Handbook of Advanced Process Control Systems and Instrumentation

Gulf Publishing Company
Book Division
Houston, London, Paris, Tokyo

Handbook of Advanced Process Control Systems and Instrumentation

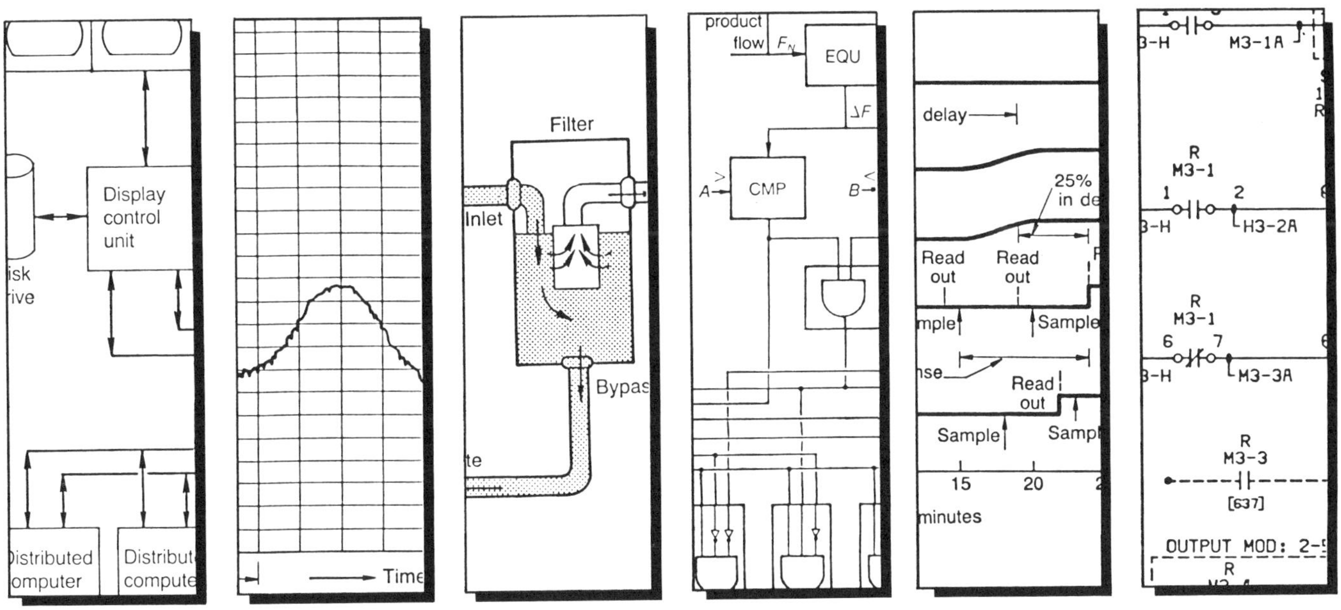

Les Kane, Editor

Handbook of

Advanced Process Control Systems and Instrumentation

Library of Congress Cataloging-in-Publication Data

Handbook of advanced process control and systems instrumentation.

Includes bibliographical references and index.
1. Chemical process control. I. Kane, Les A., 1949–
xxxx.
TP155.75.H35 1987 660.2′81 86-33584

ISBN 0-87201-721-4

Contents

PART 2—ADVANCED AND CONVENTIONAL CONTROL STRATEGIES

PART 3—ADVANCED AND CONVENTIONAL CONTROL AIDS

Preface

This book is designed to provide process control engineers, instrumentation engineers, process engineers, and management a practical guide to improving process control. The information presented is unusual for books of this type because it reflects practices used in the real world. Unlike textbooks, this book integrates problems, concerns, and actual experiences in implementing better control and relates them to the control technology.

Since processing plants vary considerably in their level of control, the book includes basic information for those just getting started, and progresses to include some of the most useful information available on the latest practical advanced control techniques. The chapters are adaptations of articles published in *Hydrocarbon Processing* magazine and reflect the progression of process control in the 1980s. They also adhere to the magazine's editorial philosophy of presenting the users' viewpoints rather than those of the manufacturers.

One important conclusion reached in review of the book is that a thorough understanding of the process to be controlled is critical to maximizing the benefits of modern control systems. Hence, considerable information is presented on specific process control strategies, process conditions, how process variables affect profitability of specific processes, etc.

I welcome any comments on the book and hope you find it useful.

Les Kane
Engineering Editor
Hydrocarbon Processing

Part 1
INSTRUMENTATION APPLICATIONS

1
Control System Design: A Brief History

B. D. Stanton

Over the past forty years the measurement and control business has changed significantly with regard to plant operation methods, plant complexity, and types of control equipment and measuring devices. Some basic things that have not changed are the need for good fundamental control systems engineering, basic single-loop control systems, and maintenance requirements. This chapter discusses the evolution of the process control system, some of the present problems we are facing with suggested solutions, and finally a look to the future.

INCENTIVES FOR NEW INSTRUMENTATION

The basic driving force of any industrial venture is the increase of profit. Profit is a function of yield of valuable products minus the cost of producing them. The effect of this interrelationship upon the process measurement and control industry can be seen by looking at examples of cost reduction and yield improvement.

Very high fuel costs have brought process heaters under close scrutiny with regard to design and measurement of efficiency. Measurements of oxygen, carbon monoxide, carbon dioxide, fuel Btu, and accurate flow are used to calculate efficiency. These measurements are of great value to plant operation, but they are even more valuable if inputted to a computer where efficiency calculations are made, safety constraints checked, and feedback signals sent to front-line controllers to bring the process to the optimum operating position. This type of operation decreases cost and thus increases profit. However, capital equipment expenditures, design engineering, software, and maintenance costs are higher.

In the area of increasing yield for greater profit, a classical example can be seen in the distillation tower. Here, computers that use on-line analyzers (themselves computer controlled) and flow meters (which are computer compensated) adjust the column to give the maximum amount of the more profitable product.

The basic requirements of computer control systems are:

- Good systems design
- Good basic measurement
- Use of computational devices to optimize the system
- Good maintenance after the system is operational

The last four decades have seen significant improvements in measurement and computational equipment. Basic measurement standards have been developed based on experience and theory. A large variety of on-line analyzers are available that do work when properly installed. On-line computational equipment that works is commercially available.

If this is true, then where do our problems exist? Problems are found in basic system design and maintenance. This chapter will present several illustrations of these problems. But first, let us look for a moment at measurement and control history; this will help us understand how we got where we are today. The following discussion is illustrated in Fig. 1 and tabulated in Table 1.

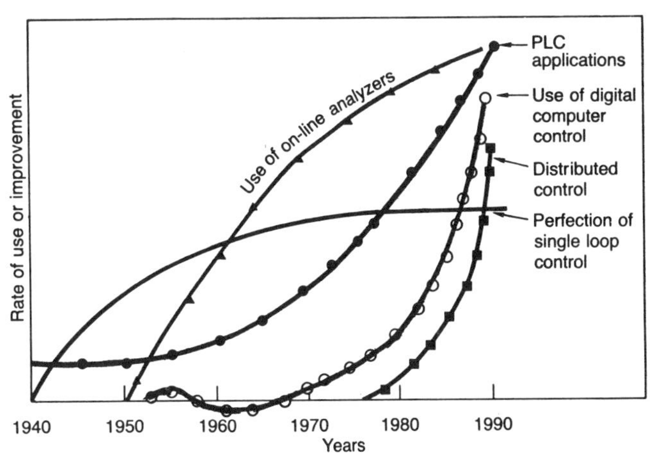

Figure 1. Trends in instrumentation.

Table 1
Control System Development

1940 - 1950	Development of single loop P.I.D. control. Start of unit control rooms.
1950 - 1960	Transmitters and transmission lines. On-line analyzers. Electronic control loops.
1960 - 1970	Control room consolidation. Digital computer control. Programable logic controllers. High density single loop controls.
1970 - 1980	Central control room, electronic instrumentation. Improved power supply systems. CRT operation distributed control. Microprocessor driven measurement devices.
1980 - 1990	Expansion of digital computer control. Intercoupling complex digital devices. More reliance on complex measurement.

PAST PRACTICES

From 1940 to 1950, operating managers were not in favor of the unit control room and they certainly did not want an air-conditioned room for the operators to work in. Instruments were scattered around in the units, and operators had to walk around looking at their process equipment and instruments. Process fluids were piped directly to the instruments. Therefore, process variable transmitters had not been developed. Single loop pressure, temperature, flow and level control loops were coming into use because the operator could not cover all the bases at once. The processes were simple, and the operator had only 10 to 20 control loops to watch over. Circular charts from flow meters were collected and sent to yield clerks, along with tank gaugings, to make up the unit materials balance reports. Operators kept rather complex hourly logs of operations. Samples were sent to a central laboratory for product composition analysis. All products went to tanks from which they were later rerun, patched, or marketed.

From 1950 to 1960, the chemical business was emerging and it requested a share of the crude barrel. Its products differed from the oil refinery's. They were sold in smaller, more cost per unit quantities and they were on specification, or they were useless. It was not so easy to rerun or to blend to specification. This accelerated development of the on-line analyzer. Also, processes were becoming more complex, and many more control loops were required. Centralization of a unit's controls into a single room was becoming essential. Development of the process variable transmitter was necessary because process fluid could not be brought safely or economically into the control room. Since control valves were pneumatically operated, the rest of the control system was pneumatic in nature. Therefore, great bundles of pneumatic tubing were required for transmitting measured signals. These massive transmission lines were the first cause of the move to electronic-type control equipment. Analyzers normally had to be made using electronic elements and had to be housed in very clumsy explosion-proof boxes.

Vacuum tubes were required in all these electronic devices, and these tubes were problematic. Great discussions were carried on in technical society meetings about the merits of pneumatic versus electronic control. Many lessons were yet to be learned about electronic devices, such as: silver migrates across insulators, shorting contacts, power supplies were not as reliable as desired, the failure rate in electronics far exceeded that of pneumatics, and maintenance men did not have good test equipment nor did they understand how to troubleshoot electronics.

In the 1960 to 1970 period, the digital computer for process control became available causing an immediate revolution in the process control business. There were many problems to be solved, and during these first few years the digital computer barely made, and in many cases lost, ground after attempts were made to apply it.

During this period conventional instruments were being miniaturized and operators were supervising larger numbers of control loops. Alarms became more important because the operator was not able to watch closely all the loops and needed aids to attract his attention to problem areas. On-line analyzers were becoming reliable and were being used as operating guides. These devices needed special installation design and maintenance. The digital computer was hungry for data to solve the problems it could resolve, and the number of back-up devices increased. Programmable-logic controllers (PLC's), which were solid state, were replacing the old relay logic systems. Control room consolidation and management information systems using digital computers were beginning to be seen in plants.

During 1970 to 1980, all of the items of the previous period were refined and used. The swing to total electronic control became strong. Microprocessor-based distributed control came into being. Control room consolidation, and in grass roots plants one central control room, became common engineering practices. Digital computers were starting to do on-line control. Operators were running plants using colored cathode ray tubes (CRT's) as their interface to the plant. Since the control room was located a long distance from the processes and the operator was confined to his CRT's, a radio link to a yard man was essential. The microprocessor began appearing in all types of instrumentation. Data links between computational devices were required. These digital devices had many types of programming languages, and communication links had all types of protocols and error-checking routines. Mixing of different manufacturers' equipment was difficult, if not impossible. Electrical power supplies became critical.

PRESENT CONTROL SYSTEM DESIGN

In our present decade of 1980 to 1990s we have arrived at process control system design as illustrated in Fig. 2. This system can be, and normally is, supplied by eight different hardware vendors. There are normally four to five different programming languages employed and as many as four different data transmission line protocols. The operator's CRT station can have up to five different CRT's. All of these devices are electrical in nature, therefore, they must continue to run no matter what happens to the main power busses. Some of the problems associated with computer control are: ground loops and isolation from stray pulses getting into digital devices; the large amount of stray electrical signals passing through the air; and, finally, most of this equipment must run in a clean relatively stable atmosphere. This puts the air-conditioning system in the reliability loop.

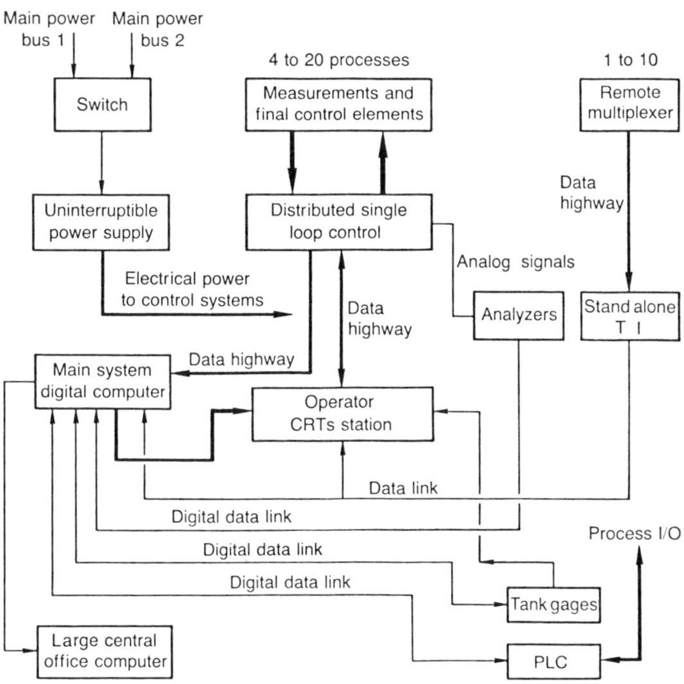

Figure 2. Present process control system design.

We can illustrate this system by making up a hypothetical system using real equipment suppliers. This will highlight the problem of complexity. Table 2 shows such a system. As examples, let us take two of the elements and run across Table 2, Number 3 - "Tank Gauges" and Number 1 - "Stand alone TI (Temperature Indicator)."

Table 2
Hypothetical Control System

Device	Typical manufacturer	Inputs	Outputs	Software
1) Stand alone TI	Transmation	Thermocouples and RTDs to five RTU's	Operator CRT. RS 232 link to computer.	Fixed by manufacturer. Configured by user.
2) Analyzer	Amscor	Hydrocarbon samples	Special data link to main computer. CRT for operator.	Fixed by Manufacturer. Configured by user.
3) Tank gauges	Varec	Level transmitter	Gray code to operator indicator and to digital computer.	None
4) PLC	Modicon	Contact closures	Contact closures. Mod bus data highway to computer & operator CRT.	Programmed by special ladder diagram technique.
5) Distributed control	Honeywell	All process measurement	Data highway to CRT to operator. Analog signals to plant.	Fixed by manufacturer. Configured by user.
6) Main digital	Modcomp	From all devices using a number of interfacing black boxes	Operator CRT Analog signals to distributed control, typers etc. Link to central computer.	Fortran
7) Measurement and final control elements	Rosemont Fisher control	Process variables 4-20 mA electrical	4-20 mA signal. Manipulation of process fluids.	None

Since the control room is centrally located and remote from the process, tank gauging from the central control room is essential. The tank status must be available to the operator to prevent overfilling or running the tank dry. Also, the computer must have these data for inventory reports. The level transmitter is electromechanical and uses a shaft encoder to convert the electrical signal to a series of pulses. These pulses go out on the line using a Gray code protocol. This code is converted to a decimal display for the operator at his station. The signal also must go to the computer and be converted to a binary format for the digital computer. Note that three different manufacturers of hardware are involved. Either the computer or the operator can request tank gauge data by going back over these data links.

Next, consider the "Stand alone TI," which we have always had to support our process control systems. The cost of running rare-metal thermocouple wire from remote units to the central control room is prohibitive. Therefore, remote multiplexers are located in the units near the measurement points. The central device, which is normally a small digital computer, handles operator or computer requests for data via a data link, by calling up the correct point in the desired remote multiplexer, checking the signal for validity, checking alarm limits, and then sending the data in engineering units to the operator via a digital display or CRT, and to the main computer in binary format via an RS 232 link. This story repeats for all devices. Therefore, the design engineer must be sure all the devices work and do the required job, and that the proper interfaces are supplied between different manufacturers' devices.

Now, how does all this complex system get put together? Figure 3 shows the type of user and user-purchased skills required. The users' manager must be sure that nothing in the interfaces is lost or forgotten. The technical experts usually define what they want and then the contractor executes their desires. There can be as many as five different contractors involved. The contractors must be checked in all phases of their work for accuracy and to verify that the user is getting what he wants. Blood, sweat and tears are usually involved in this detail design phase of the work. After this, the system check-out and field installations must be monitored carefully to make sure nothing is done to impair the operation of these complex systems. Again, although contractors have the job, the user must carefully check the work.

Now the operation and maintenance starts. Figure 4 shows the elements required. There must be a training period to teach operators to use this complex equipment. Often this is done under contract by the equipment vendors. The technicians must go through training similar to the operators' training, and also training on front-line maintenance and trouble shooting diagnostics of both hardware and software. Detailed maintenance experts on all these systems have to be supplied by the vendors. The user's technician can make front-end systems checks and sometimes repair the trouble,

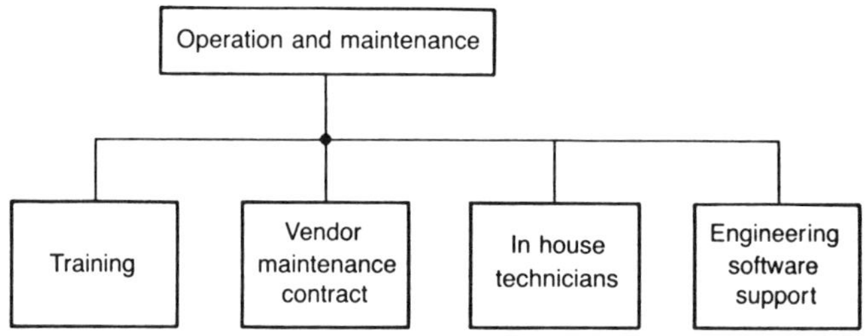

Figure 4. Required operation and maintenance elements.

but if this is not possible he must call out the vendor's expert. After operation starts, the operating department usually finds many things they don't like about the system, or changes or additions they wish to have. This requires knowledgeable in-house engineers to redesign the modifications to the system.

Examples

Now let us relate a few simple practical examples from experience that will illustrate how these complex systems can fool the designer.

Figure 5 shows the first example. As there were two sources of power it was decided to use an isolation device on this remote control loop's flow signal input to the computer. These isolators are available from many vendors. Normally, a small resistor is placed in the flow transmitter's current loop, and the current signal will generate a voltage across this resistor that the isolator passes on to the computer. The isolator was purchased with a very high accuracy specification, and extra was paid for calibration. No one asked for the small resistor

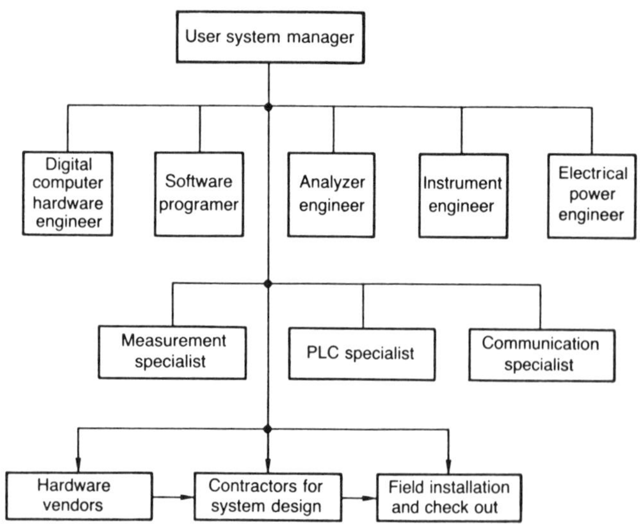

Figure 3. User and user-purchased skills required.

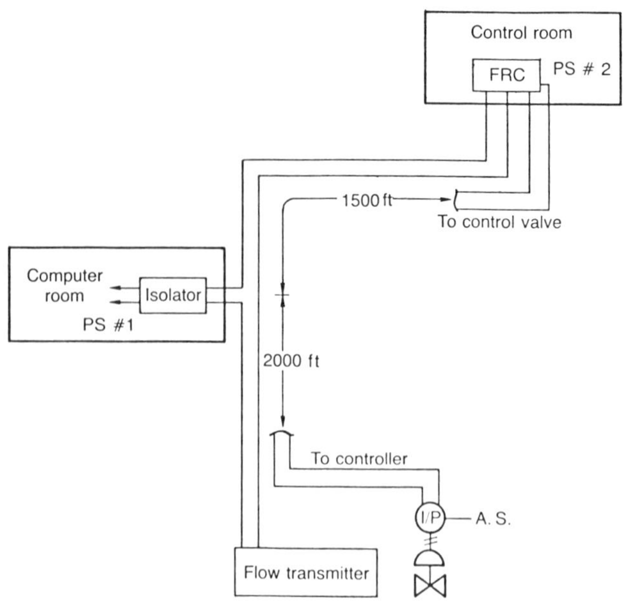

Figure 5. Isolator example.

to be placed on the plug that held the isolator. It was on the printed circuit card. The first time the technician in the computer room pulled this card to check its calibration, the current loop was broken, causing an upset to the flow control loop. The manufacturer would not guarantee the accuracy if the resistor was lifted from the card and placed on the plug. Card removal for calibration checks became very complex. The isolation cards had to be locked in their rack to prevent accidental removal. There were 40 of them in this system. Now the calibration required three sets of technicians with radios at the three equipment locations. One flow meter calibration required six men for one-half day.

Another example has to do with card replacement maintenance. A card went out on a CRT display. The mechanic replaced this with his spare and it immediately failed. Since the spare was not kept hot (on-line), the mechanic assumed it failed on startup and replaced it with his last spare, which also failed. With no spares and no way to tell what the problem was, he called in the local service man. The service man, knowing the circuits on the card, located the failure as 110 volt AC on the input, tracked it down and removed this problem and used his only spare. Now three new spare cards had to be rush-ordered at a cost of $5,000, a service charge was paid, and almost eight hours of CRT display time was lost.

A final example (these could go on forever) concerns a tank gauging system. The computer programmer put the tank gauges on a regular two-minute scan. He had no idea how these devices work and could see no reason for not putting them on a regular scan. Within a few months these electromechanical measuring heads started failing at an alarming rate. When the instrument technician located the problem he asked the computer people to put the tank gauges on an eight-hour scan rate or on demand from the operator. When the operating manager saw this, he was concerned that a tank could be overfilled between scan times. Therefore, a completely separate low-accuracy analog level measurement system had to be installed. These were on a high-speed scan with high/low alarms. There were 28 of these tanks. The low access rate but higher accuracy gauges were also retained, doubling computer tank level inputs.

In addition to many new problems that can be designed into these complex systems, there are all the old standard instrument installation, measurement point location and single loop control problems that we have resolved over many years of experience. As new people come along they seem to become enamored with the bytes, bits, nibbles, line protocols, etc., and so the basic design suffers. Many times these errors are very costly to correct. Examples of such errors are: locating orifice plates too close to elbows, placing turbine meter runs in vertical pipe runs, using asbestos-covered thermocouple wire, etc. This work, which is standard instrument engineering, must be continued along with all the new, more complex technology if jobs are to work correctly.

LOOKING TOWARD THE FUTURE

New modern control system designs are being installed, and they do work, though some do not work as well as we would wish. But there is definite improvement, and in a short time significant changes have been implemented. These projects are quite expensive and engineering-intensive, and at present there is a shortage of this type of engineer. During the present decade of 1980 to 1990, there should be a consolidation of this new technology into more workable maintenance-free, self-checking systems. A real step forward might be taken if all the systems shown in Fig. 2 could be supplied by one hardware vendor so that all interfaces could be taken care of, allowing ground loop and isolation problems to disappear. However this goes, the 1980 to 1990 period appears to be one of great opportunity for the measurement and control field.

2 How to Implement Digital Control

M. J. Sandefur

Implementing complex digital control systems requires careful planning at all stages of project execution. A systems approach to the design, specification, engineering, testing and commissioning of the digital system is essential to the success of the project.

Techniques used to engineer and build a conventional instrumentation system are not directly applicable to the digital project. For the typical analog system, discrete instruments are individually specified, purchased, and then packaged in a customized control panel. Final assembly and testing of the control panel is normally performed by the instrument vendor or a separate panel shop.

The successful digital system project follows a different set of procedures. The instrument engineer must first analyze the system's functional requirements, hardware configuration, software, operator interface and other needs. He or she must scope out the size of his system early in the project, and provide the necessary spare capacity for future expansions that may occur during project execution and later in the field. Basically, the digital control system utilizes common hardware and software to execute indication and control functions which discrete components provide in an analog system. As a result, scope changes to a digital system can be more costly and time consuming due to the interactions of hardware, software and system configuration.

PLANNING A DIGITAL SYSTEM PROJECT

Manpower Requirements. The first step in meeting manpower requirements is to assign a project engineer to the job who is familiar with the design, specification, engineering and execution of digital control systems. Depending on system size and complexity, the project engineer will be required full time during most of the project. Increased demands will be made of the project engineer during advanced stages of engineering, design review, software development, inspections and acceptance testing.

The project engineer will need assistance. For small systems with 500 process points or less, he will require a designer to handle the details of interfacing with the instrument, electrical and architectural disciplines. Numerous drawings, specifications, data base sheets, instrument loop drawings and other documentation must be developed, checked and issued for construction.

For systems with a thousand points or more, the project engineer will require an organized team of engineers, designers and technicians to successfully execute the project. Clear areas of responsibility, goals and milestones must be established for each member. Typically, engineers are assigned responsibility for the different steps in system engineering and design, while the project engineer establishes system configuration, monitors project schedules and costs, interfaces with client and vendor, and in general, maintains control of the project. In all cases, the project engineer must be willing to *delegate authority* or he soon will be lost in details of the job.

Scoping the Computer System. Scoping the system can be one of the most challenging and creative aspects of a computer system project. It is also one of the most critical steps, since everything that follows is built upon it. It is essential that the system be specified as closely as possible at the beginning of the project. Using the system specification, firm prices can be quoted, manpower assigned and manufacturing schedules established. However, none of this can be accomplished unless the project engineer and his client have a clear idea of what the system requirements are.

In general, the client usually has some concept of what is required in the way of computers or digital control systems for his plant. The client may define his requirements to the point of preparing a detailed specification from which the project engineer can obtain quotations. Typically, the client's specification must be rewritten to include additional detail on equipment types, quantities, point counts and specific operational requirements.

System configuration will be dependent on many factors. Many questions need to be answered: How many analog and digital points will be brought in? How many control loops will the computer system support? Will there be a centralized control building? How much floorspace is available? Are there suitable areas in the plant to locate remote equipment, such as remote terminal units or microprocessor controllers? Does the client wish to minimize field wiring by distributing the system, or would he prefer the computer system in one location where maintenance can be performed conveniently? Should analog instrumentation perform the primary control function and the computer act in a supervisory role? Is the client receptive to microprocessor based instrumentation for primary control instead of analog controllers? What mix of digital and analog instrumentation will be required? What degree of system security or reliability is desired? Will a backup computer be required? What is the failover philosophy if a dual computer system is implemented?

The project engineer must also establish the functional requirements of the system: How complex are the control algorithms to be implemented on the system? How much process optimization will be performed by the computer, and how will it affect system loading? Will the computer system provide process alarming, or will a hardware annunciator system be required? What are the historical data and report generation requirements (which establish bulk memory type and size)?

With small systems, some of these factors are not as critical, since the vendor's minimum standard configuration usually can handle the client's requirements. On large customized computer systems, this information is essential if the vendor is to give a realistic quotation. With large systems, some vendors may not be capable of handling the job. This will not be apparent unless the system has been thoroughly scoped and specified before sending out inquiries.

Attention must be given to the type of operator interface required. In most cases, a CRT based operator console is the preferred interface. In a few instances, the client may require digital display panels or even dedicated indicators and push-buttons, but these arrangements are usually confined to data acquisition and multiplexing systems.

Sometimes a shared video display system is required by the client. "Shared video display system" can be interpreted several ways. Some vendors offer a microprocessor based system that can be interfaced to their analog instrumentation to provide this function. If this option is not available, then a supervisory process computer is required, and the resulting interface with existing instrumentation involves extensive wiring, cabling and other expensive resources. These requirements should be identified early in the project and quoted by qualified vendors.

Steps in Projection Execution. Once the project engineer has been assigned, feasibility studies have been performed, and meetings arranged to scope out the system configuration, the project is well underway. To ensure continued progress, the project engineer should map out and schedule the various steps required to successfully execute the project. These steps are listed here and will be discussed in greater detail in the following paragraphs.

1. Prepare a detailed system specification.
2. Establish a list of qualified vendors.
3. Issue request for quotations (RFQ).
4. Evaluate bids and select vendor.
5. Issue purchase order.
6. Arrange kickoff meeting with vendor and client representatives.
   a. Establish communication channels with vendor and identify necessary information and drawings required by both parties.
   b. Establish schedules and milestones with the vendor for manufacturing, system assembly, software development, documentation and testing.
7. Assign personnel and start system engineering.
8. Establish design requirements for affected disciplines (electrical, power, grounding, air conditioning, architectural).
9. Review vendor's design documents and release for manufacturing. Conduct design reviews and visual inspections at factory as required.
10. Monitor design changes/costs and negotiate with vendor as required.
11. Generate computer data base forms and submit to vendor.
12. Define CRT displays and report formats.
13. Develop custom software.
14. Establish spare parts and maintenance requirements.
15. Develop test plan for customer acceptance of the system.
16. Conduct acceptance test at vendor's facilities. Make final arrangements for shipment and storage if required.
17. Commission system on-site.
18. Issue final revision of system specifications and affected engineering documents.

COMPUTER SYSTEM SELECTION AND PROCUREMENT

The selection and procurement of a digital or computer based system can be a major effort and require many months to complete. For small systems, it may only represent the final culmination of the project planning effort. The distinction is mainly a matter of system size and general practices followed by the engineering firm involved. On multi-million dollar computer projects, much time and effort is spent before a final system configuration is developed and a qualified vendor selected. Even after the purchase order has been released, negotiations may continue on the equipment and services to be provided, as well as on the terms and conditions of the contract.

System Specification. Before inquiries can be sent to prospective vendors, a detailed, definitive specification of the computer system must be prepared. Basically, the system specification is the final documentation of the system scope developed earlier by the project engineer and client. It is a detailed definition of system hardware, software, functional requirements and necessary services to be performed by the vendor. While the specification will be in a continuous state of revision during the life of the project, every effort should be made to make it as complete as possible initially. The system specification and resulting vendor quotation form the basis for all future negotiations between the vendor and the contracting firm.

Detail and functionality of the systems specification will be determined by several factors. System scope and definition as determined by the project engineer and client are the most obvious ones. Customer preferences for certain hardware configurations, such as dual computers or a microprocessor based distributed system, will shape the final form of the specification. Availability of *process* and *plant design data* will also have a crucial effect on the specification. Educated guesses on system size and configuration may be required so firm prices can be quoted.

Required vendor services, such as data base generation, graphic display building, custom software development and extensive acceptance testing, should be identified in the system specification so these items can be quoted. Hardware interfaces between systems supplied by different vendors should be specified in detail. Typical areas of concern are: Who will supply sensing power for analog and digital inputs/outputs? What type of interface will be required (hardwiring, connectors, special electronics, etc.)? All requirements for communications with other digital systems should be identified and necessary protocols established.

Less tangible factors, such as schedule requirements or degree of competitive bidding required, may affect time and effort spent in developing the specification. If one writes the specification around a particular vendor's hardware, or puts stringent requirements on component design, manufacturing, and testing, few vendors may respond to the inquiry.

Preparing the Request for Quotation. After the system specification is finished, a Request for Quotation (RFQ) should be prepared and sent to vendors whom the project engineer and client have determined capable of supplying the required system. This document identifies what type hardware, software, and services are to be quoted. The system specification is attached to each inquiry and forms the basis for quotation. The RFQ also identifies all necessary terms and conditions, packing and shipping instructions, system design documentation and spare parts requirements. To ensure proper response to the inquiry, it is important that the project engineer establish communications with qualifying vendors prior to release of the RFQ.

Bid Evaluation. After the bids have been received, the task of bid evaluation begins. This can be very difficult, especially when many vendors have submitted quotations. Often, all of the vendors appear capable of supplying a functional system suitable to the client's needs. The project engineer may be tempted to make a decision based on the lowest bid. However, two factors should be considered: One, all systems are not equal. Two, all "bottom lines" on comparable systems are not equal in scope of material and services provided.

It is the responsibility of the project engineer to determine which systems are acceptable from a technical and performance standpoint. It is also the project engineer's responsibility to establish his own "bottom line" for each vendor's bid. This is accomplished by tabulating each bid, breaking out hardware, software, options and services provided. Each of these categories should be based on the system specification and RFQ. The project engineer then determines if the vendor has responded to each item in the bid tabulation.

This is best accomplished when the vendors are instructed to respond to the RFQ and system specification on a *paragraph by paragraph* basis. In addition, all components, services and options should be listed and priced separately in the quotation. Many times the vendor will ignore these instructions and only quote a bottom-line figure. This is usually done to save time and present a "low bid" to the potential client. To circumvent this problem, the project engineer must maintain communications with the vendors and request *written* price clarifications. If the vendor is not cooperative, then the project engineer is wise to eliminate his bid.

As the bid tabulation is completed, the successful vendor will begin to emerge. Certain vendors will be disqualified, which simplifies the task. Other vendors may have systems which offer more attractive features. Finally, the bottom line of the bid tabulation, if prepared properly, will give a true comparison of system cost for the remaining vendors. Final selection will be a function of system features, costs, services provided, the vendor's past performance, and the client's preferences. If close attention has been paid to the system specification, request for quotations and bid tabulation, then final vendor selection can be made intelligently and the best possible system purchased.

Pre-purchase Negotiations. Depending on the size and complexity of the system, the vendor, contractor and the client may enter into negotiations before the purchase order is released. Final details on system configuration, hardware, software and services provided in the quotation are discussed and additional pricing established, if required. On large jobs, it is important that such meetings are executed prior to issuing a purchase order to ensure there are no costly misunderstandings between vendor and contractor. If an impasse is reached, the contractor still has the option of selecting another vendor without suffering cancellation charges.

Purchase Order. The purchase order is released next, with a revised copy of the system specification reflecting system configuration, equipment types, quantities and services to be provided by the vendor. Again, it is essential the specification is as detailed as possible and reflects all system requirements. As a result of pre-purchase order negotiations, there may be system configuration changes, and this should be reflected in the detailed specification.

Project Kickoff. The project engineer should schedule a "kickoff" meeting between the successful vendor, contractor and client. On smaller projects, the negotiations discussed earlier may be conducted during this meeting. However, the main function of the kickoff meeting is to establish how the work is to be done, when it will be done, and who will do it. Schedules will be developed for engineering drawings and specifications, design review and approval, manufacturing, system assembly, testing, software development, delivery, installation, and startup. Communication channels are established and *key personnel identified.* Manning requirements are established for the project.

Information requirements for the vendor and contractor need to be identified. For the vendor, this usually consists of the field terminal assignments, selected options, undefined functional requirements and final system configuration. The contractor requires dimensional drawings of the cabinets, power, grounding and heat load requirements, identification of power, grounding and signal terminations, and other particulars related to system support and interface. The contractor needs a firm date as to when final design documents will be available for approval. In turn, the vendor must know when this approval can be expected, so that manufacturing schedules can be planned in advance.

It is hoped that after the kickoff meeting, all participants will leave with a clear plan to follow. Work will start on the project and proceed according to the agreed schedule. The project engineer will have met key personnel involved, established rapport with them, and be in a good position to successfully manage the computer system project.

SYSTEM ENGINEERING AND DESIGN

Engineering and design of the computer system builds upon criteria established in the detailed specification and discussed in the kickoff meeting. Questions posed by the vendor and project engineer on hardware configuration, process interface and system functions must be answered and documented. Additional drawings and specifications, addressing specific design areas of the computer system, must be issued and transmitted to the vendor and affected design disciplines. This stage of the project represents a period of information gathering, decision making and communications with the vendor and client.

Point Count. An accurate point count must be developed to determine the total number of analog and digital inputs/outputs serviced by the system. This information is taken from the latest mechanical flowsheets and instrument index. A percentage of spare I/Os, based on live points serviced by the system, should be calculated and a total point count established. As additional points are added to the system, the point count will be revised upward to maintain the required percentage of spares.

Rack Layout. Once the point count has been established, the type and quantity of input/output cards, multiplexer chassis, controllers, terminal panels, and other supportive hardware can be identified. From this equipment list, the system racks can be identified and laid out (Fig. 1). The vendor normally performs this function, and it is important that he receive the point count, grouped according to signal type, as soon as possible. The project engineer should provide input on

Figure 1. System racks.

over-all rack layout, since this affects the electrical and architectural design disciplines.

Field Terminal Assignments. After the rack layout has been finalized and termination panels located, field terminal numbers are assigned. Although the vendor can perform this task, it is recommended that the project engineer make the field terminal assignments. This will ensure that terminal arrangements are structured in a logical and functional manner. Terminal assignments can also be transmitted immediately to the electrical design group without the delays normally associated with receiving vendor drawings.

Control Panel Interface. For supervisory control systems, the computer's multiplexer must interface with an analog control panel. This interface can be accomplished with point-to-point wiring performed in the field. To minimize installation costs, the multiplexer racks and control panels can be wired with receptacles, and multi-conductor cable assemblies can be used to make the final connection. However, coordination between different vendors on matters of signal commoning, fusing, shielding, and pin assignments can be very difficult. Control panel interface with a large computer system must be designed carefully. Errors discovered once the computer rack and control panel fabrication is underway can be very costly to correct.

Care should be taken to ensure all necessary signals are brought from each analog controller to the computer mux racks. As a minimum, the process variable, setpoint feedback, controller mode and the computer setpoint are required for each loop. Cascade loops also require the master controller output signal, the auto/cascade relay status, and the digital output from the computer mux rack to the auto/cascade relay. These signals are necessary for bumpless control transfer for all modes of operation. Should the computer fail, the setpoint must remain at its last value until the controller is switched from computer control. A special output board in the computer system, or a memory update card in the control panel, is required for this purpose.

An integrated analog controller and process computer package simplifies a supervisory control system installation considerably. All necessary interfacing is built into the equipment. Connecting the process computer system to the analog instrumentation is a matter of adding the vendor's standard interface hardware and cable assemblies.

Grounding and Shielding. A single ground reference should be provided for all logic, instrument and shield grounds. The vendor normally provides a system of bus bars and ground cables which provide signal return and ground reference for the central processor and its peripherals. All vendor recommended grounding practices should be followed.

Clear guidelines for grounding and shielding are not always available, especially for large customized computer systems. Grounding practices are further complicated in supervisory control systems when the computer must interface to an analog control panel. In such cases, an integrated approach to grounding must be followed for both the computer system and analog instrumentation. A single common ground reference must be established for both the analog and digital systems. All signal interfaces between the two systems must be analyzed to ensure ground loops are not inadvertently introduced into shields and signal returns. Analog and digital input cards for the computer system should be selected with adequate input filtering, isolation, and common mode noise rejection characteristics.

Guidelines for grounding and shielding equipment racks, power supplies, signal lines, and cable shields are as follows and should be used in conjunction with the vendor's recommended practices:

- Each equipment rack should have an instrument ground bus bar. The bus bar should be electrically isolated from the rack itself. The 115 VAC neutral, green ground wire and other chassis or safety grounds *must not be connected to the instrument ground.*

- Due to the vendor's hardware design, some computer systems and most computer peripherals have the instrument or digital ground connected to the equipment chassis. Ground connection is made through the green safety wire. Care must be taken to ensure the green wires are connected to an insulated bus bar in the power distribution panel and then grounded at the power service entrance only. Most peripherals plug into 115 VAC outlets. A green wire should be provided for each outlet and grounded in the same manner.
- Equipment racks should be grouped together and a master instrument bus bar provided for each group. Each rack bus bar should be solidly bonded to the master bus with a single #4 insulated cable. In turn, each master bus bar is solidly bonded to a single ground connection of one ohm or less to true earth. Separate 2/0 insulated cables should be used for each master bus bar. Daisy-chaining is not recommended.
- DC power supplies used to power instrumentation, digital electronics and field transmitters should be referenced to the instrument ground in the associated equipment rack.
- Individually shielded twisted pairs with an over-all cable shield should be used for field analog signals, and for cable assemblies between the computer mux racks and the analog control panels. Digital signals need not be individually shielded. However, individual twisted pairs are recommended for field digital signals and for interconnecting cable assemblies. Analog and digital signals should be assigned to separate cables.
- Signal returns must not he grounded anywhere along their length except where the final connection is made to the power supply common in the rack.
- Cable shields must be grounded at one end only. Preferably the shields should be grounded in the rack where signal power is supplied.
- When available, optical isolation should be specified for all equipment interfacing to the data highway for distributed control or multiplexer systems. Optical isolation should be specified for all field digital inputs.

Field Wiring Interface. Usually the vendor supplies the necessary terminal blocks to interface the multiplexer hardware with the field wiring. The terminal blocks mount on the edge of the multiplexer I/O boards, or in separate termination panels. Standard cable assemblies connect the terminal panels with the mux chassis. In either case, the I/O cards can be removed without disconnecting the field wiring.

Special terminal arrangements are necessary when loop fuses, zenier barriers, or interposing relays are required. To provide a clean interface with the field wiring, Wiedmuller type terminal blocks can be assembled in the rack. Special fused blocks with blown-fuse indicators can be used where required. However, additional wiring is required to connect the terminal blocks with their associated relays, zenier barriers and multiplexer equipment. Alternately, the field wiring can be terminated directly on the relay or barrier terminals, but on large systems, this may cause confusion and errors in field wiring design and installation.

Most digital outputs to the field require interposing relays with contacts rated for the applicable load. For 115 VDC service to inductive loads, relay contacts must be oversized and specifically rated for the required DC current and voltage levels, both for continuous current flow and for surge (due to inductive kick when the contact opens). It is also good practice to fuse each contact output separately to prevent a short on one output from affecting others. Fusing for analog and digital inputs is usually provided by the vendor on a per-card basis.

Signal Power and Commoning Requirements. If field wiring is connected directly to the computer system multiplexer, then the vendor normally provides the necessary DC power for energizing transmitters and sensing contact inputs. Digital outputs may be dry contacts or energized. As an example, motor status inputs normally receive sense power from the mux racks, while start/stop outputs are dry contacts. When configuring digital inputs and outputs, it is advisable to avoid commoning and run two wires for each contact input and output. This provides a clean interface and avoids wiring problems when connecting to other control systems.

With supervisory systems, multiplexer racks are connected to a control panel, usually provided by another vendor. Problems can occur concerning which vendor supplies sensing power and how signal pairs should be commoned and fused. As stated previously, it is best if signal commoning is avoided. However, most digital input cards service a group of eight, twelve, or sixteen digital inputs with a single common return. In this case, commoning should be done at the input card only.

Sense power should be provided in the mux racks for digital inputs, and dry contact outputs provided to the control panels. It is important that the vendor wires sensing power into each signal loop inside the cabinet. Otherwise, additional terminals and wiring is required when making field connections.

In all cases, interface wiring with field instruments and other vendor's control systems must be planned carefully to ensure proper operation. Criteria related to signal power, commoning, and fusing must be established, documented, and transmitted to affected vendors and design disciplines.

Data Highway Communications. Communication requirements must be established for remotely located equipment, such as remote terminal units (RTUs) or microprocessor controllers. Distributed control and data acquisition systems normally communicate over twisted pair or coaxial cable, which functions as a data highway. Two separate cables are routed through the plant to enhance system reliability. The data highway links the remote units together in a daisy chain, loop, or radial pattern (Fig. 2). Typically, total cable length is limited to about 5,000 feet for coax, and about 20,000 feet for two twisted pairs. In general, these maximum lengths are for the total length of cable on the data highway, including branches and loops.

To communicate with remote units over longer distances, modems are required. Communications is generally provided by dedicated telephone circuits in the plant. For pipeline control systems and dispatch operations, microwave channels normally link the central computer with its remotes. In either case, transmission rates must be established and circuit grades specified. It is usually the responsibility of the client's communications department to provide necessary telephone circuits. Transmission rates should be limited to a maximum of 2,400 baud to minimize demand on the communications channel and improve system performance.

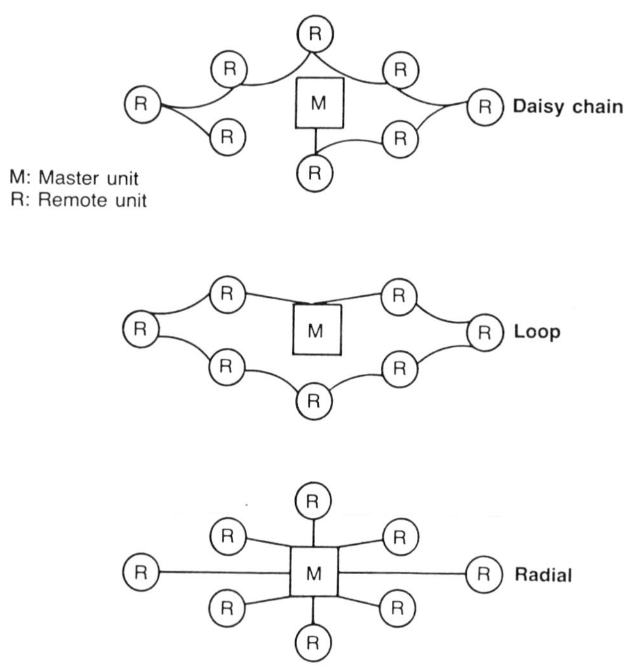

Figure 2. Data highway possibilities.

Operator Console Design. The operator console normally consists of one or more interactive CRT stations with operator keyboards. Depending on the client's preferences and the vendor's standard options, special features may be required. Pushbutton stations, pen recorders or hardwired annunciators are sometimes mounted on the operator console. A separate control console can also be installed next to the CRT station to house the hardwired instrumentation. In general, dedicated instruments should be avoided, since the CRT console provides a much more efficient and cost-effective operator interface with the process.

The computer system vendor may offer light pens, joy-sticks, or track-balls to position the cursor on the CRT screen. With some systems, cursor control provides the main operator input into the system. The choice of cursor control is mostly a matter of personal preference. There are arguments that light pens are too fragile, or that joy-sticks and track-balls are susceptible to coffee spills. In either case, the vendor usually equips his standard keyboard with pushbuttons that facilitate cursor control. On the other extreme, the vendor may supply a bank of dedicated pushbuttons that allow the operator to select a particular display, table, or report with a single keystroke. Most systems use a combination of keyboard entry, cursor control and dedicated pushbutton functions.

Printers are normally incorporated in the operator console design for logging and reporting purposes. On larger systems, it is desirable to dedicate one printer to alarm logging. Low-speed, single-character typers are adequate for this purpose. The printer can be free standing or located on the console (table top).

Depending on the vendor's limitations, modems may be required for remotely located operator consoles. This does not present a problem for the printers, but could significantly reduce CRT response time. Distributed control systems have less severe distance limitations, since the operator console can be located anywhere on the data highway.

Logic and Sequencing Requirements. Logic and sequencing requirements for programmable controller systems should be finalized, documented, and forwarded to the vendor. It is important that the logic is firmly established at the time the system specification is written. The type of programmable logic controller and amount of memory required is determined by the program's size and complexity. Finalizing the logic allows the vendor to determine if sufficient memory is available to hold the program. Twenty percent spare memory should be reserved for future sequencing requirements.

If the logic is relatively simple, a ladder diagram can be drawn up to document the logic. Usually the programmable controller has restrictions as to the number of contacts, branches, and coils in each rung. The ladder diagram should reflect these restrictions so that the program can be loaded directly.

Complex sequential logic with interlocks and branch operations is very difficult to lay out in ladder diagram form. Instead, tables should be developed depicting interlock criteria and final control element positions for each step of the sequence. The logic can also be written with input conditions, interlocks, timing functions, branch decisions and output commands identified in narrative form. However, with this approach, it is difficult to completely define the logic. Every combination of input conditions and operator commands must be anticipated and contingency logic developed for them. On large programmable controller systems, this task becomes unwieldy. Usually, system check out reveals most of the loopholes missed in the programming stage.

Logic flowcharts are best suited for defining sequential operations. Since flowcharts are sequential in nature, it may be difficult to represent static logic, such as interlocks or Boolean expressions. However, once flowcharting is mastered, it is possible to completely identify all logic requirements in a complex sequential operation. Logic flowcharts can be readily interpreted by programmers and technical personnel. It is important that controller logic requirements are communicated clearly to the programming personnel and other affected groups, such as the electrical or process design disciplines.

Backup Systems. Large computer control systems which provide critical process data acquisition, alarming and control functions may require a backup system to ensure adequate system availability. Over-all system configuration is usually established in the proposal stage, but final hardware and software functions must be determined.

The most susceptible components in a computer system are the central processor and its primary peripherals, such as disc units, magnetic tape transports, display generators, and bus interface devices. These units should be backed-up with a separate computer system. Generally, multiplexer equipment is shared by both primary and backup systems, since mux racks are modular in construction and failures are limited to the particular rack involved.

Communications between the two computers and fail-over philosophies depend on the vendor's standard configuration and the client's preferences. The data base can be shared by the two systems, using a common bulk memory storage device, such as a disc unit. Theoretically, this allows the backup system to upload the current process configuration and pick up where the failed system left off.

Sometimes the hot standby approach is used where both systems have access to the process interface devices and maintain their own data bases. However, only one system can access the operator consoles and generate control outputs. Should the primary system fail, the backup system is instantly available to take over.

In practice it can be very difficult to implement a satisfactory fail-over system that works. One approach is to program each computer to monitor the other, and to seize primary control when certain conditions occur. Depending on what failure conditions are selected and what interlocks are provided, the two computers may be locked in an endless power struggle, with control flip-flopping from one system to the other. However, if too many interlocks are incorporated in the fail-over logic, then neither system may respond when the other system fails.

Generally, the vendor has a standard approach to dual computer control that has been proven to work effectively in past applications. The fail-over system should be thoroughly exercised during customer acceptance testing. All means of failing the system should be attempted to determine the system's response. Any malfunctioning of the fail-over system should be evident as a result of the testing, and the necessary changes can then be made.

In distributed control systems, microprocessor based controllers provide primary control of the process. Controllers used for critical service should be backed-up with either a manual control station or another controller. These back-up systems can be simple patch panels, allowing the operator to switch in a backup device or controller for the failed unit. It is not necessary to provide a backup controller for each primary controller.

Some vendors offer automatic backup of their microprocessor controllers. Should a controller fail, a backup unit is switched in place automatically. All process wiring, data highway connections, device addressing and loop configuration is transferred to the backup controller. The operator sees no interruption in plant operation other than an alarm indicating which highway device failed.

Special Control Systems. Special functions, such as plant shutdown logic, may be provided by a separate control system. In this particular case, hardwired relays, solid state logic or a programmable logic controller can be used.

It is advisable to avoid distributing the logic throughout the equipment racks. All contact inputs and outputs should be brought to one area and all points labelled. Complex logic or sequencing systems should be located in a separate rack. A programable logic controller should be used to minimize cost and enhance system flexibility. If logic requirements are small, the computer vendor can implement the system with hardwired relays.

It may be advantageous to sub-contract the logic system to another vendor. However, the logic system should interface to the computer control system so the operator is aware of the PLC's operations. The interface between the logic and computer systems should be defined and communicated to the two vendors, for incorporation into their final system designs.

A mimic board will be required for all but the simplest programmable controller systems. A mimic board provides the operator with information on valve positioning, equipment status and the current sequence operation being performed. Pushbuttons, thumbwheel switches and digital readouts provide additional operator interface. When possible, the mimic board should provide a graphical representation of the process. The mimic board can be mounted on the CRT console, or control panel, depending on the operational philosophies and type of instrumentation used. It is also possible to hardwire the PLC's I/O's to the computer system and depict its operation using the computer's graphic display capabilities, if available.

Communications with Other Computers. A large refinery or chemical plant may require many digital control systems for efficient operation. Such systems would include multiple computers dedicated to major process units, computer based analyzer systems, remote multiplexer systems used for temperature readout, and microprocessor based tank gauging and royalty meter stations used for inventory control and custody transfer.

To provide coordinated control of the plant, the process computer system may be required to communicate with these other digital systems. An RS-232 interface is normally specified for digital systems within fifty feet of each other. The RS-232C interface defines the connector type, pin assignments, voltage levels and functions necessary to provide a compatible hardware interface between different makes of digital equipment. For longer distances, standard 20ma serial current loop interfaces or modems are required.

The type of interface defines hardware requirements and signal levels. However, a protocol must be established between the two computer systems. The protocol establishes data work structure, parity, interrogation commands and responses, data base structure, tag numbers and other criteria necessary for reliable communications.

It can be a real problem to establish agreement between two vendors on the protocol. Each vendor has his own way of structuring transmissions, or preferences as to which system shall be active or passive. In general, the large, more flexible computer system should adapt its protocol to the needs of the more limited multiplexer or data acquisition system. In any case, a detailed protocol must be developed and agreed upon. The project engineer may be required to meet with representatives from each vendor and lock the door until agreement is reached.

Once the protocol is established, it is a good idea to test it. This can be accomplished by using a test computer to emulate the other digital system. The two systems can also be linked together while still in the factory, using modems, acoustic couplers and a standard telephone set. A long distance connection is made, the telephone receiver on each end placed in the acoustic coupler and communications established between the two systems.

INTERFACE WITH OTHER DESIGN DISCIPLINES

Information must be communicated on a timely basis to affected disciplines associated with the computer system project. These include the electrical, HVAC (heating, ventilation, and air conditioning) and architectural staffs involved in the plant design and construction.

Electrical. Power requirements for each major component of the computer must be established. The number of 115 VAC and 220 VAC circuits must be identified. Voltage, maximum continuous current, inrush current and duration, and load power factor are required for each circuit. This information should be readily available from the vendor.

If a large number of circuits are involved, power requirements should be documented by the vendor in a written communication. This information is normally used to size an UPS (uninterruptible power supply) that provides power for the computer system and plant instrumentation. The vendor should be held to his stated power requirements so the UPS can be sized accordingly.

Requirements for power filtering, isolation transformers or motor generator sets should be identified and communicated to the electrical design group. These requirements are determined by the vendor, who usually is in a position to supply the necessary hardware. If a UPS is incorporated in design of the plant, a motor generator is not required. However, the vendor may recommend power filtering or isolation transformers to eliminate harmonics generated by the UPS inverter.

Power, ground and field wiring connections to the computer system must be communicated to the electrical design group. 115 VAC receptacle outlets in the computer room for both UPS and non-UPS loads should be identified and located. Grounding configurations must be established to ensure system performance is not degraded by noise introduced through faulty grounds. Signal and power cables should be routed in separate trays, with high and low level signals assigned to dedicated cables. Interconnecting cable quantities and sizes should be forwarded to the electrical group so cable routing can be determined and wall penetrations located.

HVAC. Operating ranges for temperature, humidity and suspended particulates must be established for the computer system. Maximum allowable concentrations of H_2S, SO_2, and other corrosive elements should be identified. Heat loads for each component are required so air conditioning equipment can be sized and bought. This information is normally supplied by the vendor.

Architectural. A floor plan for the computer equipment, instrument racks and operator consoles should be laid out. Considerations should be given to front and rear access, grouping of related equipment, future expansion and aesthetics. Room should be provided for work areas, book racks, and shelving for tape reels and disc paks. For large systems, it is desirable to locate a maintenance area next to the computer room. Operator consoles should be conveniently located in the control room, with consideration given to other control panels and instrumentation.

Maximum cable lengths should be considered when locating computer equipment. Generally, the central processor, peripheral controllers, and I/O bus hardware must be located next to each other. Line printers, mag tape units or disc drives must be located within 50 feet cable length of the central processor. Operator consoles and other CRT devices usually must be within 200 feet cable length of the central processor.

Normally, a raised floor is used in the computer and control rooms. If possible, the floor grid should be laid out with equipment rack locations in mind. Cutouts in the floor can then be minimized by lifting up a row of tiles and centering the cabinets over it. This provides easy cable entry to the equipment and allows the raised floor to be used as a plenum for directing air conditioning to the hardware.

Suitable locations must be established for remote terminal units and distributed microprocessor controllers. If possible, the equipment should be located in an air conditioned building. Electrical substations, analyzer houses, and remote control buildings are suitable locations. NEMA 4 enclosures purged with clean, dry instrument air should be used for all outdoor installations, and for dusty or corrosive indoor environments. A sun roof should be provided for outdoor installations in hot climates.

DESIGN REVIEW, CHANGES, AND COST CONTROL

The successful computer system project requires close coordination with the vendor, especially during the system design and review stages. Communication is accomplished with drawings and specifications issued by the vendor, which detail the exact hardware design and layout. The drawings are submitted to the project engineer for review and approval. Corrections are made and the drawings submitted to the vendor for reissue and final approval.

Refinements in system design, or customer requested enhancements, often result in cost overruns. If the system was thoroughly specified during the proposal stage and engineering finalized prior to hardware assembly, changes should be minor and cost impacts minimal. However, if system configuration and desired performance were not completely defined during the proposal stage, or if major scope changes have been made after release of the purchase order, significant costs over the original purchase order amount may be incurred. Some of the costs will be valid; some will be inflated and unreasonable. In such cases, the issue will ultimately come to a point where negotiations between the contracting firm, vendor and client are required. Once the negotiations are concluded, system design and cost should be finalized and future changes avoided.

Vendor Drawings. As a minimum, the vendor should submit the following drawings for approval:

- Dimensional drawings of equipment racks, cabinets, consoles, teleprinters and display panels.
- Rack layout drawings identifying electronic components, power supplies, relays, junction boxes, circuit breakers, terminal strips, panels, bus bars, mounting hardware, panduit and cable routing for each equipment rack and console.
- Wire and cable lists for all point to point wiring, internal cable assemblies, and inter-connecting cables between equipment racks and consoles.
- Field terminal drawings, identifying all process and shield drain connections by instrument tag number.
- Schematic drawings identifying AC and DC power distribution, fusing, circuit breakers, junction boxes and their locations in the rack. All power, safety ground and instrument ground connections should be identified on these drawings.
- Manuals describing system operation and diagnostics (usually applicable to custom system).

These drawings should be thoroughly reviewed to ensure inclusion of all hardware, design features and options detailed in the system specification and purchase order, and to ensure compliance with all design criteria and hardware assignments established during the system engineering phase. Copies of the marked-up drawings are submitted to the vendor for revision and final approval.

The vendor should provide a schedule and timetable of drawings he will submit for approval. The project engineer should maintain close control of vendor drawings that have been submitted, reviewed and approved for construction. It is good practice to identify changes on each drawing and assign revision numbers. When a large number of drawings are involved, it is important to ensure only one drawing revision is outstanding at any time. On large customized systems, it can become an impossible task to review system design when several versions of the drawings are outstanding. This is especially critical when the affected drawings contain information required by other vendors or design disciplines.

Design Reviews. On large customized computer systems, it is advisable to meet with the vendor and client on a periodic basis to review system design and determine project status. Schedules, design changes, change orders, test procedures and spare parts requirements are some of the items that are usually discussed. It is important to maintain close contact with the vendor to ensure that all system specifications are met and the project stays on schedule.

Whenever possible, design reviews should be combined with a visit to the vendor's plant. Early in the project, the vendor's staff requires much information on system configuration, equipment layout and terminal assignments. In most cases, it is more advantageous for the project engineer to meet at the vendor's facilities, and to work directly with the vendor's personnel. The trip also allows him the opportunity to inspect the facilities and meet key personnel involved on the project.

Inspection Trips. After system assembly has begun, inspection trips should be made on a periodic basis to ensure that the hardware meets system specifications and design criteria. Problem areas unnoticed by the vendor can be identified by on-the-spot inspections, and then corrected. This is especially important for customized systems, where some aspects of the design are not specified in detail. Usually the project engineer establishes functional requirements and attempts to specify the customized item as completely as possible. However, what the project engineer or client envisions, and what the vendor builds, are not necessarily the same. Timely inspection trips can ensure proper design and prevent costly changes later in the project.

Design Changes. Design changes and cost increases are inevitable with complex digital and computer control systems. These changes can be minimized by careful planning and detailed system specification during bid evaluation and project kickoff. The extent of design changes and cost overruns is usually determined by:

- System size, complexity and extent of customization is required.
- Degree of detail identified in the initial specification of the system.
- Thoroughness in bid tabulation and vendor evaluation. In particular, identification of services that are to be provided by the vendor, such as data base generation, loading and testing.
- Enhancements, changes, or restrictions imposed by the client after the project is underway.
- The vendor's past record on other projects in the areas of schedules, additional costs, design compliance, willingness to accommodate changes, installed system performance, system support, and services.

Design deficiencies and changes should be identified as early as possible, and cost studies initiated to alert higher level management of impending problem areas and possible cost overruns. Minor changes are normally handled by written correspondence and followed up with a change order. However, on large complex systems, these cost adders are not always resolved through normal channels, and tend to accumulate. Ultimately, a point is reached where the total cost overrun becomes unacceptable and a meeting must be scheduled to negotiate the changes and associated price increases.

Cost Tabulation. Negotiations with the vendor is best performed from a position of knowledge and strength. Therefore, a detailed breakdown of hardware, software, services and prices as originally quoted by the vendor is needed. This is best accomplished by insisting on a paragraph by paragraph response to the detailed specification *during the bid evaluation stage.* Subsequent requests for information after the contract is awarded may not accurately reflect the true price structure. The vendor should also be required to provide detailed breakdowns of any additional pricing for design changes occurring after the release of the purchase order.

All correspondence with the vendor should be gathered and organized in chronological order. A detailed tabulation of all system costs and price adders should be prepared. Each major component, option, design change and service to be provided by the vendor should be listed separately. Original purchase order prices and vendor. However, for this and other tactics to work, there must be representatives present from the contracting firm and the client with the *authority to cancel the pur-* cost increases, both contested and uncontested, are broken out for each item. The difference between contested and uncontested costs for each item is calculated and a total prepared for the entire system and services provided. From this tabulation, the project engineer can determine his exposure (percent increase in costs over the original purchase order amount), and decide what action is required.

Estimates should be made for all questionable cost adders. Extrapolations can be made using the detailed price breakdowns for similar items or services provided in the original quotations and any subsequent pricing. These estimates should be incorporated in the cost tabulation to provide a target or goal for negotiations.

Negotiations with the Vendor. Once all correspondence and pricing information has been gathered and correlated, a meeting should be set up with the vendor, contracting firm and the client. The meeting should be conducted at the contracting firm's facilities. Prior to the meeting, the vendor should be provided with a list of problem areas, an estimate of the cost differential, and notified of the seriousness of the problem.

In general, it is good practice to pursue every possible concession from the vendor and to present a well padded cost-differential in the initial stages of the meeting. Typically, the vendor inflates his pricing for changes once the contract has been awarded, and the only satisfactory approach is to pad the cost differential to be negotiated.

Negotiations normally open with a presentation of the problem by the project manager or purchasing agent. Each cost item is scrutinized and questioned when necessary. The vendor responds to each point with a defense of the quoted prices. It is at this time the project engineer draws from his calculations and cost tabulation and presents his case. The vendor should not be allowed to view the project engineer's reference material except in an abbreviated and edited form. Also, some arguments should be reserved for later sessions.

Initially, both sides present their case and some resolutions are reached. Cost reductions are achieved by demonstrating that specific items were already priced in the original quotation, or in subsequent written clarifications to the bid. In areas where an impasse is reached, a relaxation of system specifications or elimination of non-essential items can result in cost reductions. It is also helpful to present a well-padded counter bid to the vendor at the beginning of the meeting.

The counter-bid should be itemized and include uncontested cost adders to demonstrate its validity to the *chase order* in question, and to influence awarding of similar contracts in the future.

These detailed discussions on various cost adders can go on indefinitely. In fact, it is to the vendor's advantage to stay at this level of negotiations. As long as a resolution is not reached, his original pricing stands. To prevent this situation and maintain control of the meeting, it is essential that a chairman is established at the beginning, preferably a senior buyer or project manager.

A useful tactic that the chairman can exercise is to recess the meeting and call for a caucus. The point here is to break up fruitless discussion and allow each side to regroup. It is important to leave the detailed discussions, begin negotiating major items of the contract, and ultimately establish a new bottom line figure for all required hardware, software and vendor services.

For small, standard systems, the negotiations may end at this point, since the cost differential is usually small and agreement can be readily reached. However, for large, customized systems, costs and disagreements are on a much greater scale. As a result, the negotiations move into a second phase where the bottom line is negotiated. This is mainly a "horse trading" situation where the negotiators dicker, make threats, and call bluffs. Again, it is absolutely essential that the individuals present have the power to make or break the contract.

Certain tactics can be used by the negotiators to strengthen their position. The negotiators can threaten cancellation of the project if a satisfactory agreement is not reached. However, this usually is not feasible if the project is well underway, due to cancellation charges and schedule setbacks. The vendor is aware of this, but cannot be certain how resolved the customer would be if the issue is pressed. A much more effective ploy is to inform the vendor that future contract awards will be pending the outcome of the current computer system project. This should be the central theme of the negotiations, but real emphasis should be reserved for the final sessions.

Another effective tactic is to delay any scheduled partial payments. The negotiation session is a delay in itself, as a change order is usually pending the outcome of the meeting. This approach can be effective against the small firm with limited financial resources. However, if the vendor is much larger than the contracting firm or client, the situation can be reversed, since the customer may the the one with high interest payments to make.

Depending on the amount of money involved, the negotiators must be firm in their demands and should initially strive for every concession possible. Compromises are to be made in the latter stages of negotiations; not in the beginning. A "nice guy" approach, or appeals to the vendor's sense of fair play, usually have the opposite effect desired. It must be remembered that most companies are out to make money and will usually seize any opportunity to do so. A calm, firm and logical approach is best. Use of caucusing, when required, is highly recommended. If necessary, recess the meeting until the following day to allow each side time to think over their position and determine what concessions are required.

Change Order. The negotiations should be concluded with a written agreement on the final cost of the computer system and all services required. Every effort should be made to anticipate future changes and incorporate them into the agreement at this time. A change order should be issued immediately to reflect the agreement and authorize payment of additional funds.

The project usually runs smoothly from this point, but any significant changes could jeopardize results of the negotiations and open the door for more cost overruns. Changes should be avoided as much as possible. Any additional price increases, if not resolved immediately, should be allowed to accumulate. Successive negotiation sessions may be required if the cost overrun becomes significant again. However, this should not occur if the system design has been finalized in the initial negotiation meetings.

SOFTWARE CONFIGURATION

The digital or computer system must be programmed with the process information, control algorithms and operator interface instructions necessary for proper operation. Without the configuration data, the control system is incapable of performing its intended functions. System performance and pay off is ultimately determined by the software that is implemented.

Software configuration is normally initiated after the system has been purchased, and is developed concurrently with the system engineering and design. For small standard systems, the bulk of the software specification can be done in the latter stages of the project. Normally, point addresses and hardware assignments need to be defined early, as this information is required for equipment layouts, wiring lists and field terminal assignments. On larger systems, it is essential to assign the necessary personnel and start the software configuration early, due to the tremendous amount of data to be gathered and coded. Sufficient time must be allowed for loading and debugging the configuration data prior to acceptance testing.

Data Base Generation. Most microprocessor and computer based control systems come with standard, fill-in-the-blanks software for data acquisition, process control, alarming and operator displays. Data base generation requires identification of necessary process and control information for each point, and filling out appropriate forms. Information for the data base is gathered from many sources, such as piping and instrument drawings, instrument loop drawings, instrument specification sheets, instrument indexes and field terminal drawings.

The first step in developing the data base is to identify all process inputs and outputs by type (analog and digital). Each process input and any associated outputs are grouped together and called a "point." Each point record is built by coding the configuration data, such as hardware addresses, input conditioning requirements, filter constants, alarm limits, control algorithms, tuning constants and multiple loop configuration on to standard forms. Transmitter ranges, engineering units, deviation limits and tag descriptions are also required for each point for display purposes.

Data base generation for computer systems with a thousand points or more requires a major effort in data collecting, compiling and coding. Personnel familiar with computer systems and their software requirements should be dedicated to the task. Process and instrument information must be organized for each point in the system. It is important to ensure the latest data are available and new points identified. Decisions must be made as to what input processing, control algorithms, alarm types and other software functions are required for each point. Data base forms must be filled out, reviewed and updated. Depending on the type of control system, the forms are either punched on cards or entered into the system directly through a CRT console or programming device. The data base is then loaded and errors diagnosed by the system are corrected. Normally, the data base is in a constant state of revision, which continues through acceptance testing, commissioning, startup and system operation.

It may be advantageous to have the vendor perform some or all of the data base generation, including coding, keypunching, loading and debugging. This requirement should be identified early enough in the project, to be priced by the vendor during the competitive bidding stage.

Displays. Distributed control and process computer systems normally incorporate a CRT console in their design. Process indication and control are provided through a variety of displays. These displays are generated by standard software supported by a process computer, or are integral to the CRT console.

In general, the displays include plant overview, group and detailed formats (Figs. 3, 4, and 5). The user assigns points to various groups to perform indication, alarm and control functions. Once points are assigned to a particular display, other information pertaining to

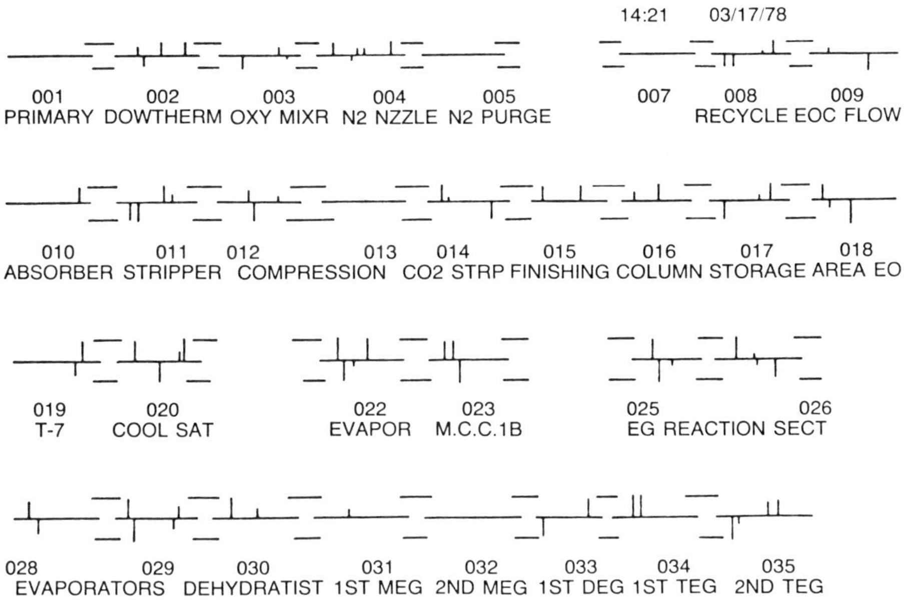

Figure 3. Overview display. (Courtesy Honeywell, Inc.)

that point can be assigned (such as controller modes, limits, scaling factors etc.).

The vendor normally supplies standard forms for display assignments. The forms are then entered directly into the operator console in an off-line mode of operation.

Some distributed control systems and most process computers offer a graphic display capability. Plant flow diagrams, schematics, one-line diagrams and other customized displays can be built. The displays are first laid out on special grid paper which establishes the field size of the CRT screen. Vessels, heaters, pumps, valves and other equipment are drawn with interconnecting process lines. Process data indication areas are defined for the display. Decisions are then made as to color, intensity, blink action and character size for each shape and process data readout.

It is good practice to keep the displays simple and avoid cluttering the screen with a lot of alphanumeric information. On some systems, tabs can be set on the display, allowing convenient paging to another display. This feature can be used on an overview display to provide a zooming effect. More information can be presented on the screen by positioning the cursor in the area of interest and using the tab function, which addresses a detailed display of the selected area.

Once the displays are laid out on paper, they are entered into the system through the CRT console. With limited graphic systems, shapes and lines are drawn using a special character set provided by the CRT display generator. With a

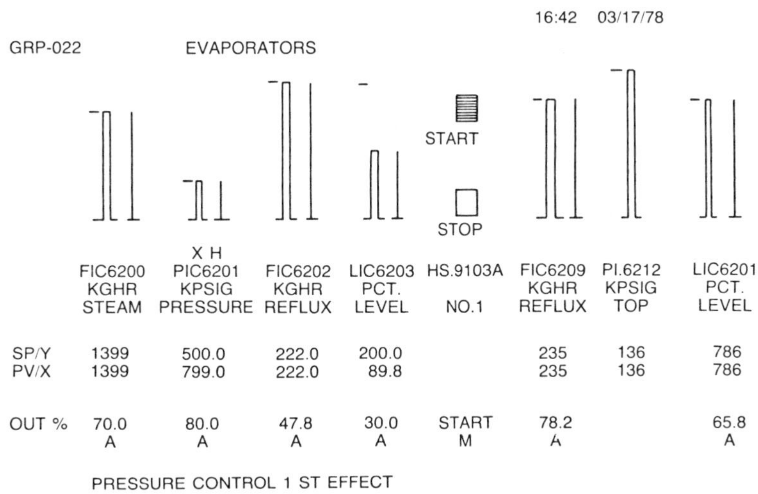

Figure 4. Group display. (Courtesy Honeywell, Inc.)

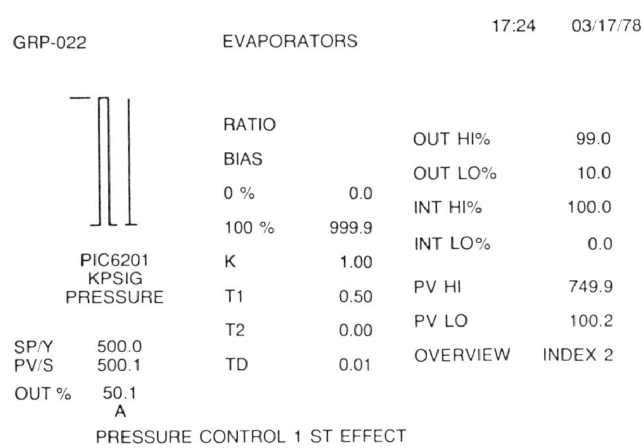

Figure 5. Detail display. (Courtesy Honeywell, Inc.)

full graphics system, circles, vector lines and other shapes are drawn by defining shapes mathematically. The computer actually draws the display using pre-programmed shape parameters, such as end points, slope, center, radius, etc.

Reports. Typical plant operations require the capability to format and print out process data, alarms, software configuration and system diagnostics. Most systems can interface to a printer, which provides a hardcopy of alphanumeric information displayed on the CRT console. Process computer systems are capable of storing historical data on selected points, and of generating reports on both a periodic and a demand basis.

The vendor normally provides standard report formats, such as summary, alarm history and operator action reports. Report types, description headings, tag numbers, point order and report scheduling are specified on standard fill-in-the-blank forms. Historical data points must be identified in the data base, as well as sample periods, process averaging and data storage requirements.

Custom reports are normally generated on computer systems that support a high level language, such as Fortran. The computer's operating system must allow Fortran programs access to the data base, and to historical files stored in bulk memory. Special formatting, point processing and data manipulation can be easily implemented using the high level programming capabilities that a process computer offers.

Custom Software. In many applications, the vendor's standard fill-in-the-blank software provides the necessary data acquisition, display and control functions. Most distributed control systems offer thirty or more control algorithms, multi-loop control, input conditioning, alarming, operator displays, multi-variable trending and logging as standard features. Process computer systems offer an even larger selection of preprogrammed functions. However, there may be special processing and control requirements that cannot be satisfied with the vendor's standard software.

Optimized operation of major equipment items and process units can require complex control schemes involving simultaneous solution of state variable equations, Z-transform transfer functions, and other forms of process modeling and multi-variable control. Material and steam balances normally are required for efficient plant operation. There may be requirements for special processing of plant information and generation of calculated variables to provide more meaningful data to the operator. In these situations, it is necessary to develop custom software.

Normally, the project engineer and client work in a joint effort to develop the required software. In some cases, the client may wish to develop custom software after the computer system is installed and operational. With large customized systems, where special software is necessary for plant operation, it may be expedient to contract out some of the software development to the vendor. Functional requirements, process models, algorithms, block diagram, and flow charts should be developed by the user and submitted to the vendor. Again, it is important to identify any software development requirements early in the project and to obtain firm pricing from the vendor.

MAINTENANCE AND SPARE PARTS

Maintenance and spare parts requirements should be addressed during the bid evaluation stage of the project. While the vendor may not quote the final cost for the desired level of support, a base line should be established from which future pricing can be developed. In general, the bid inquiry should instruct the vendors to list and price their recommended spare parts for startup and one year of operation. If the computer system is to be installed at a remote site or overseas, then the vendor should demonstrate his capability to maintain and support the equipment. Prices, terms and conditions should be quoted for onsite maintenance.

Normally, the vendor can recommend required spare parts. Type and quantity of spare parts stocked will be determined by the system's complexity, reliability of the hardware, and extent of critical functions performed by the system. The vendor may stock parts locally, which reduces spare parts requirements. As a minimum, spare circuit boards should be stocked for all major hardware items. In remote locations, or where the system operation is critical, it is also advisable to stock whole components, such as the central processor unit, controllers, CRT monitors, disc drives, and typers.

The vendor should issue the spare parts list early in the project, to allow purchase and delivery of spare parts in time for plant startup and operation. Each part should be identified with a part number, description and a reference to a maintenance manual. The manual should have an illustration of the part, identify its function, and describe how it is used to repair the system.

CUSTOMER ACCEPTANCE TEST

System testing is performed at several levels by the vendor and user. The vendor performs numerous tests on each equipment item and its individual components, as part of his quality control program. Additional inspection and testing takes place as the system is assembled. Finally, the vendor tests each input/output and exercises all components of the system.

Typically, problems are identified in the system hardware and software during the vendor's factory test. After all the bugs are worked out and the vendor is satisfied the system is fully operational, the customer is invited to conduct his own inspection and acceptance test at the factory.

Equipment Inspection. All equipment racks, operator consoles, peripheral devices and remote terminal units should be thoroughly inspected. As a *minimum,* the following items should be checked.

1. General appearance and workmanship.
2. All components specified in the purchase order for proper installation and labeling.
3. Wiring bundling and Panduit runs for neat installation.
4. Tight wire connections.
5. Cable connectors fully inserted and secured.
6. Correct tagging for all wiring and cables.
7. Operation of all cooling fans.
8. Condition of all light bulbs, LED indicators and fuses.
9. Terminal blocks and connectors labelled properly.
10. Defective relays and relay sockets.
11. Operation of all power supplies.
12. Power and instrument ground for tight connections. Also check for electrical isolation between instrument and power grounds with ohmmeter.
13. Operation of doors and latches.
14. Gaskets and sealing of NEMA type enclosures.
15. Equipment rack construction and paint.
16. CRT and keyboard operation.
17. CRT color resolution and convergence.
18. Bad keys in keyboard.
19. Hardwired instrumentation, such as pushbuttons, annunciators, trend recorders, etc. for proper installation.
20. Operation of peripheral devices, such as printers, disc units, tape drives and card readers.
21. Cable assemblies between equipment racks and peripherals. Look for loose connections, bent/broken pins, poor construction or assembly, and improper labeling.

Software Configuration. The data base, loop configuration, and displays should be checked against the data entry forms to ensure that the software configuration has been entered into the system correctly. Point tag numbers, hardware addresses, input conditioning, control algorithms, loop configuration, point descriptions, and display formats should be checked. Reports should be examined for proper descriptions and format. The software configuration can be conveniently checked by dumping a copy of all point displays, configuration data, and formatted reports on the line printer. The printout can then be placed next to the configuration forms to facilitate easy verification.

System Functional Test. Every component in the system should be exercised during the customer acceptance test. Each input and output should be tested to check wiring, point addressing, front end hardware, and shared components. Analog inputs are simulated with a resistor box or potentiometer, and a milliammeter, if transmitter power is supplied from the instrument racks. Otherwise, a 24 VDC power source is also required. Analog outputs are checked with the milliammeter. On large systems it is often expedient to use ganged potentiometers to simplify test procedures.

Digital inputs and outputs can be simulated several ways. Contact inputs can be shorted with test leads and alligator clips. An ohmmeter can be used to test contact output operation.

Programmable controller systems, having a large number of digital inputs and outputs, will require a test panel with toggle switches and lights to simulate field inputs and outputs. For systems performing complex sequencing, lights and switches can be mounted in a pegboard, with a copy of the process flowsheets pasted over the board. Lights and switches are set near valve and pump symbols to facilitate verification of the controller sequence and interlocks.

All inputs and outputs should be checked. Displays associated with each point should be called up. Point scaling, signal conditioning and controller output action should be checked. A/D calibration should be tested for each mux unit or controller by ramping the input signal and monitoring the display for proper process variable indication. Thermocouple linearization and cold junction compensation should be checked for each mux unit, using a precision millivolt source and thermocouple voltage tables. Wiring and point addressing can be checked on thermocouple inputs by shorting the terminals and reading the ambient temperature on the corresponding display.

Backup systems should be tested by failing primary components and observing the system's response. This is accomplished by powering down the device, and in some cases, removing a circuit board to simulate a specific type of failure. System diagnostics are tested by simulating failures, observing what alarms are initiated, and verifying the diagnostic information presented on the CRT console. Redundant data highways should be tested by disconnecting each cable separately, and observing the system's response. The data highway card for each highway device should be removed to test the system's response to failures in the data highway electronics.

A system availability test period should be scheduled for large computer based systems. Test duration can vary from a week to several months, and is determined by the system's critical control functions, system location (remote, overseas or local) and customer preferences. In particular, dual computer systems should be tested thoroughly to ensure that the fail-over system functions properly and that no data are lost.

At least 50 percent of the inputs should be loaded and adjusted on a periodic basis so input data are available for processing, display and verification. The availability test period should be revised backwards whenever the system fails a key test. Given a major failure and loss of data, the test period should be set back from several days to several weeks, depending on severity and duration of the failure. As with other system requirements, lengthy availability tests should be identified during bid evaluation and incorporated in the vendor's pricing.

Correcting Deficiencies. Failed components, defects in workmanship, faulty design, software bugs and errors in configuration should be corrected before signing off the system. Repairs should be made on the spot and added to the inspection checklist for approval. The checklist for complex customized systems may go through several revisions during the acceptance test. All items on the list should be inspected and approved once repairs or corrections have been made. If corrections are extensive enough, a return trip may be required to ensure all work has been completed. Final payment on the system is normally made after the system is approved. This should provide incentive for the vendor to correct deficiencies expeditiously. There may be some changes, however, which the vendor will insist are out of scope. Normally these changes can be negotiated on the spot and a change order issued. Major changes should be avoided, as they could hold up system delivery and incur large cost over-runs.

Shipping Arrangements. The project engineer should make final shipping arrangements with the vendor before concluding the acceptance test. Packing, crating, storage, and method of shipment should be finalized. The vendor should also specify what assistance he will provide during system installation and startup. The project engineer should confirm spare parts schedules to ensure delivery is in accordance with the final delivery date for the computer system.

INSTALLATION AND COMMISSIONING

The digital control system is shipped to the plant site after customer acceptance testing. The control building should be completed to the point where the computer room is enclosed, air conditioned and dust free. If the project schedule requires an early shipment of the system, then arrangements should be made to provide air conditioned storage facilities. If possible, the equipment should be vacuum packed before shipment to seal out dust and moisture.

Prior to system installation, provisions should be made for moving the equipment into the computer room. Floorplans for large systems should be laid out with a large central aisle and double doors on the end of the room, to facilitate the moving and placement of the equipment racks. If the computer room has a raised floor, field wiring, power cables, grounding conductors and cordsets should be placed in position before installing the floor tiles. It is good practice to have the vendor conduct a pre-installation inspection of the site to ensure all necessary preparations have been made.

In most cases, the vendor will provide the necessary personnel to supervise and assist in system installation. The contracting firm is responsible for making field wiring, power and ground connections to the computer system. Each step in the installation should be checked and a final inspection made to ensure all connections have been made properly. Each component is powered up and test routines performed to determine that it is functioning properly. The data base is then loaded into the system via cassette tapes, floppy disc, or disc pack. The vendor performs a system availability test to demonstrate to the customer that the computer system is fully operational and running.

Once the vendor has turned the system over to the user, a loop check is performed for each point. Copies of the process flowsheets, instrument loop drawings, and the software configuration forms should be readily available. All field instrumentation should be wired to the system's input/output racks, and the field wiring tested for shorts, ground faults, or open circuits. Each analog and digital input should be exercised, and the corresponding displays checked to verify the A/D calibration and the overall system operation.

The computer system is normally commissioned one loop at a time during plant startup. At this time, the computer system has been checked out, powered up and the data base loaded. As each loop is brought into service, the associated computer points are brought on-line and the loop is put on manual control. If the system functions properly, the system is placed on PID control and the loop is tuned.

For supervisory control systems, the analog controllers are brought on line first, and then the computer points are commissioned one loop at a time, in a data acquisition mode only. With distributed control systems, the microprocessor controllers are brought online one loop at a time. If the distributed system incorporates a process computer, then associated computer points are also commissioned simultaneously with each loop.

Normally, data base changes are required as each loop is brought on-line. Changes in tuning constants, alarm limits, control algorithms or loop configuration are often required. After necessary corrections have been made, the loop is placed on computer control. If the system is to perform direct digital control of the process, the computer control point should be turned on. If the computer system is to perform a supervisory role, the analog controller associated with the loop is switched to computer mode so setpoints and outputs can be manipulated from the system's operator console.

Once the computer system is on-line and the data base is operating properly, other system functions are commissioned, such as logs, reports, material balances and custom programs. After the system stabilizes, and operator and plant management interfaces have been established, process optimization routines are implemented and brought on-line. As new applications for the computer are found and future expansions planned, the system's hardware and software configurations will be in a constant state of change throughout its useful life.

3
How to Evaluate Distributed Computer Control Systems

S. Cocheo

Several types of companies offer distributed control systems, and the significant differences between these systems make them difficult to specify and to evaluate the wide variety of responses in the return proposals. First are the traditional suppliers of pneumatic and electronic instrumentation who have served the industry for decades, then there are the computer systems houses that have sprung up in just the last few years and offer fully integrated systems complete with hardware, and appropriate control system software. There are other suppliers who offer well-developed hardware systems but leave most of the control software in your hands.

With such a variety of offerings, you might be sold a system that is unsuitable for your plant's needs, with a programming challenge that you and your company do not have the skills to meet. There is a genuine shortage of control systems and chemical engineers who understand the process to be controlled, control system theory and practices, and who also have the computer programming and systems skills needed to make everything work together within clearly thought out targets for systems performance.

The objective of this chapter is to assist you in evaluating and establishing the desired and attainable goals that your company can accomplish for your plants using available distributed control systems.

SYSTEMS ENGINEERING COMES FIRST

To make best use of distributed computer control systems, it is more important than ever to plan carefully in a systems engineering phase. This should be done without recourse to lots of vendor contact, which can take your mind off your first goal, that is to understand fully the nature of the plant to be controlled and to determine the control strategy to be applied.

If you fail to take this step, your evaluation of various vendors' offerings will take place without much understanding of your real needs and goals. This can lead to faulty specifications, faulty evaluation, and a computer project that will require constant changes in hardware and software as you become aware of overlooked needs and goals. It is really more efficient to do your thinking first.

A computer project group should be set up consisting of engineering disciplines appropriate to the job. It should include an experienced chemical engineer with an understanding of the process and types of equipment that will be used on the job. Also, engineers familiar with control loop design and application, and with some experience in computers *are an asset.*

The plant to be controlled should be divided into functional units and their nature evaluated. This evaluation should consider any hazards associated with each unit's operation. Next, the interrelationship of these units should be studied similarly noting how performance of one unit will affect operation of previous and subsequent units.

As the plant's needs are understood, it then becomes possible to start applying the control strategy. It is not necessary to finalize the strategy at this point. However, an estimate of the number of control loops, indicators and contacts to be connected to the system is needed. This will have a bearing on the size of any vendor's distributed control system.

Next, estimate how many loops will be simple controllers or cascade, feed-forward, multi-variable, override, complex, etc. While control functions take place in software, there are some distributed control systems that cannot accommodate all the modes of control that an engineer may need. It is best to eliminate these systems from the competition early so that neither you nor the vendor wastes time and money. In addition, it is necessary to know what kinds of sensors and transmitters are to be used for each input, as some vendors provide different input cards for thermocouples, resistance temperature detectors, current transmitters, etc.

The speed with which each input can change is another important factor in system sizing. The more rapidly an input may change dictates how often it must be scanned. For example, flow rates may change very quickly, and when used in a control loop, they must be scanned at least once per second. By contrast, many temperature inputs may be scanned every 15 to 30 seconds. The less frequently the various process inputs need to be scanned, the more inputs one can "load" onto a microcomputer in the distributed-control system.

When this is complete, decide what goals you expect the computer system to achieve. There are distributed systems which can replace basic plant analog instrumentation and its associated annunciator system. These yield a little more in operating flexibility and performance than the earlier systems. If this type of equipment is what you want, select it. However, if you have more ambitious goals, equipment of greater sophistication must be specified.

As you set your goals, it is worthwhile to examine and realize the available basic functions that can be put to work in your plant by a modern digital process control system.

Within the basic area it is obvious that the computer is capable of performing the usual and unusual modes of control such as 3 mode feedback and feedforward, cascade, adaptive, override, non-linear and multi-variable. It also can perform the usual calculations previously performed by analog devices, such as adders/subtractors, multiplier/dividers, signal selectors, ratio and signal limiters.

Through cathode ray tube (CRT) video terminal and associated printers, the computer is capable of communicating the state of the process to the operator and accepting his commands within restraints imposed by plant management. Using its recording capacity, the computer can acquire and store historical data for analysis, reduction and reporting thereon. It can detect and alarm on out-of-tolerance conditions and log these conditions for later analysis. It can keep track of every command the operator has issued and thereby allow management to monitor his actions.

With proper selection of appropriate hardware and software, many additional functions become available. For example, energy consumed by your plant can be reduced significantly by the computer system provided, of course, that you know or can learn by experimentation, what process relationships affect energy consumed and generated in your plant. With additional knowledge of the process, other forms of plant unit or overall optimization become possible.

The only limits to the results that can be achieved with the computer really lie within the constraints of the equipment capabilities that are engineered within your plant and your company's ability to understand the complex, dynamic relationships of the process.

Along with evaluation of your needs and goals should be an evaluation of the personnel you have or need to implement and maintain these sophisticated control systems. The plant that relies soley on the manufacturer to maintain and repair his system is asking for trouble. Similarly, each plant should possess suitable computer-trained personnel to maintain and improve its software as knowledge of its plant grows. While these may appear to be expensive arrangements, they should really more than pay for themselves. If they do not, then the computer system was probably misapplied to the plant in the first place. Many plants are small and simple in nature and still can be handled satisfactorily with simpler systems.

Location of the plant and availability of parts and service in that location is important in your choice of equipment. If your plant is ideally located where these are easily available, then repair and maintenance difficulties are minimized. If your plant is to be located in more backward regions then you must be prepared to staff with your own qualified personnel and stock adequate spares to be self-sufficient.

Architecture Analysis

Let us now look at a "generic" style of distributed computer control system (Fig. 1). While the system architecture displayed does not reflect all those on the market, all systems must perform at least the same basic functions.

It is necessary to recognize that when one has a distributed computer architecture he has a system with "intelligence" at several locations or "nodes." What is less apparent is that there is, operating within the hardware architecture, a software architecture. Each vendor must present a system to the market that performs certain basic and, in some cases, advanced functions. The way software functions are distributed within the variety of hardware architectures can have a profound effect on how each system will perform when one or more portions of the system fail.

Part of the point of this chapter is to show the reader that various vendors perform the same functions in different nodes. The reader must understand how these differences may affect his plant.

Due to the rapid pace of evolution of hardware and software, the user constantly must be aware of new developments lest he accept a system that is becoming obsolete even as he writes his purchase order.

Each system, in its basic operation, has a time cycle associated in each of its intelligence nodes. It is necessary to examine these time cycles to ensure that they can

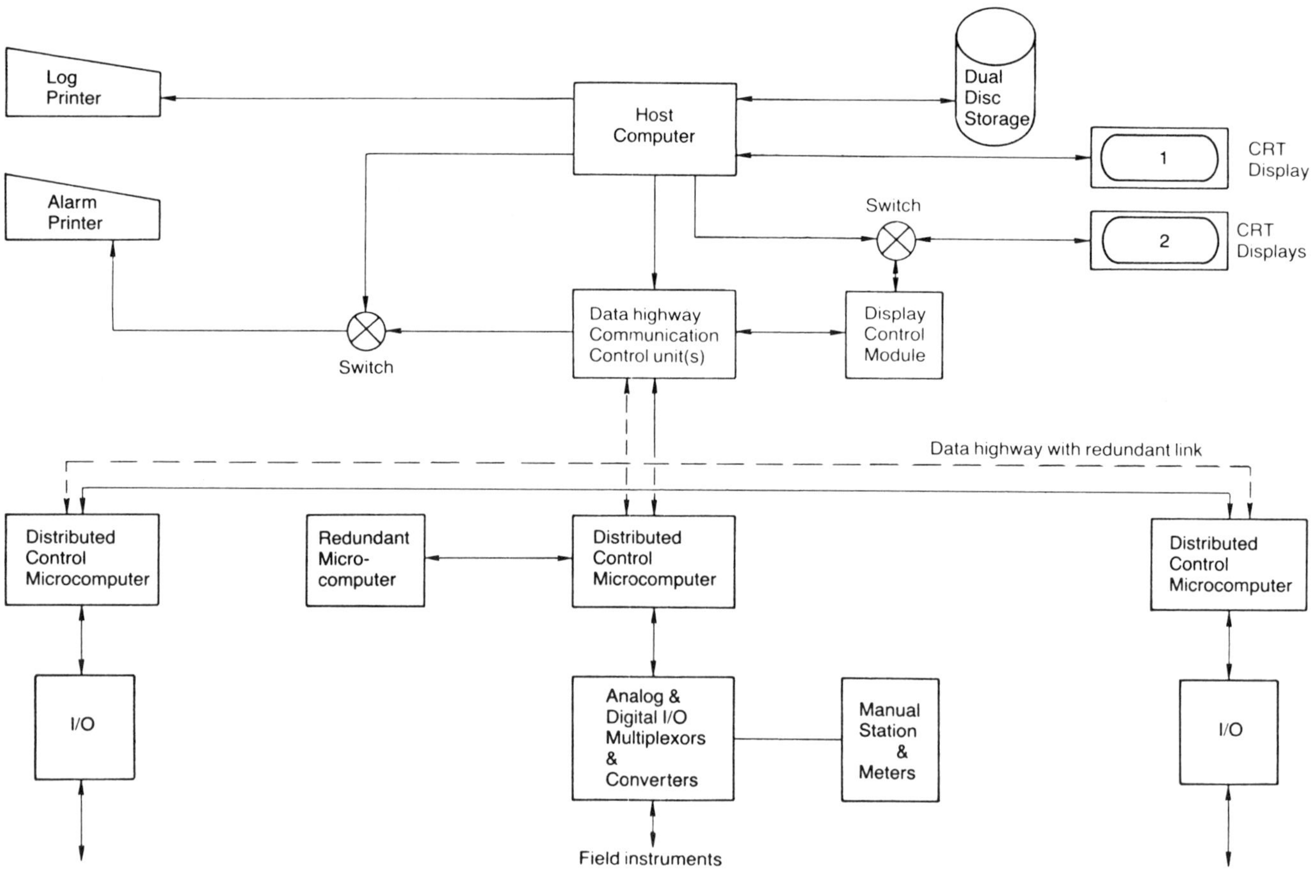

Figure 1. Typical distributed control system with host.

handle the fastest occurring events within the plant. For example, if under a plant upset condition it is necessary to detect an abnormal condition and close a control valve all within 1/2 second, then a distributed computer system with a scan cycle of one second is dangerous. In another plant, this same microcomputer might be idle for long periods of time.

Careful, critical analysis of your plant's true needs is, therefore, the key to the proper specification and evaluation of the systems offered. With the right use of this key you will spend the right amount of money on the right system that will do the job for your plant. Start-up will proceed more smoothly and continued operation of the plant will be satisfactory because pitfalls have been considered and appropriate solutions implemented.

Advantages of Distributed Systems

It would be useful at this point to briefly review the advantages of distributed computer control systems versus central computer systems as this reveals factors that must be understood for the evaluation process.

Using a central computer system exposes the plant to the risk that the system might fail. This possibility is minimized by having one or more back-up schemes available such as:

- A back-up computer
- Analog controllers
- Manual loading stations
- Various combinations of these

These back-up methods are costly. The distributed computer control system also has this need for backup. However, a distributed system also distributes the risk of failure into smaller subsections. Failure of one node should not cause other portions of the system to fail. Costs for back-up are reduced if the system architecture is designed to fail "gracefully." This type of systems design reduces the kinds and costs of redundancy and back-up needed. This enables the system to temporarily give up some of the more sophisticated aspects of computer control and fail gracefully to a less sophisticated level of control that still allows the operator to control the plant in an acceptable and safe manner. Various examples of this are pointed out in following sections.

Substantial reduction of wiring costs is another reason why distributed control systems enjoy an advantage over central computer and analog systems. Referring again to Fig. 1, visualize the distributed control microcomputers being located in various portions of the plant near the equipment each is controlling. The data highway is generally a twisted shielded pair of wires or some type of coaxial cable. This cable carries all process and computer system information necessary for efficient plant operation via a high speed digital data communication system. This cable replaces many miles of copper cables for sensor inputs and final control element outputs. In some large plants with thousands of inputs and hundreds of outputs the reduced costs of material and labor can amount to significant savings that can cost-justify a distributed control system.

Plant Hazard Analysis

As one comes to understand the capability of a distributed control system to fail gracefully, there comes a need to evaluate how each system proposed behaves in adverse circumstances. But this analysis cannot take place in a vacuum, for again relying on the system engineering phase of the project, a hazard analysis of your plant is needed first.

Is your plant a stable, non-hazardous process such as a water or sewage treatment facility? Possibly, parts of your plant might explode or in other ways damage itself, plant personnel, and the surrounding community if the control equipment failed. Most real plants have some hazard associated with some portions of the plant. You must know yours!

Each proposed system must be evaluated with your plant hazard needs in mind. It is necessary to study each system from a wary and critical viewpoint. Each system's hardware and software must be studied to learn how it will behave when each portion or node malfunctions or fails. To do this one must understand which functions are performed in which nodes. In this way, it is possible to visualize how computer system failure affects your plant.

Let us now do this type of analysis on the system shown in Fig. 1. Routine process control is performed in the distributed microcomputers with their associated input/output cards, multiplexors, analog to digital and digital to analog converters. Failure of the computer at this node will leave the control valves in their last position due to hold circuitry on the output card. If a back-up microcomputer is installed it will automatically take over control without a bump to the process. If there is no back-up microcomputer, the manual analog loading stations (Fig. 2) may be switched in to maintain control with an operator in attendance.

In many plants the microcomputer back-up is not needed as plant unit operations are not critical to itself or other units of the plant, while in other plants its presence is vital. Automatic switchover, hold circuit operation and, finally, manual loading stations are examples of failing gracefully for they each allow uninterrupted control of the plant unit at some level of security. The minimum level of security needed in each of your plant's units is determined by your analysis.

Note that failure of this microcomputer does not affect the safe functioning of any other micro, the data highway, the host, the CRT's, etc.

On the higher end of the system, observe that the CRT's are connected to the host computer. If the host

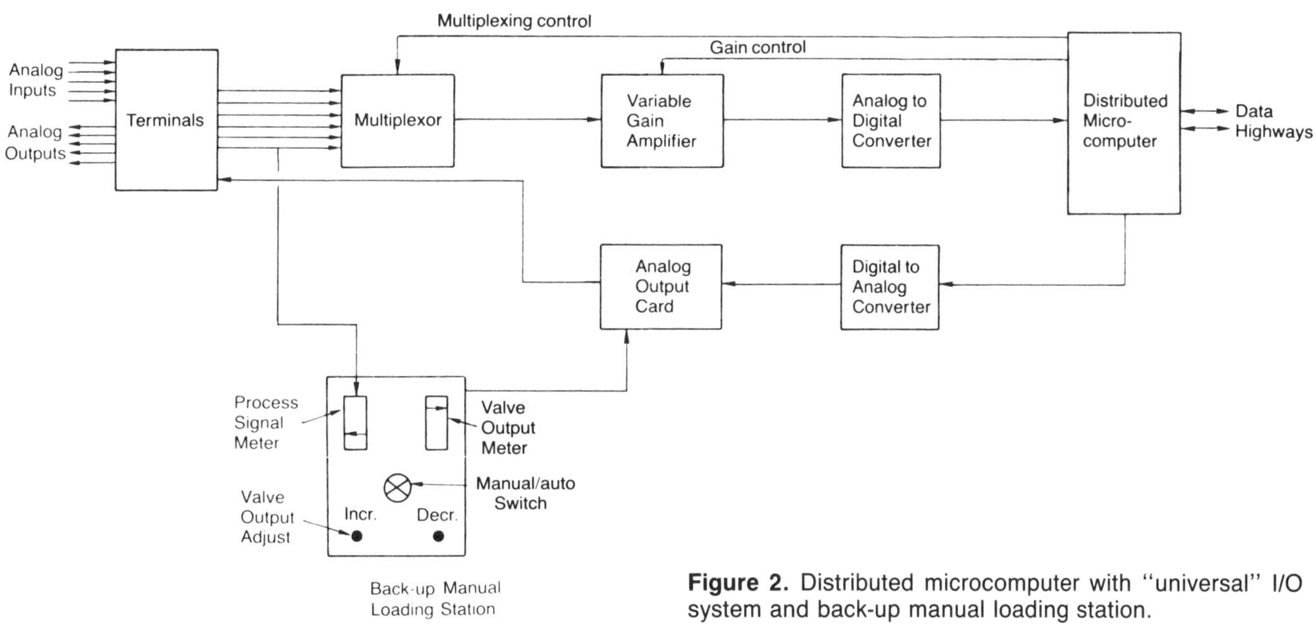

Figure 2. Distributed microcomputer with "universal" I/O system and back-up manual loading station.

computer should fail no displays could be seen. Your operator would not know what was going on except what he could see through the manual loading stations. If the distributed computers were located throughout the plant, this would be difficult.

To alleviate this problem our hypothetical system has another microcomputer to format displays for the CRT's as a back-up to the host. This unit automatically switches in when the host goes down and can provide those displays essential for operation. Many of the more sophisticated displays such as flowsheet graphics would be lost until the host was repaired. Also, other host functions such as optimization would be lost. However, operation in most plants could continue in a safe manner for the time it would take to exchange a plug-in circuit card to re-establish host operation.

In this case, a $50,000 host minicomputer was backed-up by a $15,000 microcomputer that demonstrates again the principle of failing gracefully and the lower cost of redundancy and back-up associated with distributed systems.

A similar analysis must be made for all nodes of the computer system including evaluation of whether failure of one node can affect, degrade, or block operation of other nodes. Each deterioration of computer system operation then must be evaluated versus the effect on your plant. Each competitive system similarly must be inspected and understood as to its own operation versus your needs, and also against the other proposals.

Many distributed computer systems offer several options for redundancy and back-up. The analysis just described will help you determine which options you will need with each system. Since each system is likely to be different in several ways, it is necessary to add to each system's price those options required to meet your needs and make them "equal."

HARDWARE EVALUATION

As you are beginning to see, examination of distributed control systems leads to some difficult and very detailed evaluation. To assist in performing this evaluation, it is a good idea to make an evaluation form to record your findings. This is needed as there are too many facts to keep in mind simultaneously. It is easy to mix up the facts from one vendor and apply them to another.

In a recent evaluation, the author started his analysis by briefly stating each specification or need on a form and then made columns for each vendor. This took a 17-in. by 11-in. sheet of paper for each page and the entire evaluation ran 14 pages.

As this form was gradually filled in and completed, it was remarkable how differently each vendor structured his systems hardware and software. Each vendor's original price was very close. However, as various needs were reviewed, each vendor had to add costs to bring his offering into compliance with the requirements. Their final prices were far apart. This whole process took several weeks and, without the evaluation form, it would have been difficult to keep all points clearly in mind. It also is essential to have a written basis that may be referred to at a later date.

As some systems may be eliminated along the way, you may ultimately find yourself faced with two or more vendors who meet your needs at prices that are competitive. At this point, you will need to contact previous users of each system regarding their experience with each vendor. This typically is done when evaluating equipment of substantial cost.

If the evaluators have no previous experience with any of the systems still remaining in contention, it is suggested that each remaining vendor's training course be taken as a further step in the evaluation. It is surprising how difficult some systems are to program despite the vendor's contrary claims. Time and budgets frequently will not allow this to take place and managers don't want to spend the money "unnecessarily." However, if you are going to have to live with the decision for many years, it could be time and money well spent.

The balance of this article examines the various items of hardware and software that collectively perform as a distributed computer system. The material presented is from a top-down viewpoint. The things to look for and to watch out for are included in each section.

Host Computer Evaluation

The host computer occupies the highest, most responsible position within the hierarchy. It generally consists of a 16-bit minicomputer with a reasonably fast memory access time. But before delving into these details, it should be noted that several systems currently offered have no host computer or only furnish one as an optional item to provide advanced functions. These simpler distributed systems provide a subset of the functions furnished with a more general system. They do this by putting microcomputers into each CRT display. From one viewpoint, it is as if they distributed the host computer into several nodes. The utility of this approach is again dependent on your needs and budget.

Functions of the host computer are manifold and this chapter can do no more than point out the main features. It should be realized that the host as a minicomputer is a general purpose item of computer equipment. It therefore can function in many ways depending on the collection of programs that are running it, and on the time it has available for performing them.

The host computer, along with its distributed network as used by our industry, belongs to a class of so-called "real-time" computing machines. Real-time signifies that the hardware-software combination will perform a desired function with sufficient speed so this function will be accomplished within the time constraints required by the real world. For our application this means that it will control our plants, gather the required data, display it, record it and accept commands from the operators without faltering in any essential task.

The host is capable of doing real time functions by virtue of its own internal architecture combined with the collection of properly written and organized software. The software is divided into three main parts:

- The "operating system"
- The process control programs
- User written software

These software facilities are explained in detail in later portions of this article.

The host's main work consists of the following functions:

1. Format and display on the CRT screen information needed by the operator to know the status of the plant.
2. Format and display alarm conditions on the CRT and print the same information on the alarm printer for later analysis if needed.
3. Collect data through the microcomputers and record it on magnetic media, such as a disc or drum memory, for current or later display on the CRT or output on the logging printer. This is called real-time and historical trending. The printed data can either be the raw values recorded or may be further analyzed by additional programs.
4. Format and display of graphic drawings similar to process flow-sheet type information. These graphic displays will usually have "live" process and set point data shown which will be updated every few seconds.
5. Perform calculations to provide one or more of the goals of plant or energy optimization, plant or equipment performance analysis, and similar types of analysis.

In evaluating the host, one must look at both how it functions as a whole and how it fits into the system architecture.

Host Memory Evaluation. One important aspect of the host is size and speed of its memory, for this is the main factor in estimating relative speeds of various host offerings.

Memory speed is important because the basic cycle of retrieving a value from memory is a frequent task. Once the value is moved to internal registers, calculation and other manipulation is much faster than memory retrieval due to superior circuitry. Therefore, if you are offered two machines with one microsecond and a 600 nanosecond memory access time, respectively, you can be confident that the 600 nanosecond machine will be somewhat faster.

Some machines offer a feature known as "cache memory" which can decrease significantly memory access time. Here is how it works: Cache memory is very fast and expensive. Typically, only a small quantity is purchased and installed. This is usually in the order of 2K to 8K bytes in size. The computer is organized to look, generally to cache memory, for each succeeding instruction. When the next instruction is not present in cache, the section of main memory containing it is moved into cache memory in a block transfer and operation continues. If main memory speed is 600 nanoseconds and cache memory is 100 nanoseconds, you would find that the effective average speed of operation would proceed just a little slower than cache, say 125 nanoseconds, while the cost would increase by only a few thousand dollars. Therefore, a small additional investment confers a quantum jump in performance. All host systems do not provide this option.

Memory size is important because it determines how much of the software may reside in memory versus how much must remain out on the disc to be called in on demand. The speed of a disc transfer is slow compared to a call to an internal memory location. When the required software is a vital program needed for real time operation, then excessive peripheral transfers like this can threaten to undermine the basic purpose of the machine. Some distributed systems have relatively small main memories and no capacity for further expansion or upgrading. They must constantly roll portions of their operating system software in and out of a section of main memory, called an overlay region, in an attempt to preserve orderly operation. Under adverse circumstances, this approach may cause a condition known as "thrashing." Thrashing is a circumstance where there is so much demand for different sections of the operating system or other disc resident software that most of the computer's time is spent in performing these transfers,thereby doing little productive work.

Where possible, an engineer should not order a system whose memory capacity is entirely consumed by the initial program loading. There will always be future needs that will require additional programs. Your system should either have some spare memory installed or at least have capacity for later additions. Avoid the necessity of scrapping the current system to make way for the next larger system. In the long run, it may be more cost effective to buy that larger system now and pay a 10 to 15% premium as this will preserve your company's investment. It would be foolish to find yourself with limited memory capacity and later learn that the manufacturer has furnished you with his largest capacity system.

Host's Age Evaluation. Another trap that must be avoided is the purchase of a host system that is an old design. It is wise to inquire regarding the age of the design for there comes a time in the life-cycle of any computer when the manufacturer can no longer afford to provide spare parts, updated software, and documentation. If the system is purchased close to the end of its life cycle, just how long will it be before you are forced to buy new equipment and have a major plant shut-down while it is installed?

Finally, make sure the host has suitable redundancy or other back-up within the system architecture, as described previously.

CRT Evaluation

The next system to consider is the operator/machine interface. This generally consists of several color or black and white CRT displays with operating keyboards.

Each manufacturer offers certain standard displays through which operators may observe and command plant operation. A typical arrangement is to have several CRT's, each with a different display invoked, but yet with the ability to be changed to another display as circumstances demand.

Certain display types have evolved during development of process control computer systems. In general, these have attempted to simulate the view that an operator would have if he were running a conventional plant with traditional miniature analog instrumentation. Each display type shows a view of a control board as it looks from 15 or more feet away up to one foot from a single controller or recorder. In the arrangement of these displays, a philosophy of "Management by Exception" is employed. That is, the operator is only shown enough information to know that everything is all right. When conditions depart from normal, he can see or call up displays of increasing detail depending on how much he needs to know or change to correct process upsets. The following are typical displays and their function:

- The "Overview Display" shows a complete panel board from a simulated distance of 10-15 feet. On this display there is an overview of all controllers/indicators, arranged in logical groups just sufficient for the operator to see that everything is "OK."
- The "Loop Group Display" shows a closer view (3-5 feet) of a logical group of eight simulated analog controllers/indicators. This view allows the operator to zero in on a plant section that needs attention. It also allows the operator to see greater detail on each of the "instruments", such as set point, process value, output to valve, deviation from set point, and deviation alarm values.
- The "Individual Loop Display" moves the operator still closer to view one loop. This finer view shows him all the detail he had in the eight loop group plus additional information such as tuning constants, process alarm values, and output limits.
- The "Real Time Trending Display" shows a graphical representation of the process value over the past 10 to 20 minutes (depending on manufacturer). This graph is updated so that the latest time period is always displayed. Some manufacturers allow two or more real time trends to be shown simultaneously.
- The "Historical Trend Display" shows a graphical representation of the process value over longer periods such as several hours, days or longer. Of course, this graph represents the data in averaged form over these time periods. The graph is not updated with time and that is what differentiates a real-time trend from an historical trend.
- The "Alarm Summary Display" shows, in an alphanumeric table, all active alarms and their status such as acknowledged or unacknowledged, and whether each is a warning or hazardous alarm condition. Alarm conditions are also shown on the overview, loop group, and individual loop displays by color changes to the displayed information.
- The "Graphic Display" is a special feature that shows a graphic representation of a section of a flow sheet. Special programming software allows the creation of these special displays whose only limitations are one's imagination and the physical size of the screen. Real time information, constantly updated, may be included in these graphic displays both in alphanumeric form (such as set point and process values) and graphical form (such as seeing a vessel "fill up" as its level changes).

In typical operation with a plant in steady state condition, the operator would generally set three CRT's as follows:

Overview display
Loop group display
Alarm summary display

Depending on the plant or changes in conditions, he would call up other "standard" displays. Other specialized displays can be created for unusual plant needs.

One important factor is where CRT's are located within the system architecture, and basically how they work. Some are "dumb" terminals having no computer within them. Others are "smart" as they contain a microprocessor.

Generally, smart CRT's are used in a hostless system where they are connected directly to the data highway. They do not have sufficient computer power to handle the more complex displays such as graphics. One system currently offered puts smart CRT's on the data highway for the standard displays and a less smart CRT on their optional host for graphic displays.

As with any other item of computer equipment, one must allow for CRT failure. It is a good idea to provide at least one more CRT than is absolutely needed for safe operation. Due to this precaution, interchangeability of CRT's is advisable.

The dumb CRT always must be connected to a host or other external computer. With this style architecture, a host back-up is recommended.

There is controversy as to which type of CRT is best. The dumb type run in the $7,000 to $10,000 range. The smart ones cost from $25,000 to $35,000. Actually, each has a viable place depending on whether a hosted or hostless system is needed for the plant.

The operator's keyboard is an important aspect of the CRT terminal. Some systems use a typewriterlike keyboard where the various keys are color coded and renamed to control functions as required. Other systems use a completely different layout where separate groups of keys have been arranged according to their functions. Sometimes these keys are color coded.

Each system offered should be reviewed to see that the keyboard is easy to learn and use and that the various standard CRT displays are easy to understand. In a plant upset condition the operator should not have to struggle to figure out how to go from display to display or what will happen if he pushes certain keys. One purpose of a distributed control system is to provide better control and information. It also should reduce strain on the operator by simplifying operation. Reject any system that does not provide these features. Well designed systems must be adapted to human needs.

Disc Recording System Evaluation

Host computers need magnetic recording media for medium speed memory functions. This usually is provided by rotating disc or drum memories with the former providing greater flexibility and generally is more popular.

There are two types of discs: rigid and floppy. The rigid disc pack is provided with one or more high quality magnetic surfaces that can store large amounts of data and software. Recording density depends on the nature and quality of the electronic interface package furnished with the disc. Total recording capacity is determined by recording density, number of surfaces and number of tracks per surface.

Rigid discs turn at high speeds to minimize "latency time" that is the time required for the desired information to rotate under the read/write heads. Each surface is divided into circular regions called tracks. When the computer program calls for a read or write operation to the disc, the read/write heads move from the track they are over to the track containing the information. This is a "seek" operation and time taken is called "seek time." Total access time to obtain or record information is then the sum of seek time plus latency time plus the time to transfer data to or from the computer. Compared to the computers internal speed this is a relatively slow operation.

With these basic factors in mind, what should you look for in a distributed control system that includes one or more disc drives?

A comparison should be made based on recording capacity per disc drive, number of disc drives, and their total access time.

Capacity is important because the disc system contains the set of manufacturer's supplied computer systems programs called the "operating system," which is described later. Portions of the operating system are held in central memory. The entire operating system is held on the disc for initial program loading with the unloaded portion called in as circumstances demand. All other programs that are running or can be run are stored similarly. In addition, the disc system is the medium that records data gathered during plant operation that provides the basis of the real-time and historical trending functions as well as any data analysis and reporting printouts that are desired. Total capacity is, of course, the sum of the individual drives. When comparing capacities make sure that the units quoted are the same. A five-megabyte disc would be half the capacity of a five-megaword disc if the word size was two bytes. (A byte is eight binary bits.)

Systems that provide only one disc drive can cause a problem. There is no redundancy, and disc drives need regular maintenance, for which they must be shutdown. These maintenance periods are more frequent than plant shutdowns. Any disc drive can fail and although you may be able to do without data acquisition you may not be able to do without the unloaded portions of the operating system. Secondly, disc systems occasionally are subject to a failure mode known as head crashes when the read/write heads damage the recording surface. If you only have one drive you cannot create and update a back-up disc nor have a drive to run the back-up.

Improvement of computer program speed is another reason why having more than one drive is useful. When all software and data reside on one disc, your system is limited to one disc access at a time. With your programs and data recording areas separated on two or more drives, simultaneous accesses can take place. This enhances total system thruput.

Floppy discs are soft plastic discs which turn inside a cardboard jacket. Due to their slower speed and smaller size they serve a minor use in process control applications. Some systems incorporate floppy disc drives to provide an inexpensive method of transferring programs from one location to another. They also may be used for long-term storage of acquired plant data in much the same way one might save circular or strip charts. When so used, the data are in computer readable form suitable for data analysis.

Data Highway Evaluation

As we work our way down from the host to the distributed control microcomputers we run into the data highway. Data communication is one of the more difficult areas of computer technology to master. For the general user, however, there are just a few aspects to check out to verify that the data highway system is well designed and implemented. The two main areas are the data highway cable itself, and how the hardware/software functions to control the highway.

The data highway usually consists of a shielded twisted pair or a coaxial cable. Most systems support a data highway length of 5,000 feet or more, which is usually more than sufficient to handle any real plant. If communication over greater distances is required, then different hardware/software would be required that generally would not be associated with distributed computer control system communication. It is the availability of these high-speed communication links that serves to replace miles of loop interconnecting wiring and saves substantial wire purchase and installation costs.

To ensure communication integrity, a dual data highway cable is a necessity. Each of these redundant cables should be routed to its connection points via pathways that are as widely separated as possible. This will enable the system to continue working in the event of fire, explosions and mechanical damage.

Hardware associated with this dual highway should detect failure of either side of the highway and also alert the operator. The better conceived systems switch to the second data highway on failure and actively use both paths alternatively so that the system will detect a failed highway immediately.

The data highway is really more than a wire connection for it also includes whatever hardware and software is needed to package data within suitable start and stop "Protocols" and to detect and handle transmission errors. In some systems this function is carried out by the computer equipment attached to the data highway. In the better systems a separate microcomputer is used. It is usually called the "Communication Control Unit" or "Highway Traffic Director." When the separate communication controller is not used the parts of the system responsible for communication carry a substantial load, which precludes their use for other functions. One system uses its host for communications handling. Due to this loading the host does not have sufficient time to do any optimization or energy management programs. When these functions are needed, a higher level computer is furnished with the system.

Whenever data are being transmitted a method of detecting and handling errors usually is employed. A variety of methods is in use and choice depends on application. In distributed control system data highways, the system may be transmitting up to several thousand feet with the cable running near many sources of electrical noise. All data highway systems, therefore, use one of the methods. Generally, these methods cause up to several retransmissions of the data. If this fails an alarm is initiated.

Highway speed is an important factor in evaluation. Higher speed improves the ability to function in real-time and partly determines the number of distributed nodes the system can support. In evaluating this factor it is important to understand that some systems accommodate only a few inputs in each node whereas other systems handle a larger number.

Distributed Computer Evaluation Including Input/Output Hardware

Let us now look at the variety of microcomputers that are used at the lowest rung in the distributed computer network. As in other facets that have been discussed, several approaches are currently offered.

A few systems provide individual controllers for each loop. Some have field located controllers mounted on or near their associated valves. Others have panel-board-mounted controllers similar in appearance to an analog controller.

More sophisticated and useful systems, however, usually provide from a few input/outputs per node up to a hundred or more. One recently announced system allows you to blend a combination of individual units with several types of multiple input units. This flexibility can be helpful in some control situations.

Some systems combine distributed control microcomputers with one or more types of multiplexers which handle non-control type inputs.

All of these units attach to the data highway as all the inputs must be capable of being displayed on the CRT system.

One of the most important ideas to be grasped is the concept that in most of these systems the microcomputer is only devoting a percentage of its time to any input, output, or program. In a real-time system speed of the computer, organization of the operating system, and the programs to be run must be structured properly to allow the system to get around to everything assigned without failure. It is like a juggler who keeps several objects in flight. He gives each one just enough attention to maintain it in motion. From this you can see that each distributed microcomputer and multiplexer have built-in time cycles that must be understood or they may adversely affect your process.

Systems with many individual controllers have an advantage from a time viewpoint, but many individual controllers cost more, require panels and space to mount and, in general, are more difficult to configure into complex control schemes such as a feed forward loop. Multiple loop controllers may be slower, but not necessarily. Some systems are fast enough to accommodate most real-time control situations. The point here is that you must know what you need and know that this is a factor to be specified and examined as system design and proposal evaluation proceeds.

Associated with each distributed microcomputer and multiplexer is its input/output (I/O) hardware. There are a number of systems offered. Some, working on a base of previously designed I/O boards for their analog systems, continue to offer the same equipment for their computer systems. Other companies have designed special hardware "boxes" for each type of process input. They offer one box for standard current signals, another for thermocouple signals, and still another for resistance temperature detectors.

Other companies have taken a fresh view of the input/output problem and have designed and manufactured an input card that is almost universal. Remember, what we are trying to do is convert any analog signal into an equivalent digital signal. Some of the functions that some manufacturers perform in their hardware, such as thermocouple linearization, can be more easily and cheaply accomplished in software.

In the "universal" I/O systems each input is multiplexed, one at a time into a variable gain amplifier. Gain of this amplifier is under program control. When output of the amplifier has stabilized, the signal goes into an analog to digital (A/D) converter from where the digitized signal is passed to a reserved location in computer memory. This is shown in Fig. 2. After that, the control or data acquisition programs in effect perform their respective functions. The control programs calculate an output that goes out through a digital to analog converter whose value is maintained in a holding circuit card that is connected to the final control element. This card will maintain the last signaled output even if the computer should fail as long as proper power is maintained. The data acquisition and control programs pass the process value and calculated valve output, via the data highway, to the host or smart CRT system for display, recording, trending, and printed reports as applicable.

Some systems offer a local black and white CRT display that works off of the distributed microcomputer. This can be helpful from a maintenance and operational viewpoint for the systems where the microcomputers are scattered throughout the plant in order to reduce wiring costs.

It is at this level that it is necessary to carefully evaluate and decide what types of redundancy and back-up your plant needs. There are two types of redundancy available. One is automatic switch-in of a standby microcomputer, which is the most expensive approach. Another method is to have a "hot spare" on standby. The decision as to which is needed depends on the plant requirements discussed previously and where the distributed microcomputers are physically located.

Despite the wire savings feature available with distributed systems, many smaller plants choose to keep the entire system within the control house. If this is done, the hot spare can be provided with cables to be connected quickly to the I/O equipment. It is already powered up and only needs to have the appropriate configuration program down-loaded from storage. Depending on the manufacturer this can take from a few seconds to a few minutes. In some systems a few plug-in boards are exchanged instead.

All soundly designed systems spend a small portion of their time doing self-diagnostic checks and report any malfunctions immediately. Some circuit boards contain light-emitting diodes which signal one of several types of board failure. Knowing which board to exchange then is easily determined.

Meanwhile, what is keeping the plant under control? Well, if you had automatic switch-over the redundant computer is running satisfactorily and the plant remains in control. If not, remember the holding circuit should be maintaining at least the last valve output given by the computer. Most systems use some form of analog backup as the final defense against plant upset. This can be analog controllers with or without automatic tracking of computer output or manual loading stations with a process variable and valve output indicators.

These days process control computing equipment is very reliable. In addition, well designed systems signal their own failure and tell you what needs repair. Therefore, quick repair is typical. In most systems application of digital back-up where needed, with manual analog back-up elsewhere, appears to be the most cost-effective and safest choice. It would be a poor choice to back-up a distributed control system with a complete automatic analog control system. The advantage of distributed control is distribution of risk. If the systems evaluation is done properly, then it will be easy to choose the least expensive back-up systems needed by each plant section.

Power Supply Evaluation

There is one other hardware area that can be critical to system operation and that is the power supply to the entire computer-sensor-control valve system.

Some distributed control systems have been designed around computer hardware from different manufacturers and as a result require several voltages to provide for system operation. Other systems conveniently have been designed to operate entirely on a single voltage, usually 24 VDC. While there is nothing wrong with a variety of voltages, the engineer should make certain that a redundant supply is provided for each voltage required. This should be done for both the computer system itself as well as for the transmitter and control valve supplies.

If commercial power to the system might be lost, consideration must be given to providing an uninterruptible power supply. This UPS system must transfer immediately to the back-up source without voltage fluctuation. The entire power supply system should be tied to a separate computer system ground from the plant ground. Each power input and signal input/output should be carefully protected from being connected to plant ground. If grounded tip thermocouples are used their shields must be kept isolated from other input shields. These precautions are necessary to prevent ground loops that will introduce unwanted and possibly damaging voltages into the computer system. The resulting noise at least will produce errors in the input signals and cause faulty control system responses.

SOFTWARE EVALUATION

No article on distributed control system would be complete without an examination of the software that drives the computer system. As one wag said "Software makes hardware happen!"

Software breaks down into three essential parts:

- Operating system provided by the manufacturer.
- Process control programs provided by the manufacturer or the systems house that assembled and coordinated the system.
- The programs that you, the user, writes to accommodate your special needs.

The Operating System

The operating system is one of the more esoteric subjects in the area of computer science. Its purposes are manifold. In earlier stages of computer system development, rudimentary operating systems were called by names such as "The Monitor," and "The Executive."

As this has developed in recent times, the operating system is a collection of computer systems programs whose function is to be both a master and servant of the computer system and its users.

Just as we hope to control a plant by a computer system, so it is necessary to control or govern what is happening within the computer itself. Many types of operating systems have been and are being developed. Their organization and structure depends on the end use of the computer system. Many manufacturers offer several choices of operating systems in the same machine for different applications. The goal of all operating systems is the same: to optimize management of all computer resources, to reduce time taken to perform each task, and to increase the amount of work (thruput) accomplished by the system.

In addition to providing the governance there are other programs, considered part of the operating system, that perform many useful functions and services. This is done so that you need not write a program to accomplish a particular function that is widely needed by most computer system users.

In terms of the needs of a process control computer system, the kind of operating system needed is a real-time, multi-programming system. When an operating system is a multi-programming system this means that it gives the appearance of simultaneously working on two or more programs. In practice this is not true. What really happens is that the computer will work on one program for a few fractions of a second until it encounters an instruction that will, in computer timing, take significant waiting time. It will start that instruction and then, under control of the operating system, start or resume another program. For example, when a computer operating at an instruction speed of one microsecond finds an instruction that requires data from a disc system that might require 10 milliseconds to complete, it initiates that disc access and goes on to other programs until it is completed.

The function of the operating system in all this is to perform all the bookkeeping so that it "knows" all the programs that are currently executing, where they are located in memory, at which instruction they were interrupted, and what action must be completed before each can resume. It will even perform disc access so that the specific code for performing that function need not be included in each program in each place it needs to make a disc access.

When the requirement of real-time is also imposed, then there are some additional bookkeeping and functional responsibilities built-in to the operating system. This usually is accomplished by establishing a system of priority levels. Each program that is to be run is assigned a priority. When a program of higher priority needs to run, the operating system, acting as previously described, turns the computer's attention to this program. When programs of higher priority level are satisfied, then programs of lower priority run. When there are programs of the same priority running, they each get a share of attention. Priority levels are assigned according to the relative importance of each to real time performance of the system.

Let us now briefly look at some of the other component parts and functions of an operating system.

Memory Management. This is an important function of the operating system. As different jobs need to be run, the operating system must find room, if possible, in memory for the job to fit. If there is no room, an examination of the new programs priority versus those currently executing must be made. If the new program is higher, some other program must be moved out to the disc and its current status recorded so that it may resume without error. Sometimes there is sufficient room in memory for a new program but it is broken into little areas as a result of earlier smaller programs finishing. Some operating systems can handle this by shuffling the current programs around in memory to concentrate available memory in one continuous area. Sometimes this is not wise lest it interrupt processing of a job with high priority. Therefore, these operating systems allow programs to be coded with suitable instructions that prevent their being shuffled. Memory management is a complicated topic and entire books have been written about it and the following topics. For our purposes it is necessary for you to realize that it is an area to be compared between systems.

File Management. File creation and management is another function of the operating system. These files can be source programs in FORTRAN, COBOL, or assembly languages, or their machine language translations. It may be collections of data such as the historical trend file. They could be either variable length records within a file or an entire file of records of the same length. Each of these require different handling by the computer and its operating system.

Each operating system provides routines for accessing data and programs stored or input through its various peripherals such as card readers, punched tape, magnetic tape and disc drives. It also has routines to handle this I/O depending on its own system architecture and operating system organization.

Utility Programs. Each operating system has a number of commonly needed "utility" programs. These will perform such functions as sorting and merging files, copying files, formatting new disc packs, and copying one disc pack to another for back-up.

Language Translators. Another important factor in the operating system is the language translators. In our types of systems a FORTRAN compiler and an Assembly language translator are essential for any distributed control system with a host. Some manufacturers, however, will not provide these translators at all or only on their highest priced systems. If you intend to do any optimization type programs then you must have a suitable computer language to write them in, and the system on which they are to be used must provide the necessary facilities.

Process Control Software

The next essential set of software are the special programs that perform the process control, data acquisition, and CRT display functions. Most manufacturers of distributed control systems provide these but there are some who have not yet done so. Whether your company requires these programs or would prefer to write their own depends on your current sophistication in this area. This writer believes that full advantage should be taken of previous work unless you are capable of doing it better.

Let us look at some of the functions this software serves. The most basic function is to be able to define the configuration of each indicator and control loop. This usually includes the input terminal address where the analog or digital input is to be found. It will be necessary also to decide and program how often each input will be scanned, which inputs shall have what control and calculation functions performed, and the address of the valve output. Controller set points, tuning constants, and cascade set points are also items that must be specified along with the types of alarms desired and the specific alarm values. In general, any function that you would perform with older style analog instrumentation will now be done in software plus many functions that were previously difficult or impossible to do with the older equipment.

This software configuration system generally functions in an interactive mode. In this mode, for example, you tell the system that you wish to define an analog input for your third distributed microcomputer. The system then will respond with a series of questions for this type of input which are answered one by one. As necessary, you indicate linkage to any following control and calculation functions, then answer appropriate questions for these functions.

When you have completely specified each and every loop and function, on your command, the manufacturer's software writes a machine language program for each distributed microcomputer. This will be downloaded later into the memory associated with each.

This type of programming is known as a fill-in-the-blanks language. In evaluating this software you must be sure that the manufacturer provides all the types of control that you need such as cascade, feedforward, ratio, signal select, and calculations such as add/subtract, and multiply/divide.

Some systems also provide extensive batch-oriented type software that allows you to easily specify timings, sequencings, formula changes and the like. Other manufacturers provide little if any software to perform batch functions.

Once one has a computer it should be easy to perform interlock and logic type functions. Of course, this type of programming can be done in FORTRAN or assembly language but some systems offer interlock software which further simplifies this type of programming. As before, some manufacturers provide systems which have little or no software for these functions.

While speaking of interlock type applications, it should be realized that putting any safety related interlock in a general purpose computer control system is courting serious consequences. Safety related interlocks always should be provided by an independent system. If the computer node handling safety needs is down at the instant when it is needed, the plant is in trouble.

Along with the programs that specify control configuration there must be other programs to provide for the various CRT displays. One must be able to assemble all indicating and control loops into logical groups just as he would on an analog control board. The advantage of the computer lies in the fact that the same loop can be displayed in different groupings as appropriate whereas an analog controller can only be in one place.

Part of your programming will also be devoted to specifying which inputs are to be kept in the historical trend file. Every system has some limitations as to how many inputs can be recorded in the file. There are two reasons for this.

Collection of data and averaging it over various periods of time consumes computer time. This is a limited resource. There is also a limit to the recording space on a disc drive. It can only hold up to its maximum capacity after which the new values overwrite the old unless steps are taken to transfer the information to off-line storage media. Determine the maximum number of points each offered system can handle and for how long.

Most vendors provide reporting programs to print out values held in the historical trend file. Some provide programs that will do certain simple analysis of the data. If more complex analysis is desired, it can be written in FORTRAN.

One of the most useful features on computer CRT's is provision for graphic displays. As mentioned earlier these displays can provide a convenient reminder of the process plant arrangement along with its associated controls. They can show live data versus set point values and generally can be used to get a bigger picture than loop group displays. In most systems the operator cannot direct operation of the plant from these graphic displays but must go to the loop group or individual controller displays.

Creation of graphic displays is quite time consuming from both a conceptual viewpoint as well as in programming. Remember that engineering flowsheets generally cannot be just duplicated on the screen. Significant time must be spent in making an intelligent layout of each display page. Selecting colors to represent process chemicals and plant functions also consumes time. After this the manufacturer's graphic generation package is used to program these displays. Some of them require programming in a manner similar to writing a FORTRAN program. Other graphics software works rather like a child's Etch-A-Sketch™ toy. In these you create pictures on the screen by driving the cursor and specifying colors, process variables, etc. in an interactive mode. When the picture is finished the graphic program generates the code to duplicate it on command.

User-Written Programs

One of the finer uses of process control computers is to perform process optimization. In most systems this must be done in the host computer. To perform this function a very clear idea of the complex relationships of the various plant units is needed. Alternately an understanding can be developed by appropriate experimental techniques once the plant is in operation.

With these relationships in hand, programs can be written that will calculate desirable set points for certain control loops throughout the plant. As various parameters of plant operating conditions change these calculations will continuously attempt to maintain the plant at optimum conditions.

What generally is not realized is that plant optimization seems to imply a universally recognized type of optimization goal. This is not the case. There may be a number of targets, and some of them may be mutually exclusive. For example, minimizing energy costs, maximizing product output, maximizing product quality, and reducing pollution are all valid objectives, plus others not mentioned.

However, not all of these objectives can be achieved simultaneously so the tendency would be to program for the attainable targets that would have the most beneficial effect on the bottom line of the company's financial statement. In a rapidly changing world with fluctuating market conditions the best optimizing targets may change from time to time and reprogramming may be required. Fortunately, this is where computer control can shine as changes in software are comparatively easy to accomplish when plant relationships are known. The computer, when properly used, can assist you to determine those relationships.

4
Distributed Control System Architectures

S. Cocheo

AVOIDING VULNERABLE SYSTEM ARCHITECTURE

Over the past several years I have evaluated many of the currently marketed domestic distributed control systems. During this experience, many deficiencies in some manufacturers' offerings were observed. These deficiencies can be a source of danger or plant malfunction on failure of a portion of the distributed system. Even a well conceived system design can be badly implemented by poor selection of redundant and backup equipment.

Avoiding a vulnerable system architecture involves a critical analysis of a number of system facets. These include not only the hardware and software but also the needs of the controlled process. In this analysis a careful examination is made of all data paths, instrumentation signal connections and signal processing. But most of all, it involves visualizing what will happen when various portions of the system become inoperative, how this will affect the data and signals, and ultimately how the plant process will be affected.

As a result, the system being evaluated might be rejected, or optional redundant and backup equipment added to reduce or eliminate adverse effects of failure modes.

The main steps necessary in carrying out this analysis are:

1. Acquiring a thorough understanding of the process to be controlled, and of the goals to be achieved.
2. Understanding how computer systems function.
3. Understanding the hardware functions of each component within the distributed control system architecture.
4. Understanding the software functions operating within the hardware architecture.
5. Coordinating the needs and goals of the process, along with proper selection of hardware and software, and expressing these in suitable specifications.
6. Evaluation of vendors' proposals with the wary eye and diligence of an investigative reporter.
7. Implementation of the selected vendor's system in a properly coordinated way that meets the specified needs.

Each of these steps is discussed in greater detail in the following text along with some examples of points of vulnerability in some of today's systems.

As one discusses vulnerability, it must be recognized that this negative feature is not to be compared to some idealized system that a superior vendor might supply. Rather, it is a case of designing and selecting systems of minimum vulnerability when consideration of the needs of the process and the goals set for the distributed control system are included in the conceptual phases of the project.

This approach is usually called "systems engineering." This is done so that the plant design is not a collection of process equipment to which a control system is appended in the hope that somehow it will all work. Rather, it is an integration of design so that each element is engineered in the knowledge of the assigned functions of the related elements.

Using this approach, the plant to be controlled should be divided into functional units and their nature evaluated. This evaluation should consider any hazards associated with each unit's operation. Next, the interrelationship of these units should be studied similarly noting how performance of one unit will affect operation of previous and subsequent units.

As the plant's needs are understood, it then becomes possible to start applying the control strategy. It is not necessary to finalize the strategy at this point. However, an estimate of the number of control loops, indicators and contacts to be connected to the system is needed. This will have a bearing on the size of any vendor's distributed control system.

Next, estimate how many loops will be simple controllers or cascade, feedforward, multivariable, override, complex, etc. While control functions occur in software, there are some distributed control systems that cannot accommodate all the modes of control that an engineer may need. It is best to eliminate these systems from the competition early so that neither you nor the vendor wastes time and money. In addition, it is necessary to know what kinds of sensors and transmitters are to be used for each input, as some vendors provide different input cards for thermocouples, resistance temperature detectors, current transmitters, etc.

The speed with which each input can change is another important factor in system sizing. The more rapidly an input may change dictates how often it must be scanned. For example, flow rates may change very quickly, and when used in a control loop, they must be scanned at least once per second. By contrast many temperature inputs may be scanned every 15 to 30 seconds. The less frequently the various process inputs need to be scanned, the more inputs one can "load" onto a microcomputer in the distributed control system.

When this is complete, decide what goals you expect the computer system to achieve. There are distributed systems that can replace basic plant analog instrumentation and its associated annunciator system. These yield a little more in operating flexibility and performance than the earlier systems. If this type of equipment is what you want, select it. However, if you have more ambitious goals, equipment of greater sophistication must be specified.

For example, if one of the goals is to use optimization techniques, then a distributed control system must contain a "host" minicomputer capable of performing floating point calculations to a high degree of precision. Some distributed systems make no allowance for a host computer or only provide a communication interface to a host. This latter approach leaves the software development as a heavy burden for you to handle. This is a good example that shows the need to determine the requirements first.

To properly engineer and apply distributed control systems it would be better if the reader obtained some basic education in computer science. As a minimum there are a handful of courses that will provide basic knowledge. These are:

- A high-level computer language such as BASIC, FORTRAN, Pascal or COBOL
- Computer architecture
- Operating systems

The high-level language will teach the sequential nature of computer programs plus the logic capability to jump or branch to other portions of sequential code. Computer architecture will show the structure and arrangements of the inside of the central processing unit with its registers, memory, arithmetic, logic unit, and the control unit.

The operating system course will reveal the sophisticated nature of the computer's "master program." This is the software that enables the computer system to perform several programs at the same time (multi-programming), communicate with its peripherals (such as disc drives), and to maintain orderly operation in the priority scheme required for process plant applications. The operating system also performs many other essential functions and services too numerous for the scope of this chapter.

Figure 1 shows a system architecture that represents one leading manufacturer's view of what a reliable system configuration should offer. Of course, this is only one of many possible configurations that are possible with this system since each purchased system should reflect the results of a proper systems engineering study.

SOFTWARE/HARDWARE INTERFACE

Before we begin an analysis of this system, a few basic points must be understood. As we view the architecture please note that a distributed control system is a network of several computers and auxilliary peripheral equipment. These are interconnected, in some manner, so that data and instrumentation signals may travel over communication paths. This is basic to the concept of a distributed control system for it is distributed functionally as opposed to a central control computer.

What is less obvious is that there are software programs driving and operating the hardware and this software also has its own architecture. That is, different manufacturers break up the functions performed within the system in significantly different ways. This must be clearly understood, for as failures occur in the various hardware items, each system loses different functions. This may have an adverse effect on your process that you must avoid by providing redundant and backup equipment. It also may be necessary to reject some systems, as later examples demonstrate.

The system in Fig. 1 performs all basic control functions entirely within the distributed control microcomputers. The I/O (input/output) system has its own microprocessor to control the I/O multiplexers and signal converters (analog/digital, digital/analog, digital/digital).

Data needed in other parts of the system generally travel over the high speed data highway. The highway controllers (more microprocessors) have software programs to check for transmission errors. The data on this highway can be process values for data acquisition, display on the operator's console, or for use by the host in optimization programs, etc.

The display control microcomputer formats the CRT screen displays using data gathered from the highway, prints alarm messages and log reports, and sends new operator commands, such as new set points, down to the distributed controllers.

The optional host computer can perform all the functions of the display control microcomputer and several other important functions such as:

- Store large quantities of data for reports, trend recording and historical data analysis.
- Monitor performance of various portions of the plant or items of equipment such as efficiency of a boiler, or power consumption of large motors driving pumps and turbines.
- Run process optimization programs that improve the economic performance of the process or improve its product quality.
- Generate management information to report on the economic performance of the process and its units.

As can be seen the functions performed by the system are complex, interrelated and or varying degrees of importance. The software that performs the work falls into four main categories:

1. Operating systems in each computer.
2. Process control and data acquisition programs that must be tailored to the particular process.
3. Special programs that must be written to perform many of the functions enumerated above for the host computer.
4. Special programs the manufacturer provides for overall operation of the system such as the programs to drive the data highway, format the CRT displays, and to generate a selection of standard reports. Some of this last group is frequently "transparent" to the user. That is, you rely on their working but you cannot access them or change them.

The manufacturer of the system designs both the hardware and software architectures as an integrated whole. Our job, in evaluating the vulnerability of the system, is to learn whether his choices might impact the safety and operability of the process.

STRENGTHS AND WEAKNESSES

Referring again to Fig. 1, let us review some of the architecture's strengths and possible weaknesses. In doing so we will look at hardware failure modes and assess their impact. Some factors that we will examine are:

- Can the operator always know the status of functional portions of the process?
- Can any essential functions be lost, and how can we avoid, or reduce the chances of this happening?
- What functions can we afford to lose rather than incur excessive costs to avoid the loss? In addition to the extra cost, we do not want to engineer greater complexity into the system than necessary.

Starting with the operator's stations we observe that they operate off both the display control and host computers. We have operator security then for this configuration. Since the host is optional in this system, what would be the vulnerability if there were no host? It would be high, for failure of the display control micro would mean losing our only window on the process and would prevent the operator, through his keyboard, from affecting plant operation.

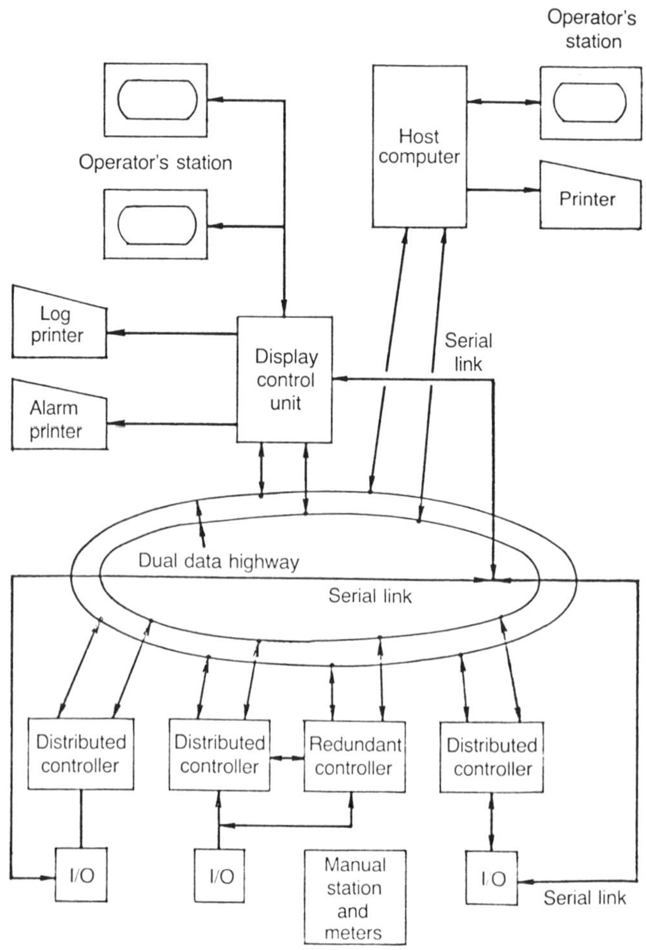

Figure 1. Control architecture from vendor one.

In this latter case a second display control micro with its own operator's stations might be a wise investment, depending on the size and safety requirements of the process. In larger processes additional operators would be needed in any case, and their stations should operate off independent computers. With some vendors these redundant stations are very expensive and that may make their proposal too costly. You must be wary of proposals that offer too few operator's stations to minimize price.

Another factor of concern is operator's stations that are not interchangeable. In some hosted systems the operator's stations connected to the host cannot be connected to the microcomputer display control unit. Since operator's stations also can fail, you may not wish to own a system where a failed host, simultaneous with a failed operator's station on the display control unit, leaves you blind. Interchangeability is a good feature.

Good systems design will assign essential control functions to the distributed control microcomputers and nonessential functions to the host. In this way failure of the host or data highway will not suddenly plunge the entire process into an uncontrolled state. An example of this is given later.

Data highway design is a critical factor in system performance. The system in Fig. 1 shows a redundant data highway and is essential. Data highways also need redundant highway controllers for greatest security. In some systems this is not available at all. In others it is either optional or standard.

The best systems offer is "four-way redundancy," which allows either redundant highway to operate with either redundant highway controller. This provides four combinations and will maintain the integrity of the data highway system if either highway *and* either controller fails. Some systems offer only two-way redundancy, which is still better than those which offer redundancy only of the highway cable.

The distributed network is characterized by several competing philosophies. One philosophy employs a combination of distributed controllers coupled with several types of multiplexers for noncontrolled variables. The type of multiplexer used depends on the variety of instrument signals connected to it, such as current transmitters, thermocouples, resistance temperature detectors, etc. Other systems bring all varieties of signals into one system, through appropriate interface cards, and, under input program control, manipulate the gain of their multiplexed input amplifier. These latter systems are more flexible in practical applications for they tend to require fewer items of equipment to install, maintain and provide with spare parts.

The number of loops handled in each distributed controller or multiplexer also varies widely. Some systems consist of only single loop units, some with 8 or 16 loops, and others with 32, 48, and higher. In selecting the systems for your process, you must analyze the number of loops in each process unit of your plant and decide which will require which type of redundancy or backup.

In Fig. 1, you will note that one of the distributed microcomputers has a redundant system in case of its failure (redundancy). There is also a set of manual loading stations with process variable input and valve output meters shown (backup).

This backup unit is not required if the serial link is implemented. This link is a direct digital communication path between the operator's stations, through the display controller, and the microprocessor in the input/output section. The operator can view the state of the process and manually manipulate the outputs via this link. Usually, he or she would do this only if automatic control were lost. Other vendors offer similar capability in their systems; however, some of them do so over their data highway. This, of course, requires that the data highway and the communications portion of the distributed computer operate at the same time the backup means is to be used.

As the capabilities of the distributed network are reviewed, it would be wise to examine the types of control methods each provides. All systems accommodate simple indicating, three mode and cascade control methods. Many systems, however, do not provide for ratio, high-low signal select, feedforward, signal limiting, arithmetic calculations, etc. One very popular system provides nothing that allows you to configure the simplest interlock logic while others provide significant facilities in this area. Some systems provide for batch oriented programs, while others are limited to continuous control. In some systems the more complicated control structures may be passed to the host computer. If the host computer failed, would your process be in danger?

Speed of operation in these systems also must be observed. Some systems have a scan cycle speed as slow as two seconds per input scan. Others perform a similar scan in 1/4 of a second. This scan rate must be compared to the fastest occurring event in your process for compatibility. For example, if an antisurge control system were to be implemented in one of these systems, it is obvious that a two-second time frame might allow a compressor to go into surge before the following scan. If your process has only slow moving variables, however, a two-second scan would be acceptable.

EXAMPLES OF VULNERABLE ARCHITECTURES

Let us now review some architectures for a few examples of vulnerability. These are based on actual proposal offerings. Since review must be based on the realities of the process involved, a brief statement of the needs and goals is in order.

The process is for manufacture of a product using a batch processing technique. The plant has several independent trains with interconnections to provide for flexibility of operations and equipment redundancies. The process is new and still in a stage of final evolution.

Due to this last factor, the system to be selected requires a host computer for extensive data collection and numerous analytical reports of process performance. At a later date the client wishes to implement process optimization for energy savings and supervision of product quality.

Certain portions of the process equipment may be damaged if control is lost during particular phases of the batch cycle. Full redundancy at the distributed computer level, the data highway and the operator's console is desired.

The job is moderately large consisting of over 1,000 digital (contact) inputs and outputs for such uses as process status switches; valve position verification; and motor starting, stopping and running verification. There are also several hundred continuous control and indicating loops, many of which only function during particular portions of the batch process. The system must be expandable to cover additional trains.

A systems architecture similar to Fig. 1 was proposed by its manufacturer. It differed in that two display controllers were configured for both redundancy and the need for two plant operators. The distributed controllers were all provided with a redundant companion. The communication controllers (which are hidden within each item of highway connected equipment) were also made redundant for four-way redundancy.

All operators' consoles were eight-color CRT's having full graphics capabilities to display simulated piping and instrument diagrams (P&ID's). Since "live" process data and set points may be included in these graphic displays, the operators may control the process in an "interactive" manner while viewing them.

Another vendor proposed a very similar system which differed in some details. One involved a larger quantity of distributed controllers needed by his system. His data highway provided only two-way redundancy, and he provided no alternative communication path to the input/output equipment.

The software presented was of high quality in both systems. They each had well-developed batch control software through which the engineer could configure all batch/sequential and continuous control functions. These languages are engineering oriented using the Instrument Society's concept of a "fill-in-the-blanks" language. In this approach, the computer will respond to a command to configure a particular type of loop with a menu or series of questions. The engineer responds to these questions with engineering or English information such as tag number, transmitter range, engineering units, set point, etc.

When the process is running, all configured control programs reside and operate entirely within the distributed controllers. The display, data acquisition, reporting, alarming and optimization functions are assigned to the host and/or the display controllers.

Let us now contrast these well-conceived systems with three others having a higher degree of vulnerability or whose costs to eliminate that vulnerability were sufficiently high so that their proposals were noncompetitive.

Figure 2 presents an architecture that is somewhat different from Fig. 1 in the hardware architecture but, as discussed previously, is significantly different in the software architecture.

While this system has distributed controllers, which presumably would control the process, in fact they only perform some of the control functions. The batch software resides only in the host computer. Continuous control and digital input/output reside in the distributed controllers and input multiplexers. All batch/sequential functions must originate in the host and thereafter manipulate the process through the distributed units.

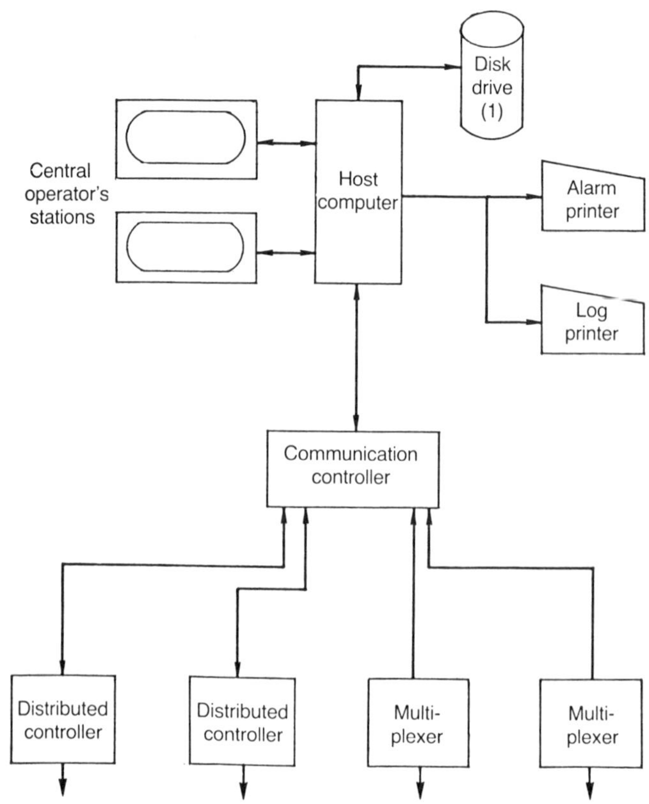

Figure 2. Control architecture from vendor two.

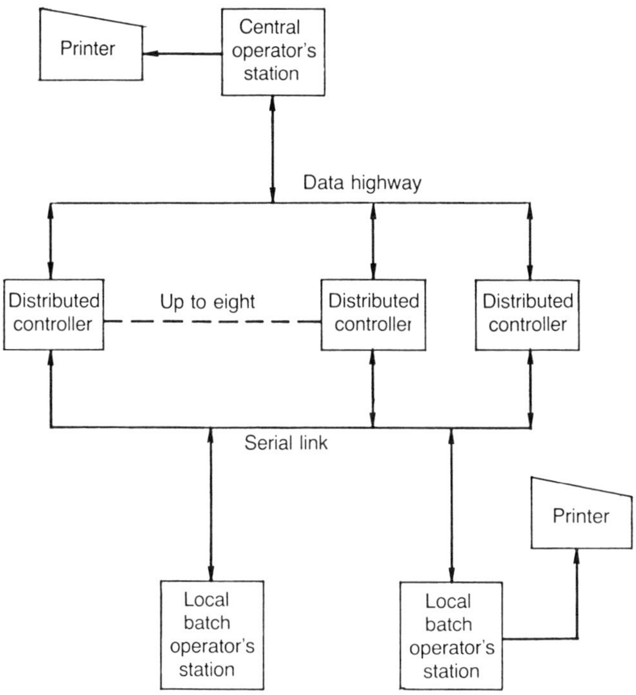

Figure 3. Control architecture from vendor three.

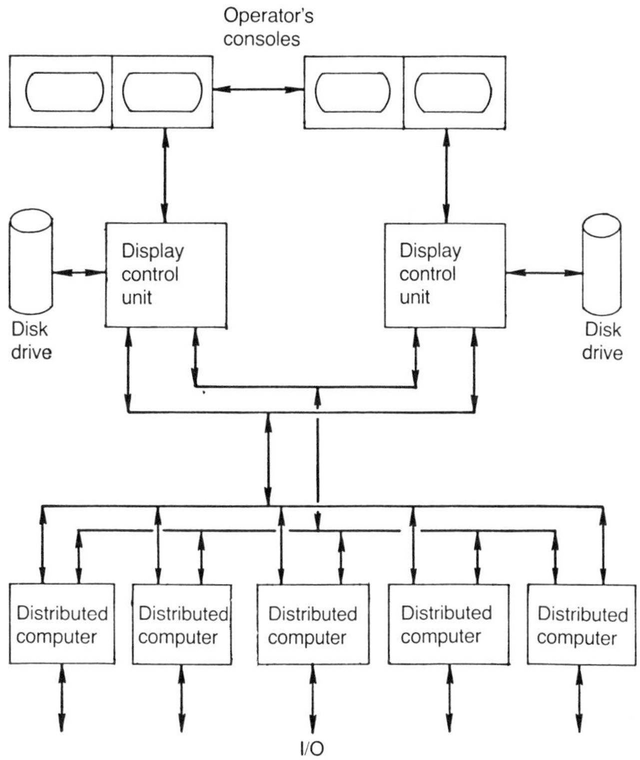

Figure 4. Control architecture from vendor four.

Also note that the operator consoles, the disc drive and systems printers all depend on the host computer for continued operation. When the host fails, virtually all essential functions are lost. This failure would be a very serious consequence to many processes.

In optimization applications, it is generally the host computer that manipulates the process in this manner. The difference to be noted is that optimization, however important, is not an essential control function. The safety of the process or human life is not dependent on its continued operation.

Figure 3 shows another proposed configuration. The lack of a host computer is immediately obvious. The vendor stated that a host could be purchased at a later date when optimization was desired. Their central operator station was capable of performing sufficient data acquisition and reporting functions for immediate needs.

In addition, as this architecture is analyzed there are four main areas of concern:

1. The local batch operator's station is a black and white alphanumeric terminal only. In other words, it cannot show graphic displays of any color.
2. The central operator station has all required capabilities but each additional unit is very expensive.
3. The distributed controllers cannot be expanded to cover additional trains.
4. One-for-one redundancy for each distributed controller is lacking. This system provides one redundant controller to backup several controllers.

Most system specifications can be met by purchasing sufficient quantities of the various components. This makes this offering noncompetitive.

Figure 4 shows the final system for evaluation. Similar to Fig. 3, this system also offers no host. The vendor expects to have a host available in the future.

The data highway and its control are a serious cause for concern in this configuration. The system has no independent data highway controller but rather places the communications load on the operator's console display control unit. This unit is a minicomputer. It is of sufficient capacity to be used by others as a host computer.

Unfortunately, communication represents a significant load in distributed systems. When this load is combined with the loads of operator console displays, data storage on the disk drive, and log report printing the system is being used close to capacity.

There is no redundancy of data highway control if either display control unit is down. The system would simultaneously lose communication redundancy along with operator displays, data acquisition and printing capability.

The system provides only one disk drive, which is another vulnerability. The display control unit will shut down within a few minutes if the disk drive fails or is turned off. Disk drives require regular maintenance and must be shut down to perform some of the maintenance work. When this is done the system is deprived of all the functions of that display control unit. If the drive fails, it may be many hours before repairs can be accomplished.

The proposed system has little capacity for the future expansion desired. This limitation again involves the data highway system for its load on the display control unit and its data rate limit the system to processes that are moderate in size.

As each engineer analyzes the needs and goals of his assigned process, he must establish a hierarchy of targets based on a priority level. Some functions are essential, others important and some just desirable. He must be prepared to evaluate each proposed system with these priorities clearly in mind. When he does this, he can engineer and select the most suitable system to meet the requirements at a competitive price. The system he selects must be chosen for minimum impact on his process when various components fail.

5

Operating Experiences with Distributed Digital Control

A. Ollila

Neste Oy chose distributed digital instrumentation for the first time in 1979 for the MTBE-unit. Today Neste has six distributed digital systems in operation. These include small and large systems; the largest involving about 300 control loops and 12 video stations in the control room of the phenol plant.

Today we are able to analyze and to make some conclusions from the experiences we have had in Finland and especially at Neste Oy. Those experiences have been compiled and give us an over-all view of what the typical features are and the positive and negative qualities of these systems, as well as how we can improve the level of control.

Special attention has been paid to the control room technique, because this seems to be the area where we do not have generally accepted solutions. The characteristics of the human element are essential in this connection, and its basic aspects are presented. Finally, some conclusions and suggestions are made, especially concerning the control room technique and disturbance situations. Computer control techniques are not dealt with in this chapter.

EXPERIENCES WITH MICROPROCESSOR-BASED SYSTEMS

There is now plenty of experience concerning this new automation technique. Everything is not valid in every case, but quite a clear picture can be given of the overall situation.

Positive Experience

- Hardware is reliable and systems are down very seldom. We can forget the first prejudice which the operation personnel had regarding operational security.
- Operation via video terminals works after a few days of training.
- Trend records on the screen are useful because measurements other than analog can be easily recorded.
- Operators can supervise larger areas.
- Testing before installation can be very thorough. Startups have been very short and without surprises.
- The maintenance personnel can be easily trained for new jobs. Only some instrumentation specialists need to be trained for making software changes. The main part of maintenance work belongs to field instrumentation repairing.
- Failures can be quickly located.
- No delivery problems.

Problems or Development Targets

- Correcting software errors does not usually succeed the first time.
- Disturbance situations are not handled as fluently as in analog control systems because all the control circuits needed are not necessarily within sight.
- Alarms on the screen can be easily overlooked by operators because they are not graphic.
- Too many alarms.
- Process graphics in the panel are usually useful.
- Trend records should be more than 64 h long because otherwise weekend situations are lost.

HOW TO IMPROVE THE CONTROL LEVEL

Automation systems offer a wide variety of special features. The majority of systems include special control algorithms which can be put into effect only through programming efforts.

It is quite clear that all the positive qualities should be gained. Neste Oy has had this approach from the very start. Our experiences are as follows:

- The feedforward control principle is used widely. Trimming of controls succeeds the same way as feedback control: through trial and error.
- Calculation of various aid quantities is very frequent. The pure PID control circuit number is nowadays about 20%, earlier it was about 80%.
- Temperature and pressure compensations are frequently used for flow measurements.
- Override controls can be carried out in a more advanced way than with analog technology.
- Nonlinear level controls have been useful, especially in distillation train applications for smoothing flows from units.
- In interlocking situations operators may have extra help from the automation system which can switch certain control circuits to manual control automatically and give a desired valve position.
- Special tricks can be carried out in the control circuits because software technique is more flexible than analog technique and the costs may be unimportant.

Special attention should be paid to the documentation technique when planning controls. If the microprocessor-based instrumentation is presented by normal instrumentation symbols in P&I drawings, there is a great danger that the control level will be lower than the technique presumes. This is a result of the fact that there is not enough space and symbols.

Neste Oy has developed its own documentation technique. In P&I diagrams, microprocessor based instrumentation is presented by square symbols and only connections are visible (Fig. 1). Each control circuit is presented in the control circuit diagram so accurately that it can be programmed from these documents (Fig. 2).

Because the new technique has changed the control room completely, some problems have arisen. The role of operator and control room design deserve closer study.

CHARACTERISTICS OF THE HUMAN ELEMENT

The human element is essential in process automation. Even in the most sophisticated automation systems one can find a person who communicates with a machine. To plan and construct a good interface between these two elements—a person and a machine—we must know each's characteristics.

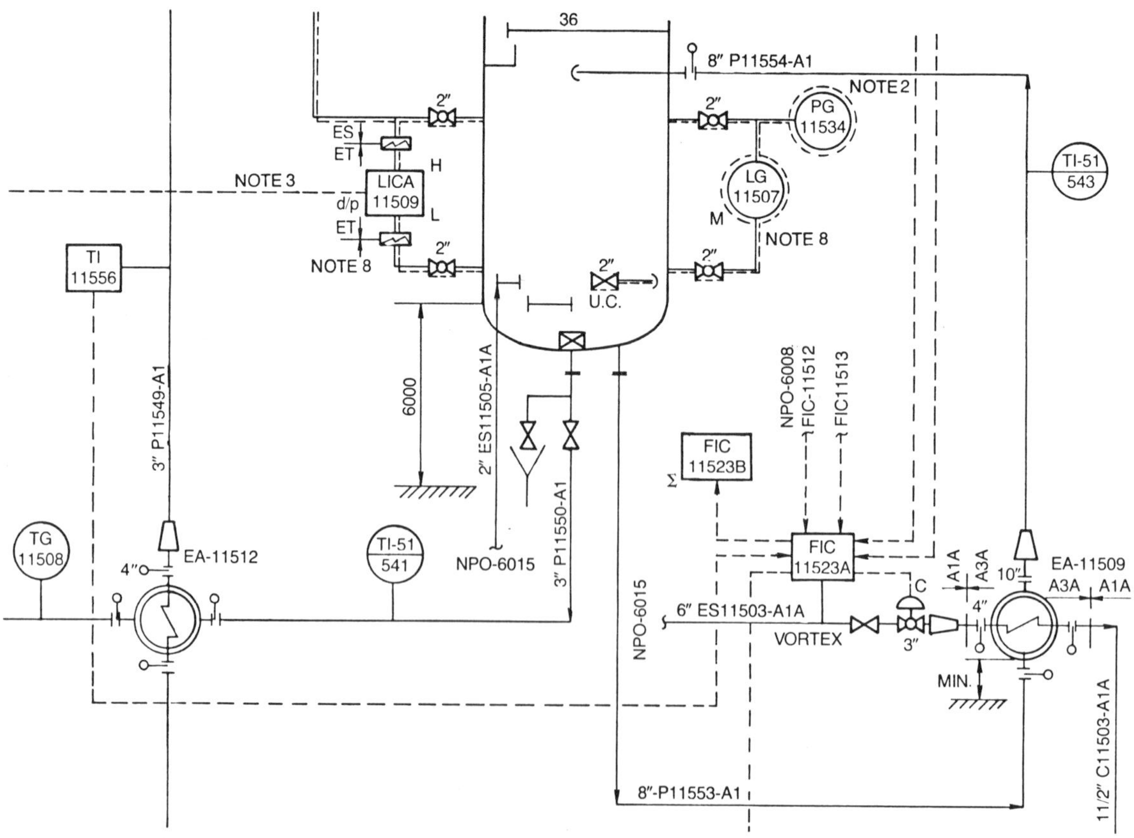

Figure 1. In P&I diagrams distributed instrumentation is represented by squares and only connections are visible.

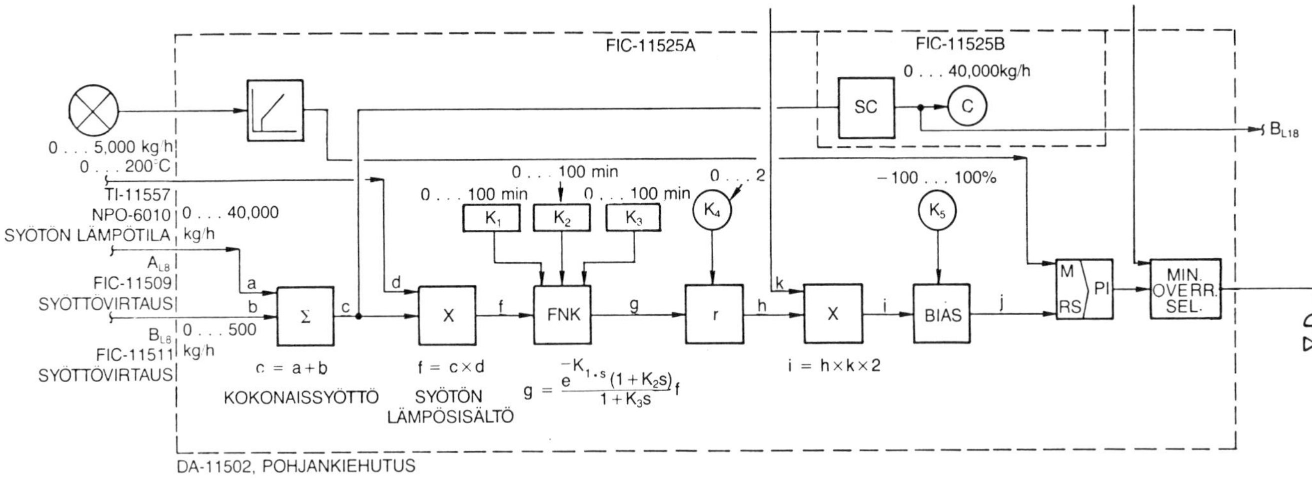

Figure 2. Control circuits are presented so accurately that they can be configured from these documents.

A man or woman knows every property of the machine or the system he or she has created. But man does not know man, because human qualities vary to a great extent and humans react in unpredictable ways. Numerous factors affect human beings and make them different from one another.

Certain basic qualities are characteristic concerning human performance and qualities. The most important ones in this connection are:

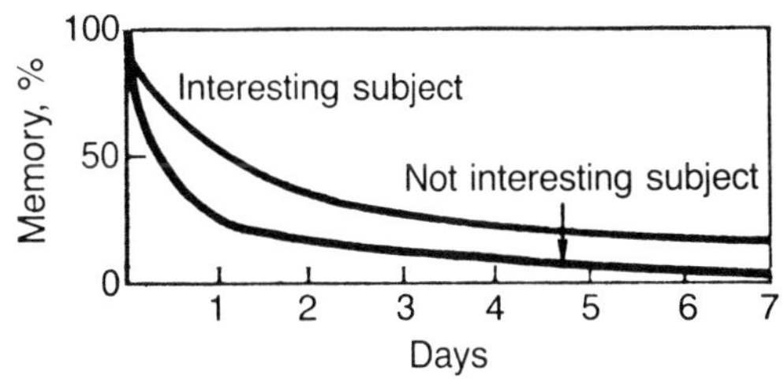

Figure 3. The memory of man.

Man's memory is short. Figure 3 shows that man has a very short memory. There is a saying among teachers; what you hear you forget; what you see you remember; what you do you are able to do.

If we put this into practice, it means that an operator is able to handle those things competently which he handles almost every day. If some odd things happen we can assume that an operator will only remember some basic instructions.

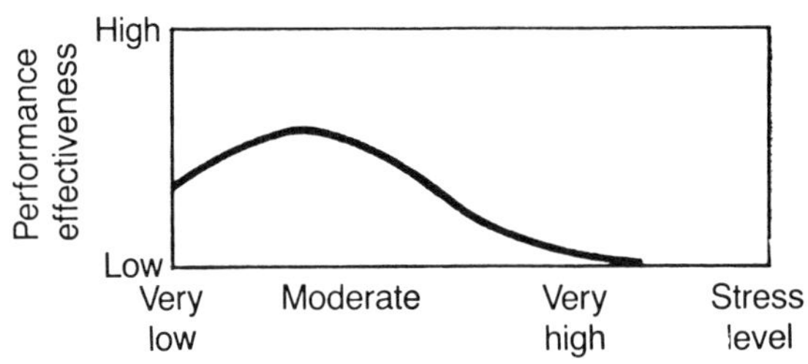

Figure 4. Hypothetical relationship between performance and stress.

Stress level affects human performance. Figure 4 shows us that a very moderate level of stress increases performance, but after a certain limit the performance will decrease rapidly.[2]

In the control room environment, this means that a fair amount of stimuli (actions, alarms) has a positive effect. In panic situations the performance of an operator may collapse.

Man cannot handle complex situations at once. Man can work effectively when solving only one problem at a time. A closely related phenomenon is that man can only keep four to six situations active in his brain simultaneously. A complicated logical problem must be divided into sub-tasks. This makes problem solving slow.

Man is not perfect. From experience we can point out that man makes errors. Human beings are lucky because in our world most errors are not fatal. Hence, we can usually correct our errors in time.

In process control situations extreme conditions can be so serious that they must be handled automatically.

There have been some attempts to make man unerring. In the USA there was in the 1960s a program called Zero Defects Program.[2] The aim was to eliminate all operator errors by selecting the right individuals, by training and motivation. Because man is subject to mistakes this program did not succeed.

After this failure another approach was taken. The basis came from the experience of ergonomic personnel that most of the errors are due to design parameters and not to human qualities. This program is called the Work Situation Approach.

If we want to make any progress in eliminating errors, we must admit that there are some shortcomings in the work situation.

There are basically two techniques which can be used in analysis work: MMSA (Man—Machine Systems Analysis) and THERP (The Technique for Human Error Rate Prediction). In normal process control situations they are probably too heavy tools; atomic power plants are good exceptions.

A very useful method in practice is an error likely situation report. If the operators and other personnel could freely report observed potential risk situations (without fear of being fired) many dangerous working situations could be revealed and improved.

Man can perceive patterns and generalize them. This item favors man compared to machines. Humans possess this ability even though the patterns may change in size or position, or may be partly masked by noise.[2]

Man is a creative being. Humans are creative in many various ways and this skill is lacking in machines.[2] Man has the ability to use judgment, to improvise, to profit from experience, to find a new solution and to reason inductively.

A good man/machine interface can be planned by recognizing these basic elements.

EXPERIENCES OF NEW CONTROL ROOM TECHNIQUE

New control room technique based on video terminals has now been in use in Finland for about 10 years. After such a long time we are able to see things as they are; both positive and negative.

The new technique seems to cause two kinds of reactions:

- Users and nonprofessional people have prejudices. After training and operating experiences they accept new solutions surprisingly fast.
- Among professional people there seems to be "a steam engine effect." Opposite opinions concerning application principles or technical solutions are suppressed or are not expressed at all. In technical papers, these negative features of new techniques are getting very little attention.

Based on experiences gained from six different types of microprocessor based systems in Finland, the following generalized conclusions can be made:

1. Operation via video and push buttons works well after a few days.
2. Graphic drawings are well liked by operators.
3. In disturbance situations troubles have arisen because all controllers needed were not in the same picture.
4. Operators do not appreciate general or overview displays (Fig. 5). They tend to look at group displays (Fig. 6).
5. If alarms are presented on the screen only as lists (Fig. 7), they are often experienced as the weakest points of the system.
6. Process graphs are still said to be necessary in spite of video graphics.
7. Experiences from analog back-up controllers have pointed out that operators had forgotten how to operate in disturbance situations.

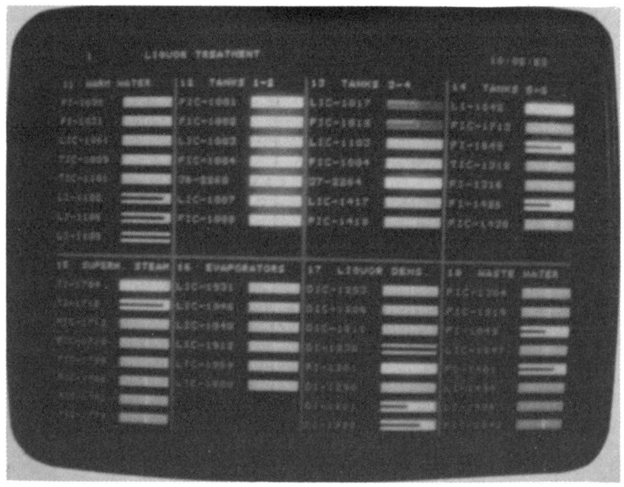

Figure 5. Overview displays[3] are not appreciated by operators because of high density and difficulty in relating them to the process.

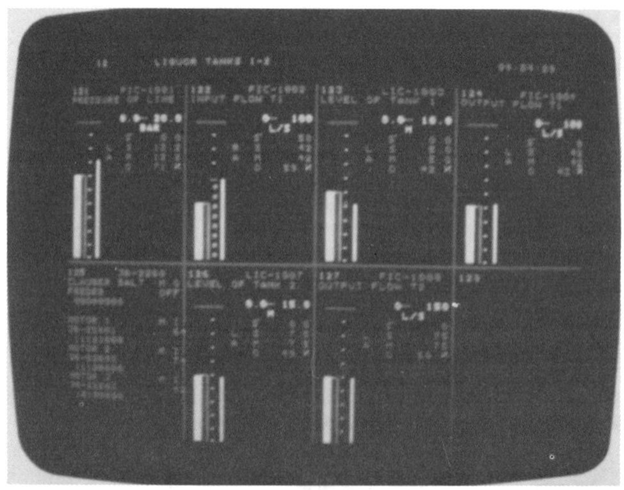

Figure 6. Operators tend to look at group displays[3].

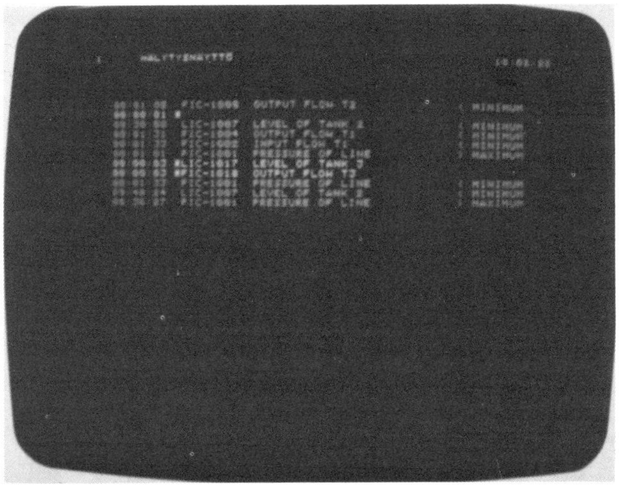

Figure 7. Alarm displays[3] in the form of lists are the weakest point of the distributed system.

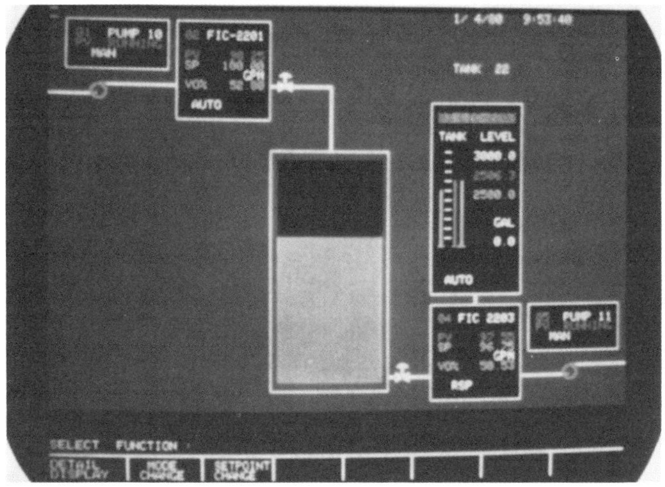

Figure 8. Process graphics[4] that closely resemble the real process are well liked. Limited space is the main problem.

Many of these experiences can be explained by human characteristics given in the previous section. For example, in point 7 the explanation is that the operators really forget the connection between the controllers and the process. Stress in disturbance situations still degrades the performance of an operator.

HOW TO APPLY NEW CONTROL ROOM TECHNIQUE

Graphic displays are popular among operators. This is explained by the observation that most men have an image memory.

A study made in the USA revealed that the process panel, where controllers and other panel instruments were situated in the right places in the process graphics (Fig. 8), was clearly appraised as the best solution. And why so? Because it so closely resembled the real process. Operators need minimum conversion from the image to the expression in the control room.

Especially alarm and disturbance situations seem to be the weakest points in the new technique. If alarms are only presented on the screen as lists they are ungraphic.

- Operators must think where the disturbance is located.
- Many alarms at the same time make it very difficult to determine how they relate to each other.
- Operators must remember tag numbers.

Also, we must remember that in this work situation there are two strong negative factors affecting operators: stress and operating instructions are not in their minds in detail.

If we think about this kind of a work situation we must admit that there are several shortcomings. In fact, it is very probable that the operators make errors in bad disturbance situations.

SUGGESTIONS FOR OPERATING IMPROVEMENT

Can we make any improvements by combining the best of the old and new techniques? Some proposals:

Video terminals are suitable for control operation. Operation on the screen via push buttons normally works well. Special attention should be paid to disturbance situations by choosing the right controllers in group displays. Because the same controller can be put in many displays, special displays are sometimes very useful. Analog back-up instrumentation should be minimized.

Line recorders are usually needed and must be located close to operators otherwise the operators will not use them often enough.

Process graphics on the screen are useful. Because process graphics resemble the real process, they are very valuable to operators. Controller operation directly via process graphic displays best resembles the old-fashioned control panels with controllers and graphics together and this may be the best solution in the future.

In video graphics it is so easy to use colors that they, as a rule, are used incorrectly which can be very confusing for operators. First of all we must remember that some colors possess a special task:

- Yellow is a deviation from the normal.
- Red is bad deviation, i.e., real danger.
- Green is all right.

It is an old tradition in the U.S. to use green and red for valve positions. This is an illogical choice and should not be in use any longer. It would be better to use a neutral color such as white instead of red. The shape images are probably better than lights of different colors (Fig. 9).

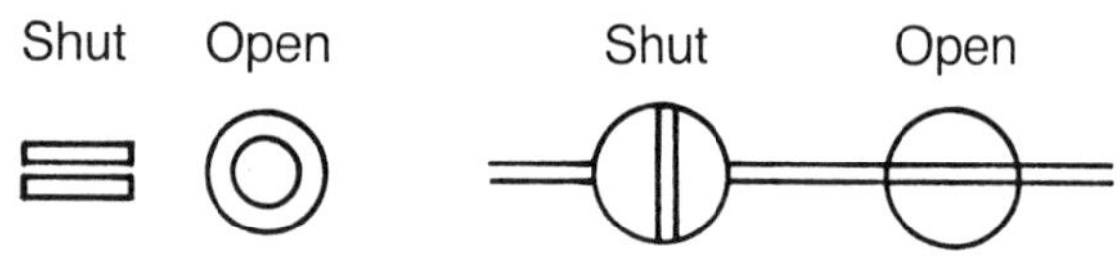

Figure 9. Different shapes for valve indicators.

Analog process graphics (Fig. 10) are still needed. Analog process graphics have a great positive advantage because they are "on" all the time. They are especially suitable to shown alarm indications because:

- Overall situation is seen all the time.
- An alarm situation shows at once where it is located and how it is connected to other alarms.
- Operators do not need to remember tag numbers.
- Color and situation tells the quality of the disturbance.

The right grouping of alarms is essential. The operators in the new control rooms easily have too much invaluable information because it is so easy to configure new displays and alarms. An alarm must be shown in two places only: (1) one amend for operators; (2) the other place is an alarm printer that should not be too close to the operators. If the same alarm is shown on the panel and on the screen it demands extra attention from the operator.

Figure 10. Analog process graphs are especially suitable for indicating alarms and showing the overall situation.

Alarms should be grouped according to their severity.

- Process alarms and interlocking alarms on the panel process graphics
- Deviation alarms (measurement deviation from a set point) and control circuit alarms (taking place very seldom) on the screen.

Operating instructions automatically on the screen in disturbance situations. As shown before it is very probable that the operators do not remember operating instructions for rare situations and that stress worsens the work situation. It is a very promising technique to program operating instructions for certain disturbance situations which are thrown on the screen automatically or manually by the alarm.

We can use the best qualities of computers and microprocessors to store and sort out the data to help operators. This technique is used by Neste Oy's phenol plant in the most serious disturbance situations.

Automation offers many opportunities for oil refineries to improve profitability. Computer control is suitable for optimizing product flow relationships in a distillation tower and between separate units. Microprocessor-based automation systems offer more advanced control strategies for stabilizing control. Feedward control can be used widely to solve distillation control problems.

The new control room technique is very flexible and possesses a great information density. It is possible to make wrong design solutions in this technique. Compared to the old analog technique, the process graphics seem to be very useful especially for indicating alarms in disturbance situations and for showing the overall situation.

References

1. "The New Process Automation System V"—Operation and Maintenance Experiences.
2. Alan D. Swain, "Design Techniques for Improving Human Performance in Production," Lutab, July 1977.
3. Damatic, A Brief Description. The brochure of Valmet Oy.
4. Provox instrumentation system. Bulletin 4:001 of Fisher Controls.

6
Microprocessor-Based Controllers in Refineries*

G. F. Hall

Plant economic considerations, environmental constraints, and fluctuating markets are three major factors compelling the refiner to optimize process unit performance. In an effort to do so, it's become a common practice to supplement basic proportional-plus-integral-plus-derivative (PID) control with advanced control strategies designed to increase energy efficiency, throughput, or yield.

For the past 15 years or so, much has been published on more sophisticated control strategies implemented with process computers. Unfortunately, the process computer with its input/output subsystem has also been burdened with various mediocre tasks that hinder its more effective use as a heavy-duty "number cruncher" and central information source and distributor. These include signal conditioning and characterization, minor logic operations, and multiterm calculations.

The advent of the microprocessor in the early '70s brought about the evolution of console-based control systems and controllers capable of providing more than basic control. Microprocessor-based signal multiplexers, input conditioners, and communications links have significantly increased data handling efficiency and reduced material and installation costs. As an option, these and other system components are frequently available in redundant configurations to minimize process interruptions because of control system failures. Further, system component integrity is interrogated regularly by self-diagnostic tests, the negative outcome of which will force a transfer to the backup mode.

UNIT CONTROLLERS

A unit controller is a "shared" controller capable of handling a quantity of process instrument loops. These microprocessor-based controllers reside in a distributed system or are self-contained. Both offer a variety of control functions, full input/output flexibility, and computer compatibility.

As with a computer, the most apparent benefits of a unit controller include high accuracy, much better configuration flexibility, and a considerably lower cost per loop on a multiprocess unit basis. Accuracy is maximized because very few signal conversions and manipulations are necessary.

The lower cost of implementing a multivariable control scheme becomes more evident when an addition is to be made to existing instrumentation. If, for example, the operations department wants to add heater pass balancing to the vacuum unit charge heater to aid in the reduction of tube coking, the strategy shown in Fig. 1 could be applied. Implementing this control in a distributed system with redundant unit controllers would cost less than a quarter of that for conventional analog control of a four-pass heater and less than a fifth for that of an eight-pass heater. Labor and additional installation material costs would offset the comparisons even more.

UNIT CONTROLLER CONFIGURATIONS

Unit controllers have built-in, predefined algorithms that perform specific, configurable tasks. The algorithms reside in a variety of "blocks," the function of which describes their types (Table 1). Note that the blocks are being used to show typical functionality; i.e., block count and type are not necessarily accurate. For example, an input block may handle multiple inputs; input and output blocks are usually not necessary with PID blocks; and logic and equation blocks are multifunctional, thereby reducing block quantity.

Figure 2 illustrates a configuration of function blocks for a distributed system unit controller that will provide

* The author wishes to thank C. W. Piper and M. J. Greenwood, both of The Foxboro Company, for their assistance with Fig. 8.

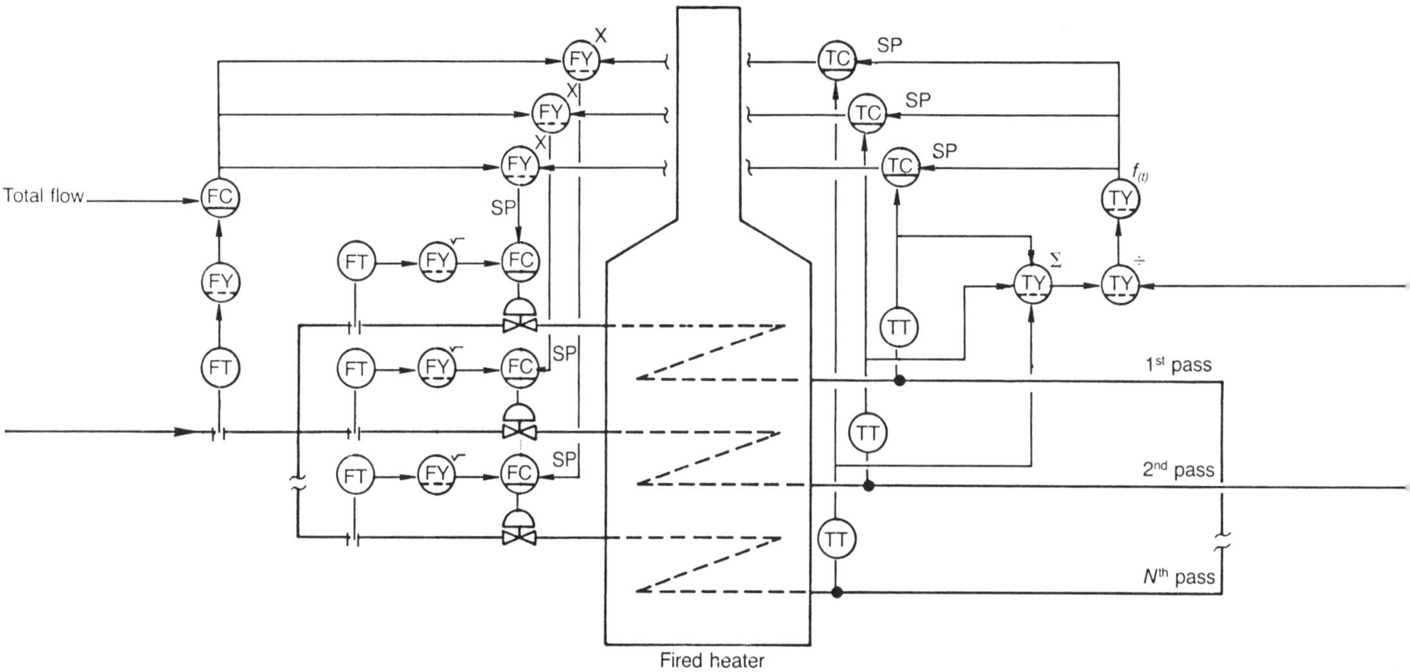

Figure 1. Fired heater pass balancing by conventional analog control.

Table 1
Unit Controller Function Blocks

Symbol	Block function
AI	**Analog Input:** does signal conditioning and characterization of analog inputs.
AO	**Analog Output:** conditions signal output to external hardware.
CI	**Contact Input:** passes the state of an external contact to the controller or another block.
CMP	**Compare:** performs a relational comparison of two inputs.
CO	**Contact Output:** forwards switch status commands to external hardware.
CON	**Constant:** permits input of a conversion factor or a standard value.
D-T	**Dead-Time:** delays controller reaction to a process change.
EQU	**Equation:** performs mathematical operations.
LAG	**Lag:** causes a gradual output change in response to a changing input signal.
[LOGIC]	**Logic:** executes Boolean tasks utilizing the logic operators shown.
NLN	**Nonlinear:** provides nonlinear signal characterization capability.
PID	**Controller:** furnishes proportional, integral, and derivative control actions and absolute, deviation, and output alarms.
S/H	**Sample/Hold:** signal output tracks the input as long as the track command is present; otherwise, the output is held at the level of the last input sample.
SEQ	**Sequence:** produces digital outputs in a stepped numerical sequence.
TMR	**Timer:** provides digital outputs for a specified time.

Notes: A digital input to a block other than a logic block has been represented as:

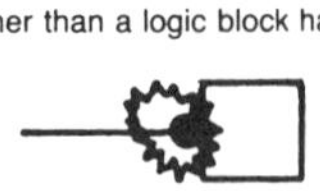

TGL indicates a "toggle" signal.

feedforward manipulation (based on feed rate) of a distillation tower reflux flow and bottom heat input. Dynamic compensation is provided to further minimize or eliminate tower upset caused by feed-rate fluctuations. Control block (PID) features such as set point tracking, output limiting, and external feedback aid the operator in applying multivariable controls and increase plant operational security. Since measurement signals are scaled and linearized by analog input blocks (AI), transmitters for millivolt or resistance conversions are not required.

Integrating a process computer and a unit controller produces the ultimate system for advanced control. However, in accordance with good engineering practice, caution must be exercised when configuring a control strategy to ensure loop integrity in the event of computer failure.

The marriage of a computer and a distributed system unit controller for advanced control of a crude unit atmospheric tower pumparound stream is shown in Fig. 3. Primary control is not merely pumparound flow, but rather pumparound heat removal rate (duty). Constants for the duty calculation and the duty controller set point are *updated* by the computer. Should the computer go off line, only the updating would cease. New values could be entered by an operator.

ANALYZERS AND UNIT CONTROLLERS

Microprocessor-enhanced programmers and diagnostic tests have tremendously strengthened the security thence acceptance) of analyzer-directed control. Improved control security can be attained when unit controllers are interfaced to the analyzers.

Control of the mole percentage of normal butane allowed to escape in debutanizer bottoms is detailed in Fig. 4. In this distributed system configuration, the composition controller will make adjustments to the bottoms temperature to maintain a desired normal butane slippage. A status change to "off" directed by the chromatograph programmer will transfer the composition controller to the manual mode. An operator can reset the controller to automatic once the programmer has cleared the hold.

An effective, economical approach for expanding the capability of panel instruments is to add the power of a self-contained unit controller as shown in the hydrogen recycle flow control example of Fig. 5. Information from the naphtha and hydrogen flow transmitters and the hydrogen gravitometer is used in calculations to furnish engineering data and to maintain the hydrogen/hydrocarbon mole ratio set by the operator. The deviation alarm and output clamps of a control block augment

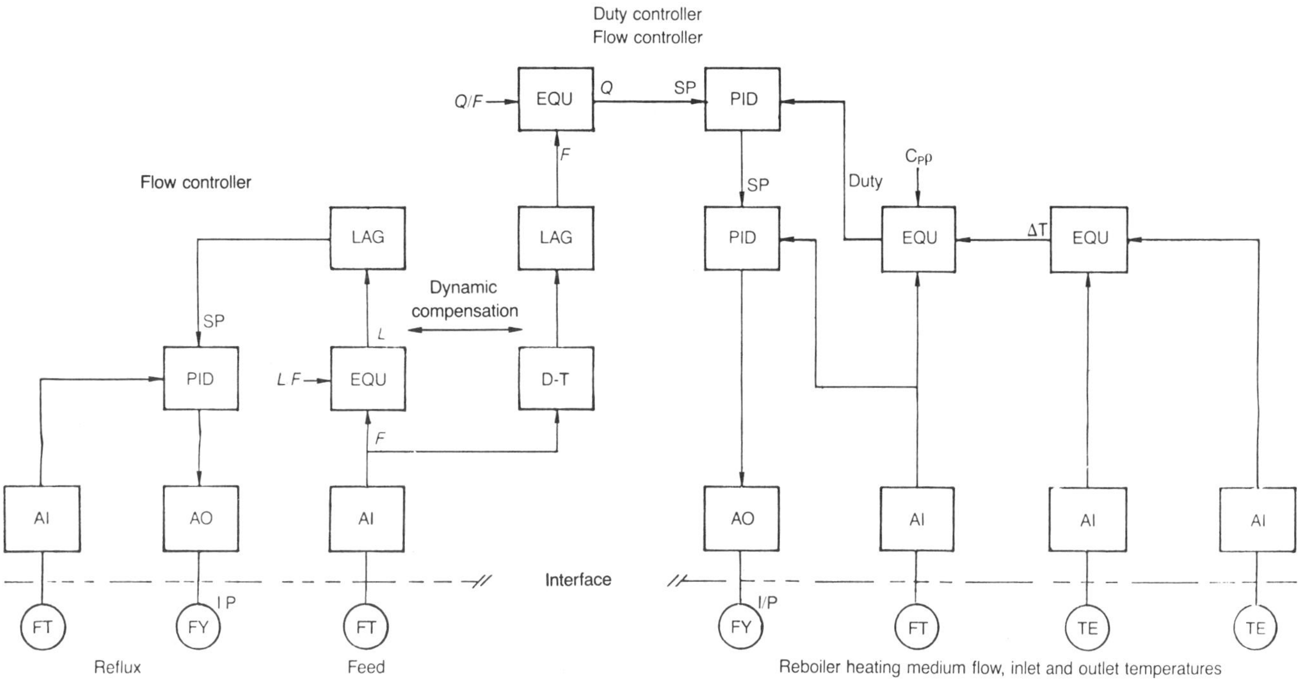

Figure 2. A feedforward scheme for distillation tower heat control.

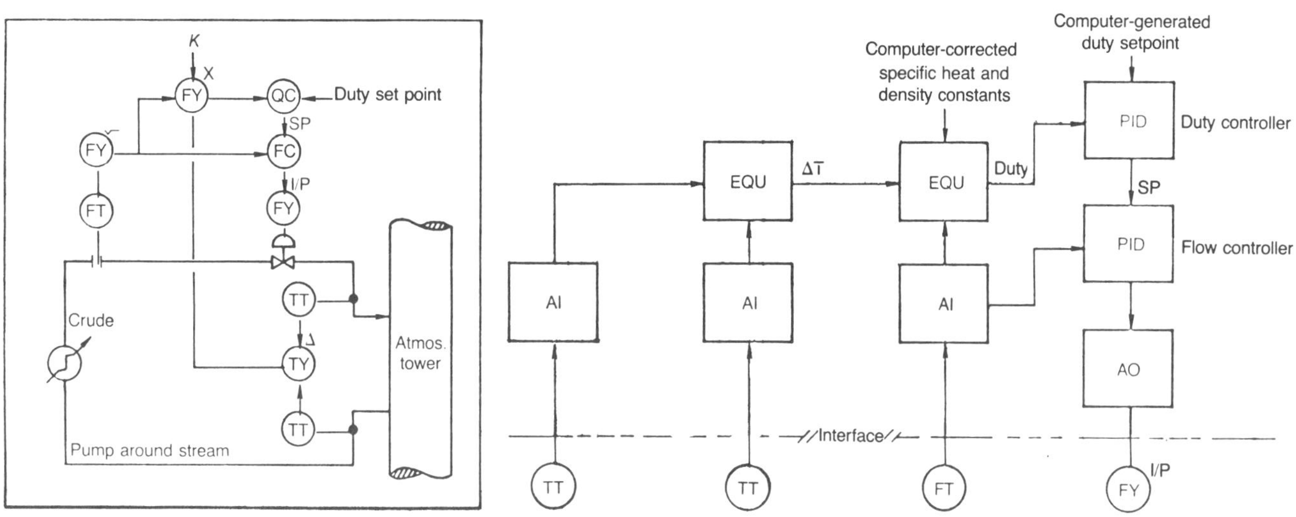

Figure 3. Process computer/distributed control system control hierarchy.

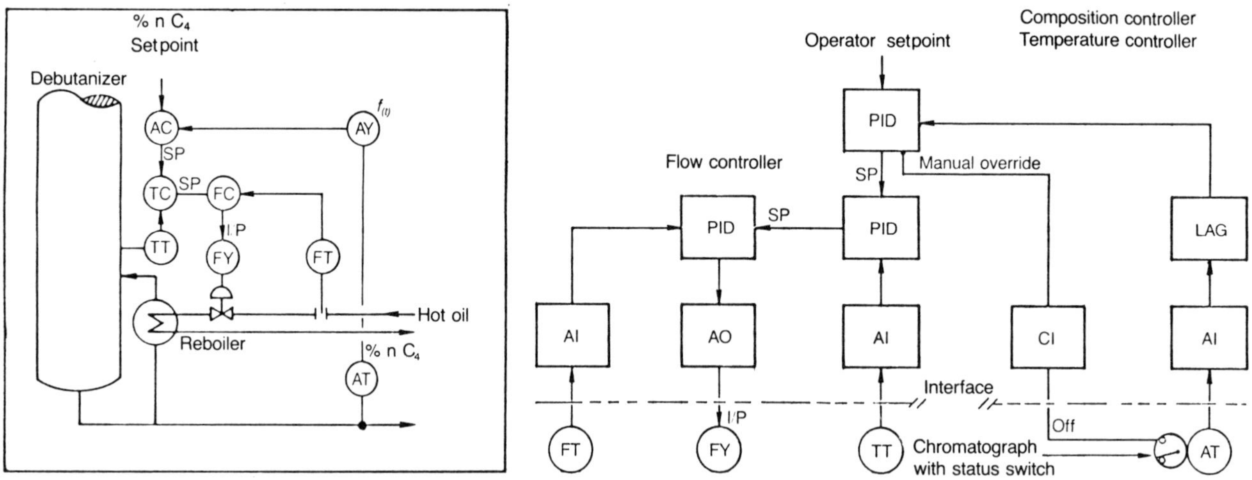

Figure 4. Debutanizer bottoms composition control.

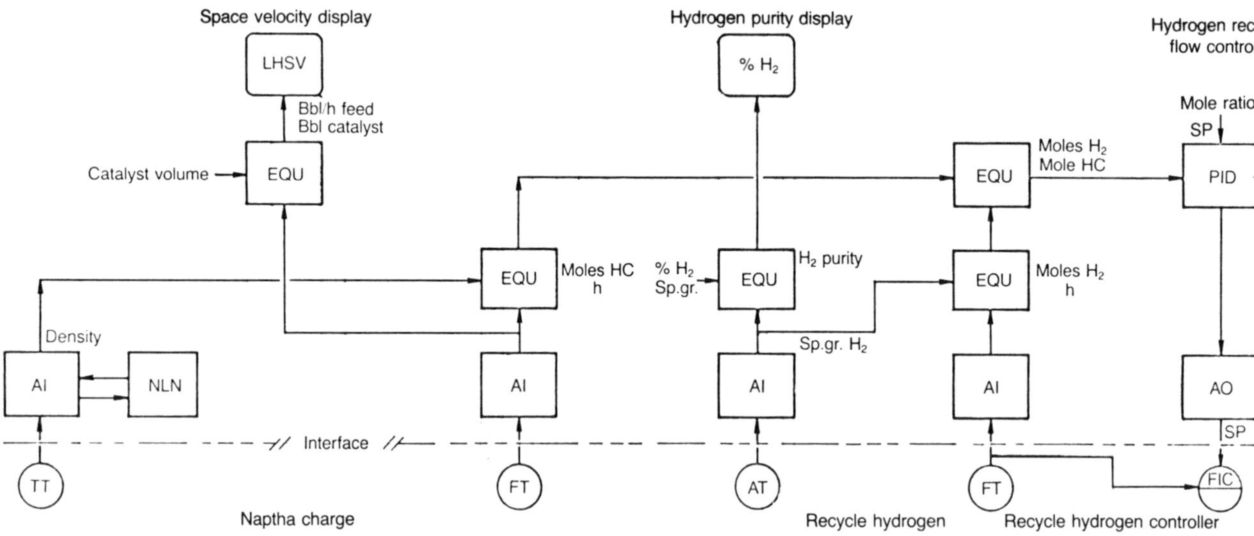

Figure 5. Hydrogen-to-hydrocarbon ratio control for a catalytic reformer.

control security in the event of a measurement failure. Independent output boards equipped with fail-safe options provide a sound unit controller/panel controller interface.

The more sophisticated analyzer interface presented in the scheme of Fig. 6 includes a dependable safeguard against using poor analyzer information for control. In the event of an analyzer failure or calibration check, the octane controller would transfer to the manual mode where it would remain until the condition cleared and the operator again selected automatic control. The self-contained unit controller manipulates panel reactor inlet temperature controllers to maintain a set reformate octane quality.

USING LOGIC FUNCTIONS

Typically, logical operations are performed by electromechanical or solid-state logic devices, programable logic controllers, and process computers. However, when it is necessary to incorporate a few logic functions in a continuous control scheme (such as those employed in analyzer-directed control), it becomes more practical to use unit controllers. They offer simple logic circuit configuration and modification, and an accurate and reliable logical operation/instrument loop interface. The logic circuits of Figs. 7 and 8 demonstrate the power available in a unit controller to perform logical operations.

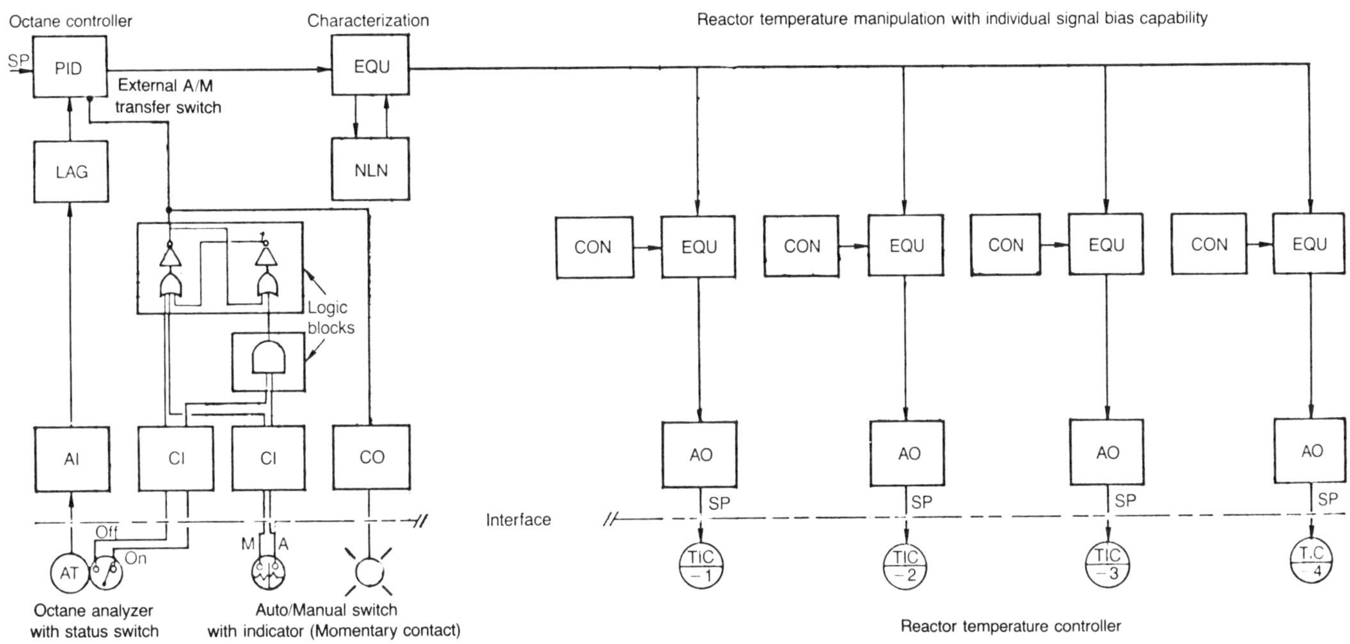

Figure 6. Octane control for a catalytic reformer.

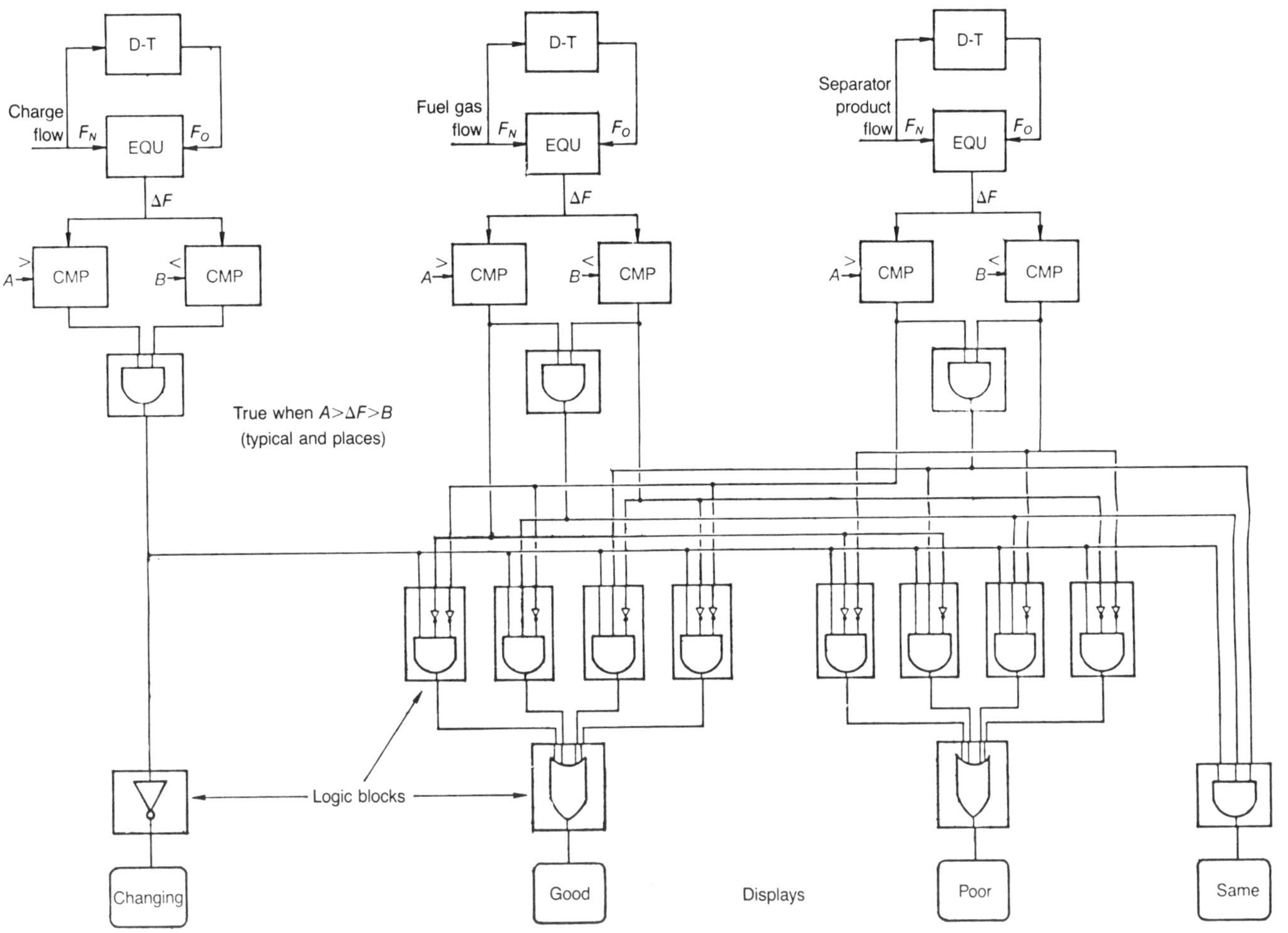

Figure 7. Logic circuit for a catalytic reformer energy conservation performance test.

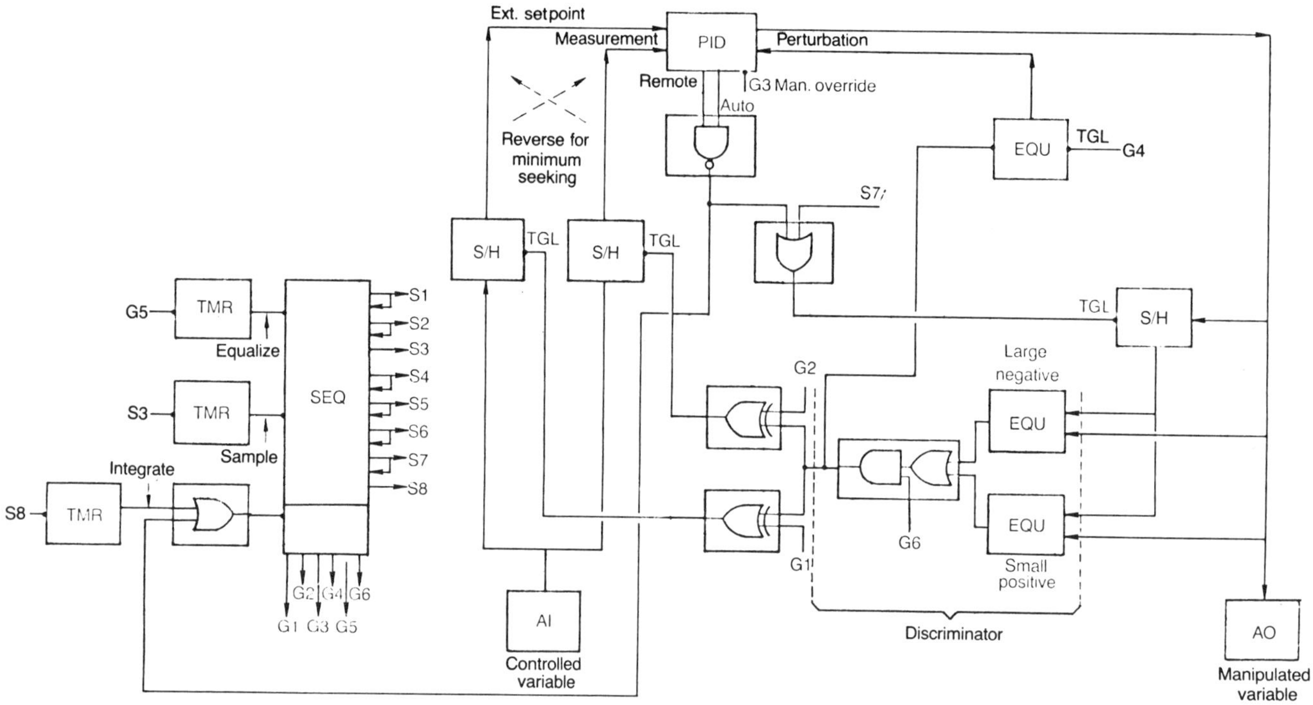

Figure 8. Distributed control system self-optimizing controller.

Table 2 describes an energy conservation performance test that can be used as a guide to an operator as he adjusts the reactor inlet temperatures of a catalytic reformer (constant blocks in Fig. 6). Fig. 7 depicts the logic instruction for the performance test as it would be configured in a unit controller.

Table 2
Catalytic Reformer Performance Test (Based on Flow Rates)

Charge	Fuel gas		Separator product	Performance
+/−		← All cases →		Changing
=	=		+	Good
=	−		=	Good
=	−		+	Good
=	+		+	Good
=	=		−	Poor
=	+		=	Poor
=	+		−	Poor
=	−		−	Poor
=	=		=	Same

Where:
(=) is no change in flow rate
(−) is a decrease in flow rate
(+) is an increase in flow rate

If a refiner would like more than off-line guidance of reactor inlet temperature manipulations, the self-optimizing controller described in Fig. 8 could be applied to the octane control scheme of Fig. 6. It would continuously evaluate, bias, and compensate the reactor inlet temperature manipulations as necessary. The fuel consumption is minimized while the octane quality is constrained to the desired value. Examples of other suitable applications for the self-optimizing controller include furnace combustion control, compressor station prime-mover loading management and multiple heat exchanger duty balancing.

SUMMARY

The unit controller has the ability to execute many of the more complex controls and engineering calculations once exclusively assigned to a process computer. Available in either self-contained or console-based designs, the unit controller provides accurate data conversion and characterization, a high degree of loop security, computer and analyzer compatibility and the maximum in loop configuration flexibility. The use of unit controllers to enhance process performance with more sophisticated control strategies affords improved operator efficiency and, generally, a good return on investment.

7
Programmable Controller Problems

E. A. DeVries

Programmable controllers have been around for about fifteen years and have come into widespread use during the last five. They have many advantages over old-style relay logic systems, including lower cost per number of inputs and outputs, smaller size for the same logic control, increased flexibility for easy changes and less power consumption. The newer PC's have many expanded capabilities over that of their relay predecessors, including mathematical and computing capabilities, process control algorithms, report generation and information handling, and they will interface to another computer.

However, with any new device, no matter how wonderful that device is, there comes a new set of problems. In applying PC's, one quickly discovers that these new devices require cleaner power supplies, and are more noise sensitive than their electromechanical counterparts. However, hardware problems account for only a small percentage of programmable controller problems. A great many problems can be traced to the PC software.

Hardware problems can normally be found quickly. Experienced technicians can isolate a problem electronic board, replace it with a spare, and get the PC back on line within a few hours. Generally, hardware failures result from line transients, excessive heat, corrosion, or noise, with the power supply being the cause much of the time. For this reason, a PC system should be installed with a line conditioner, isolation transformers, or uninterruptible power supply (UPS). Many users also prefer redundant hardware to minimize the effects of hardware failure.

However, processor failures are not uncommon, and these are generally more difficult to diagnose. Many PC's have self-diagnostics to help isolate these processor problems. Sometimes watchdog timers are used to monitor processor performance.

The most difficult problems to diagnose, however, are software problems. Some software problems may take days to find. Good documentation is an absolute necessity for troubleshooting PC software. At a minimum, software documentation should include a ladder diagram, an address assignment list, and input and output schematics (Figs. 1, 2A and 2B). Also, a program listing, cross-reference sheets, and a narrative type operating description are recommended for easier startup, training and troubleshooting. In addition to these documents, some operating companies require binary type logic diagrams in accordance with industry standards

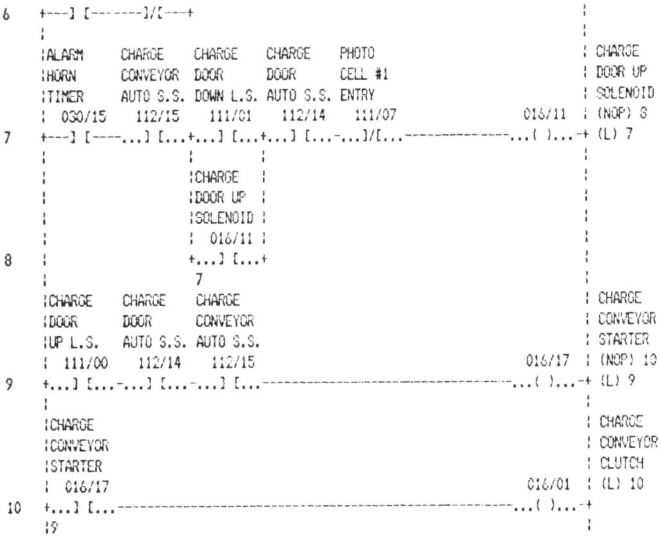

Figure 1. Typical ladder diagram. Note line numbers, address assignment numbers, verbal descriptions and cross-references.

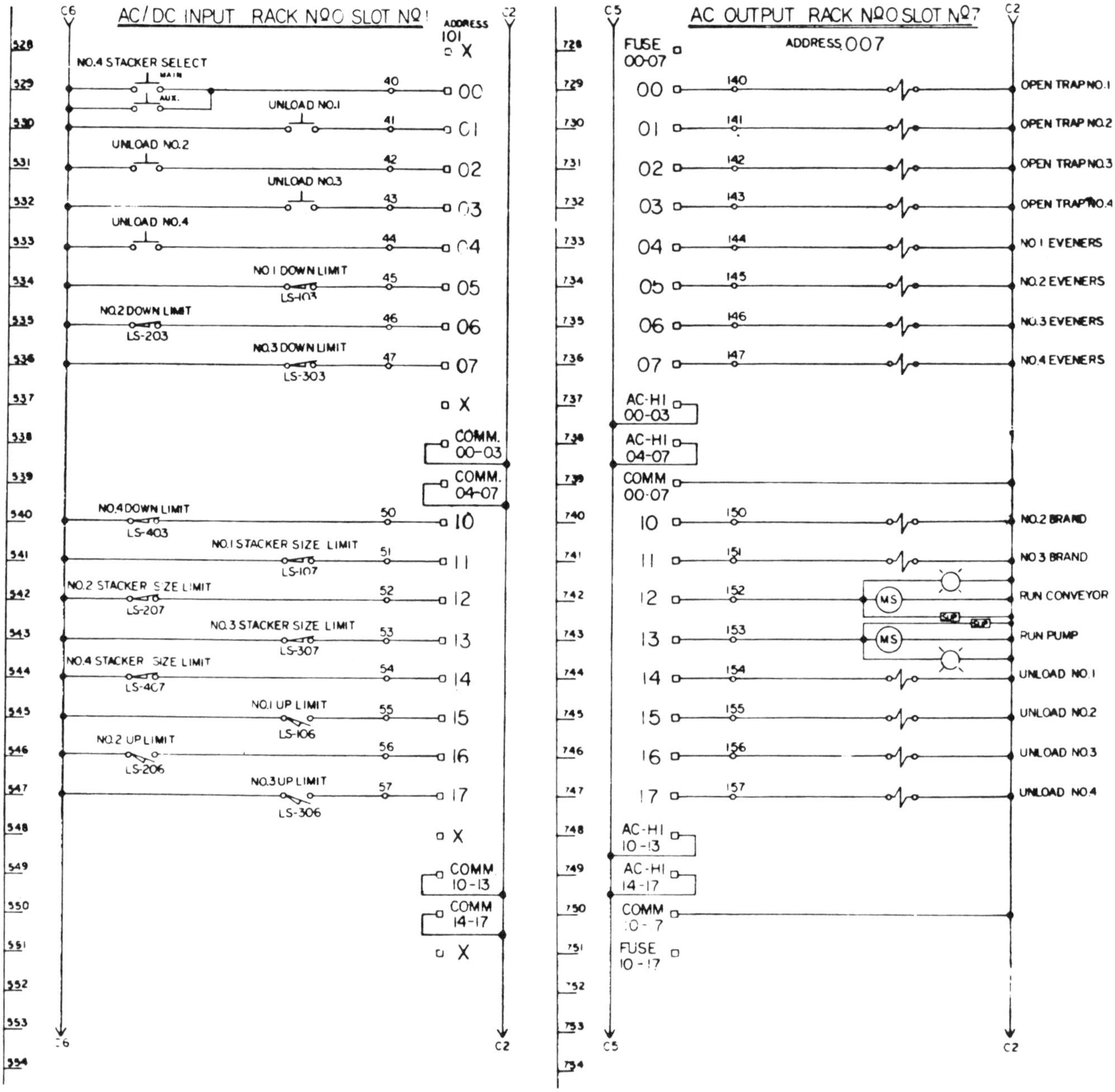

Figure 2A. Typical input and output schematics (Courtesy Allen-Bradley Co.).

ISA-S5.2 or ANSI/IEEE 91-1973 (Figs. 3A[10] and 3B). The latter standard has been adopted by the Department of Defense for logic systems associated with military equipment. Old-style logic documentation such as truth tables or sequence charts commonly used for relay systems are not very useful for programmable controllers.

Once the problem has been diagnosed as being in the software, the documentation comes into play. The most common software problems are an improper end statement, incorrect address assignments, or duplication of address assignments. It is easy to see why a program listing and cross-reference sheets are recommended.

There are more difficult PC problems one sometimes encounters, and most of these are time duration phenomena. These time-related problems include contact bounce, rapid intermittant signals, improper timer settings and program racing. These more difficult problems can sometimes take weeks to properly diagnose and correct. Tools needed to trace these problems include binary logic diagrams and CRT terminals that have error reporting, diagnostics, and contact histograms.

A contact histogram records the history of a bit in memory over time duration. If the bit is changing its status more than that of its input device, contact bounce may be the problem.

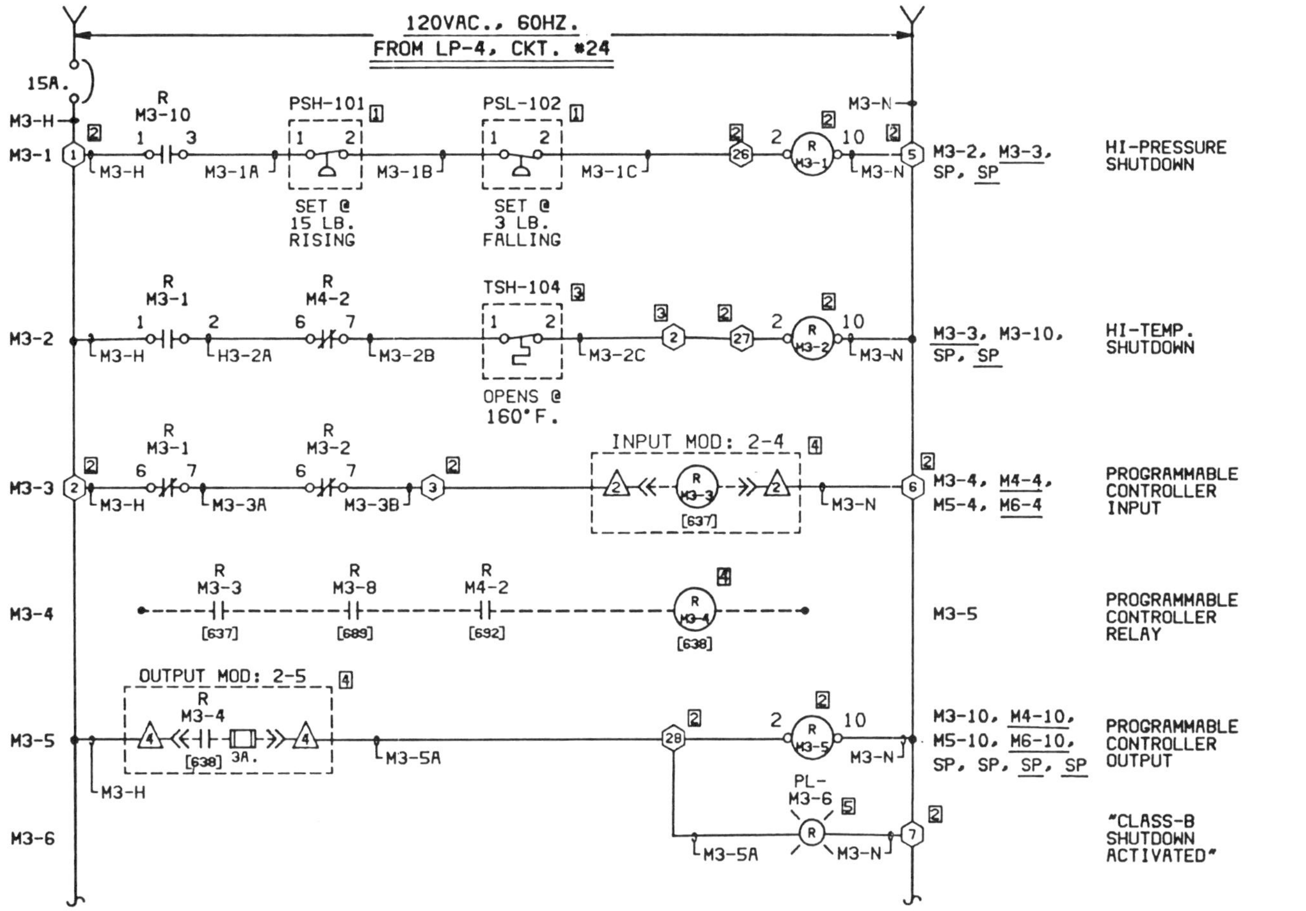

Figure 2B. Many users prefer schematics and ladder diagrams combined in one document (Courtesy the John West Co.).

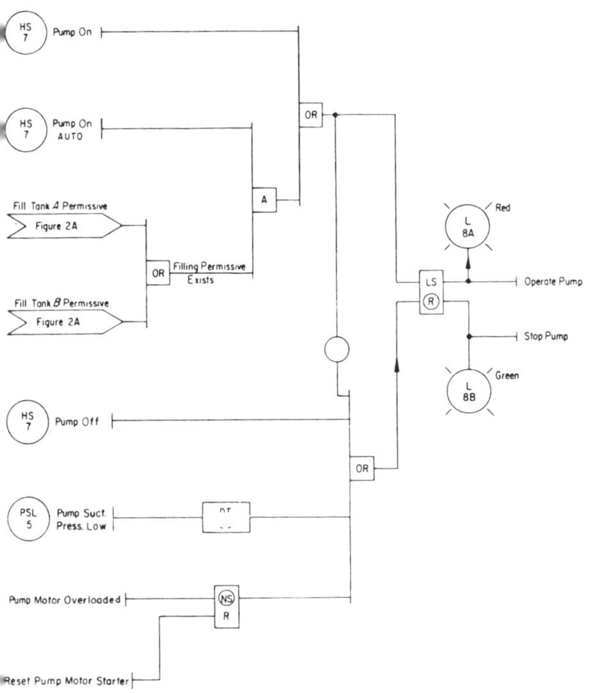

Figure 3A. Example from ISA Binary Logic Standard.[10]

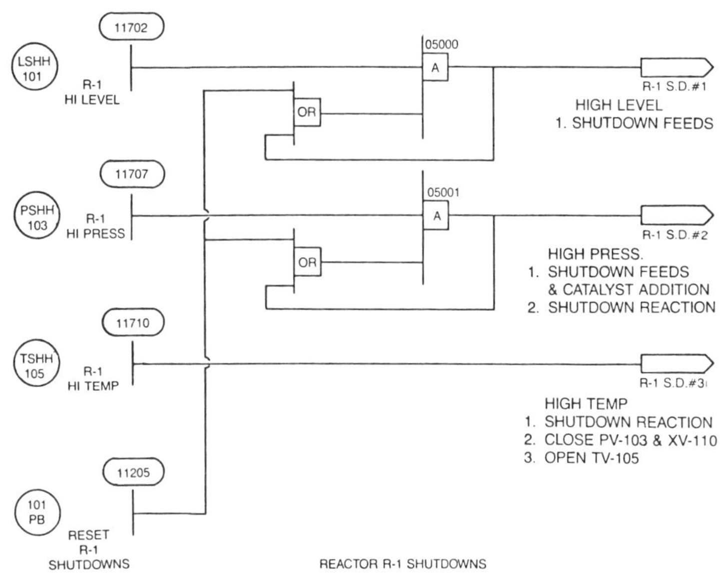

Figure 3B. Binary logic diagram used to document PC logic software.

Program racing may happen when the scan time for the PC must be faster than the process reaction time. It is entirely possible that the PC is not able to keep up with a very fast process. However, the first thing to check using these CRTs is the program end statement for proper format.

Proper address assignments are a necessity if the program is to work properly. Internal storage locations and interlocks seem to be the most problematic of these. Many people feel the binary logic diagram best shows the relationship of interlocks and internal storage locations. However, good assignment documents and cross-reference sheets are important also.

There are several computer-generated programmable controller documentation systems on the market, and various companies offer documentation packages using these systems (Fig. 4 and Table 1). The references to wire numbers and other drawings are very helpful in troubleshooting. Many of these systems have the advantage of checking the assignments for duplication, or for improper use. Many people have found this feature useful. Many times these computer-generated packages can be purchased cheaper than a company can make their own. A further advantage of these computer systems is that some can generate documentation from a program tape of an existing programmable controller. This is particularly useful for older PC installations whose documentation may not be up to date.

```
Address     PC operation and references
--------    --------------+------------+---------
016/01       CHARGE     CONVEYOR  CLUTCH
             (L) 10
016/02       OIL        PUMP      STARTER
             (L) 33
016/03       OIL TANK   ROLLS     STARTER
             (NOP) 37,45
             (L) 36
016/04       OIL        ELEVATOR  SOLENOID
             (L) 43
016/05       ALARM      BEACON    FLASH
             (L) 5
016/06       ALARM      HORN      ALERT
             (L) 3
016/07       WATER      ELEVATOR  SOLENOID
             (L) 26
```

Figure 4. Typical computer generated cross-reference list.

Table 1
Assignment List and Cross-Reference Sheet

PLC no.	Wire no.	Schem. dwg. no.	Schem. line no.	Logic dwg. no.	Logic line no.	Description
		D-3-10				
110 00	502	8030	100	B-10001	10	START P.B.
110 01	503	8030	101	B-10001	10,9, 38A	STOP P.B.
110 02	508	8030	105	B-10001	3,14, 29,30	FLAME FAILURE
110 03	509	8030	106	B-10001	26,42	ZSH-914 F.D. FAN BACK-UP TURBINE STEAM VALVE
110 04	510	8030	107	B-10001	6	LSLL-910 BOILER LOW WATER
110 05	511	8030	108	B-10001	1	PSHH-916 MAIN GAS HIGH PRESS.
110 06	512	8030	109	B-10001	2	PSLL-916 MAIN GAS LO PRESS.
110 07	513	8030	110	B-10001	14	ZSH-916 MAIN GAS FLOW CONTROL VALVE

Following good hardware installation practices and careful software documentation as illustrated here will help minimize PC problems and aid troubleshooting.

References

1. Barron, G. R., "Application of Programmable Controllers in Process Interlocks and Shutdowns," ISA Joint Spring Symposium, Houston, Texas, April 19, 1983.
2. Allen-Bradley, "Industrial Terminal System User's Manual," Ohio, June 1981.
3. Allen-Bradley, Systems Group, "PLC School Outline," 747 Alpha Dr., Highland Heights, Ohio, December 1978.
4. Beck, W. A., and Carver, T. E., "Document Your Process Control Software," *Instrument Technology,* September 1982.
5. McCann, B., "Troubleshooting Programmable Controllers," *Plant Engineering,* October 14, 1982.
6. Penz, D. A., "Organizing PC Software Development (Part 1)," *Instruments and Control Systems,* February 1982.
7. Penz, D. A., "Organizing PC Software Development (Part 2)," *Instruments and Control Systems,* March 1982.
8. Saitta, G. N., "PC Manages Vapor Recovery System at Alaska Pipeline Terminal," *Instrument Technology,* May 1978.
9. Togino, K., and Furuta, K., "A Computer Route to Ladder Diagrams and PC Programs," *Instrument Technology,* September 1978.
10. "Reprinted with permission from ANSI/ISA-S5.2-1976 (R1981), "Binary Logic Diagrams for Process Operations," Instruments Society of America 1981. Copies available from ISA, P.O. Box 12277, Research Triangle Park, NC 27709.

8

Convert to Microprocessor Controls Without a Shutdown*

M. E. Leister
R. R. Sanders

Converting all process control and other instrumentation, while upgrading existing field instrument transmitters, to a modern, centralized microprocessor-based distributed control system with the latest generation electronic transmitters obviously requires careful planning and coordination. Making the conversion while a unit is operating requires even greater concentration on planning and coordination. However, careful planning and a loop-by-loop approach allowed this conversion to occur without any disruption of operations. How this conversion project was planned, organized, and implemented is presented.

BACKGROUND

Prior to the move to a centralized control room, Marathon Oil Company's Robinson Refinery consisted of 12 major units which operated out of six separate control rooms. The newest units were built in 1968 to 1970 and used mostly 10 to 50 mA electronic transmitters, recorders and controllers. These units included a Crude Unit, Unicracker, Ultraformer, Alkylation Unit, Saturate Gas Plant and Sulfur Plant. The remaining units were instrumented with pneumatic and a few older 4 to 20 mA transmitters, recorders and controllers. These included an older Crude Unit, Cat Cracker, Light Ends Plant, Regular Delayed Coker, Special Delayed Coker, and a Thermal Cracker.

Marathon's experience at its Garyville Refinery revealed that use of a microprocessor-based, shared display control system could result in a substantial operational improvement over traditional control systems. Consolidating the refinery process controls into a central control room also allowed the incorporation of supervisory computer control. The combination of these two factors led to the decision to consolidate all refinery operations into one central control room with six "Complexes" or major unit groupings utilizing a microprocessor-based control system.

The decision to consolidate control rooms required other changes in refinery operations. A new radio system was necessary to maintain communications between the central control room and each unit. The Ultraformer regeneration control panel, which consisted of 320 relays, was replaced by a programmable controller that interfaced with the distributed control system. All unit shutdowns and shutdown bypasses were also incorporated into the programmable controller. Unit reactor bed temperatures were transmitted to the central control room via a thermocouple multiplexing system. In addition, the decision was made to upgrade all field instrument transmitters which interfaced with the new shared display system. The old transmitters could have been adapted to the new control system, but the latest generation of electronic transmitters has a much faster response time, and the old electronic and pneumatic transmitters were nearing the end of their useful lives.

PLANNING

The requirement of doing work on line with a refinery in operation without endangering personnel and/or equipment, shutting down any process and at the same time gaining operator acceptance of the new system was a large, intricate project that required the total efforts of many dedicated individuals and organizations. Prior to starting the actual conversion effort, several areas of prime concern had to be studied, potential risks evaluated against expected gains and a thorough strategy for implementation of the work was developed. A "steering

* Originally presented at the National Petroleum Refiners Association Computer Conference, San Francisco, Calif., Oct. 24–27, 1982.

committee" comprised of plant and project management, members of Refinery Operations and Maintenance Departments, with a conversion team leader was established to set guidelines, assure standardization of approaches and solutions to problems, and help the project stay on schedule and within budgeted costs. Areas addressed during pre-conversion planning were:

Safety. Each refinery unit and process was reviewed for potential unsafe conditions, such as hazardous materials, consequences of a process upset or shutdown caused by conversion activities, fire, and environmental considerations. From this review, safety procedures and safe working practices were established for the conversion team, and their responsibilities in the event of an abnormal condition were established. Safety became the guideline for every conversion activity.

Operator acceptance. No matter how good a new control system is, operator acceptance is essential to gain all the desired benefits. Fear of change and "the unknown," along with the apprehension of operating a unit from a CRT, cause even experienced operators to have grave concerns. These apprehensions must be combatted as soon as possible to prevent misinformation from dooming the project to failure. As soon as the decision to convert to a microprocessor-based system was made, demonstration units were brought on site for three weeks to allow operators, maintenance personnel and engineers a small amount of "hands-on" time to familiarize themselves with the new hardware. This was followed by an in-depth training program, both in-house and at vendor facilities. Foremen and chief operators were trained by project personnel, and they, in turn, trained the boardmen. Maintenance personnel received both in-house training and formal training at vendor facilities. Foremen, unit engineers and maintenance personnel were included in project meetings and given active roles in the conversion team effort. This participation in and with the project team and project personnel helped build trust, confidence, and working relationships between the project and plant people. Especially helpful in our acceptance efforts was the use of longtime operations and maintenance individuals on the project team. These people were trusted and established expert resources within the plant community.

Definition of Work. Defining "what really exists in a unit" is essential as a starting point for work to be done and seems simple to determine. However, units 20 years old or older do not usually have the same level of documentation as modern units. Over the years, many changes in operational philosophy have occurred. Control valves and vessels which once had important functions may be bypassed or no longer in service. The cost of converting these points must be weighed against the likelihood of their future use. Extensive reviews with the foremen and unit engineers are necessary to establish firm definition of the current status of each field device. Piping and instrument drawings, electrical wiring and logic drawings and instrument data sheets must be reviewed for accuracy. Areas where information is missing or undependable must be field surveyed to bring the documentation up to date. Where possible, past maintenance problems must be reviewed to prevent future ones. All possible sources of information must be reviewed to get an accurate definition of the work to be done.

Speed of Conversion. As with any good project, justification for the expenditure is a major consideration and the faster the project is completed, the sooner the benefits and rewards are realized. In the case of online conversion, rapid completion becomes a necessity since units, processes and even individual control loops can become divided, i.e., controlled and monitored from two separate locations which are vastly different in their modes of operation. Therefore, the operating group from each plant was asked to identify "critical" loops; those that must have priority when converted or must be completed in a given sequence or at the same time. Next a "trial" conversion was conducted on out-of-service loops. This gave the project team an idea of the actual time required to go to manual control or bypass (simulated), remove the existing transmitter, mount the new transmitter, rework process tubing, steam tracing, insulation, and install winterization enclosures. The time trial data and the critical loop list were the basis of conversion sequence schedules which, after refinement and detailed planning by the conversion team, later became the controlling document for the conversion portion of the project. Time saving installation techniques were developed. In the process units utilizing pneumatic instrumentation, all conduit, wire, etc. was installed and checked out prior to conversion. In existing early generation electronic units, signal wire and cables were field checked prior to conversion to assure their suitability for reuse. Data links from remote instrument interface units to the central operator consoles were verified. Maximum preassembly of transmitters, manifolds, winterized enclosures and fittings was accomplished, allowing 50% of the piping work to be done in the shop. All instruments were bench calibrated and tested so only a functional test and verification of calibration test were required in the field. Then, all instrumentation and materials not pre-installed were clearly identified and marshalled to facilitate sequential access during conversion.

ORGANIZATION

The organization of the conversion team was as complex as any of the problems faced during the conversion effort. The myriad interfaces between construction, engineering, operations, maintenance, the project team, vendor service representatives, etc. can lead to an organization much too large and bulky to respond to the needs of loop-by-loop conversion. The organization must be as streamlined as possible, with clear cut channels of responsibility. The problems of working in operating units were numerous—who would be in charge of a process during conversion, operations or the conversion team? How and when can a loop be taken out of service and then put back into service? Even worker density was a concern since working space around a single transmitter may be limited. To solve these problems and to organize the conversion team, the first step was to divide the conversion effort into activities or steps. With each step defined, numbers of people required were estimated and assigned responsibilities. The conversion was divided into three steps as shown in Table 1.

Table 1

1. Pre-conversion:	Complete detailed conversion sequence schedule. Develop installation task lists. Preassembly and shop calibration. Marshall materials. Train and inform operations personnel. Acquire and update all documentation.
2. Actual conversion:	Isolate loops. Site clear existing transmitter. Install new transmitter, process air and steam. Fit tubing. Calibration check and loop check. Process on and function test. Functional acceptance by operations and maintenance.
3. Post-conversion:	Steam trace. Insulate. Clean up. Final acceptance by the complex foreman.

After defining each step of conversion, it was obvious the team had to be made up of individuals from more than one group—construction, plant maintenance, operations and engineering all had to be represented and take active roles, while some roles needed to be duplicated with counterparts in the contractor and client organizations to assure checking, cross-checking and proper coordination took place. An organizational chart of the conversion team is shown in Fig. 1. The following roles were defined:

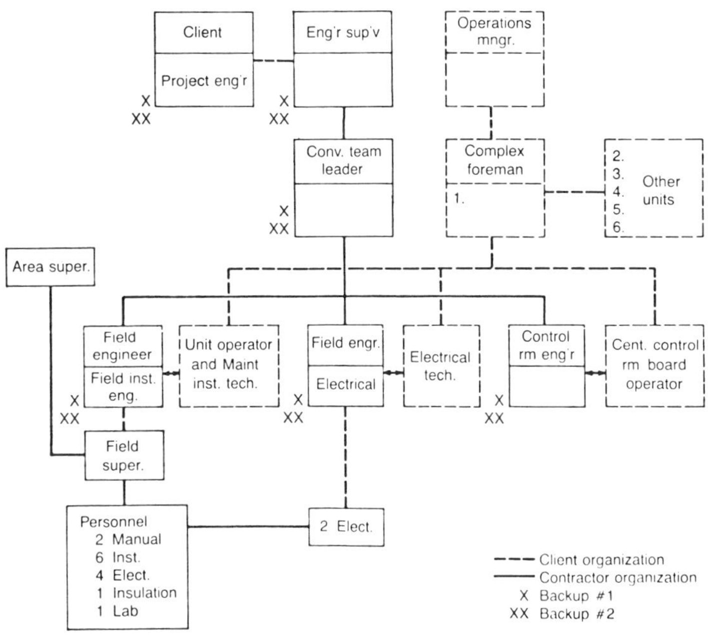

Figure 1. Conversion organizational chart.

Conversion team leader—In charge of the contractor's organization, he had overall responsibility for the conversion effort, including organization coordination and actual execution, and seeing established procedures were followed.

Complex foreman—The conversion team leader's counterpart in the Marathon organization, he was responsible for his personnel being trained and ready for the conversion. During conversion, he was the go/no-go decision maker for each loop and for the overall conversion effort in his complex, based on safety and operating conditions. Each day's work was reviewed one day ahead of conversion for approval.

Control room engineer—A member of the contractor's organization, he manned the CRT operator station on the console of the complex being converted. He confirmed loop operation and functionality and commissioned loop to client upon field acceptance by operations and maintenance.

Complex console operator—Marathon's unit operator took possession of commissioned loops and operated the plant from the new control room.

Field operator—Marathon's outside operator, his prime responsibility was to remove the loops from service and maintain stable unit operation by manual control in the field, where necessary. He also witnessed the field functional test.

Instrument field engineer—The contractor's outside team leader, he supervised and was responsible for quality control of the change out or isolation of instruments, installation of new process instruments, tubing, steam tracing and insulation, testing and field calibration checks.

Instrument technician—Marathon's counterpart of instrument field engineer witnessed calibration and functional field tests and accepted field operation of the loop.

Electrical field engineer—This person supervised and coordinated wire continuity and polarity checks, wire transfers in field and local control house junction boxes to new instrument interface units. He checked and coordinated all I/O satellite rack operations.

Electrical technician—He was the Marathon counterpart to the electrical field engineer. He witnessed electrical tests and was an available resource for field problems.

Manual crafts people—Was a composite trade work force made up of instrument fitters and electricians with insulators and laborers supporting the composite crews to do all physical installation and site clearance. It should be noted that the trade unions agreed to work as a composite (different trades) team and to work through breaks, lunch and on overtime as required to complete any loop started, to endure adverse weather conditions and to take direction from conversion engineers rather than the normal general foreman-to-foreman chain of command. Without this cooperation by the crafts people, a safe, efficient conversion effort would not have been possible.

Once the positions were identified, the task of filling them with top notch people began. Some positions were filled from project people on site; others, such as the conversion team leader, were brought in from other contractor jobs or offices to supply the caliber of people required to undertake on line conversion. Plant Operations and Maintenance Departments identified members to fill their part of the team, and contractor manual crafts people that had demonstrated above average performance during preconversion installation were utilized to complete the conversion team.

IMPLEMENTATION

Once the conversion team was organized and guidelines for conducting on line conversion were established, the process of execution began. First, a preconversion meeting was held with complex operations and maintenance people, and the preliminary conversion schedule was given to them to identify critical and non-critical loops. Non-critical loops were those not used in actual unit operation, such as accounting or indicating loops that could be converted at any time without interfering with the unit's operation.

Critical loops were identified as follows: (1) must be done alone, (2) must be done in conjunction with another loop, (3) cannot be done on the same day as another loop, and (4) requires special precautions for safety or process stability. With the loops identified, a revised conversion list was made with the date and approximate time conversion of each loop. Since normal starting time was 7 a.m., a 6:30 meeting was held each morning to review the previous and present days' work. With all supervision in attendance, it was at this meeting that the operations foreman approved the day's work, rearranged the schedule because of process needs, and sped up or slowed down the conversion effort. With everyone in agreement as to the work to be completed for the day, crews with their materials were sent into the units with the instrument and electrical conversion engineers to prepare for the day's work. Once in the field, the team was in constant contact with each other via portable radio and sound-powered telephones in the I/O rack areas. With everyone in place, the field operator made one last check with the existing control board operator and the new console operator to make sure the instrument could be removed from service. With their OK, the field operator took manual control of the valve and blocked off the transmitter process taps and steam tracing as required. The field manual crews removed the transmitter, tracing, tubing and insulation, as instructed by the instrument engineer, and the mounting of the new transmitter started. During this time, the electrical crews checked wiring one last time with the new transmitter in place. A calibration check (actual calibration had been done earlier in the shop) took place with the instrument technician witnessing the tests, and the field operator checked with the console engineer for the OK to bring the instrument back on line. The console engineer then would complete a functional test of the loop, and the process would be turned back on. A last check of system stability was made by the console engineer, resetting any variables required. With the instrument stable and operating, the console engineer passed the instrument over to the console operator, and the instrument was again a part of the operating plant. In cases where the valves could not be stroked manually, I/P (current to pneumatic) transducers were installed, and the loop was simulated without air connected to the valve operator.

In this manner, the conversion team moved from eight loops a day in the beginning of the conversion effort to as many as 23 in a single day during the closing weeks of conversion.

DOCUMENTATION

While a complete design package, including P&ID's, electrical equipment layouts, electrical raceway plans, junction box details, connection diagrams and vendor configuration drawings, was generated and issued as a part of the project's engineering effort, the instrument loop connection diagram (or loop sheet) was the data base for conversion. A loop drawing was made for every instrument interfacing with the central control consoles. Each loop sheet had all the configuration documentation as part of the drawing format and included all information required by the field and/or console engineer for conversion. To assure the refinery that all procedures for conversion were followed, the sign-off took place on the original loop diagram drawing. The instrument technician signed for the calibration and functional field check, the electrical technician for electrical completeness, the field operator for functionability and completeness of installation, and the console operator that the loop was operational from the console. The complex foreman reviewed these signatures and made the final acceptance of each loop. All "as-built" changes resulting from conversion were transferred from the copies of the loops used for conversion to the original drawings, and the drawings were reviewed by the project engineer for engineering completeness and signed.

RESULTS, BENEFITS, AND LESSONS

The results of the conversion project were outstanding. No process shutdowns due to conversions took place. The actual conversion started May 14, 1981, and ended December 16, 1981, some 14 weeks ahead of schedule. During this time, about 3,000 loops were converted, 1,000 transmitters installed, and nearly 100 shutdown switches or bypasses moved to the central control room. Six hundred of the loops were controllers which directly manipulated control valves. The refinery operators have accepted the shared display systems, and a considerable improvement in operating performance has been realized.

Looking back, there were several lessons learned:

1. Materials: Scheduling of materials is critical. Delivery of instrument transmitters, at one point, almost shut the project down. Specialty and long delivery items should be well on their way or in hand before starting a conversion effort. Delivery one day with installation scheduled for the next makes for many sleepless nights.
2. People: Top quality people are the key to success. The time and effort must be spent to put together a good team. It pays big dividends. Utilize plant people as much as practical with the engineering team.
3. Union agreements: The agreement made with the local trade unions to work composite crews, to take directions from the conversion engineers, and to work until all loops started each day were complete was essential to our successful efforts.
4. Spend the time and effort to find out what's really existing in each unit and how it works. Make new drawings if none exist or if they are so out of date they are useless. The new drawings made from field investigations turned out to be invaluable during conversion of the shutdowns and trip tests and for later maintenance efforts.

Experience indicates that on line conversion loop-by-loop is preferred in most cases over the conventional shutdown-convert-startup method. Problems are encountered one point at a time and can be readily identified and corrected. Loops are turned over to the newly trained operator one or two at a time rather than all at once during a unit startup. And finally, a loop-by-loop approach allows the unit to continue to operate in a stable manner. The shutdown-convert-startup approach requires the unit to go through the upset conditions of startup at a time when the operator is learning a new control system and while the errors made in converting the control system are being discovered.

9
Retrofitting Process Controls

D. J. Krompak
D. L. Bishop

The concepts and strategies contained within this chapter were employed by the writers at a 120,000 barrel-per-day oil refinery. The retrofit project involved nearly one-third of the entire facility; approximately 800 instrument loops.

A flow chart depicting the phases and steps involved in retrofitting a facility is outlined in Table 1. This chapter follows the flow chart, thus, taking the reader through an entire project.

PHASE I—PROJECT IDENTIFICATION

Justification for pursuing an alternative control system must be established first. This task is accomplished by performing economic comparisons of the existing systems and state-of-the-art systems. Since many refineries and other industries' facilities are at least three decades old, often their analog control systems are somewhat obsolete and replacement parts may not be available. In addition, many central control rooms suffer from overcrowding, making process expansion unfeasible with the existing system. Often, too, a need for advanced controls for system improvements is determined by efficiency studies. Details on justifying a new control system are available elsewhere and are out of the scope of this article.

Once the justification is made, a design philosophy must be adopted to properly choose the system that best fits the requirements. This philosophy should include considerations such as failsafe operation, centralization of the control room, advanced control scheme requirements, management information requirements, system expandability, environmental compatibility and operator interface requirements.

The philosophy is formalized through engineering studies that culminate in a recommendation to pursue certain specified systems.

Often, the conclusion reached by the studies is to pursue distributed digital control, either as a direct replacement of an existing system or as part of a total control system enhancement. In many cases, a high level process computer interface is included to obtain process op-

Table 1
Project Flow Chart

Phase I Project identification	Phase II Project planning	Phase III Project implementation
1. Identify a need and justify 2. Initiate a project team 3. Investigate options available 4. Pursue best alternative (if best alternative includes a distributed control system, proceed) 5. Document existing system 6. Establish data base 7. Identify improvements needed 8. Develop preliminary specifications	1. Finalize project objectives 2. Select a design team 3. Produce preliminary design package 4. Produce budgetary cost estimate (if required)* 5. Finalize specifications • Distributed control system • Field instruments • Rack instruments • Power devices • Interconnection materials • Facility items 6. Prepare project schedule	1. Select a DSC and remainder of material 2. Produce a construction bid package 3. Produce definitive cost estimate (if required)* 4. Purchase distributed control system and remainder of major material 5. Select contractors(s) 6. Configure DCS 7. Complete engineering/design package 8. Approve engineering/design package and issue for construction 9. Begin construction • Facility • Power • Control systems 10. Commission and start up 11. Document system

* Client obtains approval to install the system at one of these two points

timization. Once this decision is made, the project is born.

Detailed identification and documentation of the existing control system is the initial step. Engineers and process personnel work cooperatively to determine the various types of control and monitor loops and the quantities of each. This usually involves reviewing existing drawings and specifications and comparing this documentation to the actual installations. This step is imperative since it serves as the base line data for the entire project. Accurate process and instrumentation diagrams are an essential product of this survey.

Once the existing system is completely defined, the team is ready to look at improvements. Existing control loops are individually and systematically scrutinized. At this point, it is determined, for example, whether or not to cascade certain loops, add feed-forward control, or, perhaps, add an optimization package. Detailed studies may be required that always "bottom line" with return on investment figures.

Once determined, the proposed control system is ready for preliminary specification. This usually takes the form of a document written by the engineering team members with input from process and operations personnel. This document, once approved (see Table 2 for outline of project manager's approval requirements), is added to the base line documentation and is the first formal step in selecting a distributed control system.

Table 2
Design Approval Requirements

Project identification	Client formally approves preliminary specifications
Project planning	1. Client formally approves preliminary design package 2. Client formally approves budgetary cost estimate 3. Client formally approves final specifications 4. Client formally approves project schedule
Project implementation	1. Client formally approves construction bid package 2. Client formally approves definitive cost estimate 3. Client formally approves DSC configuration 4. Client formally approves engineering/ design package 5. Client formally approves final documentation

The following outline serves as both a checklist and the basis for the preliminary specification for a distributed digital control system with CRT. Not all the items listed are required for each process under consideration.

General

Purpose. To define the hardware configuration and performance requirements of the proposed system.

Vendor responsibilities. Complete system, power information, grounding information, environmental information, isolation information, code conformance.

Functional criteria. Effective process control, data acquisition, process connections & interface, analog system replacement capabilities, altering control schemes, high level computer interface, back up & redundancy, maintenance considerations.

Data Base Definition

System description. The data base is the configuration information stored within the system's microprocessor driven memory. It is pertinent to the process interface and control strategy. The strategy may include the following parameters: signal conditioning, operational mode status, range, limits, algorithms, control constants.

Process Signals

- Analog signals: high level (4 to 20mA), low level (millivolt, such as from thermocouples, analyzers, and RTDs.
- Digital signals (on/off): alarms, start or stop pumps and compressors, reversible motor drive.

Algorithms. PID (all combinations), high/low select, sum/multiply/divide, ratio/bias/manual load, lead/lag, nonlinear PID, differential gap.

Control files. These files shall be microprocessor based with the capability of transmitting analog signals.

Controller considerations. Self startup (without the assistance of any higher-level devices), reconfiguration, local or console configuration, reduced performance capabilities, effect of failure of other devices connected to the communication system, power failure and memory retention, removal capabilities, console and rack tuning, back up system, bumpless transfer, reset wind-up, trend memory, manual loading, built-in service manual.

Data acquisition and accessibility. High level process I/O, low level process inputs, remote low level process inputs.

Operating Center Description

Console functions. Indication of variables, manipulation of control loops, recording (real time and historical), alarm annunciation and display, tables, start/stop station for motor operation, status indications.

Console hardware. CRT monitor (quantity, size and color requirements), keyboard(s), electronics, disc system, pen recorder, printer, videocopier.

Display requirements. Overview, group, trend (single loop), trend (multi-loop), communication system status, clock time, tag number summary, summary of group titles, communication link status, system status, library link status, averages (by designated time, internal), diagnostics.

Process Signal Alarm Monitoring

Analog points. High and low absolute value of the process variable, high and low deviation value from the setpoint of the process variable.

Digital inputs. Alarm upon a change in field contact status.

Alarm monitoring action sequence.

Alarm in display. Alarm summary, alarm group display, system diagnostic alarms.

Hard Copy Outputs

Real-time trending, historical trending, tabular printout, CRT screen print-out, automatic alarm print-out, trend recording, dedicated recorders.

System Interface

Communications between I/O racks & operator's station, circuitry separation, system diagnostics.

Vendor Services

Project management, training, engineering, configuration, documentation, project scheduling, shipment, delivery, installation/troubleshooting, servicing, and maintenance.

Testing & Acceptance

General requirements, testing at vendor's facility hardware and software, shipping approval, field acceptance test.

Base Line Information

(Attach system identification and documentation), system strategy, I/O lists & summary, control room layout, process flow diagrams or process and instrument drawings.

PHASE II—PROJECT PLANNING

Project planning overlaps and, in fact, encompasses the identification process just described. They are separated because identification is preliminary planning that is generally developed ahead of the active planning stage.

The initial function of active planning is to fully define all the objectives of the project. This allows the existing project team to obtain a good overall picture of the system requirements, summarizes work done to date, and gives project management insight into obtaining sufficient human resources needed for implementation.

Once the objectives are firmly established, project management selects a team of engineers and technicians. This design team is a source of personnel who should be experienced in general process control as well as distributed control systems. The design team must, to even a greater extent, be able to coordinate well the many-faceted aspects of a job of this type. They will assist the project manager throughout the remainder of the project. Their functions include the following:

- Produce a preliminary design package
- Produce a budgetary cost estimate (if required)
- Produce additional studies (if required)
- Finalize specifications
- Assist in preparation of project schedule
- Assist in selection of DCS (digital control system)
- Produce a construction bid package
- Prepare a definitive cost estimate (if required)
- Assist in configuration of DCS
- Produce a construction engineering/design package
- Provide construction assistance
- Provide commissioning/startup assistance
- Document system.

The preliminary design package consists of drawings detailed enough to show the proposed locations of all major equipment, routings of all interconnecting wires and tubing, current process and instrumentation drawings, an instrument list and a basic scope of work. Some typical interconnection drawings are sometimes added for further definition.

A budgetary cost estimate is often required by a client to show formal justification to obtain necessary funding for the project. This step is, of course, critical. It is imperative that all data supporting actual need for system replacement or enhancement be included as an integral part of this estimate. The preliminary design package also serves as back up data by showing visually the direction and form the project will take.

Upon completion of the budgetary cost estimate, the design team produces detailed specifications as identified in Item 5, Table 1. The preliminary design provides the design team with enough process definition to finalize all specifications for a distributed control system, associated instrumentation, and all sundry hardware.

The specifications are approved by the client, then sent out to approved vendors for price quotations.

The project schedule identifies and dates all major tasks to be completed in the areas of approvals, engineering, purchasing, construction, and technical assistance through commissioning. It is prepared by the design team with input received from project management.

Sometimes, this schedule is produced in phases concurrent with the design packages and cost estimates. For instance, a preliminary project schedule may be part of a preliminary design package to serve as back up for a budgetary cost estimate. Later, a final project schedule is produced in conjunction with the construction bid package and definitive cost estimate.

PHASE III—PROJECT IMPLEMENTATION

When the project manager receives the quotations from the various vendors, he establishes reviews of each quotation with his engineering and purchasing personnel. Quotations for the DCS, field and rack instruments, power and facility items, and interconnections are individually scrutinized by the review team. They have the task of evaluating the quotes for conformity to the specifications while objectively evaluating cost.

Ultimately, they select a system that optimizes such variables as configuration, operation, maintenance, flexibility, expandability, ease of installation and price.

By this time, all material suppliers have been selected (pending project approval, if required), and firm prices are established. The next task is to obtain a firm labor price for installing the system. The most consistent and accurate way to obtain realistic labor figures is to produce a construction bid package and obtain bids from selected contractors who are experienced in distributed control.

The construction bid package is produced by the design team and uses the preliminary design package as a basis. The bid package is actually the first step in preparing the final construction engineering/design package. As a minimum, it should contain the following items:

1. A detailed construction scope of work (narrative installation requirements and procedures for all work).
2. Final control room and rack room equipment layout drawing(s) and details.
3. Final tubing and/or cable routing drawing(s).
4. Final process and instrumentation drawings.
5. Final instrument list.
6. Final field plan drawings (if required).
7. Facilities and/or demolition drawing(s).
8. Identification and quantities of all terminations.
9. Construction schedule.

The construction bid package is sent to the selected contractors after receiving approval by the client.

It is suggested that the contractors review the package, then meet collectively with the client and design team at a prebid meeting before submitting their bids. The meeting should take place at the client's facility where the system is to be installed. At this time, all questions and concerns of the contractors are reviewed, and everyone obtains common description of the project objectives. The meeting should culminate with a guided tour of the job site for the contractors so they can see firsthand the conditions they will face.

This bid procedure provides the client with the accurate figures he needs to forecast total project cost.

The project manager reviews the bids he has received from the contractors and selects a contractor (pending project approval, if required).

The client may or may not require a cost estimate based on the firm material and labor costs provided by the quotations and the bids. If such a definitive cost estimate is required, it is prepared by the design team. The sample table of contents shown in Table 3 summarizes items addressed in a typical definitive cost estimate.

Upon completion, the definitive cost estimate is sent to the project manager for his approval. He, in turn, submits it for final project approval. If approved, the project becomes reality.

All material, especially the distributed control system (which may already have a tentative purchase agreement), is placed on order with the selected vendors. Scheduling requirements may have required the DCS to be placed on order ahead of receiving final approval. This is generally a tentative contract that is subject to cancellation charges if the project is not approved. Delivery dates established by the vendor quotations must be reinforced at this time by the project manager to maintain continuity with the project schedule.

Table 3
Definitive Cost Estimate Sample Table of Contents

	Page(s)
1.0 Purpose of cost estimate	1
2.0 Project cost	
2.1 Summary	2 & 3
2.2 Instrument quotations	4–72
2.2.1 Contractors' bids	
2.2.2 Distributed control system	
2.2.3 Electronic transmitters	
2.2.4 Pneumatic transmitters	
2.2.5 I/P converters	
2.2.6 P/I converters	
2.2.7 Indicators	
2.2.8 EMF/I transducers	
2.3 Electric/facility quotations	73–77
2.3.1 Contractors' bids	
2.3.2 Power conditioners	
2.3.3 Miscellaneous electrical equipment	
2.3.4 Facility equipment	
2.4 Current area labor rates	78–81
3.0 Construction scope of work & drawing list	
3.1 Instrumentation contractor scope of work	82–108
3.2 Instrumentation drawing list	109–111
3.3 Electrical/facility contractor scope of work	112–119
3.4 Electrical/facility drawing list	120
4.0 Detailed engineering cost estimate	121–135
5.0 Construction schedule	136
6.0 Attachments	
P&IDS	137–139

Concurrent with the material purchases, the project manager formally awards the installation contract to the selected contractor(s).

Immediately following the purchase of the distributed control system, the design team begins system configuration in conjunction with the DCS supplier's engineering personnel. The configuration consists of translating all data, as defined by the process requirements, into software compatible with the system selected. This task requires a great degree of coordination and communication between design team members, the client, and the vendor, since there are numerous loops involved, many of which are critical to plant operation.

When system configuration is complete or in progress, the design team begins work on the detailed engineering/design package. This will be the drawing package used for construction. The package consists of all drawings and documents included in the construction bid package, additional drawings showing each termination point as well as each configuration assignment, and individual loop interconnection diagrams. Additional detail drawings also may be added to clarify difficult areas of construction.

The client reviews and approves the engineering/design package. It is then issued to the contractor for construction.

Construction is performed in accordance with the requirements of the project schedule. The contractor mobilizes his crews and work begins in specified phases. Facility changes and instrument power feeders are usually scheduled first. This is followed by major equipment installation, interconnections and commissioning. Any removals, such as obsolete analog instruments and panels, are scheduled last.

The importance of good planning and scheduling becomes quickly apparent during construction. This is particularly true if the plant is to remain in operation during the changeover to distributed control. Process and operations personnel must be directly involved with the construction crew to assure that no process upsets occur. This is efficiently accomplished by assigning a field representative to act as a liaison between all disciplines and to enforce and update the project schedule as required.

The field representative should be someone familiar with all phases of the project. He should come from the project management or design team and must be very familiar with the job site. This provides continuity between the design package and construction. It also enhances the need to follow the project objectives.

The approaches taken to construction may vary greatly. The key is to perform sufficient up front planning to provide the contractor with a realistic project schedule geared to guide him in the direction needed to meet the objectives of the project.

Loop commissioning occurs continuously during the instrument construction phase for jobs performed with the plant in operation. If the plant is shut down, loops are commissioned as part of the startup function. In either case, loop commissioning involves getting an instrument loop operational and tuning the loop to perform its specified function(s) to meet process requirements.

Loop commissioning, like all other construction tasks, must closely involve the operations personnel. They must become familiar with the system and be confident it is functional. Including them in the commissioning builds both familiarity and confidence.

The final task of documenting the system takes place after all loops are commissioned and the system is operating as required. Final documentation involves a complete review of the drawing package with revisions provided as required to show the "as built" conditions of the system. This task is accomplished by the design team in conjunction with the field representative. The as built set assures the client he has complete and accurate drawings from which to work pending a future change or expansion in the needs of production.

Bibliography

Albanese, R., *Management: Toward Accountability for Performance,* Richard D. Irwin, Inc., 1975.

Burriss, C. B., and Ketcham, R. S., "Retrofitting Industrial Boilers with Digital Controls," ISA Paper, Discover the new world of Instrumentation, Proceedings of the 1982 Joint Symposium, Columbus, Ohio, April 13-15, 1982.

D'Abred, F. A., "Replacing Installed Process Control Computers," *In Tech,* 29, July 1982, Instrument Society of America, 1982, pp. 57-59.

Garret, L. T., and McHenry, J. M., "Analyzing Costs of Digital and Analog Control Systems," *Hydrocarbon Processing,* Dec. 1981, pp. 103-108.

Liptak, B. G., "Application of Process Computers," *Instrument Engineers Handbook Volume II, Process Control,* Philadelphia, Chilton Book Co., 1970, pp. 1,109-1,113.

McConnell, C. R., *Economics,* 7th ed, New York, McGraw-Hill, Inc., 1978, pp. 135 and 469.

Sikula, A. F., *Personnel Administration & Human Resources Management,* New York, John Wiley & Sons, Inc., 1976.

10

Control Graphics

E. A. DeVries

The CRT (cathode ray tube) has become the major focal point in distributed digital control systems, which constitute most of the new control systems being purchased. Much has been written about these CRTs being "The Window On The Process," and this is all very good. However, many problems still arise in control graphic design that could be avoided if we remember their primary purpose: to provide quick recognition, interpretation and manipulation of the process by an operator. The goal of this chapter is to outline some ways to achieve this purpose and avoid some common problems.

Our reference point is the operator, a typical human. His or her limitations may become the limitations of our process control system. For example, some people are forgetful or easily confused. However, we may be able to capitalize on their operating experience, anticipation of necessary adjustments, and ability to make decisions during extraordinary events. Therefore, it becomes our job to provide a system that allows the operator to perform to the best of his ability without putting stumbling blocks in his path. For example, we have learned that if we deprive the operator of a normal level of interaction with the system, he may become disassociated with the process. This sensory deprivation may cause boredom

Figure 1. Finding the optimum stimulation level.

Figure 2. Full graphic panels resembled the plant.

(Fig. 1). On the other hand, if we flood the operator with too much information too fast, we may overload him and cause him to react improperly. Avoiding sensory overload becomes very important during process upsets. Therefore, we must design our graphics to keep the operator properly involved with the process at all times.

GRAPHICS EVOLUTION

But this is nothing new. Providing a control format that is quick to interpret, easy to manipulate, and one that holds the operator's interest has always been the goal. Years ago we built full graphic panels that looked very much like the process (Fig. 2). The controllers were located in the panel corresponding to the control points in the process. Often these panels were located on the operating deck. This made the controls both easy to in-

Figure 3. Semi-graphic panels grouped controllers.

terpret and to associate with the real plant. However, the spread out controls were difficult to keep up with during upsets. Often it was necessary to manually manipulate controllers at opposite ends of the panel. To keep the operator from running himself to death, semi-graphic panels were introduced (Fig. 3). The controllers were grouped together, but the graphic resembling the plant was still there for reference and training. These semi-graphic panels had now moved into control rooms. However, they were still very large and generally required several operators, particularly during upsets.

Figure 4. CRT graphics allow multi-unit control from a single console.

CRT GRAPHICS

Microprocessor-based CRT graphics and distributed control systems have taken over the place of the full graphic and semi-graphic panels. One operator is able to control one or more units without having to leave his console (Fig. 4). These graphics contain live process data at strategic locations representing the process, in much the same way as the old full graphic panels did. Also, most distributed control graphics have displays of controller groupings that appear very much like controller groupings on the old semi-graphic panels (Fig. 5).

This brings us to our first problem with modern graphics. Changing the CRT screen from one display to another can be just as bad as an operator having to run from one end of the control panel to the other. As always, this problem is particularly acute during a process upset. Therefore, the first method of improving graphic system design is to minimize changes of screen displays. Some ways of approaching this problem have included increasing the amount of dynamic or live information on the primary display (eight points are good, 16 are better), minimizing the number of controller displays by increasing the number of controllers on each display (eight on a screen are good; 12 are better), and having dedicated display select keys to quickly change displays (Fig. 6).

As previously mentioned, all the system limitations are most acute during process upsets. This is the time when both the system and the operator must handle the greatest amount of process information most efficiently. Color graphics have improved this situation. Some studies indicate that information displayed in color is four times easier to understand and assimilate than information displayed in black and white. However, as previously mentioned, we do not want to overwhelm the operator with too much information. Experience has shown that the most overwhelming aspect of process upsets is the myriad of alarms that the operator must deal with. There are a few good approaches to the problem of an avalanche of alarms that should be included in the graphic design. I recommend that the following be considered.

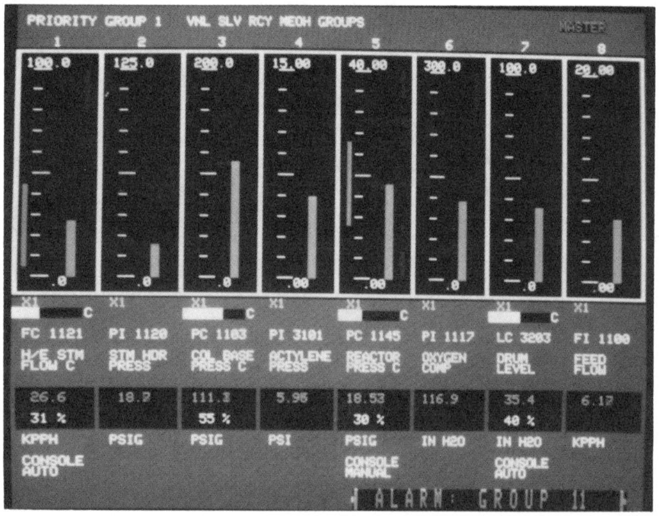

Figure 5. Associated controllers grouped similarly to semi-graphic panels.

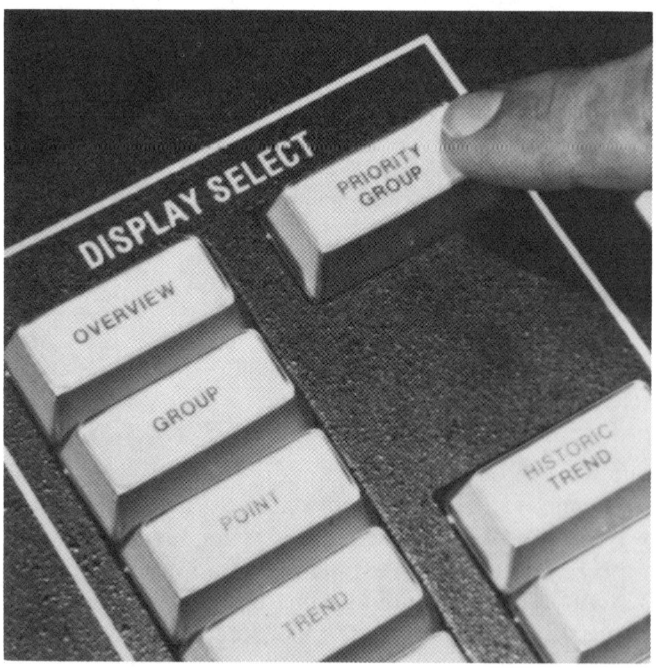

Figure 6. Dedicated keys speed display changes.

ALARM GRAPHICS

Most companies that sell CRT graphic terminals and control systems will demonstrate wild, bright circus-like graphic displays. This is not what we want in front of our operators. Graphic designs need to employ carefully planned, judicious use of color, starting with our alarms. Some military studies recommend that for best recognition and response, red should be used only for high-priority alarms, shutdowns and emergency conditions; and that yellow be used only for medium-priority alarms and cautionary conditions. Further recommendations of many experts are that low-priority alarms (such as maintenance flags) should be blue, and that a normal or safe condition should be green. Most CRTs have only eight colors available. Sometimes removing colors for dedicated alarm priorities from the available repertoire leaves the graphic designer with a real challenge for assigning the remaining colors to the process lines. The best solution to this problem is to create a system of cross hatched or alternatingly colored lines (Fig. 7). This effectively squares the number of available colors to use for this purpose. Another approach to managing alarms

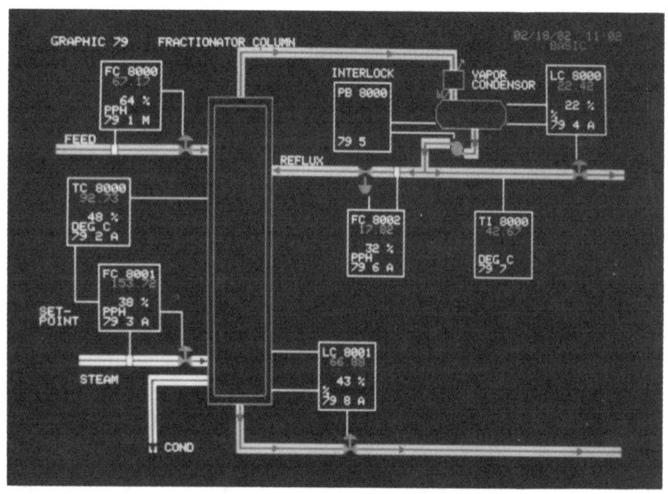

Figure 7. Combining colors expands color possibilities.

is to have the low priority alarms go to a printer only, so that they can be dealt with as time permits, and thus do not flood the screen and become a nuisance during critical times.

NEW IMPROVEMENTS

Graphics hardware is continually improving and manufacturers are providing graphic designers with better solutions to old problems. Information updating and display changing is getting faster. Better resolution on the screen helps to provide better graphic details and symbols, giving the operator a better understanding of the graphic representations of his process. After all, the more accurately and realistically the graphic represents the process, the less we must depend upon the operator's intermediate thought processes to interpret what he sees on the display.

Modern displays are becoming more dynamic and interactive with the operator. This direction of hardware and software improvements tends to better hold the operator's interest and understanding of what he is trying to achieve. Symbols that change color with their state, and vessels with moving levels, help the operator identify with the changes in his process. The new touch-screen type displays help the operator to easily access his equipment and their controls, without having to manipulate intermediate hardware. It is sometimes advantageous to eliminate an intermediate step like this, as this helps to reduce operator error.

In conclusion, control graphics help our plant operators to more easily learn, understand and control the process that they are entrusted to manage. If the graphics designer employs a few guidelines to aid the operator in his task, particularly with process upsets in mind, results such as improved productivity with reduced losses can be expected.

Bibliography

Bader, F. P., "Custom CRT Graphics Enhance Operator - Process Interaction," *Control Engineering,* February 1982.

Bailey, S. J., "From Desktop To Plant Floor, A CRT Is The Control Operator's Window On The Process," *Control Engineering,* June 1984.

Brough, W. R., "Human Factors - The Role Of The Operator," *ISA Advances In Instrumentation,* Volume 38, Part 1, 1983.

Herb, S. M., "Technology Improves Process Control Displays," *I & CS,* May 1984.

Miller, T. J., "Ease Of Use, Simplicity, Flexibility Are Key Requirements For Process Control Graphics," *Control Engineering,* June 1983.

Ollila, A., "Operating Experience With Distributed Digital Control," *Hydrocarbon Processing,* October 1983.

Schellekens, P. L.,"Alarm Management In Distributed Control Systems," *Control Engineering,* December 1984.

11
Windup Protection and Initialization for Advanced Digital Control*

S. E. Gallun, C. W. Matthews, C. P. Senyard, and B. Slater

Proper windup and initialization processing, along with provisions for handling bad measurements, are essential to smooth operation of advanced digital control systems. Procedures are discussed for simple feedback and additive feedforward control structures using PID, ADD, and lead-lag algorithms. Other algorithms used in control paths such as ratio, high/low select and output characterization should also provide such procedures.

A new algorithm class that has attributes of full position and incremental (velocity) algorithms is discussed. This new class of algorithm, called pseudo incremental, allows feedback and additive feedforward to be implemented in cascade structures with rigorous handling of windup. Algorithm initialization and bad data response also are addressed.

INTRODUCTION

Digital controller windup and initialization have received minimal attention in engineering literature. The basic concepts of PID windup protection and control algorithm initialization are well known, but the details necessary for application in commercial control systems have not been discussed. Implementation of these features in commercial digital control systems varies from "primitive" to "adequate." The application of these concepts to feedforward control algorithms is even less appreciated than for feedback control. Only a handful of commercially available digital control systems attempt to provide these features for feedforward algorithms.

The concept of windup due to integral action in output-constrained PID controllers has received some attention in the literature. Shinskey[1] discussed reset windup in analog PID controllers and proposed a technique to handle the problem. Shinskey's technique, external reset feedback, is also applicable to digital PID controllers that are mathematical images of pneumatic or electronic controllers. The technique does not apply to digital PID algorithms that generate integral action by summation of an error term. Smith[2] discusses such an algorithm and indicates that reset windup can be prevented by suspending summation when the output of the PID is constrained. This idea appears to have been first suggested by Fertik and Ross.[3] Shinskey[4] notes that in PID cascade systems, reset windup of a primary controller can occur if proportional action forces the primary output to a limit. Again the prescribed solution is dependent on external reset feedback. Khandheria and Luyben[5] experimentally evaluated several algorithms to prevent reset windup in a supervisory control setting. Their control configuration consisted of a supervisory digital PID cascaded to the set point of an analog flow controller. The recommended windup algorithm was a modification of the external reset feedback technique.

State-of-the-art block-structured or fill-in-the-blank digital control systems, with algorithms designed to exploit digital computation power, can employ sophisticated windup prevention and automatic initialization procedures. These procedures should be integral to the algorithms. This means that windup and initialization problems will not plague the process operator and further, since the procedures are inherent in the algorithms, the control engineer is not required to design and implement windup and initialization logic for minor loop he builds. A proven technique to accomplish windup protection and algorithm initialization in a set of full-position algorithms will be discussed. Although not discussed in this chapter, the basic technique has been also applied to velocity algorithms.

* The authors wish to acknowledge the early work done by Mr. Bobby Ewing of Exxon USA and Mr. Mel Ambuehl of Exxon Chemicals Central Engineering Division in the application of windup and initialization concepts in DDC and supervisory computers.

WHAT IS WINDUP?

Traditionally, windup has been considered to be a PID controller problem. However, the concept of windup and its prevention is much broader. For example, in simple additive feedback-feedforward system shown in Fig. 1, windup can be generated by both the feedback and feedforward algorithms. If the output of the feedforward controller increases such that the secondary PID output reaches a limit (cause of classical windup), windup will occur in the primary and secondary PIDs unless preventive measures are taken. Further, if the feedforward signal continues to increase, the set point of the secondary controller will be increased (wound up) after the secondary PID is no longer able to respond to set-point increases. One component of the windup prevention procedure must limit the outputs of both the feedback and feedforward primary controllers. The feedforward block output must be limited in a manner that preserves the feedforward reference level.

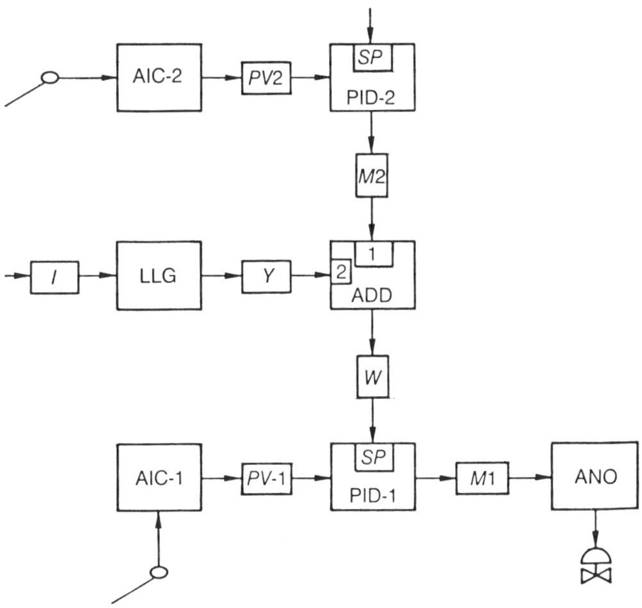

Figure 1. Feedback and additive feedforward cascade.

CONTROL SYSTEM STRUCTURE

The ideas to be discussed are quite general, but the details of their implementation will vary depending on the precise structure of a control system's fill-in-the-blank algorithm set. We will define a particular algorithm structure to tie the discussion to. An algorithm and its output will be indicated as in Fig. 2. The algorithm block can have multiple inputs, if required, that

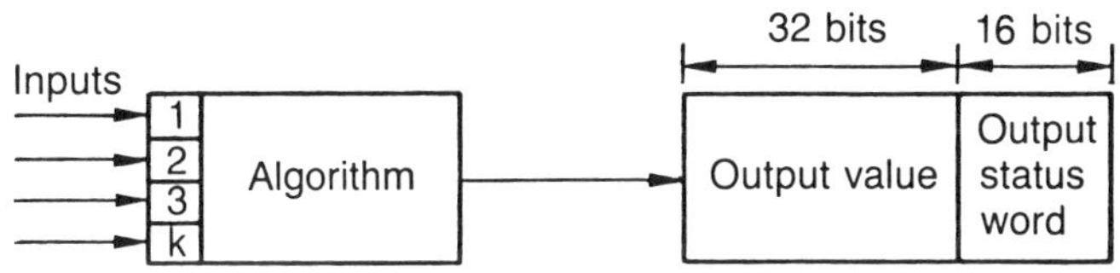

Figure 2. Algorithm block and its output.

are the outputs of other algorithm blocks or are constants. Some algorithm blocks read from or write to memory locations that receive outputs from analog-to-digital converters or provide inputs to digital-to-analog converters. The algorithm output will be a 32-bit real value that is the result of the algorithm's calculation and a 16-bit word that contains algorithm status and control information. The 16 bits will be called the algorithm status word and the 32-bit real value the algorithm output. Six of the algorithm status bits will be of interest in this discussion. The six bits are the active/inactive bit, the auto/manual bit, the data quality bit, the initialization bit, and two bits defining the windup status of the algorithm. In general, active/inactive, data quality, and initialization bits have meaning for all algorithms. The auto/manual and the windup bits are meaningful only when algorithms are configured in a feedback and/or feedforward control path.

Control path refers to the ultimate use of an algorithm's output as a component in the calculation of a secondary PID's set point or a secondary analog output (as opposed to use in a controller's measurement or process variable, PV). Fig. 1 shows a typical cascade structure in which a feedback controller, PID-2, and a lead-lag feedforward algo, LLG, provide the set point to a PID (normally just PI) flow controller. The path leading from the analog input, AIC-1, to the measurement or PV of PID-1 is a calculation path. The paths leading from the outputs of PID-2 and LLG, through the ADD to the set point of PID-1, and eventually through the analog output algorithm, ANO, to the valve are control paths. Windup has meaning with regard to control paths and not calculation paths.

The flow of information required in a control structure such as Fig. 1 is bidirectional. Windup and initialization information will be passed up control paths towards the ultimate primary controllers. Quality information and algorithm output values are passed down control paths towards a valve and from analog inputs along calculation paths that typically terminate as PID measurements. Proper algorithm initialization procedures will be shown to require passing outputs up control paths. Again, it should be emphasized that the ideas to be discussed are general and not specific to the particular algorithm structures we have defined.

DIGITAL ANTIRESET WINDUP (ATW) CODE

Classical reset windup involves integrating or summing a value that will cause an internal control calculation and/or the output of the control calculation to move in a direction that cannot result in response by a field device (i.e., a valve or a compressor speed controller). The windup status of a controller and its output will be one of four states, which we will designate ATW status 0, 1, 2 or 3. We will define ATW = 0 to mean that the controller output is unconstrained and that both increases and decreases will ultimately result in some field action. ATW = 1 will be defined to mean that the controller output is constrained on the low side and that further decreases in controller output will not result in field action. ATW = 2 will mean that the controller output is constrained high and that further increases in controller output will not result in action by the field element. ATW = 3 will mean that the algorithm has no path to a valve and that changes in the output in either direction will not result in any response by a field element. Clearly, four numbers define all possible constraint states of an algorithm output. Thus windup or constraint status can be represented in two bits of binary information.

The windup information stored in the algo status word is stored by the next block down the control path. In Fig. 1, the windup bits of the ADD algo's status word are set by PID-1 each time it executes. The windup bits of the LLG algo and PID-2 status words are set by the ADD algo when it executes. An algorithm never sets its own output status word ATW bits. ATW information is in counterflow with control algorithm output values.

The mechanism of ATW passing needs to be further enhanced to account for constraints generated within the algorithm and to account for controller action. For example, suppose the process requires that the output of PID-1 never exceed 75% valve opening out of a possible range of 0 to 100% valve opening. In some systems this would be accomplished by putting a high output clamp of 75 percent on PID-1. The analog output algo, ANO, would never generate and pass an ATW = 2 if it were ranged to cover the full valve opening of 0 to 100 percent, as it should be. PID-1 must look at its own clamps, determine if it has generated an internal constraint, and combine this information with the ATW information passed to its output status word and determine the ATW status to pass up the control path to the ADD algorithm. Whether the high clamp causes an ATW = 1 or 2 to be passed to the ADD block depends on the reverse/direct-acting configuration of the PID. If PID-1 is passed ATW = 3, then it should pass the 3 up its control path without modification. If the control path is not connected to the process (ATW = 3) clamps and other constraints have no meaning.

The ATW status passed up a control path exactly indicates the direction that a PID's set point can be moved and result in a controller output than can affect a final control element.

FULL-POSITION PID ALGORITHM

A simplified full-position digital PID algorithm is defined by Eqs. 1 through 3. Derivative action is calculated from the process variable rather than from the error. The option of proportional action from the process variable rather than error is useful for ultimate primary controllers that are tuned to respond to load disturbances. The sign of the gain determines direct or reverse action and is important in determining proper windup propagation. The PID algorithm defined by Eqs. 1 through 3 would thus have a positive gain for direct action and negative for reverse action.

$$E_i = SP_i - PV_i \tag{1}$$

$$G_n = \begin{cases} E_n & \text{if proportional is from the error} \\ -PV_n & \text{if proportional is from the PV} \end{cases} \tag{2}$$

$$M_n = K_n \cdot [G_n + (PV_{n-1} - PV_n) \cdot TD/T] + \sum_{i=0}^{n} (T \cdot K_i \cdot E_i/TI) + M_0 \tag{3}$$

The PID algorithm described by Eqs. 1 through 3 normally would be implemented using Eqs. 4 through 6, assuming E_n and G_n have been determined by Eqs. 1 and 2, and the gain, K_n, calculated (if variable) and assigned the proper sign for reverse or direct action.

$$D = K_n \cdot E_n \cdot T/TI \tag{4}$$

$$\overline{S}_n = S_{n-1} + D \tag{5}$$

$$\overline{M}_n = \overline{S}_n + K_n \cdot (G_n + (PV_{n-1} - PV_n) \cdot TD/T) \tag{6}$$

The logic defined by Eqs. 7 through 13 is required in the implementation of Eqs. 4 through 6 to provide windup protection and propagation of windup status up the control path. It will be assumed that the PID has high and low output clamps as discussed previously. The tentative internal summation, $\overline{S}_n$, and the tentative PID output $\overline{M}_n$, are calculated with Eqs. 5 and 6. The windup bits of the output status word are examined and stored as a number in the variable ATW. Internal values of the clamps are calculated using Eqs. 7 and 8. The PID output, the integral summations, and the ATW code to pass up the control path are determined by sequential application of Eqs. 9 through 13.

$$\overline{HOC} = \begin{cases} HOC & \text{if ATW} = 0 \text{ or } 1 \\ Min(M_{n-1}, HOC) & \text{if ATW} = 2 \end{cases} \qquad (7)$$

$$\overline{LOC} = \begin{cases} LOC & \text{if ATW} = 0 \text{ or } 2 \\ Max(M_{n-1}, LOC) & \text{if ATW} = 1 \end{cases} \qquad (8)$$

$$\overline{M}_n = Min(\overline{HOC}, \overline{M}_n) \qquad (9)$$

$$M_n = Max(\overline{LOC}, \overline{M}_n) \qquad (10)$$

$$S_n = \begin{cases} S_{n-1} & \text{if } M_n = \overline{HOC} \text{ and } D > 0 \\ S_{n-1} & \text{if } M_n = \overline{LOC} \text{ and } D < 0 \\ \overline{S}_n & \text{all other conditions} \end{cases} \qquad (11)$$

$$\overline{ATWP} = \begin{cases} 0 & \text{if } \overline{LOC} < \underline{M_n} < \overline{HOC} \\ 1 & \text{if } M_n = \underline{LOC} \\ 2 & \text{if } M_n = \underline{HOC} \\ 3 & \text{if ATW} = 3 \end{cases} \qquad (12)$$

$$ATWP = \begin{cases} \overline{ATWP} & \text{if } \overline{ATWP} = 0, 3 \text{ or direct acting} \\ 1 & \text{if } \overline{ATWP} = 2 \text{ and reverse acting} \\ 2 & \text{if } \overline{ATWP} = 1 \text{ and reverse acting} \end{cases} \qquad (13)$$

Equations 7 through 13 show that the PID output is never moved in a direction that violates an existing constraint or beyond its output clamps. Further, and most important, the summation of the integral term is suspended when the sign of the current increment indicates that the integral summation would be moved in the direction of an existing or generated constraint. The windup code generated (ATWP) indicates what direction or directions the set point of the PID can be moved and still affect the process. This windup code is passed up the control path and should be displayed on the operator's CRT meter face. Several control systems indicate the windup status of the PID algorithm by displaying <, >, or * next to the CRT meter face set point for ATW codes 1, 2, and 3, respectively. The lack of a character indicates the controller is unconstrained and connected to the process.

It is important to calculate and display the ATWP code even if the PID has no primary controller and is operating on a local set point supplied by the operator. In this case the operator is the primary controller and should be told in which direction he can move the PID's set point and achieve a process response.

PID INITIALIZATION

Automatic and bumpless initialization is easily implemented for the PID algorithm. Generally, initialization should occur when the output status word ATW code changes from 3 to either 0, 1, or 2, when the output status word initialization bit is set, when the PID changes from manual to auto, from inactive to active, and finally when the PID's measurement becomes "good" as indicated by the quality bit of the PV block's status word changing from "on" (bad data) to "off" (good data).

PID initialization involves two possible stages of initialization. A well designed set of fill-in-the-blank algorithms should allow the option to configure the PID to equate its set point to the current measurement during initialization. Normally, a PID will be configured to be set point initializing when it is a secondary in a cascade, and set point noninitializing when it is the ultimate primary. The PID set point initialization configuration status should be available for display through the operator's CRT so that the operator can determine when set points will be changed by the control system. Set point initialization, if configured, should occur before the initialization of the PID algorithm. The initialization process is defined by Eqs. 14 through 18. The current PID output, M_0, is the basis of the initialization process.

$$SP_0 = \begin{cases} SP_0 & \text{if not set point initializing} \\ PV_0 & \text{if set point initializing} \end{cases} \qquad (14)$$

$$G_0 = \begin{cases} SP_0 - PV_0 & \text{if proportional from error} \\ -PV_0 & \text{if proportional from PV} \end{cases} \qquad (15)$$

$$\overline{M}_0 = Min(HOC, M_0) \qquad (16)$$

$$M_0 = Max(LOC, \overline{M}_0) \qquad (17)$$

$$S_0 = M_0 - K_0 \cdot G_0 \qquad (18)$$

The initialization procedure recognizes that a full-position algorithm replaces its output each execution. Eqs. 16 and 17 clamp the output at the specified limits prior to calculating the reference value by Eq. 18. This step is necessary because a control scheme is initialized up a control path with each secondary writing over the output of its primary, setting the primary's status word initialization bit and passing the appropriate ATW code to the primary's status word. In the process of writing to the primary's output, the primary's output clamps are not necessarily respected.

The final step in PID initialization is the passage of its set point to its primary's output and passing the ATW and initialization bits to the primary's output status word. If no primary is configured, this step is bypassed.

ADDITIONAL SECURITY FEATURES

The basic flow control loop of Fig. 1 consists of three distinct algorithm blocks. Reading hardware to determine the flow value is allocated to the AIC algorithm. The AIC algorithm should have options for input checking, square rooting, filtering and pressure and/or temperature compensation. The PID provides the control calculation and the ANO algorithm converts the real valued PID output to an integer input to the digital-to-analog converter. An example of an AIC algo data-checking function is detecting "bad" data from a field transmitter. Suppose a 0 to 20-milliamp signal is input to a 12-bit analog-to-digital converter. The converter output would be an integer ranging from 0 to 4,095. Normally 4 milliamps is the bottom range of a current transmitter. A failed transmitter will normally produce a signal of less than 2 milliamps. If the the AIC detects a transmitter output of less than 3 milliamps (leeway for poor calibration), indicating transmitter failure, the AIC's output should not be updated. The quality bit of the AIC's status word should be set on to indicate bad data.

The PID using this AIC algorithm's output should detect the transmitter failure by examining the quality bit of its PV input and take appropriate control action. Similarly, if the AIC algorithm is made inactive by the operator it should hold its current value in the algorithm's output location and set the algorithm status word's quality bit to indicate bad data.

The appropriate action for the PID when it detects a bad PV is to hold its output constant, and generate an ATW code of 3 to pass up its control path and display on the CRT meter face. When the AIC determines that the transmitter has recovered, or when the AIC algo is made active (with good input data), the AIC's quality bit should be cleared. The PID should initialize when it detects a transition of its PV from bad to good. This security feature, and others like it, should be inherent in the algorithm set. When an AIC or PID algorithm is configured, the user should automatically be availed of the sophisticated control security features incorporated in the control system. Incorporating these features should not depend on the knowledge of the person configuring the system.

The second PID of Fig. 3 has been configured to have three possible set point sources. The sources are local-operator-entered set point, cascade set point from within the distributed control system, and supervisory computer cascade set point from a host computer. When the PID is switched to local set point, both the control path to PID-2 and that to the host computer should receive an ATW code of 3 (no path to process element). Similarly, if in cascade, the nonselected cascade set point

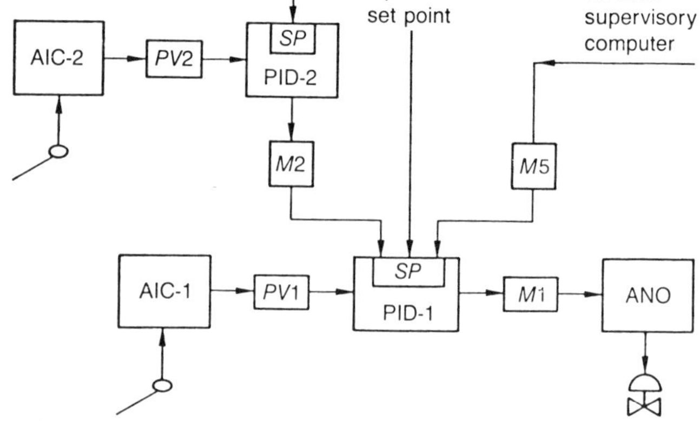

Figure 3. Local, cascade, and supervisory PID set points.

path should be passed an ATW code of 3. The ATWP code determined by Eq. 13 should only be passed up the selected control path and used to generate a character for display on the operator CRT.

ADDITION OF FULL POSITION OUTPUTS IN A CONTROL PATH

A full-position control algorithm operates on its inputs and calculates an output, which it stores in its algo output location. The ADD algorithm of Fig. 1 gets the algo outputs of PID-2 and the LLG algorithm, adds them together, and places the result in its algo output location. Full-position algorithms use full-position inputs to produce full-position outputs. Full-position algorithms do not operate on changes or produce changes as incremental or velocity mode algorithms do. Each type of algorithm implementation, full-position or incremental, has its own set of advantages, disadvantages, advocates and opponents. It has been the authors' collective experience that either approach is adequate if the algorithm set has been properly designed. Ideally, an algorithm set would provide both types of algorithms.

When an ADD algorithm is an element of a control path, some consideration must be given to its utilization of windup and initialization information and the generation of additional windup information from its own output clamps. The functions of an ADD algorithm will differ in a control path relative to a calculation path. The features required in a calculation path will not be discussed, but a well designed algorithm set will permit the engineer to obtain either set of features by merely specifying that the ADD block is in a control or a calculation path. An ADD algorithm in a control path will use the windup and initialization information passed to its algo status word and pass initialization, and possibly modified windup information, to the algo status words of its inputs.

Normal ADD algorithm processing begins by calculating a tentative output using Eq. 19. The final ADD algo output is calculated by using Eqs. 7, 8, 9, 10, and 12 (replacing M with W). It is not necessary to modify the windup coded by Eq. 13 as reverse acting has no meaning for the ADD algorithm. If instead a subtract algorithm that subtracted the values of inputs 2 through K from input number 1 was used, then an equation analogous to Eq. 13 would be required for inputs 2 through K.

$$\overline{W}_{n} = I^1 + I^2 + \ldots\ldots + I^k \quad (19)$$

The initialization of an ADD algorithm should be bumpless, transparent, automatic and propagated up the control path. The ADD initialization procedure, for control paths, should be designed with the idea that Fig. 1 represents a typical control configuration. The feedforward shown in Fig. 1 should be viewed as an enhancement of pure feedback control. This implies that PID-2 should be able to manipulate the set point of PID-1 across its full range and achieve stable but degraded control relative to the feedback-feedforward system. On this basis the following initialization procedure can be used for the ADD algorithm. The procedure begins by obtaining the current ADD algorithm output, W_0, and using it in place of M_{n-1} in Eqs. 7 and 8 to determine temporary high and low clamp values. A clamped value of the algo output is determined using Eqs. 9 and 10, where M_n on the right side of Eq. 9 is the current ADD output, W_0. The windup code to pass to all variable ADD inputs is determined by Eq. 12, again recognizing that reverse action has no meaning in an ADD algorithm.

The values to write to each of the ADD algorithm's inputs are calculated by subtracting all constant inputs from the modified ADD output just determined. The result of this calculation is passed to the number 1 input and zero is passed to all other nonconstant ADD inputs. The intent is that the feedback control signal will be the number 1 ADD input and the remaining inputs will be feedforward inputs and constants. The fill-in-the-blank editor should require that the number 1 input of an ADD in a control path be a variable and that the ADD have at least two inputs.

WINDUP AND INITIALIZATION IN FEEDFORWARD CONTROLLERS

The feedforward control algorithm of Fig. 1 is specified to be a lead-lag element. The lead-lag is a common feedforward algorithm and as shown by Luyben[6], provides perfect feedforward when the process is first order with respect to the feedforward variable and the feedback manipulated variable. The following discussion will use the lead-lag feedforward controller for illustrative purposes. The procedure developed will be shown to apply to general feedforward algorithms if their outputs are passed through a special algorithm to be discussed.

The lead-lag transfer function relating the input, I, to output, X, is given by Eq. 20. A common digital realization of Eq. 20 is obtained by converting the transfer function to the time domain and numerically integrating the resulting differential equation across one sample time using the trapezoidal rule. Eq. 21 is the result of this process. The coefficients of Eq. 21 are defined by Eqs. 22 through 24.

$$X(s)/I(s) = K \cdot (T1 \cdot s + 1)/(T2 \cdot s + 1) \quad (20)$$

$$X_n = A \cdot X_{n-1} + B \cdot I_n + C \cdot I_{n-1} \quad (21)$$

$$A = (2 \cdot T2 - T)/(2 \cdot T2 + T) \quad (22)$$

$$B = K \cdot (2 \cdot T1 - T)/2 \cdot T2 + T) \quad (23)$$

$$C = K \cdot (T - 2 \cdot T1)/(2 \cdot T2 + T) \quad (24)$$

Before proceeding further it is important to appreciate the basis upon which Eq. 20 represents a feedforward control algorithm. Luyben's[6] derivation is based on process transfer functions which implies that I(s) and X(s) in Eq. 20 are Laplace transforms of perturbation variables. In the real full-position physical world the meaning of Eq. 20 is that if the process is near the point where K, T1, and T2 were evaluated, that a change in I will produce a change in the process output that can be cancelled by making a change in the manipulated variable. The key idea is that changes in I are counterbalanced by making changes in the manipulated variable in an anticipatory manner. If the manipulated variable is constrained such that it cannot move in a particular direction, changes in this direction cannot be physically achieved. If the feedforward control algorithm were allowed to continue to move its output in the direction of the constraint, the feedforward algorithm would wind up in the sense that it will output values that cannot be responded to by the process. Referring to Fig. 1, if PID-1 passes an ATW code of 2 up its control paths, both PID-2 and the LLG algorithm should be prohibited from increasing their outputs. In the case of the LLG algorithm, it is also necessary to ensure that the algorithm is not saving up its unrealized increases in output to wallop the set point of PID-1 if it momentarily passes the LLG an ATW of 0 or 1.

If the output of a feedforward algorithm, such as the LLG algorithm, is not properly handled in a constrained situation, the secondary set point using this output could be driven to a point that holds the output of the secondary PID at a constraint and eventually causes a disturbance in the primary feedback controller measurement of PID-2. The potential exists then that PID-2 will have to negate the effect of the feedforward by integral action to move PID-1 off a constraint. This situation is exactly analogous to windup in a PID to PID cascade.

The desired windup protection in full-position additive feedforward control configurations can be obtained by defining a new class of algorithm which we will call pseudo-incremental. The digital lead-lag algorithm defined by Eq. 21 can be implemented as a pseudo-incremental algorithm using Eqs. 25 through 28.

$$\overline{\mathrm{DEL}} = X_n - X_{n-1} \tag{25}$$

$$\mathrm{DEL} = \begin{cases} \overline{\mathrm{DEL}} & \text{if ATW} = 0 \\ 0 & \text{if } \overline{\mathrm{DEL}} > 0 \text{ and ATW} = 2 \\ 0 & \text{if } \overline{\mathrm{DEL}} < 0 \text{ and ATW} = 1 \end{cases} \tag{26}$$

$$Y_n = Y_{n-1} + \mathrm{DEL} \tag{27}$$

$$\overline{\mathrm{ATWP}} = \begin{cases} \mathrm{ATW} & \text{if ATW} = 0, 3 \text{ or } K > 0 \\ 1 & \text{if ATW} = 2 \text{ and } K < 0 \\ 2 & \text{if ATW} = 1 \text{ and } K < 0 \end{cases} \tag{28}$$

The X_n in Eq. 25 is calculated using Eq. 21. The X_{n-1} was the result of this same calculation the previous sample time. X_n is saved internal to the algorithm regardless of the windup code passed to the algorithm status word. Thus X_n follows all variations in the input, I, and is always the correct value to use in Eq. 21 next sample time. The pseudo-incremental lead-lag algorithm output is a full position output that is built up from a series of increments.

The initialization procedure for a lead-lag in a control path must be set up to be bumpless and consistent with the initialization procedures of other control path algorithms. Lead-lag initialization should be self-initiated when its input changes from bad to good, as indicated by the quality bit of its input status word. Initialization should also occur when the lead-lag's status word initialization bit is set or if its status word ATW changes from a 3. Initialization basically consists of getting the current input, I_0, and calculating X_0 by Eq. 29. X_0 and I_0 are then saved and used as X_{n-1} and I_{n-1} next algorithm execution. The tentative windup code that *may* be passed to the lead-lag input is determined using Eq. 28. The output of the pseudo-incremental lead-lag algorithm is left at its current value.

$$X_0 = K \cdot I_0 \tag{29}$$

It should be noted that if lead-lag algorithm initialization is triggered by a quality bit transition, the current algorithm output normally will not be zero. If the initialization is triggered by the algorithm using the lead-lag output, the lead-lag output will have been set by the downstream block. In Fig. 1 the ADD block would have set the lead-lag output to zero. The implications of this are that if the lead-lag input, I, goes to steady state, the lead-lag output generally will not be K•I, as one would expect if only Eqs. 20 and 21 are considered. This is the proper response and results from initializing at an arbitrary output value or not outputting increments due to ATW codes of 1 or 2. Eq. 20 was derived in the unconstrained world of perturbation variables and does not carry over exactly to the real full-position world.

PASSING ATW AND INITIALIZATION IN FEEDFORWARD PATHS

The normal calculation procedure for the lead-lag algorithm defined by Eqs. 21 through 28 determined the tentative windup status to pass up the control path, but stops short of passing the windup status to the input's output status words. The 2 by 2 decoupler shown in Fig. 4 is a simple example of a situation where feedforward algorithms should use windup and initialization information but not pass this information up the control path. Suppose ADD-2 passes an ATW of 2 to LLG-1 and PID-2. This is the correct action, as it will prevent PID-2 and LLG-1 from increasing their outputs. Clearly it would make no sense for LLG-1 to pass an ATW of 2 to PID-1 and constrain its output and inter-

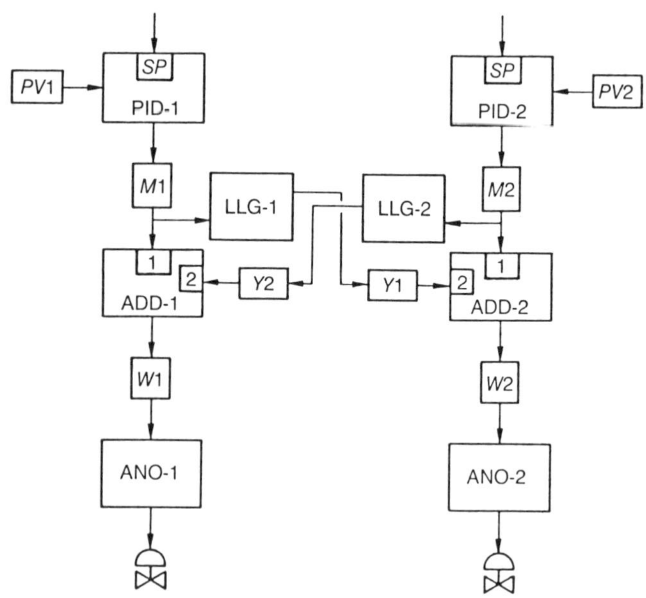

Figure 4. Feedforward decoupling.

fere with feedback control of PV1. The same considerations apply to LLG-1 passing initialization information to PID-1 by setting its status word initialization bit. Other control configurations require the passing of ATW and initialization status by lead-lag algorithms in feedforward paths.

The result is that passing initialization and ATW status information should be an algorithm option specified by the engineer. Feedforward control algorithms will be ATW and initialization users but not necessarily passers.

FEEDFORWARD CONTROL SECURITY

The idea of a feedforward algorithm using, but not passing, ATW and initialization information is one aspect of feedforward control security. A second aspect is the response of the combined feedback-feedforward control structure to failure of an instrument providing feedforward measurements. Suppose the input to the LLG algorithm of Fig. 1 is the output of an analog algorithm that is holding its output constant with the quality bit set indicating bad data. If the LLG algorithm continues to follow its normal calculation sequence, its output, Y, will reach a constant value and remain there until the AIC recovers. Changes in I will occur only when the AIC outputs good data.

The ADD, PID-1, and PID-2 do not need to notice that Y has gone constant. Feedforward control has been gracefully eliminated from the control structure. Feedback through PID-2 has not been degraded. When the LLG input, I, changes from bad data to good, the LLG algorithm will follow its prescribed initialization procedure and begin feedforward control from the current value of Y and the new value of I.

The situation is no more complicated when I is the result of a calculation involving several analog inputs as long as data quality is propagated through the calculation sequence that determines I.

GENERAL FEEDFORWARD ALGORITHM INTERFACE

The discussion of feedforward algorithm initialization and windup utilized the lead-lag algorithm. The ideas developed for this algorithm could be incorporated in other calculation-type algorithms, or a special algorithm could be designed to convert the full-position output of any calculation algorithm to pseudo-incremental. Further, this general algorithm could be designed to use windup and initialization but pass or not pass it as required. Incorporation of a general algorithm to interface calculation algorithms to feedforward control paths provides system flexibility and does not require the calculation algorithm designer to envision how the algorithm will fit into a feedforward system.

The general feedforward interface algorithm should basically operate as lead-lag with zero lead, zero lag and unity gain. If these values are introduced into Eqs. 21 through 28, the required algorithm results. Normally, a special interface algorithm block should be designed to incorporate calculational efficiencies possible with the specified lead-lag parameter values. This has been the approach taken by the authors in systems where they defined the algorithm set.

SUMMARY

Windup and initialization processing and propagation, along with a benign response to bad measurements, have been discussed for simple feedback and additive feedforward control structures employing PID, ADD, and lead-lag algorithms. Other algorithms used in control paths such as ratio, high/low select, and characterization algorithms used to linearize controller outputs to achieve constant loop gain should also provide such procedures.

Windup, initialization and bad data-handling procedures should be an integral part of the fill-in-the-blank algorithm set provided by a distributed control system. The utility of these features is such that they are applicable to the vast majority of process control situations. It is far better to design the algorithm set to automatically supply these features and take special action on the rare control loop where they are not totally applicable, rather than to require the control engineer to construct these features for each loop.

NOMENCLATURE

ATW	- Windup status in algorithm status word.
ATWP	- Windup status to pass up a control path to status word of primary algorithm.
ADD	- Digital addition algorithm.
AIC	- Analog input algorithm.
ANO	- Analog output algorithm.
D	- Increment related to PID integral action.
DEL	- Increment related to lead-lag output.
E	- Controller error.
G	- Component of proportional term of PID.
HOC	- High clamp on output of a control algorithm.
I, I^k	- An input to an algorithm.
K	- PID or lead-lag gain.
LLG	- Digital lead-lag algorithm.

LOC	- Low clamp on output of control algorithm.
M, M1, M2	- Control algorithm output. Usually a PID output.
MS	- Supervisory computer algorithm output.
MIN, MAX	- Operators that determine minimum or maximum values of an argument list.
PID	- Digital algorithm for proportional, integral, derivative control.
PV	- Process measurement, usually measurement input to PID.
s	- Laplace variable.
S	- Integral summation term for PID.
SP	- PID set point.
T	- Time between algorithm executions, sample time.
T1	- Lead time constant for lead-lag algorithm.
T2	- Lag time constant for lead-lag algorithm.
TD	- Derivative time of PID.
TI	- Integral time of PID.
W, W1, W2	- Add algorithm output.
X	- Unconstrained lead-lag algorithm output.
Y	- Output of pseudo-incremental lead-lag algorithm.

NOTE: Subscripts on variables refer to sample number, the nth or ith sample. A ¯ over a variable means an intermediate or tentative value of the variable before the final value is determined and stored.

References

1. Shinskey, F. G., *Process-Control Systems Application/Design/Adjustment,* McGraw-Hill, 1967.
2. Smith, C. L., *Digital Computer Process Control,* Intext Educational Publishers, 1972.
3. Fertik, H. A., and Ross, C. W., "Direct Digital Control Algorithms with Anti-Windup Feature," *ISA Transactions,* Vol. 6, No. 4, pp. 317–328, 1967.
4. Shinskey, F. G., *Process-Control Systems Application/Design/Adjustment,* Second Edition, McGraw-Hill, 1979.
5. Khandheria, J., and Luyben, W. J., "Experimental Evaluation of Digital Algorithms for Antireset Windup," *Ind. Eng. Chem., Process Des. Dev.,* Vol. 15, No. 2, pp. 278–285, 1976.
6. Luyben, W. L., *Process Modeling, Simulation, and Control for Chemical Engineers,* McGraw-Hill, 1973.

12

Specifying Process Chromatography Systems

A. Mukerji

Control of process plants by instruments providing measurements that infer product compositions is giving way to on-line chromatographs that provide actual composition data. Incentives include improved product yields and conservation of feeds and energy. Specifying chromatography systems involves several considerations including sample conditioning, analyzer shelters, documentation and drawings, inspecting and testing, piping and electrical requirements, vendor responsibility, etc. Presented are typical specifications intended to aid accurate and complete specification of chromatography systems.

SPECIFICATION REFERENCES

Several engineering references must be adhered to in the design of a chromatography system. A listing is given for applications in the U.S.:

All engineering-design shall be based on references listed below and current good analyzer system design practices. The analyzer vendor shall be responsible for the soundness and accuracy of his designs. Approval by the user does not relieve him of this responsibility.

1. API RP 550 "Installation of Refinery Instruments and Control Systems, Part II-Process Stream Analyzers."
2. API RP 550 "Installation of Refinery Instruments and Control Systems, Part I-Process Instrumentation and Control."
3. Users engineering standard instrument electrical.
4. Users engineering-standard instrument tubing/piping.
5. Users engineering-pipe specifications on cooling-water, steam, instrument air, chilled water.
6. API RP 520 "Recommended Practice for the Design and Installation of Pressure Relieving-Systems in Refineries," Part I-Design, Part II-Installation.
7. Electronic Industries Association.
8. Latest National Electrical Code.
9. RS-232 C Standard for Interface Between Data Terminal Equipment and Data Communication Equipment.

SYSTEM DESCRIPTION

This section should contain a brief system description outlining the salient features of the system package. A typical description may be as follows:

The analyzer system package shall consist of an analyzer shelter, eight chromatographic analyzers, and microprocessor-based programmers. The analyzers shall be in a centralized location. The programmer(s) shall be located in the main control room on a free standing-cabinet. All these items will be furnished by the analyzer vendor.

The programmer shall be tied in to a host computer. The programmer shall also display trend for each measured component on an analog trend recorder (furnished by user).

The programmer shall incorporate the following features:

1. Modular plug-in electronics.
2. Electronic auto-zero.
3. Battery back up for volatile memory.
4. Audio-visual alarms for detector temperature deviation and zero offset.
5. On analyzer power failure or communication failure between analyzer and programmer, the analyzer shall revert to a safe condition (back flush).
6. Programmer output to host computer shall represent actual concentration of measured component in ppm or weight percent as called out in the General Data Section. Area integration technique

shall be used to determine component concentration.

7. All program data tables shall be capable of field modification without user knowledge of any higher level programming. Initial program data tables shall be supplied as tape cassettes and capable of being loaded using standard voice grade cassette recorder/player.
8. Maintenance diagnostics shall be included in the software system to allow rapid trouble shooting in the event of system malfunction.
9. The following outputs shall be available:
   a) 4 to 20 maDC continuous trend-type output signal for each component measured. Trend will be displayed on analog-recorders
   b) RS232C, full duplex asynchronous serial communication interface for communication with host computer
   c) 20 ma current loop interface for communication with host computer
   d) 0 to 10 volts DC chromatogram output for potentiometric recorder

The potentiometric recorder shall be a single speed, single pen strip chart type. This recorder shall be furnished by the analyzer vendor.

The programmers and potentiometric recorder shall be suitable for flush panel mounting. These units shall be mounted on the panel by the selected panel vendor.

INSTRUMENTATION, PIPING, AND ELECTRICAL REQUIREMENTS

Typical specifications covering these items are as follows:

1. Measured variables shall be expressed in the following units:
   a) Temp - °F
   b) Positive pressure - psig
   c) Vacuum - psia
   d) Differential pressure - psi or inches water
   e) Liquid flowrate - gpm at
   f) Gas flowrate - scfm.
2. A pressure switch shall be provided inside the analyzer to shut-off power to the detector in the event carrier gas flow is lost, where such loss might result in damage to the detector.
3. If the sample conditioning system contains precut columns, an intercolumn thermal detector shall be provided to monitor separation efficiency. This shall be displayed on the programmer.
4. Electrical wiring and hardware shall comply with the latest national electrical code and users attached engineering standards for electrical design practices.
5. Shutdown and/or interlock devices which operate under emergency conditions shall be electrically isolated from alarms.
6. All interconnecting wires shall be numbered and these numbers shall appear on the respective interconnecting wiring drawings furnished by the analyzer vendor.
7. All terminal blocks shall be clearly identified and each terminal shall be clearly marked with the same identification shown on the wiring drawings as on the hardware.
8. Devices that operate alarms shall be energized under normal operating conditions.
9. Electrical classifications are as follows: Outside analyzer shelter: Class 1, Group D, Division 1 Inside analyzer shelter: Class 1, Group D, Division 2 Control room: General purpose non-hazardous.
10. Analyzer system shall operate within specifications at 120V $\pm$ 10 percent and 60Hz $\pm$ 5 percent.
11. All connected piping to the analyzer and its sample conditioning module shall be identified by permanently attached labels marked with the same designations as used on the documentation.
12. All tubing, piping hardware, and its material of construction shall be in accordance with users piping standards.

SAMPLE CONDITIONING SYSTEM

The sample conditioning system is a very important element of any analytical system. This is especially true in gas chromatography since the chromatograph only analyzes that portion of the process stream that is brought to it by the sampling system; analytical information provided by the chromatograph can be no better than the sampling system that transports the sample from the process stream to the analyzer. Actually, success or failure of an installation can be determined by this portion of the overall system. A sample conditioning system may by required to accomplish several or all of the following:

- Take a representative sample from a vessel or process line.
- Transport it to the instrument and without excessive delay.
- Filter out particulate matter, mists, and tar.
- Vaporize liquid samples.
- Control sample temperature.
- Measure and control pressure and flow.

- Scrub or treat the sample to remove water or corrosive materials.
- Switch streams when more than one stream is analyzed.
- Introduce calibrating mixture.

In most applications, the sample conditioning system is designed by the analyzer vendor due to his broad exposure and extensive expertise in this field. In rare cases, a user may have better expertise than the analyzer vendor. Whatever the case may be, the user must completely define the vendors scope of work, specify physical property and composition data on streams to be analyzed and define his requirements. Typical specifications may be as follows:

1. Analyzer vendor shall furnish a completely prefabricated and installed sample conditioning system. The system shall include all equipments and instruments necessary to separately condition each stream prior to injection into the analyzer. Typical equipments may be vaporizers, dryers, filters, heaters, coolers, scrubbers, demisters, coalescers, separators, pumps, etc. Typical instruments may be pressure reducers, vaporizing regulators, rotameters, pressure and temperature gauges, thermostats, temperature switches, relief valves, solenoids, etc. There shall be provisions for connecting a calibration standard cylinder for each stream and also a valved connection for grab sample. The sampling system shall include block valves on all process sample lines.
2. At the inlet to each analyzer, the following local indicators shall be provided: a) Sample flow rate b) Sample pressure c) Sample temperature.
3. A pressure gauge shall be provided on each side of the pressure regulator.
4. For multistream analysis, the conditioning system shall be designed to prevent contamination of one stream by another stream by valve leakage, diffusion from dead volumes, or any other source while the sample is being injected for analysis.
5. Sampling system shall transport the sample rapidly and with a minimum of memory effects which would tend to smooth out composition changes that the instrument shall record. Typical transport time from entry to sample conditioner until entry to analyzer shall be within 30 seconds.
6. The sample conditioning system shall be located within the analyzer shelter.
7. User shall specify the acceptable material of construction for all the sample wetted components of the sample conditioning system.
8. Since the majority of analysis applications are on hydrocarbon streams which are explosive and hazardous, it will be unsafe to discharge or vent sample vapors inside the shelter. As such, all vapors from each sampling system shall be vented outside the shelter in a flare system. Analyzer vendor shall terminate these vents on bulkhead fittings on the shelter wall. Tie in between vent on bulkhead fittings and the flare header shall be by the user.
9. If the sample stream pressure is such that it can over-pressure the analyzer due to sample stream regulator failure, malfunction or high pressure setting, a relief valve will be required to protect the analyzer. Vendor shall size and furnish this valve. Vendor shall also terminate all relief valve discharges on bulkhead fittings on the shelter wall. User shall tie these bulkheads to the flare header.

ANALYZER SHELTER SPECIFICATIONS

All the analyzers and their associated sample conditioning system, carrier gas cylinders, and containers carrying liquid calibration standards should be installed inside an analyzer shelter. The analyzer vendor should furnish a self-framing, free-standing house as a completely assembled package suitable for installation on a concrete pad and suitable for maintenance of all sample conditioning systems and their associated analyzers. All sample conditioning systems, instruments and related equipments, various cylinders and containers should be installed and interconnected inside the shelter with all necessary utility lines and accessories provided. The shelter should facilitate transportation by motor freight and permit the crane to set it in place without distortion.

The following minimum specifications apply to analyzer shelters:

1. Minimum height shall be 8 ft 6 in.
2. Vendor shall establish depth and length and obtain user's approval prior to fabrication.
3. Shelter roof shall be shed type.
4. Shelter wall material shall be 20 gauge (minimum) galvanized steel fiberglass insulated with foil backing towards the inside of the building.
5. Material of construction for hinges, bolts, washers and nuts shall be at least cadmium plated steel. Rivets can be aluminum.
6. The shelter shall be assembled on a welded galvanized structural steel. Structural steel also can be used for mounting internal equipment and instruments.
7. Access doors shall be provided at each end of the house. These doors shall be galvanized steel 3 ft.

wide by 7 ft high and of double wall construction. Each door shall incorporate a panic bar, window with reinforced glass and bottom adjustable vent indoor door closer. The glass window will be 12 in. by 12 in. The door shall open outward.

8. All electrical/electronic instruments inside the shelter shall meet electrical classification.
9. Signal and power supply wiring from each analyzer shall be brought to a common signal termination box and a common power termination box.
10. Vendor shall layout and provide fluorescent lighting with gas tight fittings and seal inside the shelter. Lighting illumination level measured horizontally at a height of 3 ft above the shelter floor shall be 75 foot candles. The light fixtures shall meet shelter (inside) electrical classification. All lighting wires shall be terminated on a common lighting termination box located on the shelter exterior wall. One 115 V 60Hz service receptacle will be provided at each end of the shelter. These receptacles shall also be wired to a common lighting termination box.
11. All wiring inside the shelter up to various termination boxes (located on shelter exterior wall) shall be done by the analyzer vendor. Wires shall be run in rigid screwed conduits. Analyzer vendor shall furnish and install all termination boxes.
12. One AC double pole disconnect switch shall be provided for each analyzer. Near each door inside the shelter, a lighting-on-off switch shall be provided.
13. Signal wiring, power supply wiring and lighting wiring shall be run in separate conduits.
14. Wire size, conduit size, insulation, termination boxes, disconnect switches etc. and wiring shall be in accordance with users attached electrical standards and latest national electrical code.
15. All the piping, tubing and associated valves inside the shelter used for sample utilities (steam, air, water, etc.) and vent shall be furnished, installed and terminated at the bulkhead fittings on the shelter wall/walls by the vendor. Each bulkhead fitting shall be tagged. Typical tagging shall be as follows:
    a) Sample in analyzer #1
    b) Sample return analyzer #1
    c) Air supply
    d) Sample vent analyzer #1.
16. Analyzer shelter shall meet the following codes:
    a) Uniform building
    b) AISC (American Institute of Steel Construction)
    c) AISI (American Iron and Steel Institute)
    d) MBMA (Metal Building and Mfr's Association)
    e) ANSI (American National Standards Institute)
    f) API Recommended Practices (American Petroleum Institute)
17. The following design loads shall be applicable to the analyzer shelter:
    a) Wind load per ANSI A58-1 1972 Table 5.
       100 year mean recurrence
       100 mph basic wind speed
    b) Live load 20 lb/ft^2 of horizontal projected area
18. For each analyzer, the vendor shall furnish two size 1A cylinders of certified carrier gas and eight gallons of liquid calibration standard in a carrier gas pressurized container. Cylinder will be furnished with all necessary auxiliaries such as fittings and two stage regulators. With gauges, shut off and purge valves, the carrier gas cylinders for each analyzer shall be manifolded together. The manifold shall be arranged so that a bottle may be replaced while the system remains in operation on the remaining bottle. The carrier gas from the common manifold shall be tubed to the liquid container via an isolation valve and gauge for pressurization. The liquid container shall be tubed to the sample conditioning system at the sample inlet via an isolation valve. The carrier gas cylinders manifold shall also be tubed directly to the analyzer.
19. All the cylinders and containers shall be installed inside the shelter in mounting racks. Mounting racks shall be furnished and installed by the vendor.
20. Vendor shall also furnish and install all tubing between cylinders, containers and analyzers.
21. Composition of the calibration standard shall be approximately the same as normal composition of the sample being analyzed.
22. Each cylinder and container shall be furnished with a certificate of title transfer.
23. Temperature inside the shelter shall be maintained at 70°F year round. Vendor shall furnish and install outside the shelter a heating and air conditioning package.
24. The shelter shall be pressurized to maintain + 1/2 in. WC gauge air pressure to prevent a hazardous environment inside the shelter. A pressure switch set at + 1/4 in. WC gauge air pressure shall be provided to detect loss of air inside the shelter. This will be alarmed on the main panel. Analyzer vendor shall furnish and install the switch and run the wire from the switch to a termination box on the shelter external wall. The wiring from box to main panel shall be done by the user.

25. The suction air duct to the heating and air conditioning package shall be run by the user. Vendor shall advise duct size. Duct air intake shall be at an elevation of 25 ft.
26. User shall provide a foundation or concrete pad for installation of the package. Vendor shall provide package foundation and loading details.
27. The ambient dry bulb temperature and relative humidity variations shall be 0 to 100°F and 10 to 80 percent. Package shall be designed for above cnvironmental conditions.

GENERAL DATA

This section should provide information on the sample stream and available utilities.

1. Analyzer tag no: AT-1
2. Flow sheet: A-1
3. Number of steams analyzed: 2
4. Objective: To measure concentration of components X & Y in reactor R-1A & B product streams.
5. Analyzer location: Inside the shelter
6. Sample process connection: User shall provide a ½-in. valve (as per pipe specs) at the sample take-off point and run the sample supply line to the bulkhead fitting on the shelter wall. Sample probe with internal cut-off valve shall be furnished by the vendor. The sample probe connector shall be adjustable over the entire probe length. Material of construction of the probe and cut-off valve shall be as per pipe spec.
7. Sample return: User shall run sample return line from the shelter bulk heat fitting to the sample return point.
8. Phase: Liquid
9. Temperature: T°F
10. Pressure: P psig
11. Available press drop: Δ psi between sample take-off and return
12. Sample supply line length: L_1ft
13. Sample return line length: L_2ft
14. Heat tracing: None
15. Sample tubing outside the shelter furnished by the user shall be ½ in. OD seamless SS tubing of 0.035-in. wall thickness, unless larger OD tubing is required due to insufficient available pressure drop between the sample take off and return points. Analyzer vendor shall recommend sample flow rates in gpm in primary and secondary (if required) sample loops. All instrumentation for both loops shall be furnished and installed inside the shelter by the analyzer vendor.

Stream composition

Components	Molecular weight	N-boiling point in °F	Composition in wt.%	Component of interest	Range in wt.%
C_1	M	T_1	W_1	---	---
C_2	M_2	T_2	W_2	C_2	0 to W_2[1]
C_3	M_3	T_3	W_3	---	---
C_4	M_4	T_4	W_4	C_4	0 to W_4[1]
C_5	M_5	T_5	W_5	C_5	0 to W_5[1]

Available utilities at shelter

1) Power supply:	115V 60Hz, 1 phase 460V 60Hz, 3 phase
Power consumption:	by vendor
2) Instrument quality air:	80 to 100 psig
Air consumption:	by vendor
3) Cooling water:	50 psig at 85°F
Usage:	by vendor
4) Chilled water:	50 psig at 45°F
Usage:	by vendor
5) Saturated steam:	50 psig
Usuage:	by vendor

ANALYZER SYSTEM DATA AND DOCUMENTATION

For each analyzer the vendor shall furnish the following data:

1. Sample size.
2. Analyzer temperature.
3. Column and valve arrangement.
4. Column packing material—stationary phase and support.
5. Column life at analyzer temperature.
6. Column head pressure and flow rate.
7. Cycle time.
8. Type of detector (thermal conductivity or flame ionization).
9. Number of detectors per unit (two or four elements).
10. Detector settings, e.g. voltage and current.
11. Programmer settings in actual time.
12. Component attenuator settings.
13. Chromatograph showing—component separations.
14. A full specification on carrier gas.

Analyzer vendor shall furnish the following documents and drawings:

1. Overall installation drawings for each process analyzer system. Drawings shall describe in detail interconnecting piping and wiring of sample conditioning system, process analyzer instrumentation and receivers.
2. Shelter plan drawing which shall show the physical location of each sample conditioning system, each

process analyzer, all calibration and carrier gas cylinders and racks, all utility lines, vent lines, each sample supply line, each sample bypass line, drains, lighting fixtures, disconnect switches, doorways, termination boxes, all relief valves discharge lines, etc.

3. Shelter bulkhead detail drawing which shall show the physical location tagging and size of each bulk head fitting on the shelter wall/walls.
4. Interconnecting piping drawings which shall describe in detail tubing of the sample conditioning system to analyzer and process stream, piping of utilities to analyzer, physical location of each instrument associated with each analyzer system, tubing and manifolding of carrier gas cylinders to all analyzers inside the shelter, tubing of calibration gas and or liquid calibration standard to each analyzer etc.

   All the tubing and piping associated with each analyzer and sample conditioning system shall be tagged using analyzer tagnumber and service. All tubing and piping connections on the analyzer and sample conditioning system also shall be tagged using the analyzer tagnumber and service.
5. Interconnecting wiring drawings which shall indicate all connections, wire color coding or numbering and terminal identifications as it actually exists for each process analyzer system and termination boxes. The drawings shall also show wiring between various termination boxes and analyzers, sample conditioning systems, lighting fixtures, disconnect switches, pressure switch, etc.
6. Instructions and operating manuals shall cover all components and include all wiring diagrams, analyzers and sample conditioning system piping and tubing details, detailed installation instructions, calibration and trouble shooting procedures, complete parts list, vendor prints on all instruments and recommended spare parts list for one year operation.

INSPECTION AND TESTING

1. Each analyzer system shall require a functional test at the vendor's shop. Vendor shall be responsible for demonstrating system operation in accordance with users' requirements. Users' representatives shall inspect all the hardwares and witness and approve all tests.
2. A leakage test using a leak detection fluid shall be conducted on each analyzer system. No visible leakage is permitted.
3. To verify that the instrumentation sample conditioning system and utilities are properly interconnected, a functional test of all piping, tubing and system components shall be conducted with instrument air. Power shall be applied to the system to verify proper electrical interconnections and actuation of solenoid valves, alarms, etc.
4. System response shall be demonstrated by upscale deflection with span gas and down scale movement with zero gas.
5. Calibration shall be demonstrated with zero and span and with test samples giving approximately half scale and three quarter scale readings for the component of interest.

13
Analyzers and Reformer Efficiency*

W. A. Bajek

General goals of the HPI are reduction of utilities consumption and improving conversion of feedstocks to desirable products while ensuring continued on target product quality. In process fractionation sections, reduction of fuel consumption can be rather significant, whereas, in catalytic reaction areas, it is often more significant to maximize conversion of feedstocks to the desired product. These goals can most readily be achieved by monitoring and control of parameters with control systems that have analytical devices as their primary transducers. Recently developed on-line boiling point and octane analyzers are two examples. Their application with respect to improving the efficiency of certain aspects of crude fractionation and catalytic reforming is presented.

CONTROL STRATEGIES—CATALYTIC REFORMER

A processing scheme common to many large refineries is shown in Fig. 1. In the normal fuels refinery situation, gasoline production is of primary importance. Numerous processing possibilities exist for production of various grades of gasoline, but common to almost all schemes is a catalytic reformer. This process is important not only in its ability to convert low-octane naphtha to a high-octane blending component, but also in its ability to produce reformate of a widely variable octane. Other components such as alkylate, FCC gasoline, "poly" gasoline, and straight run are valuable gasoline blending components, but their octane may be varied little, if at all.

Current crude supply problems necessitate frequent feedstock changes in many refineries. This situation makes it desirable or necessary to monitor and control quality of feed to the catalytic reformer and product reformate octane.

Reformer Feed Quality

There are three major reasons for closely controlling feedstock distillation characteristics to the catalytic reformer. Endpoint control is necessary to prevent production of reformate materials which boil beyond the specified gasoline endpoint. Since reformate endpoint can increase 10–40°F above the feed naphtha endpoint, depending on reformer severity, close control of this parameter is very important. A second reason for naphtha endpoint control relates to the effect heavy naphtha has on reforming catalyst coking rate. Since catalyst coking rate increases in an exponential manner with naphtha endpoint, processing an overly heavy naphtha can prove very costly. Reformer feed naphtha initial boiling characteristics is a third feed control criterion. Since pentanes and lighter materials gain little benefit from being passed through the reformer, the most efficient scheme is to carefully exclude these materials in the feed fractionation.

Fig. 2 illustrates a typical naphtha catalytic reforming operation with its associated feed fractionation train. In this figure a boiling point monitor (BPM) is shown as the primary control transducer on the naphtha overhead cut on the crude facility. Rather than relying on product quality information inferred through column vapor temperature measurements, the boiling point monitor produces actual naphtha quality information. Newly developed boiling point monitors are capable of producing a complete distillation analysis that correlates either with ASTM D-86, ASTM D-2887 or a custom correlation. Analytical cycle time of approximately two minutes on 450°F EP naphtha makes this apparatus ideally suited for this control scheme.

In this scheme, the monitor controls product endpoint by resetting the crude column top temperature controller. A strip chart, Fig. 3, shows a readout of the distillation characteristics which, in this installation, were correlated to ASTM D-86 (Engler). Nine boiling points (115°, 130°, 150°, 210°, 250°, 290°, 340°, 360°, 375°F) were selected as being important to characterize the complete distillation curve. Irrespective of the num-

* Based on a paper originally presented at the Energy Sources Technology Conference and Exhibit, New Orleans, Feb. 3–7, 1980.

Figure 1. Refinery base scheme.

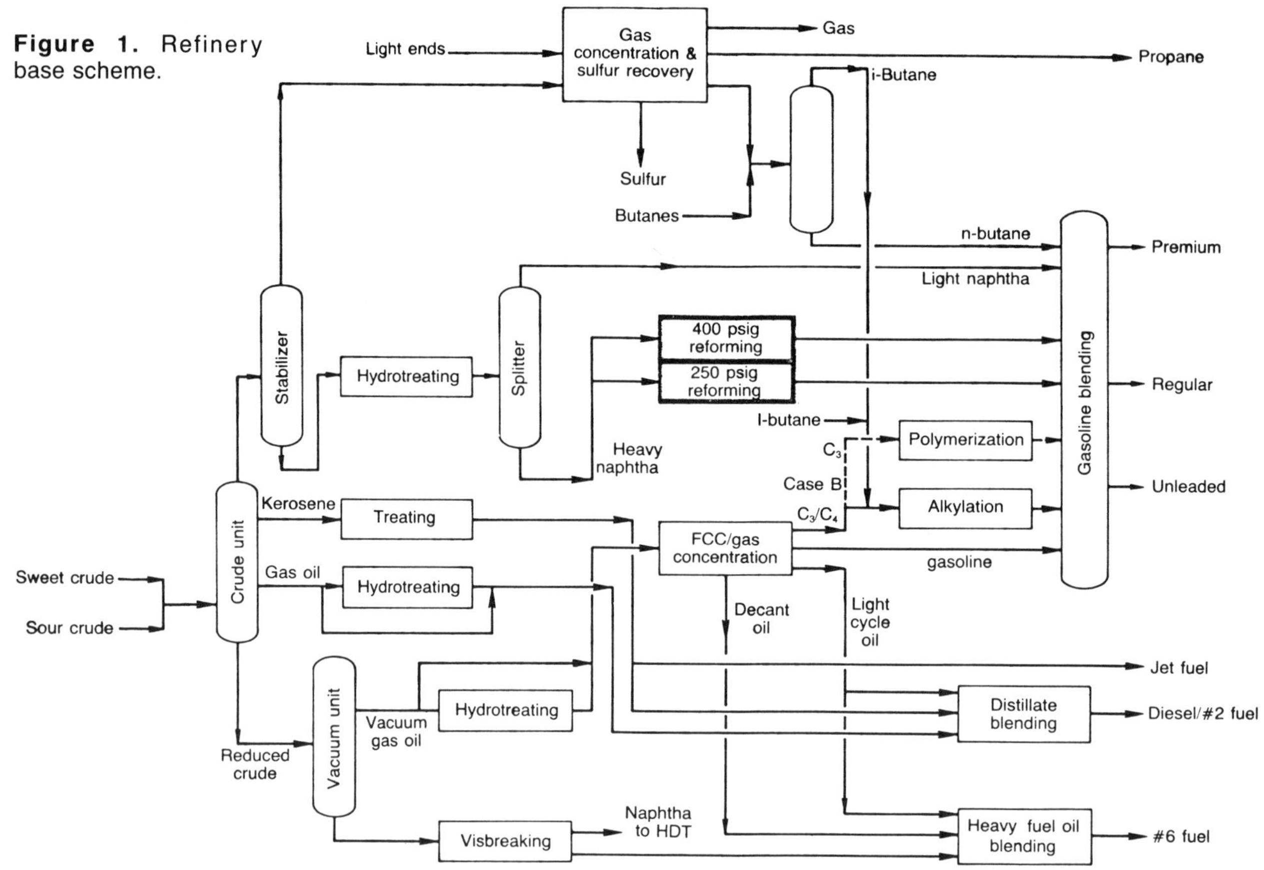

Note: Unit hydrogen and gas streams not shown.

Table 1
Catalytic Reformer Within One RON of Target Operation

			Target		
RON	94.0	94.5	95.0	95.5	96.0
Operating cost, $/barrel charge	1.488	1.504	1.52	1.548	1.576
Δ Operating cost, $/barrel charge		0.032		0.056	
Δ Operating cost running ± 1 RON of target			$0.024/barrel charge		
Δ Operating cost running ± 1/2 RON of target			$0.012/barrel charge		

Example: Costs incurred due to small variations from target RON operation. A 20,000 bpd reformer fluctuating one octane in 24-hour period.

(10,000) (0.024) = $240/day
$240.00/day x 30 = $7,200/month
$7,200.00/mo. x 12 = $86,400/year

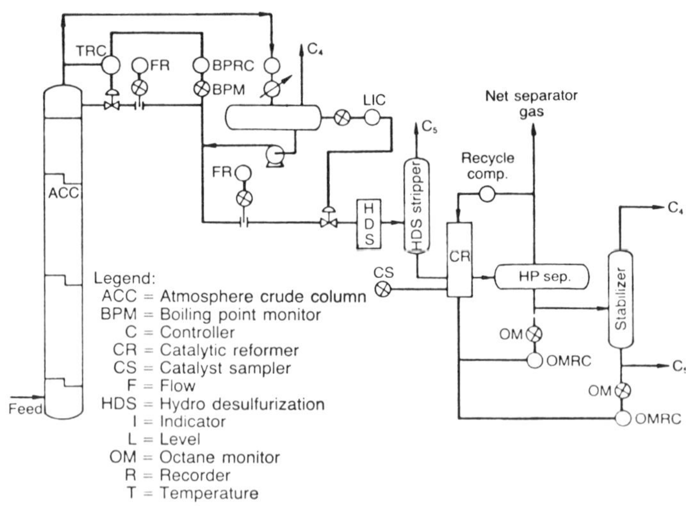

Figure 2. Catalytic reformer with on-line analyzers.

ber of points, the monitor has the inherent capability to generate a complete distillation curve in approximately two minutes, with a reproducibility within 1 percent of each boiling point selected. This cannot be achieved consistently by a laboratory and accuracy of control is improved. More importantly, the low response time of the boiling point monitor allows its direct use in a real time control system which is impossible with laboratory data.

Other crude column product streams such as straight run gasoline, kerosene, and light and heavy naphthas also may be monitored and controlled to improve over-

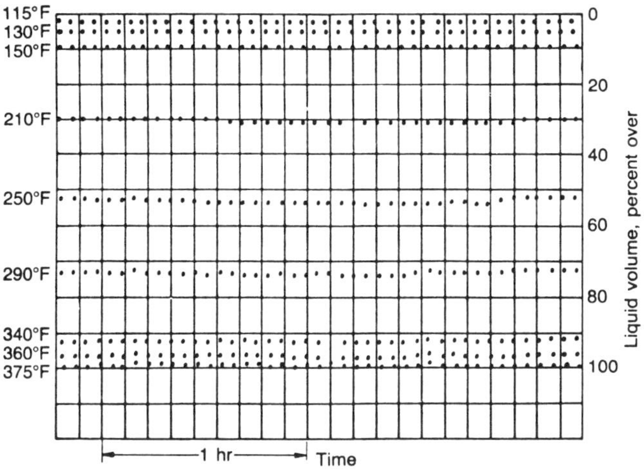

Figure 3. Boiling point analyzer readout of catalytic reformer feed.

all crude column operation. These applications will not be addressed in this discussion.

Reformer Product Quality

As previously mentioned, the catalytic reformer is one of the few refinery processes that allows variation in product octane. This process, therefore, provides an invaluable "knob" that the operator can turn to meet product quality needs. This process variable is not without its cost, however. Since catalytic reforming is a highly endothermic operation, tremendous amounts of heat are required to effect desired conversions. It is not uncommon for typical naphthas to require 300–400 Btu's per lb. of feed to produce high-octane reformate. Also, the higher the octane, the higher the endothermic heat of reaction. Therefore, if one operates the reformer beyond the necessary conversion level, energy is wasted.

A more serious consequence of operating at a higher than required octane is loss in reformate yield. The reforming reaction (actually many competing reactions) results in an overall volumetric yield loss. The higher the octane, the lower the volumetric yield. Over the typical range of octanes and feedstocks, 0.6 to 1.4 liquid volume percent of reformate yield may be lost per single octane increase. An example of the impact of this variable can be seen in Table 2. Assuming conservatively that continuous octane control can allow an overall reduction in average octane of 0.2 RON-0 (Research Octane Number, unleaded) compared to conventional control strategy, a savings of greater than $100,000 per year can be realized on a typical 20,000 bpsd reformer.

Commercial analytical devices are now available which continuously measure and record the knock rating (octane number) of a fuel. The monitor is not dependent on cyclic action or automatic calibration checks in

Table 2
Effect of Close Octane Control on Catalytic Reformer Yield

Basis:

a) Δ value between reformate and fuel gas = $7.50/barrel
b) Feed rate = 20,000 bpsd
c) Δ RON = 0.2
d) Annual operation = 330 stream days
e) Δ C_5+/RON = 1.1 liquid volume percent

$/year = 20,000 x (0.011 x 0.2) x $7.50 x 330 = $108,900
or for 0.5 RON = $272,250.00

Note: Example ignores delta hydrogen value and delta operating costs.

its operation and has the inherent characteristic of low dead time. This makes the instrument ideal for online monitoring of a gasoline process scream, and its output can be used as the input signal for process control by computer or standard analog instruments. A conventional control application for which the monitor is ideally suited is monitoring product from a catalytic reformer. In many catalytic reformers, the monitor is installed in the stabilized reformate line. Fig. 4 is a strip chart of such an installation. However, where the stabilizer handles other streams in addition to the reformer, or where minimum response time is desired, octane control is accomplished by locating the monitor in the high-pressure separator liquid line. Since separator liquids have high vapor pressures, direct rating of octane cannot be obtained with conventional laboratory knocktest engines because of light material loss at normal carburetion pressures. Rating high vapor pressure samples by the octane monitor is easily accomplished without the need for any additional equipment as the process stream sample does not fall below its vapor pressure until after it is introduced into the reaction zone of the analyzer.

The chart, Fig. 4, shows octane monitor readout from a startup of a catalytic reformer where stabilized refor-

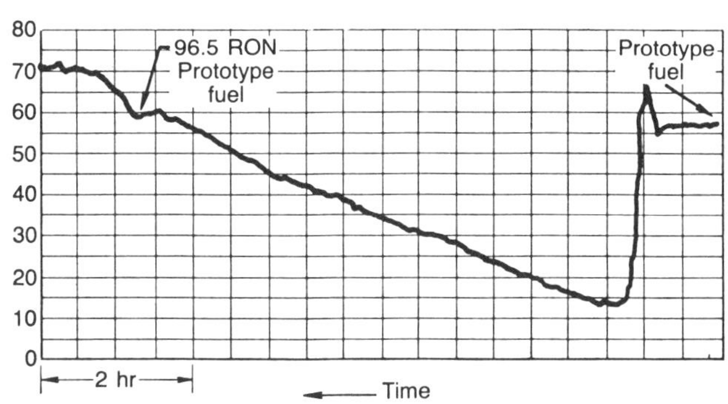

Figure 4. Response from stabilized reformate for catalytic reformer.

mate was being monitored. The octane monitor was initially operated on a prototype fuel that represented the product from the process. The monitor was switched from prototype fuel to the process sample, and the reformer reactor temperatures were raised according to prescribed procedures. To demonstrate that the monitor was performing properly, the prototype fuel was reintroduced. Satisfied with the repeatability of the prototype fuel, the refiner switched the monitor back to the process and reactor temperatures were raised to meet his new target octane. In this operation, target octane was reached within eight hours. It should be apparent that with the monitor on-line, the refiner now has a tool that will allow him to produce target octane in a minimal time frame within constraints of the process.

The chart in Fig. 5 is from a catalytic reformer employing a swing reactor. The process is on computer control with the octane monitor providing the octane input to the computer. Benefits derived with this system are that the octane swing on startup of regenerated reactors, as well as the period of deviation from target production, are minimized. Monitoring and control of catalytic reformers can best be achieved with a continuous monitoring system providing input to either analog or computer type control systems.

Response time comparison between a continuous monitor and a hatch type monitor with an approximate five-minute cycle time is diagrammed in Fig. 6. Stabilized reformate from a 15,000 bpd catalytic reformer typically has a process delay of approximately 15 minutes as shown. If a step change is introduced in the process line A, the octane effect of the stabilized reformate will occur as shown in line B. With a continuous monitor having negligible response time, the equivalent octane readout line C will be similar to line B. With a batch type monitor, assuming a five-minute cycle time, the resulting octane level inputted into the control system could be similar to line D in the worst case, and similar to line E in an intermediate case.

The step change output response associated with batch type monitoring requires special control limits and algorithms. This system can only approach the results inherently characteristic of a control system utilizing real time octane input. Conversely, the output from the continuous monitoring system can be used directly with standard process control techniques. This tighter control enables the refiner to obtain maximum performance and associated economic benefits from the reforming unit.

ECONOMIC JUSTIFICATION

To generally determine the economics for each refiner's operation is an impossible task. However, a simplified economic basis can be developed which the refiner can use as a starting point to find the costs for his particular situation. Economic factors that should be considered in evaluating performance of a reforming unit include catalyst (cost, interest, replacement and regeneration), utilities (energy) and labor. While the effect of these factors is known and accepted by the industry, what is not recognized is the effect they have on yield.

We have monitored reformate produced by catalytic reformers and have found that regardless of the type of unit, octane number of the product is not constant, even though the laboratory knock rating of product samples taken every twenty-four hours indicates relative stability; i.e, variations of approximately 0.5 RON which is close to the accepted deviation for the analysis. It is not

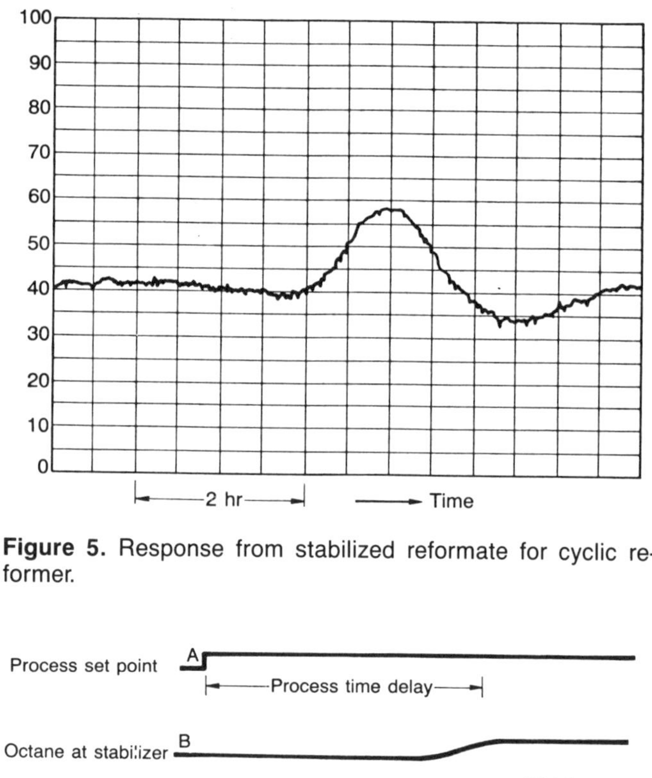

Figure 5. Response from stabilized reformate for cyclic reformer.

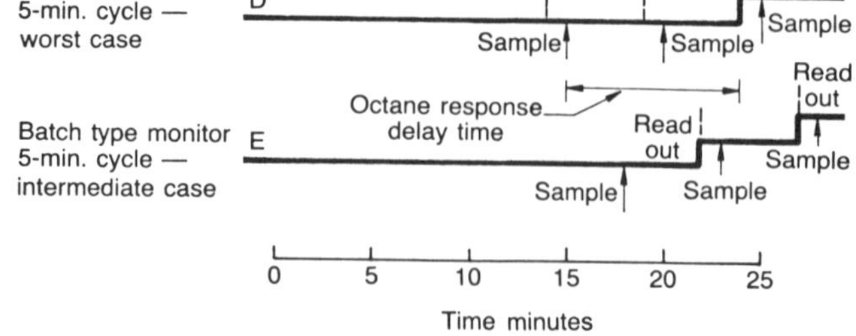

Figure 6. Comparative octane analyzer response.

uncommon to observe a 0.5 to 1.0 RON swing between day and night operation as well as similar swings caused by undetermined process variables.

The following example of reformer costs is examined to exemplify the relationship between target octane variation and operating costs. Consider a catalytic reformer operating at 95.0 RON. Operating costs for this operation have been determined to be $1.52/barrel charge. If the reformer operates part of the time at 94.0 RON, the operating costs will be $1.48/barrel charge. Conversely, operation at 96.0 RON will cost $1.56/barrel charge. For simplicity of comparison with steady operation at 95.0 RON, we have assumed the unit operated half of the time at 94.0 RON and half of the time at 96.0 RON, even though we know this will result in an average RON of the total product slightly below target because yield at 96.0 RON is less than yield at 94.0 RON. Table 1 shows the operating costs in dollars per barrel charge for production of 94.0, 94.5, 95.0, 95.5 and 96.0 RON reformate. The delta operating cost from 95.0 RON to 94.0 RON is $0.032 and from 95.0 RON to 96.0 RON is $0.056. The delta cost of operating within ±1.0 RON of 95.0 RON target is $0.024/barrel charge and within ±0.5 RON is $0.012/barrel charge. Therefore, a 20,000 bpd reformer fluctuating one octane number in a twenty-four hour period will incur a cost of $240.00/day or $7,200.00 on a monthly basis and $86,400 in one year. Using the graph, Fig. 7, which assumes linear variation of operating cost with product RON, the effect of octane deviation on operating costs for various reformer thruputs can be seen.

In addition to increased operating costs, there will be an overall reduction in the volume of the highest octane reformate produced. The example shown in Table 2 indicates that if an overall reduction in actual produced octane number of 0.2 RON can be achieved by the

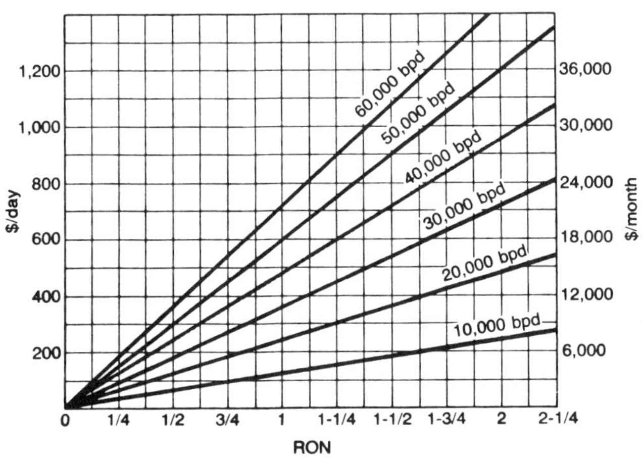

Figure 7. Catalytic reformer incurred cost due to deviation from target RON production.

continuous monitoring strategy, an improvement in C_5+ yield of 40 barrels per day can be expected. While this quantity seems small, it results in a product value savings of over $100,000 per year.

It should be noted that yield differences increase rapidly at higher target RON reformer operation.

CONCLUSION

Continuous on-line monitors provide the refiner with a tool to obtain tighter process control, which minimizes dependency on laboratory analysis with their inherent lag time. The improved efficiency of the operation allows the refiner to achieve conservation of feedstocks and energy. Each refiner must determine savings obtainable by on-line product and feedstock monitoring. The simplified examples discussed here provide a basis for such a study. While they do not take into account the effect of many variables individual to one unit, they show the magnitude of savings attainable. Further savings can be achieved in the refiner's motor fuel blending operation since the octane quality of the reformate, which is the prime octane control, is known and closely controlled. Ultimate savings are achieved by continuous on-line monitoring and control using recently developed monitors in the blender.

References

1. Bajek, W. A., Sampson, R. W., and McLaughlin, J. H., "An On-Stream Boiling Monitor Employing a Microcomputer," 8th IMECO Congress, Moscow, USSR, May 21-27, 1979.
2. Sampson, R. W., Bajek, W. A., and McLaughlin, J. H., "On-Stream Monitoring for Hydrocarbon Boiling Characteristics Employing a Microcomputer," ISA, Analysis Instrumentation Division Conference, Charleston, West Virginia, May 17-19, 1977.
3. McLaughlin, J. H., and Bajek, W. A., "Continuous Octane Monitoring," 24th Annual ISA Conference, Houston, Texas, October, 1969.
4. Board, T. G., Addison, G. E., Conser, R. E., and Salazar, J. R., "The Next Octane Increment: Where From and At What Cost?," API 44th Refining Midyear Meeting, San Francisco, Calif., May 14-17, 1979.
5. Kuchar, P. J., Addison, G. E., "Platforming: A Tool For Tomorrow's Fuels," UOP 1974 Technology Conference.
6. Kuchar, P. J., Braun, A. O., Shelton, J. B., and Olson, R. K., "Platforming Developments for the Energy Conscious Era," UOP 1978 Technology Conference.

14

Computer Model of Analyzer Payouts

J. W. Black

Studies performed on a refinery computer model developed for the U.S. Department of Energy indicate large payouts can be obtained by applying on-line boiling point, pour point, Reid vapor pressure and/or viscosity analyzers to refinery operations. Studies were done for a 60,000 bpd and a 200,000 bpd refinery by Turner Mason & Associates. The results have been extrapolated to provide estimates for any refinery. Maintenance costs were included as described in the modeling procedure for each analyzer application.

THE MODEL

The increases in refinery profitability were calculated from a series of linear programming (LP) runs of a computer model developed by Turner, Mason & Associates for the U.S. Department of Energy. This model describes refining operations along the Texas Gulf Coast and has been updated repeatedly as changes in feedstock, product qualities or processing flexibility have occurred. Although calculated for a Gulf Coast location, the results can be adjusted and used for other refinery locations.

The current model (pricing, specifications and processing configuration) describes a typical refinery on the Texas Gulf Coast for 1980. This model is constructed to carefully calculate blending requirements for finished product specifications. In motor gasoline blending, research and motor octane rating, distillation, and RVP are calculated to assure compliance with specification qualities.

In No. 2 fuel oil blending, gravity, sulfur content, pour point, flash point and distillation points are calculated to assure compliance with specification qualities.

The model also compensates for changes in other refinery units as a result of changes in the analyzed parameters. Hence, it calculates a more accurate profitability than would be obtained by hand calculation of the improvement of a variable change on one unit. Processing units and other pertinent data included in the model are listed in the Tables.

BOILING POINT ANALYZER

Three different boiling point analyzer cases were evaluated:

- A package of four analyzers on a crude unit.
- One analyzer on a crude unit.
- One analyzer on a fluid catalytic cracker (FCC) main fractionator.

The following paragraphs discuss one boiling point analyzer application on an FCC unit and illustrate how we determined the typical 3°F improvement in cut point attainable from using an analyzer. Normally, the cut point between the overhead (FCC gasoline) and first side cut (light cycle oil) from a FCC main fractionator is maintained by a temperature controller. This controller normally regulates the main fractionator reflux rate. Routine ASTM D-86 laboratory distillation results are used to adjust the setpoint on the temperature controller. Studies conducted on this method of controlling the cut point (similar to the ASTM 95 percent point) reveal a cycling.[1]

To avoid exceeding the target cut point, and to meet required unfinished and finished product quality constraints (FCC gasoline 95 percent point and end point and finished gasolines end point) the cut point is set below the desired target. This results in "downgrading" FCC gasoline to light cycle oil (No. 2 fuel) a majority of the time. This product "downgrading" costs the refiner because he makes less of his more valuable product (gasoline) and more of his less valuable product (No. 2 fuel).

An alternate cut point control relies on a continuous boiling point analyzer. With continuous analyzer control, the 95 percent point cycling is reduced greatly. Hence, the cut point can be maintained about 3°F closer to target.[1] This 3°F improvement also was supported by discussions with several refiners. This improvement reduces the product "downgrading" compared to the conventional temperature control adjusted from lab analyses. The increased profitability is attributable primarily to less "downgrading" of gasoline to No. 2 fuel and the difference in value between finished gasoline and No. 2 fuel.

This type of "downgrading" also occurs for overhead gasoline from a crude unit or other refinery process column. Similarly, unfinished products taken as sidecuts from atmospheric crude units or other refinery process fractionators normally are controlled or adjusted by sidedraw temperature to approach the target cut points. These unfinished sidecut products also are affected by cut point cycling that forces product "downgrading" to meet required unfinished and finished product quality constraints. On a typical crude unit the following avoidable unfinished product "downgrading" might occur without boiling point analyzers:

- Reformer feed to light straight run gasoline
- Reformer feed to kerosene jet fuel
- Kerosene jet fuel to No. 2 fuel base
- FCC feed to No. 2 fuel base

Analyzer Costs and Profitability

To estimate the potential profitability of installing boiling point analyzers, a base case was simulated with the LP, assuming laboratory analyses are used to adjust cut points. This resulted in a base profitability posture. Then, for comparison, we ran the three other cases (which we evaluated) using the same raw materials as in the base, but allowing one or more unfinished product cut points to be controlled "tighter" to approximate the use of on-stream boiling point analyzers.

Changing unfinished product cut points also affects finished product qualities. Our LP optimized the product blending to continue to meet all quality specifications in the most profitable manner. As a result, when four boiling point analyzers were used in the crude unit, the LP slightly changed reforming feed rate and severity, FCC conversion, isomerization feed, and alkylation product. When only one boiling point analyzer was used, the changes were less extensive, but still taken into account. Hence, the LP calculated a more accurate profitability for these cut point improvements than one would obtain by a simple hand calculation.

Table 1 describes the overall effects of these constraints on a 60,000 bpd refinery for four boiling point analyzers. It is apparent that the four "analyzers" case

Table 1
Effects of Four Boiling Point Analyzers on Crude Unit Material Balance, Processing and Profitability

	Base case lab analyses	Proposed case with 4 analyzers	Analyzer Increment
Material balance, bpd			
Crude feed	49,767	49,767	—
Other feeds	2,957	2,957	—
Total feeds	52,724	52,724	—
Gasolines	26,074	26,234	160
Kerosene jet fuel	3,956	4,206	250
No. 2 fuel	10,079	9,704	(375)
No. 6 fuels	3,380	3,384	4
Other products	7,143	7,143	—
Propane	457	405	(52)
Plant fuel burned	2,374	2,387	13
Refinery loss (gain)	(739)	(739)	—
Total products	52,724	52,724	—
Processing			
Fluid catalytic cracker			
Fresh feed, bpd	14,570	14,570	—
Conversion, %	71.4	71.7	0.3
Catalytic reformer			
Feed, bpd	11,097	11,205	108
Research octane, clear	96.7	96.6	(0.1)
Hydrocracker feed, bpd	2,187	2,187	—
Delayed coker feed, bpd	1,504	1,504	—
Pen/hex isomerization feed, bpd	2,122	2,029	(93)
Alkylation product, bpd	1,474	1,487	13
Profitability			
Over base, $/year	Base	562,000	562,000

generates considerably higher profits than the "lab analyses" case. Tables 3 and 5 provide similar information for adding one boiling point analyzer to the crude unit and the FCC respectively. Material balances, processing rates and severities, and profitability are detailed separately on these tables.

In addition to costs of about $23,000 for each boiling point analyzer and sample conditioning system we estimated installation costs of about $22,000 for a total capital cost of about $45,000 for each analyzer. The package of four analyzers would have a capital cost of about $180,000.

The maintenance cost of about $7,000/year for each analyzer has been subtracted from the incremental analyzer earnings to get analyzer profit (Table 2). This maintenance cost was estimated on the high side to be conservative. This cost assumes a refinery performing its own maintenance would have an expert "instrument man" who makes about $25,000/year and has fringe benefits that cost the company another 40 percent. We estimated that this person could service about seven instruments of this complexity on a full-time basis. This

Table 2
Economics of Four Boiling Point Analyzers on Crude Unit

Investment	
Final boiling point analyzer[1] and sample system	$ 23,000
Installation, prototype system, shelter	22,000
Total—per unit	$ 45,000
Total—package of four units	$180,000
Economics	
Incremental analyzer earnings, $/year	$562,000
Analyzer maintenance, $/year	27,000
Net analyzer earnings, $/year	535,000
Before-tax payout, years	0.34
Rate of return (discounted cash flow), %	>300

[1] Initial boiling point analyzer has same cost as final boiling point analyzer. Proposed package of four analyzers includes three final boiling point and one initial boiling point.

Table 3
Effects of Single Boiling Point Analyzer on Crude Unit Material Balance, Processing and Profitability

	Base case lab analyses	Proposed case with analyzer	Analyzer increment
Material balance, bpd			
Crude feed	49,767	49,767	—
Other feeds	2,957	2,957	—
Total feeds	52,724	52,724	—
Gasolines	26,074	26,074	—
Kerosene jet fuel	3,956	4,206	250
No. 2 fuel	10,079	9,829	(250)
No. 6 fuels	3,380	3,380	—
Other products	9,974	9,974	—
Refinery loss (gain)	(739)	(739)	—
Total products	52,724	52,724	—
Processing			
Fluid catalytic cracker			
Fresh feed, bpd	14,570	14,570	—
Conversion, %	71.4	71.4	—
Catalytic reformer			
Feed, bpd	11,097	11,097	—
Research octane, clear	96.7	96.7	—
Hydrocracker feed, bpd	2,187	2,187	—
Delayed coker feed, bpd	1,504	1,504	—
Pen/hex isomerization feed, bpd	2,122	2,122	—
Alkylation product, bpd	1,474	1,474	—
Profitability			
Over base, $/year	Base	227,000	227,000

manpower, plus minor maintenance materials cost, results in an average cost of about $7,000/year or about 15 percent of capital costs/year for upkeep of each analyzer, which is much greater than the typical 3 percent of capital costs/year for process units.

Table 4
Economics of Single Boiling Point Analyzer on Crude Unit

Investment	
Final boiling point analyzer and sample system	$ 23,000
Installation, prototype system, shelter	22,000
Total	$ 45,000
Economics	
Incremental analyzer earnings, $/year	$227,000
Analyzer maintenance, $/year	7,000
Net analyzer earnings, $/year	220,000
Before-tax payout, years	0.21
Rate of return (discounted cash flow), %	>600

Table 5
Effects of Boiling Point Analyzer on FCC Material Balance, Processing and Profitability

	Base case lab analyses	Proposed case with analyzer	Analyzer increment
Material balance, bpd			
Crude feed	49,767	49,767	—
Other feeds	2,957	2,957	—
Total feeds	52,724	52,724	—
Gasolines	26,074	26,138	64
Kerosene jet fuel	3,956	3,986	30
No. 2 fuel	10,079	9,888	(91)
No. 6 fuels	3,380	3,380	—
Other products	7,143	7,143	—
Propane	457	455	(2)
Plant fuel burned	2,374	2,373	(1)
Refinery loss (gain)	(739)	(739)	—
Total products	52,724	52,724	—
Processing			
Fluid catalytic cracker			
Fresh feed, bpd	14,570	14,570	—
Conversion, %	71.4	71.4	—
Catalytic reformer			
Feed, bpd	11,097	11,097	—
Research octane, clear	96.7	96.7	—
Hydrocracker feed, bpd	2,187	2,187	—
Delayed coker feed, bpd	1,504	1,504	—
Pen/hex isomerization feed, bpd	2,122	2,122	—
Alkylation product, bpd	1,474	1,474	—
Profitability			
Over base, $/year	Base	148,000	148,000

Extrapolation to Other Users

Tables 2, 4, and 6 present the economics for each of the three applications on a 60,000 bpd refinery. Results of the studies done on a 200,000 bpd refinery were much more impressive, as indicated in Figs. 1-3. These figures also are useful in estimating the profitability for users in other than a typical 60,000 or 200,000 bpd Gulf Coast refinery. The figures were all derived by:

- Using three differentials between pertinent major product prices
- Reducing profitability in our modeled case to the basis of ¢/barrel of feed rate

Table 6
Economics of Boiling Point Analyzer on FCC

Investment	
Final boiling point analyzer and sample system	$ 23,000
Installation, prototype system, shelter	22,000
Total	$ 45,000
Economics	
Incremental analyzer earnings, $/year	$148,000
Analyzer maintenance, $/year	7,000
Net analyzer earnings, $/year	141,000
Before-tax payout, years	0.32
Rate of return (discounted cash flow), %	>300

- Extrapolating that profitability to other refinery sizes—on the basis of unit feed rates.

These figures can be used by any refiner to estimate the potential profitability of installing boiling point analyzers in his plant.

POUR POINT ANALYZER

No. 2 fuel pour point specifications normally are 10 to 30°F lower in the four winter months than in the rest of the year. They are set lower so the fuel will not solidify at the colder winter temperatures. During the winter, many refineries are producing No. 2 fuel oil at its maximum pour point. Only a few refineries are producing No. 2 fuel at its maximum pour point during the remainder of the year. During the summer, No. 2 fuel normally is limited by the distillation 10 percent point maximum instead of the pour point maximum. Either of these specifications can set the minimum kerosene jet fuel in the No. 2 fuel blend. Hence, there is normally summer pour point "give away" that is unavoidable. However, in the winter, the lower pour point specifica-

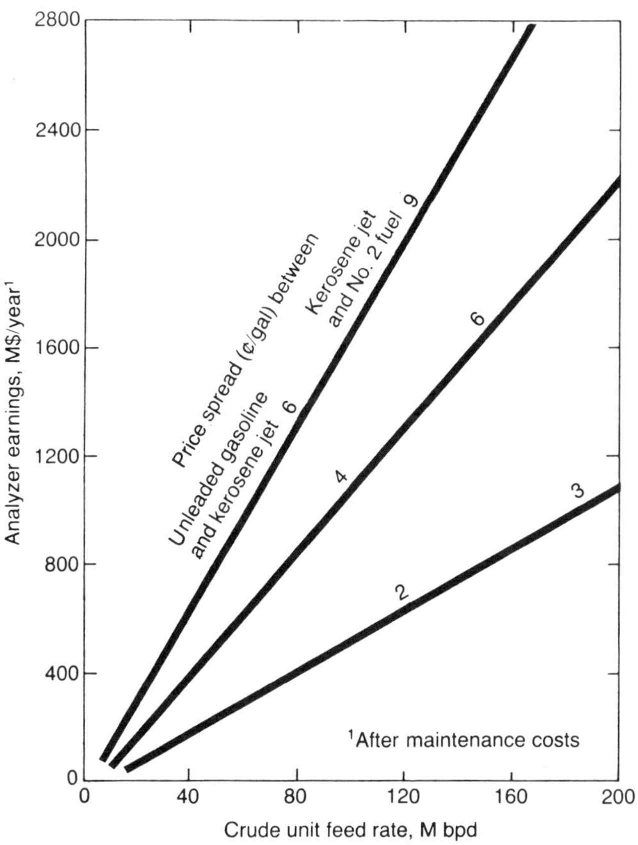

Figure 1. Estimated profitability four boiling point analyzers on crude unit.

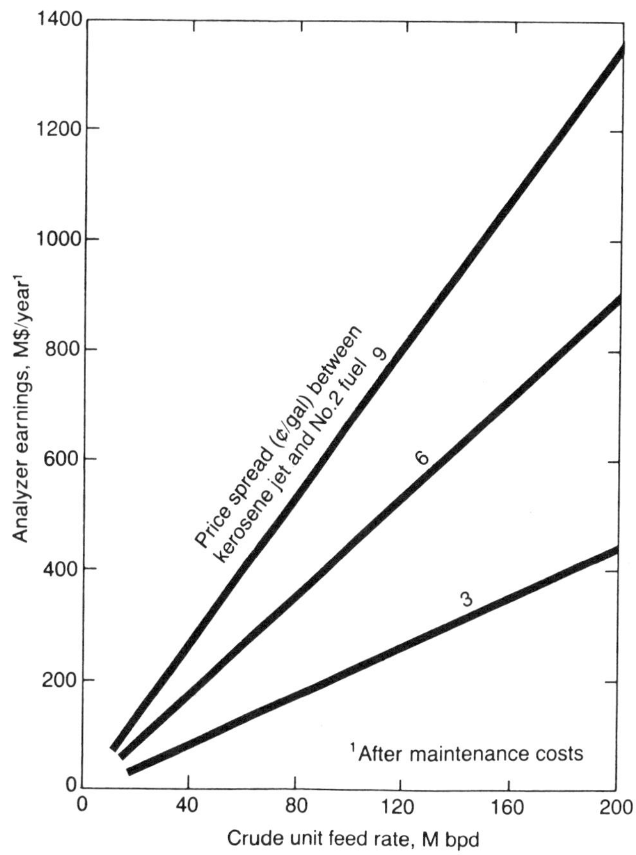

Figure 2. Estimated profitability single boiling point analyzer on crude unit.

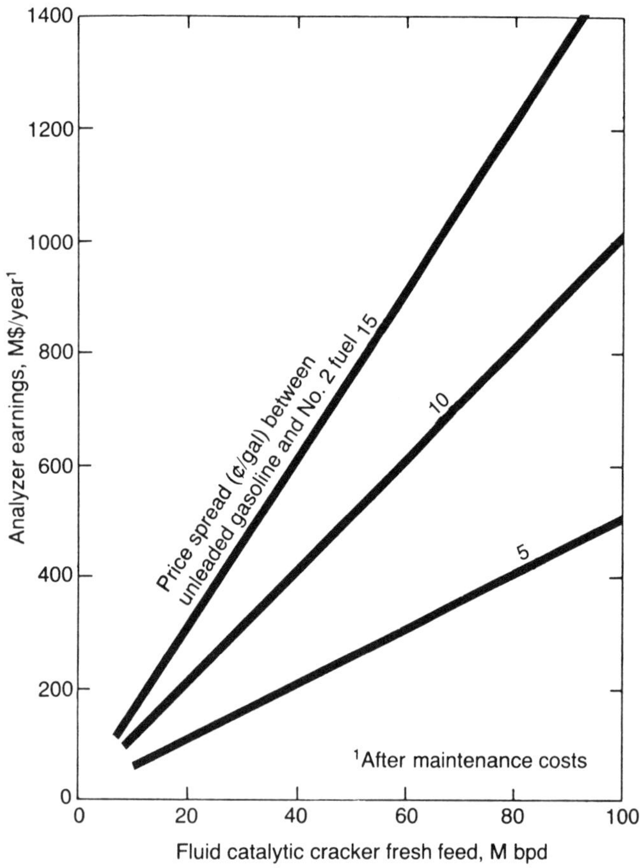

Figure 3. Estimated profitability single boiling point analyzer FCC unit.

tion normally sets the minimum kerosene jet fuel in the No. 2 blend. Hence, winter pour point "give away" is avoidable.

This analysis applies to the refinery whose cracking units are operating at capacity. In this situation the refiner cannot reduce No. 2 fuel pour point during the winter by the much cheaper option of undercutting salable No. 2 fuel base stock and routing it to cracker feed. Hence, his only means of reducing No. 2 fuel pour point during the winter is increasing kerosene jet fuel content.

Pour point of a No. 2 fuel blend can be determined using ASTM test method D-97. This laboratory method has an established statistical repeatability of ±5°F. This implies that a refiner must "target" his blending operations to a pour point value 5°F *lower* than he is obligated to sell to *always* meet his sales specification. This results in his "giving away" sales specification a majority of the time. Pour point "give away" costs the refiner because he has added a profitable blend component (kerosene jet fuel) into his No. 2 fuel at a higher-than-optimum rate.

Instead of relying upon set laboratory pour point analyses to set his No. 2 fuel blend composition in the winter, the refiner could rely upon a continuous pour point analyzer. These analyzers typically demonstrate a statistical repeatability of ±2°F. Consequently, the refiner who uses an analyzer can blend his No. 2 fuel closer to the winter sales specification and "give away" less product specification than his competitor who still relies upon spot laboratory analyses. The increased profitability is attributable primarily to lower usage of kerosene jet fuel in his No. 2 fuel blending recipe in the winter, and the difference between the value of the kerosene jet fuel and the value of the finished No. 2 fuel.

Analyzer Costs and Profitability

To estimate the potential profitability of installing a pour point analyzer, a base case was simulated with the LP, assuming laboratory analyses are used on individual No. 2 fuel blends. This resulted in a base profitability posture. Then, for comparison, we ran another case using the same raw materials as in the base, but allowing the pour point specification to be blended "tighter" to approximate the use of a continuous pour point analyzer.

Increasing No. 2 fuel pour point also affects its flash point, gravity, sulfur content and distillation points. Our LP optimized the No. 2 fuel blending to continue to meet these other quality specifications in the most profitable manner. Hence, it calculated a more accurate profitability for this pour point improvement than one would obtain by a simple hand calculation based on the kerosene jet fuel addition to the No. 2 fuel.

Table 7 describes the overall effects of these constraints on a 200,000 bpd refinery which illustrates the processing scheme of the 200,000 bpd refinery model vs. the 60,000 bpd model illustrated in the boiling point analyzer study. Table 8 shows the economics with maintenance costs again estimated at 15 percent of capital cost per year.

Extrapolation to Other Users

To demonstrate the profitability of the pour point analyzer for users other than the typical 200,000 bpd Gulf Coast refinery, we created Fig. 4. This figure was derived by:

- Using four differentials between kerosene jet fuel and No. 2 fuel values
- Reducing profitability in our modeled case to the basis of No. 2 fuel production rate
- Extrapolating that profitability to other refinery sizes—on the basis of No. 2 fuel production rate.

Figure 4 can be used by any refiner, whose cracking capacity is fully utilized, to estimate the potential profit-

Table 7
Pour Point Analyzer Effects on Material Balance Processing, No. 2 Fuel Quality and Profitability

	Base case lab analyses	Proposed case with analyzer	Analyzer increment
Material balance, bpd			
Crude feed	165,889	165,889	—
Other feeds	9,858	9,858	—
Total feeds	175,747	175,747	—
Gasolines	86,913	86,913	—
Kerosene jet fuel	13,187	17,987	4,800
No. 2 fuel	33,597	28,797	(4,800)
No. 6 fuels	11,268	11,268	—
Other products	33,246	33,246	—
Refinery loss (gain)	(2,464)	(2,464)	—
Total products	175,747	175,747	—
Processing			
Fluid catalytic cracker			
Fresh feed, bpd	48,567	48,567	—
Conversion, %	71.4	71.4	—
Catalytic reformer			
Feed, bpd	36,991	36,991	—
Research octane, clear	96.7	96.7	—
Hydrocracker feed, bpd	7,290	7,290	—
Delayed coker feed, bpd	5,015	5,015	—
Pen/hex isomerization feed, bpd	7,074	7,074	—
Alkylation product, bpd	4,913	4,913	—
No. 2 fuel quality			
Pour point, °F	−5	−2	3
Distillation, 10% pt, °F	407	418	11
Sulfur, wt%	0.16	0.21	.05
Gravity, °API	36.5	36.6	0.1
Flash pt., °F	150	156	6
Profitability			
Over base, $/year	Base	1,454,000[1]	1,454,000[1]

[1] Based on benefits during four winter months only. Pour point not limiting during rest of year.

Table 8
Economics of Pour Point Analyzer

Investment	
General pour pointer and sample system	$ 27,000
Installation, prototype system and shelter	23,000
Total	$ 50,000
Economics	
Incremental analyzer earnings, $/year	$1,454,000
Analyzer maintenance, $/year	8,000
Net analyzer earnings, $/year	1,446,000
Before-tax payout, years	0.03
Rate of return (discounted cash flow), %	>1,000

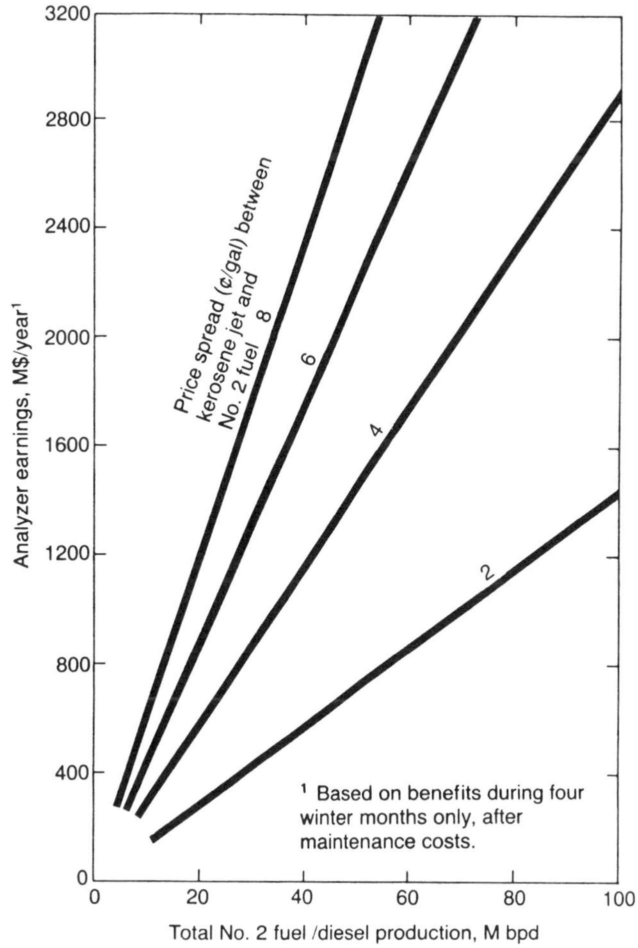

Figure 4. Estimated profitability pour point analyzer.

ability of installing a pour point analyzer in his plant for No. 2 fuel blending.

REID VAPOR PRESSURE ANALYZER

RVP of a gasoline blend can be determined using ASTM test method D-323. This laboratory method has an established statistical reproducibility of ±0.3 psi. This implies that a refiner must "target" his blending operations to an RVP value 0.3 psi *lower* than he is obligated to sell to *always* meet his sales specification, with testing performed by the customer's laboratory using similar equipment. This results in his "giving away" sales specification a majority of the time. RVP "give-away" costs the refiner, because he has not added a profitable blend component (butane) into his gasoline pool at the optimum level.

Alternately, the refiner could blend his gasoline product while relying upon a continuous RVP analyzer. These analyzers typically demonstrate a statistical reproducibility of ±0.1 psi. Consequently, the refiner who uses an analyzer can blend his gasoline closer to the sales specification and "give away" less product specification than his competitor who still relies upon spot laboratory analyses. The increased profitability primarily is attributable to the greater usage of butane in his blending recipe, and the difference between the cost of the butane and the value of the finished gasoline.

Analyzer Costs and Profitability

To estimate the potential profitability of installing an RVP analyzer, a base case was simulated with the LP, assuming laboratory analyses are used on the individual gasoline blends. This resulted in a base profitability posture. Then, for comparison, we ran another case using the same raw materials (except butane) as in the base, but allowing the RVP specification to be blended "tighter" to approximate the use of an on-stream RVP analyzer.

Increasing gasoline RVP also affects its octane and distillation qualities. Our LP optimized the gasoline blending to continue to meet these other quality specifications in the most profitable manner. As a result, it slightly changed reforming severity to meet the same maximum lead limit. Hence, it calculated a more accurate profitability for the RVP improvement than one would obtain by a simple hand calculation based on the butane addition to the gasolines.

Table 9 describes the overall effects of these constraints on a 60,000 bpd refinery. Table 10 shows the economics using the same maintenance criteria as the other examples.

Extrapolation to Other Users

To demonstrate the profitability of the RVP analyzer for users other than the typical 60,000 bpd Gulf Coast refinery, we created Fig. 5. This figure was derived by:

- Using 6 sets of differentials between butane cost and motor gasoline value.
- Reducing profitability in our modeled case to the basis of gasoline production rate.
- Extrapolating that profitability to other refinery sizes—on the basis of gasoline production.

Figure 5 can be used by any refiner to estimate the potential profitability of installing a Reid vapor pressure analyzer in his plant for gasoline blending.

Table 9
RVP Analyzer Effects on Material Balance Processing, Gasoline Quality and Profitability

	lab analyses	case RVP analyzer	Analyzer increment
Material balance, bpd			
Crude feed	49,767	49,767	—
Butane feed	859	957	98
Other feeds	2,098	2,098	—
Total feeds	52,724	52,822	98
Gasolines	26,074	26,173	99
Kerosene jet fuel	3,956	3,956	—
No. 2 fuel	10,079	10,079	—
No. 6 fuels	3,380	3,380	—
Other products	9,974	9,974	—
Refinery loss (gain)	(739)	(740)	(1)
Total products	52,724	52,822	98
Processing			
Fluid catalytic cracker			
Fresh feed, bpd	14,570	14,570	—
Conversion, %	71.4	71.4	—
Catalytic reformer			
Feed, bpd	11,097	11,088	(9)
Research octane, clear	96.7	96.7	—
Hydrocracker feed, bpd	2,187	2,187	—
Delayed coker feed, bpd	1,504	1,504	—
Pen/hex isomerization feed, bpd	2,122	2,122	—
Alkylation product, bpd	1,474	1,474	—
Gasoline qualities			
Reid vapor pressure, psi	10.9	11.1	0.2
Lead level, gms/gal			
Regular	1.34	1.35	0.01
Premium	2.52	2.51	(0.01)
Gasoline pool	0.80	0.80	—
Research octane			
Unleaded	93.6	93.6	—
Regular	93.4	93.4	—
Premium	98.1	98.1	—
Profitability			
Over base, $/year	Base	295,000	295,000

Table 10
Economics of RVP Analyzer

Investment	
Reid vapor pressure monitor and sample system	$ 30,000
Installation, prototype system, shelter	30,000
Total	$ 60,000
Economics	
Incremental analyzer earnings, $/year	$295,000
Analyzer maintenance, $/year	9,000
Net analyzer earnings, $/year	286,000
Before-tax payout, years	0.21
Rate of return (discounted cash flow), %	>600

VISCOSITY ANALYZER

The application of a viscosity analyzer on vacuum resid should reduce the "downgrading" of vacuum gas oil to vacuum resid by about 0.1 percent of crude run. This benefit was indicated by refiners we contacted who had added viscosity analyzers on vacuum resid. This vacuum gas oil "downgrading" occurs primarily because of frequent, minor variations in the vacuum tower operating conditions and crude composition. These variations cannot be followed as closely with routine lab analyses as they can with an on-stream analyzer. Hence, the analyzer will permit running closer to the "target" cut point and minimize vacuum gas oil "downgrading." In addition, using an analyzer avoids larger quantities of vacuum gas oil "downgrading" during major upsets and large crude composition swings. Low viscosity vac-

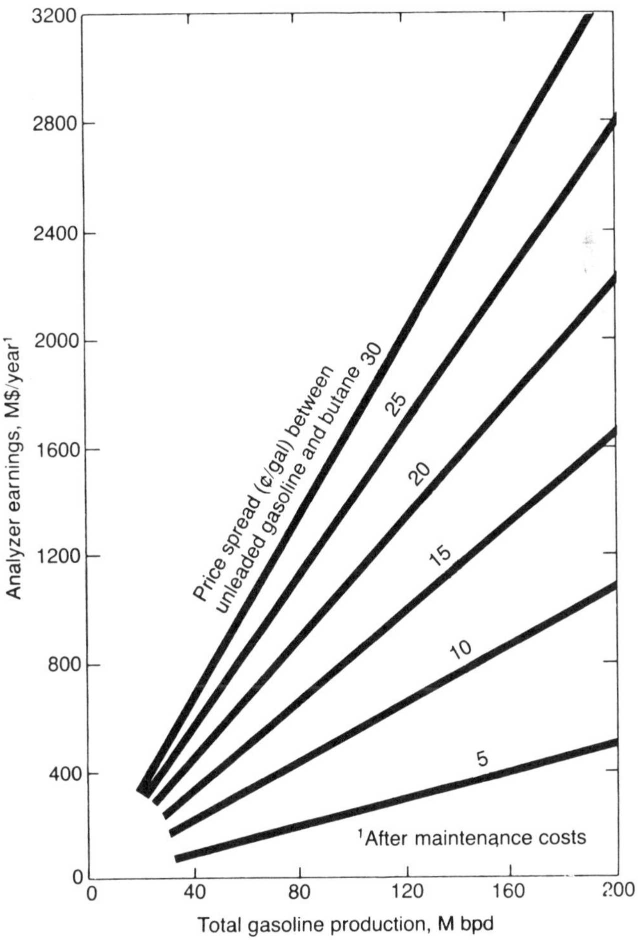

Figure 5. Estimated profitability RVP analyzer.

uum resid, which contains excessive vacuum gas oil, then can be detected quickly and diverted to recovered oil for profitable reprocessing.

Analyzer Costs and Profitability

To estimate the potential profitability of installing a viscosity analyzer, a base case was simulated with the LP, assuming laboratory analyses are used to monitor vacuum resid. This resulted in a base profitability posture. Then, for comparison, we ran another case using the same raw materials as in the base, but allowing the vacuum resid cut point to be controlled "tighter" to approximate the use of an on-stream viscosity analyzer.

Changing vacuum resid cut point also affects downstream processing and finished product rates and qualities. Our LP optimized processing and product blending to continue to meet all quality specifications in the most profitable manner. As a result, when a viscosity analyzer was used in the crude vacuum unit, the LP slightly changed fluid catalytic cracker feed composition and the feed rates to the coker, reformer and alkylation units. Hence, the LP calculated a more accurate profitability for the analyzer than one would obtain by a simple hand calculation. The increased profitability is attributable primarily to the higher production rate of No. 2 fuel, lower production rate of No. 6 fuel, and the difference in value of these products.

Table 11 describes the overall effects of these constraints on a 200,000 bpd refinery. Table 12 shows the

Table 11
Effects of Viscosity Analyzer on Crude Vacuum Unit Material Balance, Processing and Profitability

	Base case lab analyses	Proposed case with analyzer	Analyzer increment
Material balance, bpd			
Crude feed	165,889	165,889	—
Other feeds	9,858	9,858	—
Total feeds	175,747	175,747	—
Gasolines	86,913	86,962	49
Kerosene jet fuel	13,187	13,117	(70)
No. 2 fuel	33,597	33,750	153
No. 6 fuels	11,268	11,159	(109)
Other products	23,809	23,809	—
Propane	1,523	1,508	(15)
Plant fuel burned	7,914	7,913	(1)
Refinery loss (gain)	(2,464)	(2,471)	(7)
Total products	175,747	175,747	—
Processing			
Fluid catalytic cracker			
Fresh feed, bpd	48,567	48,567	—
Conversion, %	71.4	71.5	0.1
Catalytic reformer			
Feed, bpd	36,991	36,973	(18)
Research octane, clear	96.7	96.7	—
Hydrocracker feed, bpd	7,290	7,290	—
Delayed coker feed, bpd	5,015	4,945	(70)
Pen/hex isomerization feed, bpd	7,074	7,073	(1)
Alkylation product, bpd	4,913	4,927	14
Profitability			
Over base, $/year	Base	688,000	688,000

Table 12
Economics of Viscosity Analyzer on Crude Vacuum Unit

Investment	
Continuous viscosimeter	$ 20,000
Installation, prototype system, shelter	20,000
Total	$ 40,000
Economics	
Incremental analyzer earnings, $/year	$688,000
Analyzer maintenance, $/year	6,000
Net analyzer earnings, $/year	682,000
Before-tax payout, years	0.06
Rate of return (discounted cash flow), %	>1,000

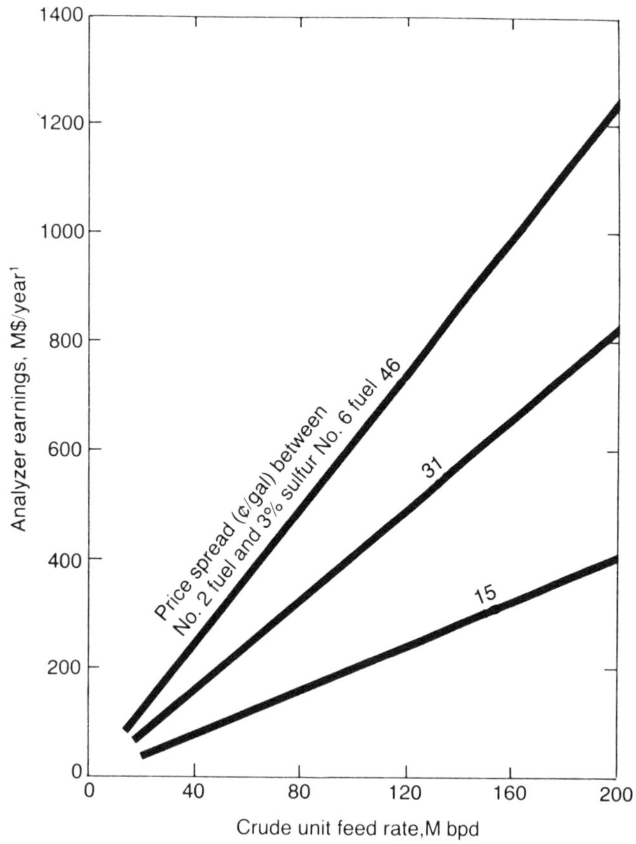

¹After maintenance costs

Figure 6. Estimated profitability viscosity analyzer on crude unit.

economics using the same maintenance criteria as the other examples. These economics assume only one large vacuum unit in the refinery. The payout would be longer, but still very attractive if the refinery has several vacuum units.

Extrapolation to Other Users

To demonstrate the profitability of the viscosity analyzer for users other than the typical 200,000 bpd Gulf Coast refinery, we created Fig. 6. This figure was derived by:

- Using three differentials between No. 2 fuel and No. 6 fuel.
- Reducing profitability in our modeled case to the basis of ¢/barrel of crude unit feed rate.
- Extrapolating that profitability to other refinery sizes—on the basis of crude unit feed rate.

Figure 6 can be used by any refiner to estimate the potential profitability of installing a viscosity analyzer on his crude vacuum units.

Reference

1. Grillo, Peter H., "Practical Boiling Point Analyzer Applications," *Instrument Society of America Transactions,* Volume 18, Number 1, pages 33–39, ISA Symposium, Houston, 1978.

15

How Boiling Point Monitors Can Increase Profits*

W. A. Bajek, P. J. Kuchar and A. A. B. Remec

INTRODUCTION

Boiling characteristics of product streams are regularly checked by the refinery laboratory. While the analysis of the sample may only take a few minutes, the results of the most recent sample may not be transmitted back to the operator for an hour or longer. This delay may be due to sample delivery or its priority in the laboratory, preparation of the sample, analysis itself and, finally, conveyance of the results to the column operator. Meanwhile, he operates on the "safe side" of the specifications, to guard against producing off-specification products, caused by some unforeseen disturbance. If the operator had an instantaneous, real-time report of the streams that would also be accurate, he could reduce the "safety range" to zero.

The boiling point monitor (BPM) is an analytical device that continuously monitors the boiling characteristics of a hydrocarbon stream, using gas-liquid chromatography (GLC), which simulates distillation. Use of a conventional chromatographic column achieves full component separation. Summing the areas under the peaks and recording the cumulative sum produces an analogy to a true boiling point curve. Typical laboratory techniques may take 20 to 45 minutes or longer to perform the same analysis. The BPM, using the GLC approach, compresses the analysis time and provides a separation that gives good correlation with the ASTM D 86[1] or ASTM D 2887[4].

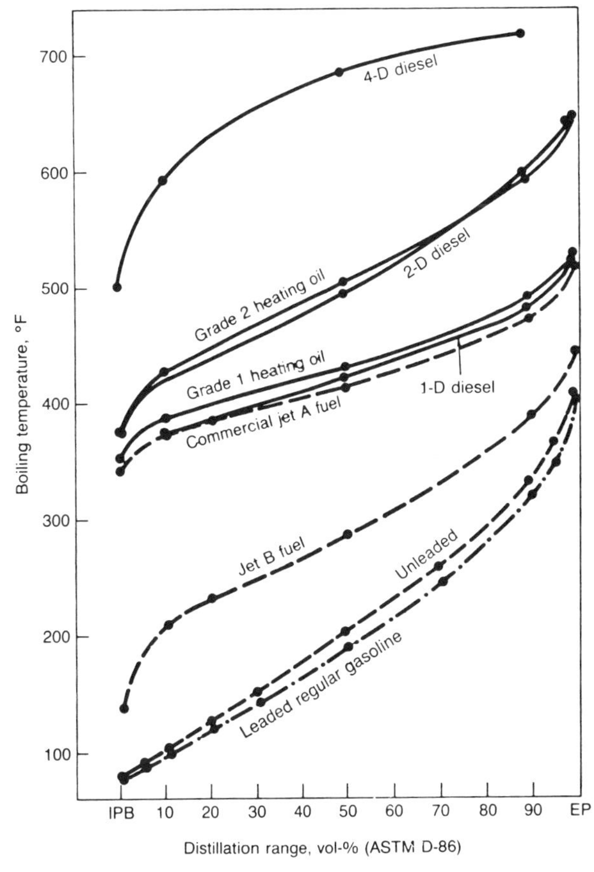

Figure 1. Boiling ranges of refinery products.

CRUDE COLUMN APPLICATIONS

Boiling ranges of refinery products are graphically presented in Fig. 1. These graphs show the boiling curve shapes, the temperature spreads between them and identify the products within a common boiling range. The graph is also useful in understanding the crude distillation control points to obtain these products or cuts. The following discussion characterizes the various fractions obtained from a typical crude unit and where BPMs can be profitable.

* Adapted from a paper copyrighted by the Instrument Society of America and presented at the Multinational Instrument Conference, Shanghai, China, April 1983.

Crude Column Overhead Product

The overhead naphtha product with a 360 to 390°F end point range is frequently the charge for fixed bed catalytic reformers. Even small amounts of contained heavier ends will be significant in causing premature catalyst deactivation as well as production of reformate with an end point exceeding specification.

Considering a single boiling point detector in this service, it will approximately meet the end point specifications by controlling near the fractionation end point. However, depending on the slope of the boiling curve, the actual end point will be either on the low or high side of the specification. The end point deviations are of serious concern to the catalytic reformer operator. The shape of the final portion of the distillation curve, not just the single point, can affect downstream operations.

Very accurate distillation quality control can be attained with a distillation analyzer which continuously analyzes the complete end range of the boiling curve. The control set point of the top TRC (Fig. 2) thus can be adjusted either by the reference point and the slope on the curve, or by the slope of several boiling points monitored in the end portion of the boiling curve.

Kerosine Cut

Currently, kerosine demand is mostly for jet fuel. It often commands a premium price over naphtha and diesel.

Typical jet fuel specifications[6,7] include as many as four distillation points. In the case of jet A fuel, ASTM specifications specify distillation temperatures of 10% recovered and end point, with requirement to report temperatures of the 50% and 90% recovered. The commercial jet B fuel specifies maximum distillation temperatures at 20%, 50% and 90% recovered. This fuel is also known as low freeze point jet fuel (Fig. 1). It is specified for use in extremely cold temperatures.

If jet fuel is to be maximized due to its premium value, at the expense of heavy ends from naphtha and lighter ends from diesel, the use of a full range distillation analyzer will assure reliable quality control over the full boiling range of the product. In selecting the maximum attainable boiling range for the kerosine, it is, of course, necessary to consider the other specifications, including flash point, freeze point, smoke point, etc., which will require different boiling ranges to be selected depending on crude source.

With real-time monitoring of the full range kerosine cut, the initial boiling range of the jet fuel is maximized by lowering the set points of the crude column top TRC, calling for lower overhead temperatures. This allows the heavy ends of naphtha to slump into the side cut kerosine stream. Also, the kerosine end point can be maximized by increasing the set point of the kerosine draw rate controller (FRC). Increasing the draw rate reduces the internal reflux below the kerosine draw and allows more of the lighter diesel components into the kerosine cut.

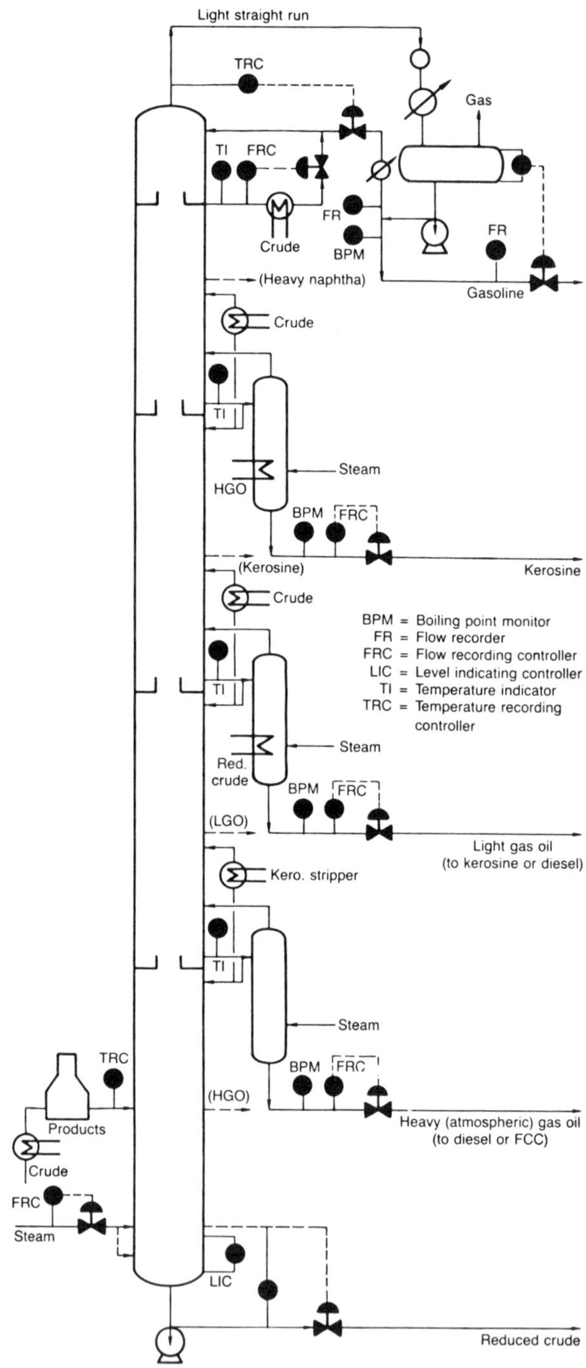

Figure 2. Crude fractionator.

Diesel Cut

The product stream from the next lower side cut location, adjacent to the kerosine cut, is the next heavier or higher boiling product, the light atmospheric gas oil (LGO). The diesel, No. 1 and No. 2 fuel oils fall into this same boiling range (Fig. 2).

Diesel fuel products have a variety of distillation or boiling range specifications.[6,7] For example, the ASTM specifications for No. 2 diesel include a minimum and a maximum temperature at the 90% point specification. Other specifications that define this product are a minimum flash point, a viscosity range, cloud point, carbon residue and cetane number. In addition, the diesel boiling range and other characteristics are subject to seasonal change and geographic area.

As with kerosine, the diesel cut must also be quality controlled, not for one but several specified boiling points during its fractionation. Here, again, a real-time full distillation range analyzer provides instant information, allows continuous control to maintain quality, eliminates time delays inherent with laboratory analysis and, thus, minimizes product loss.

If the light gas oil stream is to be maximized with kerosine components, the set point of the kerosine flow controller can be decreased manually by the operator, or automatically by the monitor. This allows the heavy ends from kerosine to slump into the diesel stream. Similarly, increasing the diesel draw rate (FRC) will lift the light ends from the adjacent (lower) heavy gas oil stream into the diesel cut.

If kerosine is concurrently maximized, at the expense of diesel, the only components that can maximize diesel cut must come from HGO. However, their amount in diesel may be limited, not by the boiling range, but cloud and pour points or other specifications. Therefore, only a carefully controlled amount of HGO could be used in such diesel maximization.

Atmospheric Heavy Gas Oil

Usually the lowest side cut location on the crude column is that of atmospheric heavy gas oil (HGO). Its boiling range includes heavy diesel and No. 4 fuel oil (Fig. 1). The value of this stream is lower than that of the LGO. The HGO stream is often later upgraded by fluid catalytic cracking (FCC) or by hydrocracking.

The initial boiling range of HGO depends on the diesel draw rate (FRC). Ideally, diesel components should be completely recovered. A boiling range analyzer installed in the HGO stream will detect diesel components and suggest their recovery.

The HGO end point is controlled by its draw rate (FRC), similar to the control of other side-cut products. Normally, the draw rate is adjusted for removal of approximately 650 to 700°F material from the reduced crude, the crude column bottom product. A real-time distillation analyzer in control of the HGO stream can help maximize HGO from bottoms and stabilize the HGO end point. A stable end point, in turn, allows FCC or other upgrading units to operate consistently and maximum end point materials, since end point peaks can thus be eliminated.

POTENTIAL SAVINGS

Improved fractionation and especially maximization of a high-value product are examples of easily discernible savings. These are due to improved process control. By using a real-time full boiling range analyzer, the useful components from the adjacent products are monitored continuously in the maximized stream to minimize losses to the less valuable products. Thus, the value of the useful components is upgraded and a better value obtained for them. The loss of these valuable components, the "give-away," in adjacent products is thereby minimized.

Harder to estimate are the earnings related to, or de- . rived from, improved fractionation control. With the aid of a real-time analyzer, the fractionation products can be brought within specifications much sooner after a startup, a major feed or process variable change, or after an upset in fractionation. The reduction of lineout time reduces the quantities of recycle and minimizes production loss.

Maximizing Jet Fuel Kerosine

The demand for this fuel may dictate a premium in a particular location, as it does in the U.S.A. For a period of several months, jet fuel kerosine commanded, on the average, six U.S. cents per gallon more in price than regular gasoline or diesel fuel.[8]

As previously discussed, maximization of kerosine cut requires control of not only the end point, but also the initial portion of the boiling range. For such control it is essential that a full boiling range analyzer be used. Inherent benefits of a monitor, which include the realtime display of multiple boiling points, far outweigh those obtained from installation of two or more single boiling point type analyzers, which can monitor only a single point of the front and end portions of the boiling range while offering no other information.

The percentages of recoverable naphtha or diesel components can be estimated from a temperature plot of certain boiling points from the ASTM D86 laboratory distillation reports of adjoining products. Examples of these plots are Figs. 3 through 6 depicting, on one hand, end ranges of naphtha against pertinent portions of kerosine and, on the other hand, the initial ranges of LGO (diesel) against the pertinent portions of kerosine for two different refineries. These data were obtained from refiners during several weeks of actual crude unit operations.

The percent "recoverable naphtha", Fig. 3, is only a portion of the "temperature overlap" in the area of naphtha end range and kerosine initial range. The "full overlap" amount is not recoverable due to fractionation inefficiencies and laboratory distillation test precision. When 10% of naphtha end range (from 90% to end point [EP] of naphtha) is within the initial kerosine range (from kerosine initial boiling point [IBP] to 10%) then the final 5% of naphtha (95% to EP) could be recoverable (Fig. 5), provided the crude column is capable of a better split.

The recoverable portion of diesel components, for kerosine maximization, is estimated similarly. In Figure 4, the recoverable portion was assumed at 2%, (from IBP of LGO to the 10% BP of kerosine) even though more of the LGO materials could have been assumed recoverable. (Note that the LGO temperature range, from IBP to 5% BP line, extends beyond the 10% kerosine specification line, at 400°F maximum.) In Fig. 6, only 1% of LGO was assumed recoverable, between IBP of LGO and 90% BP of kerosine, to allow for some

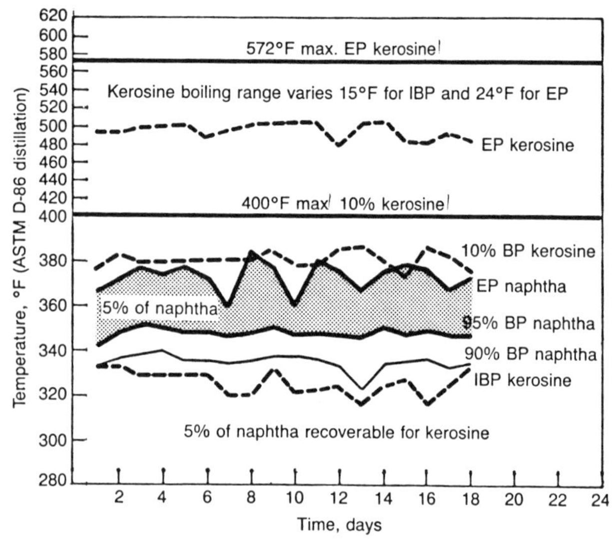

Figure 3. Recoverable naphtha (refinery A).

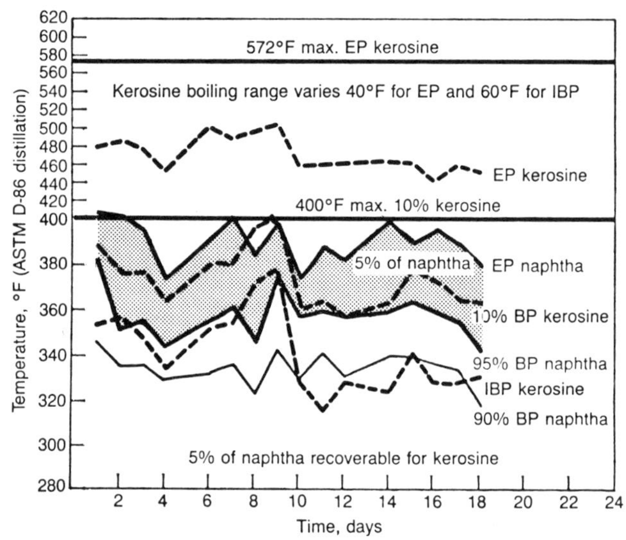

Figure 5. Recoverable naphtha (refinery B).

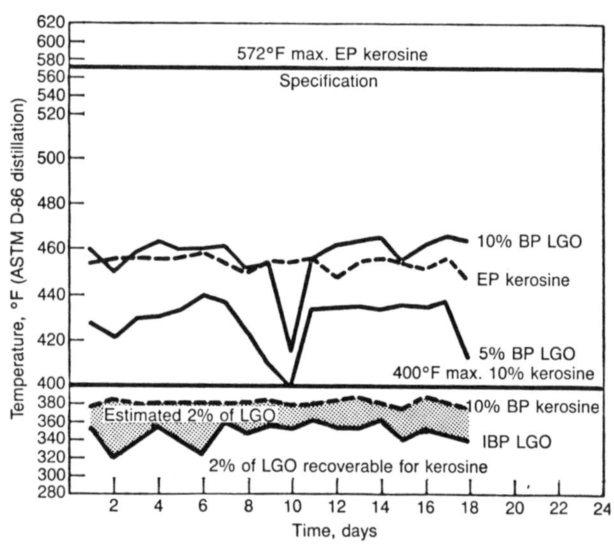

Figure 4. Recoverable LGO (refinery A).

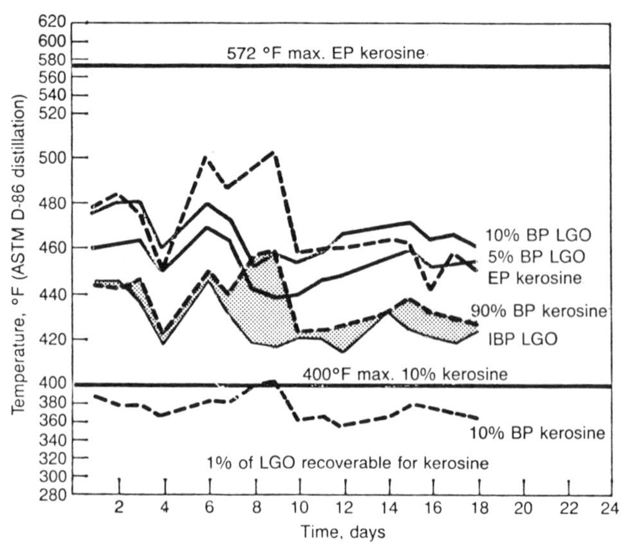

Figure 6. Recoverable LGO (refinery B).

temperature overlap between the products. Fig. 6 shows a better product split than Fig. 4, which represents a "poor" split.

Worthy of note are the fluctuations of kerosine boiling point temperatures in Fig. 5. The IBP fluctuates 60°F for IBP and 60°F for EP temperatures between lows and highs, respectively. These fluctuations could be reduced significantly with a real-time boiling range monitor.

Maximizing Diesel

If No. 2 diesel or No. 2 fuel oil is to be maximized, the same considerations apply as for kerosine maximization. A control system for front and end of boiling range is necessary. Use of a real-time full boiling range analyzer is most desirable, rather than the use of two or more single boiling point type analyzers because of inherent full range information capability from the analyzer, as discussed earlier. The savings, of course, depend on the narrowing of the "operator safety margin" and the amount of components saved from the adjacent lower value cuts.

Control of Side-Cut Naphtha

Usually, naphtha is stabilized and split in stabilizer and splitter columns, downstream of the crude column. If heavy naphtha is taken as a side cut from the crude column, the case for maximization becomes similar to that of kerosine. The initial portion of the heavy naphtha boiling range is closely controlled for the C_5 and C_6 split, and the end portion is closely controlled for maximum recovery of the aromatic precursors. Here again, the preferred approach is to monitor the boiling range instead of just one point.

Control of Naphtha as Top Product

An analyzer is justified by the differential savings derived from improved reformer operations, e.g., slower increases in catalyst temperature which compensate for carbon deposits formed by periodic occurrence of heavy ends in the charge. Thus, the tighter end point control helps extend reformer catalyst life.

Energy Saving Operation

Still other easily discernible earnings are related to the savings in heat energy due to reduction of reflux rates. Generally, reduction of reflux rates tends to increase the temperature "overlap range" of adjacent products. A certain amount of overlap is acceptable, of course. The use of real-time analyzer control allows cutbacks of reflux in addition and beyond the ones already instituted with laboratory control. These additional cuts may amount to as much as 5% of design rate.

Note, however, that reduction of reflux for energy conservation runs counter to the benefits derivable from improved, high resolution fractionation splits. A better split between product cuts usually requires increasing amounts of reflux or more fractionation trays.

FINANCIAL CONSIDERATIONS

Favorable factors for payout are as follows:

- Large fractionation capacity of the crude column.
- A favorable composition of crude oil containing more LGO and lighter components with fewer heavier ones, producing larger product rates of more desirable streams.
- High differential prices for one or more product streams.
- A wide overlap of boiling ranges between adjacent products, indicating a "poor split" or a low component resolution which would be improved by better fractionation, e. g., by tighter, onstream control of side cut and/or reflux rates.

These factors vary independently of each other. Each must be investigated before a final decision can be reached regarding justification of a project or its payout.

Several examples of differential savings for a kerosine maximization case are presented in Table 1. A similar table, of course, could be prepared for naphtha, diesel and HGO as well. In this table, the crude column capacity is evaluated with regard to type of crude, percent fraction saved from naphtha and diesel streams at various price differentials.

This differential savings can quickly be estimated graphically using Figure 7. A 2% kerosine recoverability from naphtha (naphtha rate at 8,100 bpd) is 162 bpd. The incremental value of kerosine over naphtha (at 6¢/gal) amounts to U.S. $143,000 over a one-year period. Similarly, a 1% kerosine recoverability from diesel (LGO rate at 2,700 bpd) is 27 bpd, for an incremental value of kerosine (over diesel at 2¢/gal) of 8,000/year, for a potential increased profitability of $151,000. This compares to $150,822 from Column 3 of Table 1, which is accurate enough for a quick estimate.

Differential streams for other product cuts can be estimated similarly.

Energy savings can be effected by safe reduction of column top reflux rates, recovering more heat in the

Table 1
Savings Possible From BPM Installation to Maximize Kerosine

Column number	1	2	3	4	5	6	7	8
Crude unit capacity, bpd	**20,000**	**20,000**	**30,000**	**30,000**	**40,000**	**40,000**	**60,000**	**60,000**
Approx. type crude	Heavy Arabian	Light Arabian	Medium Arabian	Brass River Nigeria	Maya Mexico	Bonny Light Nigeria	Bachequero Venezuela	Amna Libya
Naphtha, vol-%	25	30	27	40	20	35	10	24
bpd	5,000	6,000	8,100	12,000	8,000	14,000	6,000	14,400
Kerosine, vol-%	10	15	12	15	8	13	6	10
bpd	2,000	3,000	3,600	4,500	3,200	5,200	3,600	6,000
Diesel, vol-%	10	10	9	10	6	15	10	10
bpd	2,000	2,000	2,700	3,000	2,400	6,000	6,000	6,000
Gas oil, vol-%	10	15	10	10	8	10	10	9
bpd	2,000	2,000	3,000	3,000	3,200	4,000	6,000	5,400
Reduced crude, vol-%	45	30	42	25	58	27	64	47
bpd	9,000	6,000	12,600	7,500	23,200	10,800	38,400	28,200
Differential value of:								
Kerosine to naphtha, ¢/gal	6	6	6	4	6	4	4	4
Kerosine to diesel, ¢/gal	2	2	2	1	2	1	1	1
Kerosine increment:								
From naphtha, vol-%	1.5	2.0	2.0	1.5	1.5	1.5	1.5	1.5
bpd	75	120	162	180	120	210	90	216
$/D	189	302.40	408.24	302.40	302.40	352.80	151.2	362.88
From diesel, vol-%	1.0	1.0	1.0	1.0	1.0	1.0	1.0	1.0
bpd	20	20	27	30	24	60	60	60
$/D	16.80	16.80	22.68	12.60	10.08	25.20	25.20	25.20
Incremental earnings, $/D	205.80	319.20	430.92	315.00	312.48	378.00	176.4	388.08
$/Year*	72,030	111,720	150,822	110,250	109,368	132,300	61,740	135,828

* 350 stream days per year.

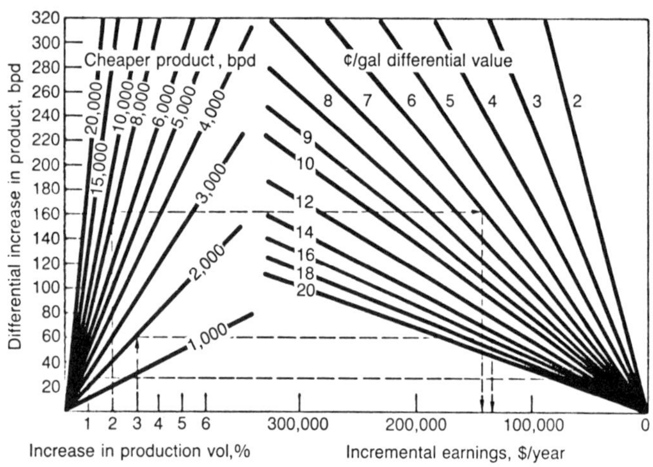

Figure 7. Estimated savings from improved operations.

pumparound intermediate reflux streams, lower down the column.

As an example, consider a 50,000 bpd crude unit. An additional 5% reflux reduction may amount to 1,500 bpd of naphtha reflux. For a 58.5°API gravity reflux material, and using 135 Btu/lb vaporization heat for 275°F average boiling range reflux, $5.50 per million

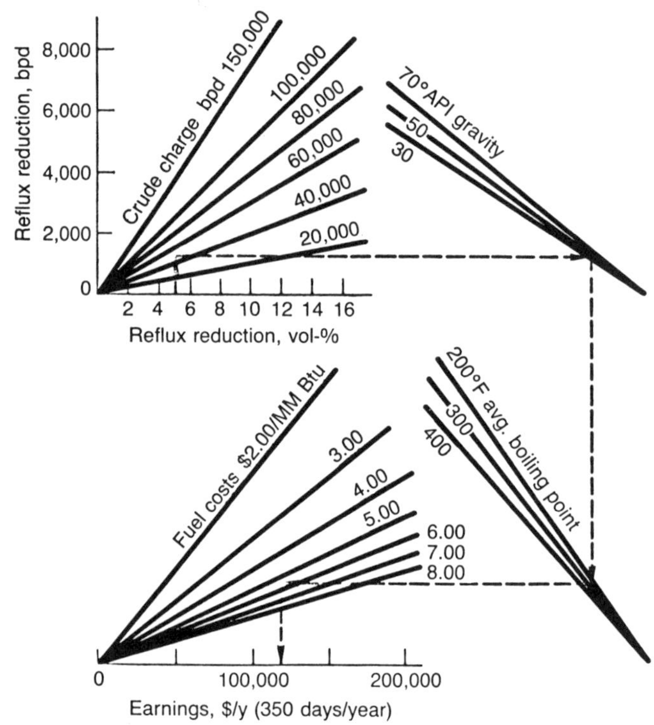

Figure 8. Estimated savings from reduced heat energy costs.

Btu fuel, and 85% greater efficiency, the heat savings due to this reflux reduction amount to $115,000 per year. The potential savings may be read conveniently off Fig. 8.

References

1. ASTM Method D 86-78, "Standard Method of Test for Distillation of Petroleum Product," 1981 Annual Book of ASTM Standards, Part 23, p. 8, American Society for Testing and Materials, Philadelphia, Pa., 19103 (1981).
2. McLaughlin, J. H., Bajek, W. A., and Sampson, R. W., "Continuous Onstream Analysis of Boiling Characteristics of Petroleum Fractions," ASTM Committee D-2, Dallas, TX, Dec. 6, 1973.
3. Sampson, R. W., Bajek, W. A., and McLaughlin, J. H., "Onstream Monitoring for Hydrocarbon Boiling Characteristics Employing a Microcomputer," Instrument Society of America, Analysis Instrumentation Division Conference, Charleston, W. Va., May 17-19, 1977.
4. ASTM Method D 2877-73, "Test for Boiling Range Distribution of Petroleum Fractions by Gas Chromatography," 1981 Annual Book of ASTM Standards, Part 24, p. 789, American Society for Testing and Materials, Philadelphia, Pa. 19103 (1981).
5. Bajek, W. A., "Energy Conservation Through the Use of Continuous On-Line Process Stream Analyzers," Instrument Society of America, New Orleans, La., Feb. 3-7, 1980.
6. Petroleum Products Surveys, National Technical Information Service, U.S. Dept. of Commerce, 5285 Port Royal Rd., Springfield, Va. 22161.
7. Kirk and Othmer, Encyclopedia of Chemical Technology, Vol. 17, Third Edition, John Wiley & Sons, Inc., pp. 257-271.
8. "Platt's Oilgram Price report" for the period from May 15, 1981 to May 28, 1982.

16

Interfacing Analyzers to Digital Systems

M. G. Thor

Several factors have focused attention on interfacing analyzers to digital systems. Microprocessor-based distributed control systems are now widely accepted. Almost every new or revamped control room employs a DCS, and the decision to use one is made almost automatically. The hows and whys of connecting process analyzers to these digital systems are the topics of this chapter.

Revamps of the controls in existing plants are proliferating, due to the need to make existing plants run more efficiently. Also, the high cost of borrowing money has made it more advantageous to revamp existing plants rather than build new ones. Interfacing older existing analyzers to new digital systems is quite a challenge since, in many cases, these analyzers were designed at the dawn of the application of process control computers.

The application of advanced control strategies is also on the rise. Often these schemes include composition control, which is another reason for the increased emphasis on the topic of this article. These advanced control strategies may require some of the newer types of analyzers such as mass spectrometers, and nuclear magnetic resonance (NMR) spectrometers, as well as conventional analyzers including gas chromatographs, zirconium oxide analyzers and infrared analyzers.

The term digital systems is meant to include minicomputers, distributed control systems, and integrated plant-wide systems comprising a combination of distributed control systems and area and plant-wide computers. For the purpose of interfacing, analyzers can be divided into two categories; continuous and discontinuous. An analyzer that provides a single continuous signal, such as an O_2 analyzer, is considered to be continuous. A gas chromatograph, which analyzes one or more streams, each with one or more components, is considered to be discontinuous. As shown in this chapter, the latter type of analyzer presents the greatest challenge to the systems engineer.

CONTINUOUS ANALYZERS

The output of a continuous analyzer is usually an analog signal, typically 4 to 20 mA or 0 to 1 volt DC, which represents a measurement and is often proportional to that measurement.

A typical hookup of a continuous analyzer with a current output in a plant with existing electronic analog panel board instrumentation is shown in Fig. 1. Resistors R1 and R2 are located in a terminal board cabinet, where they will not be disturbed during maintenance of the equipment. In this way, the digital system analog input card can be pulled from its slot for servicing and the measurement circuit will not open up. Similarly, the panel board receiver instrument, such as a controller or indicator, can be disconnected without affecting the digital system reading. Also, a short across R1 or R2 will not affect the other reading. It is imperative that this circuit be grounded in only one place, i.e., at a connection

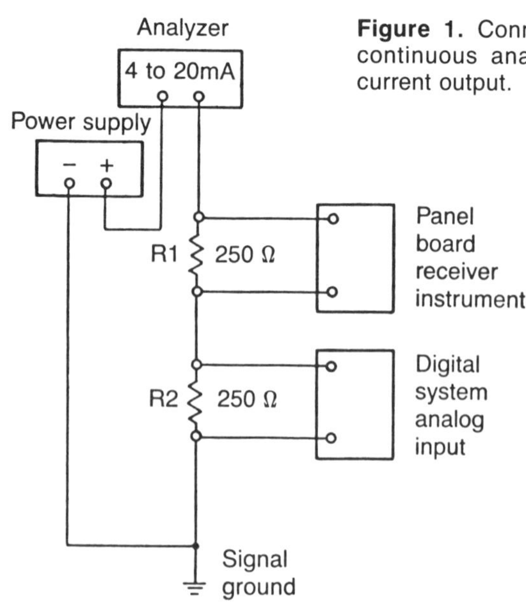

Figure 1. Connection for a continuous analyzer with a current output.

to signal ground. Signal ground should be a separate ground connection running into the plant grid or several feet into the ground. The wire connecting the signal to earth ground at the equipment should be large enough to present a low impedance. (In applications where intrinsic safety is required, barriers should be employed between the field transmitter and the control room/computer room equipment.)

GAS CHROMATOGRAPHS

Before we can discuss how to connect a gas chromatograph (GC) to a digital system, we must first briefly discuss its operation. A GC measures the amounts of various components in a process stream. For example, a GC may be used to measure the amount of methane (CH_4), ethylene (C_2H_4) and propylene (C_3H_6.) These components have molecular weights of 16, 36, and 42, respectively. A sample of the stream is injected into the GC, which contains a column used to separate the components. Since methane is the lightest, it will reach the end of the column first. The output from the GC signal detector may look like Fig. 2. The time after sample injection at which each peak occurs remains constant. The concentration of each component can be determined from either the height of the peak or the area under the curve for each component.

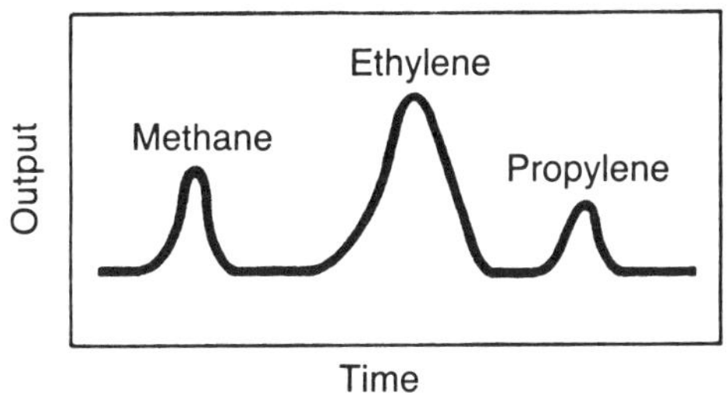

Figure 2. GC detector output.

In older GC's, a mechanical programmer is used to control the timing of the analyzer. The programmer starts an analysis by injecting a sample and triggers the peak picker or curve integrator at the proper moment. If more than one stream is being analyzed, the GC programmer also controls the switching of streams to be sampled. The signals that must be connected to a computer are shown in Fig. 3. (This is an example of a parallel interface, to be discussed later in this chapter.)

After an analysis has begun and the concentration of the first or lightest component has been determined, the programmer sends a digital pulse signal to the computer indicating that analysis results are ready. This signal is generally connected to an interrupt in the computer. Since a GC analysis of a stream takes on the order of one

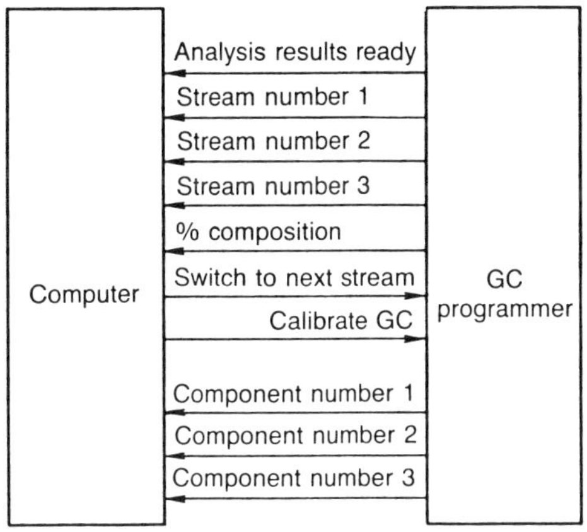

Figure 3. GC digital interface.

to ten minutes to complete, there are generally at least several seconds between the analysis of each component. Once the computer recognizes the interrupt, it reads the stream number, component number and the percent composition of that component. In our example, if this stream is assigned as stream number 1, and methane, ethylene and propylene are assigned as components 1, 2 and 3, then the computer would read stream 1, component 1 and know that the percent concentration being read is the concentration of methane. When the next component has been analyzed, an interrupt occurs. The programmer will have incremented the component number to 2, and the computer will read the concentration of ethylene, and so on.

The stream number is normally sent as digital contact signals. Usually there are at most three streams connected to a GC. Therefore, there is usually one contact dedicated to each stream, with only one contact active at a given time. A similar method is used to indicate the component number. The percent composition is sent as a 4 to 20 mA analog signal.

In some cases, the GC programmer is capable of accepting external commands to switch to the next stream, or to inject a known gas sample for calibration purposes.

Newer GC's use microprocessor-based programmers. This gives more accurate results and allows for a two-wire serial digital interface, such as RS-232C. (Digital interfaces are described later in this chapter.)

Another method is to provide a dedicated sample-and-hold circuit for each component in each stream. As a component is analyzed, a sample-and-hold circuit stores the value of that component. The output of each circuit is a 4 to 20-mA signal that is connected to an analog input in the computer. Thus, the discontinuous analyzer can be interfaced as if each analysis were coming from a continuous analyzer (Fig. 4). This approach entails the use of more analog input and sample-and-hold circuits than does the digital interface method. While

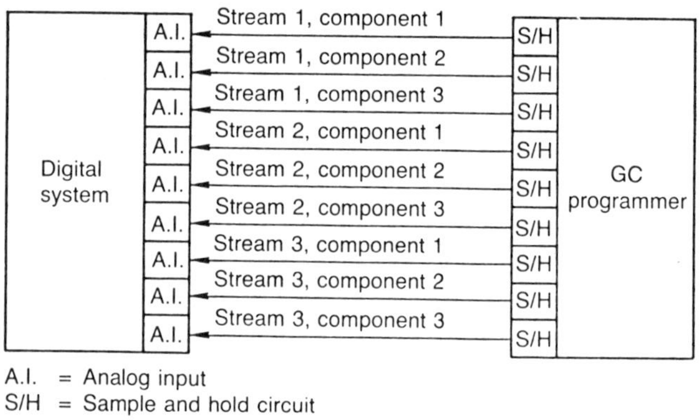

Figure 4. GC analog interface.

the hardware cost is higher, the software cost is much less since no special software handler is required in the computer for the data link to the GC. Most microprocessor-based instrumentation (i.e., distributed control) systems do not provide the software to accommodate digital communications to a GC. Therefore, GC's are being interfaced with these systems predominantly by the analog signal method.

DIGITAL INTERFACES

Most newer analyzers provide digital outputs. The fact that a device has a digital output does not necessarily make it easier to interface to a computer than an analog output device. Often special programs must be written for the computer and these programs can be complicated and expensive. If these development costs can be amortized over several applications, a digital interface often will be less expensive.

Digital interfaces can be divided into two categories: parallel and serial.

Figure 5 shows a typical parallel interface for 16 bits of information. The data to be transferred to the computer are held in a register. The 16 bits of data each have dedicated signal paths. During the period when the register is being updated, the data ready signal will be false. When the register contains valid data, the data-ready signal will be true. The computer should be programmed to read the data whenever the data-ready signal makes a transition from false to true. This will allow the computer to retrieve information as soon as it is available. Also, the computer should disregard any information read when the data-ready signal is false.

Sometimes a parity check bit is included. A parity check is the simplest but least secure method of error checking. This involves adding an extra bit, called a parity bit, to the message. This bit is set or reset so that the total number of binary ones in the message is always odd. (This is referred to as "odd parity." Similarly some systems use "even parity," where the number of binary ones in a message is always even.)

A serial interface is depicted in Fig. 6. Here only one communications path is used, as opposed to the parallel interface, where a separate path is needed for each bit. In the serial interface, the message is sent a bit at a time, with the bits following in order. The parallel interface is, of course, much faster, as an entire block of data (17 bits in this example) can be sent in the same time as one bit can be sent over a serial interface. But the parallel interface requires a separate communications channel for each bit and therefore is more expensive. For communications over long distances, the serial interface is preferred, because communications become expensive. Parallel interfaces are used for short distances where high speed is essential. Serial interfaces are gaining in popularity even for short distances because data rates of 1 million bits per second can be achieved today over a communications channel. This is more than adequate to meet the interfacing requirements of any transducer. Also, using a serial interface allows the use of the same software for interfacing with a digital transducer, re-

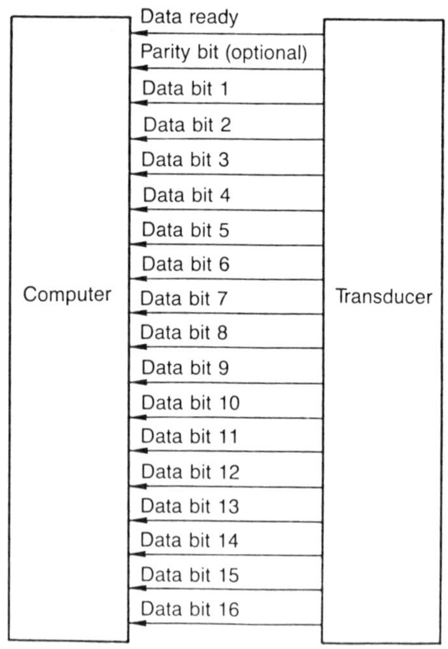

Figure 5. Parallel digital interface.

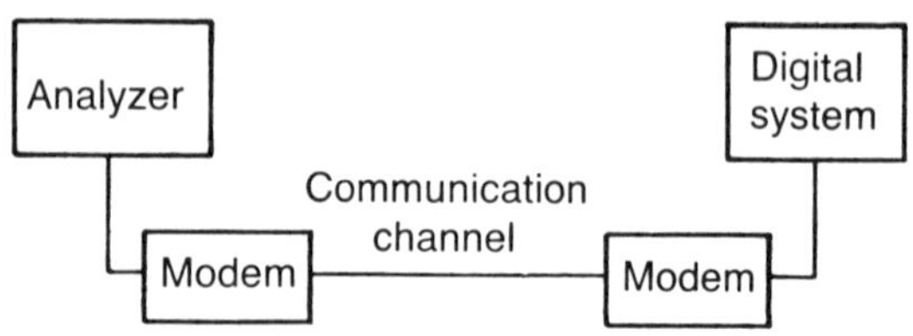

Figure 6. Serial digital interface.

gardless of how far from the computer it may be. For long distances, greater than a few hundred feet, digital data are normally transmitted over the communications channel as modulated signals. At the receiving end, the signals are demodulated back into digital signals. The transformation of digital logic to modulated signals and the transformation from modulated signals to digital logic is performed by a modulator-demodulator, commonly referred to as a modem or data set. A modem is not required for short distances.

More sophisticated methods of error detection are used when a communications channel is employed. This is because a communications channel, such as a telephone line, microwave link, etc., is highly susceptible to errors due to noise on the signal.

Most discontinuous analyzers designed today provide a digital output compatible with the RS-232C standard. This standard defines the signals used in the interface, the voltage levels used for these signals, the cable connectors to be used, and the location of the signals on the pins in the connectors. It is a popular standard for interfacing terminals to a computer.

CONTACTS

A contact can be used to represent the on/off status of a device. Also, a group of contacts can be used to express numerical information in binary arithmetic form. A contact should be isolated from ground and all other signals, except those from the computer. This is sometimes referred to as a "dry contact." A typical connection for a contact computer input is shown in Fig. 7. The contact input, or digital input, is connected to a multiplexer within the digital system. Note that the computer sends out a signal (a ground in this example) and senses if the signal is returned. If the multiplexer were to be disconnected from the contact, the contact would be completely isolated. One should also make certain that the contact is rated for the voltage and current of the computer sensing signal although, in general, these voltages and currents are not very high.

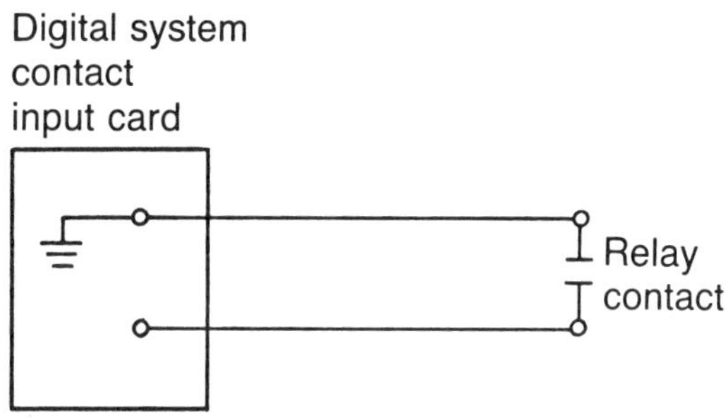

Figure 7. Connection for a contact input.

Wiring Practices

Signal cables must be run in trays that are separate from trays containing power wiring. Where signal cables must cross power wiring, it should be done at right angles, with as much separation as is practical. Twisted pair cable with an overall shield should be used for long signal runs. Shields should be grounded at one end only.

17

Filtering Samples to On-Line Analyzers

R. Strauss

Increasingly precise process control strategies, often linked to computer control, have accelerated the use of sophisticated on-line chemical composition analyzers in plant applications. Instruments such as gas and liquid chromatographs, ion chromatographs, laser optic instruments, atomic absorption instruments and specific ion analyzers, which were rarities in laboratories ten years ago, are now found routinely in plant settings. While continuing miniaturization and "ruggedization" of the electronics are making the instrument circuitry more tolerant of the plant environment, a problem that will not diminish is the level of contamination in plant samples compared to laboratory samples.

Factors in plant operation that magnify the difficulty of delivering acceptable samples to on-line process analyzers are:

- The sample stream must be processed and fed to the analyzer continuously, rather than the batchwise method permissible in the laboratory.
- Test frequency in the plant is far greater than in the laboratory. Where a laboratory analyzer might perform 100 tests a month, an on-line plant analyzer could do 100 tests a day. In the high frequency plant usage, trace contaminants that would not be a problem in the laboratory can rapidly build up and cause instrument failure.

Filter Requirements

It is not surprising that contaminants in the samples are reported to be the most frequent cause of problems with on-line process analyzers. While the requirement for an effective filter in the sampling line is generally recognized, it is also important to recognize that it is usually necessary to use a filter specifically designed for sampling applications, rather than trying to make do with a general purpose or homemade filter. The characteristics that a sample filter should have, in addition to filtering out contaminants, are:

- The filter must not change the composition of the sample, other than to remove unwanted impurities. Therefore, the choice of filter media generally is limited to a few chemically inert materials: glass, stainless steel and Teflon.
- Since the sample filter often is in a remote or inconvenient location, it must be capable of operating for a reasonable period between scheduled maintenance checks. Even more important, it should not be susceptible to unscheduled problems, such as filter element plugging or rupturing, between regular maintenance checks.
- Sample filter maintenance in the field usually is performed under adverse conditions by personnel who are not trained chemists; therefore, the filter should be designed for easy and uncomplicated maintenance. Filter elements should be rugged and not susceptible to handling damage; the unassembled housing should have a minimum number of loose parts, and the housing should be designed so that it is virtually impossible to install a filter element incorrectly.
- A filter should introduce minimum lag time into the system. Lag time can be dealt with in the sample system design (slipstream sampling, for example), but a sizable dead volume in the filter housing should be avoided. Since large reservoir volume is desirable in many filter applications—such as compressed air or water filters—filters not specifically designed for sampling usually are not suited for analyzers.

Requirements for sample filters range so widely that specifying a filter is best done on a case-by-case basis. There is, however, one generalization that applies to all sample filter requirements: the filter must be able to separate efficiently a noncontinuous phase contaminant from the continuous sample stream phase. Specifically, the filter must be able to make the following separations, in addition to removing solid particles:

- Gas samples—remove liquid droplets
- Liquid samples—remove immiscible liquid droplets and gas bubbles.

Most filter media will do an adequate job of removing solid particles from liquids or gases, but the only practical commercial media that will separate liquids from gases, gas bubbles from liquids, and two immiscible liquids is resin-bonded glass microfiber media. All recommendations in this article are based on resin-bonded glass fiber media.

Separating Liquids from Gases

Glass microfiber filter tubes efficiently separate suspended liquids from gases. The filter fibers capture the fine liquid droplets suspended in the gas and cause the droplets to run together to form large drops within the depth of the filter tube. The large drops are then forced by the gas flow to the downstream surface of the filter tube, from which the liquid drains by gravity. This process is called "coalescing." Since the coalesced liquid drains from the tubes at the same rate that liquid droplets enter the tubes, the tubes have an unlimited life when coalescing liquids from relatively clean gases, and the filters operate at their initial retention efficiency even when wet with liquid. The flow direction is inside-to-outside to permit the liquid to drip from the outside of the filter to the housing drain.

The filter tube grade should be selected for maximum liquid drainage rate, rather than maximum filtration efficiency rating. Since liquid drainage rate *decreases* with increasing filtration efficiency rating, this is a clear case where overspecifying filter efficiency will lead to unsatisfactory results.

If liquid is carried into the filter in slugs rather than dispersed as droplets in the gas, a filter that is properly sized for steady-state conditions can be flooded and permit liquid carryover. If slugging of liquid is expected, a filter with a relatively large bowl should be selected to provide adequate liquid holding capacity, and positive provisions should be made to drain the liquid automatically from the bowl of the housing as fast as it accumulates. An automatic float drain can be used if the pressure is in the 10 to 400 psig range. Above 400 psig, the upper limit for commercially-available float drains, the possibilities are: a constant bleed drain, a valve with an automatic timed actuator, or an external reservoir with manual valves (Fig. 1). The external reservoir can be constructed of pipe or tubing with sufficient volume to hold all the liquid that is expected to be collected during any period of unattended operation. To drain liquid while the filter is operating at pressure or vacuum conditions, valve #1 is closed and valve #2 is opened.

If the filter is under vacuum, the external reservoir is a practical method of collecting coalesced liquid for manual draining periodically. Alternatively, if any external vacuum source is available, such as an aspirator, the liquid may be drained continuously from the housing drain port.

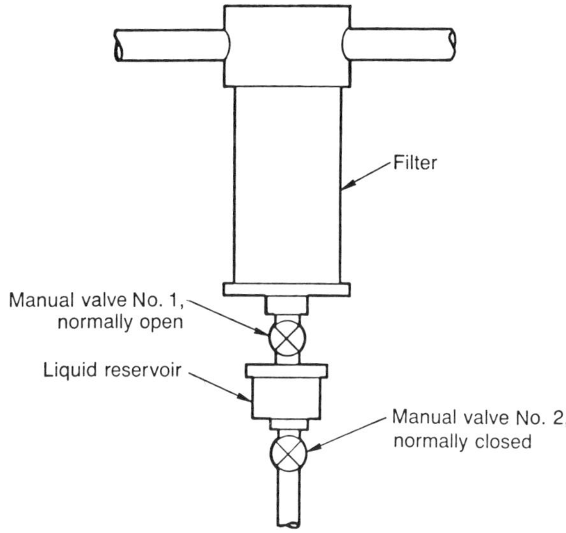

Figure 1. External reservoir for high pressure applications.

Separating Two Liquid Phases

In principle, glass microfiber filter tubes separate suspended droplets of a liquid which is immiscible in another liquid by the same process as they separate droplets of liquid from a gas. The liquid droplets suspended in the continuous liquid phase are trapped on the fibers and run together to form large drops, which are then forced through the filter to the downstream surface. The large drops separate from the continuous liquid phase by gravity difference, settling if heavier than the continuous phase and rising if lighter. The coalescing action of glass microfiber filters is effective with aqueous droplets suspended in oil or other hydrocarbons, and also with oil in water suspensions.

In practice, however, liquid-liquid separations are much more difficult than liquid-gas separations. The specific gravity difference between two liquids is always less than between a liquid and a gas, and therefore, a longer phase separation time is needed. Either the filter housing must be oversized or the flow rate greatly reduced to avoid carryover of the coalesced phase. As a rule of thumb, flow rate for liquid-liquid separation should be no more than one-fifth the flow rate for solid-liquid separation. Even at low flow rates, if the specific gravity difference between the two liquids is less than 0.1 unit (for example, if an oil suspended in water has a specific gravity between 0.9 and 1.1), the separation time for the coalesced phase may be impractically long. In that case, if there is only a small quantity of suspended liquid, the filter tube can be used until saturated with the suspended liquid and then changed.

Another practical problem with liquid-liquid separations is that small quantities of impurities can act as surface-active agents and interfere with the coalescing action. For that reason it is not possible to predict accurately the performance of a liquid-liquid coalescing filter, and each system must be tested on site. The general guidelines for the system to start testing are to use 25-micron filter tubes and flow inside-out at very low flow rates. If the suspended liquid is lighter than the continuous phase the housing should be oriented so that the drain port is up.

Removing Gas Bubbles from Liquids

Glass microfiber filter tubes readily remove suspended gas bubbles from liquid, eliminating the need for deaeration tanks, baffles or other separation devices. Flow direction through the filter is outside-to-inside, and the separated gas bubbles rise to the top of the housing and are vented through the drain port. If slipstream sampling is used (see below) the separated bubbles are swept out of the housing with the bypassed liquid. Filter tubes rated at 25 microns are a good choice for gas bubble separation.

Slipstream or Bypass Sampling

Instrument sample use rates are invariably quite low, yet it is essential to minimize lag time in the sample system. Since analyzers often are located some distance from the sampling point, samples usually are transported to the analyzer at a relatively high flow rate to minimize lag time. The sample is divided at the analyzer, with the analyzer using the portion it requires (usually a very small fraction of the total sample), and the balance recycled to the process or vented.

If the sample filter is located in the low-flow line to the analyzer, it will have good life between filter element changes because the solids loading rate is very low; however, the filter must be carefully selected to avoid introducing unacceptable lag time. If the filter is located in the high-flow portion of the sample system, its effect on sample lag time can be relatively low, but the life between filter changes may be inconveniently short because the element is filtering a much greater volume of material than the analyzer is using.

Ideally, a filter should be located at the point where the low-flow stream is withdrawn to the analyzer (Fig. 2). This arrangement permits the main volume of the filter to be swept continuously by the high flow rate system, thus minimizing lag time; at the same time, only the low-flow stream to the analyzer is filtered, thus maximizing filter life.

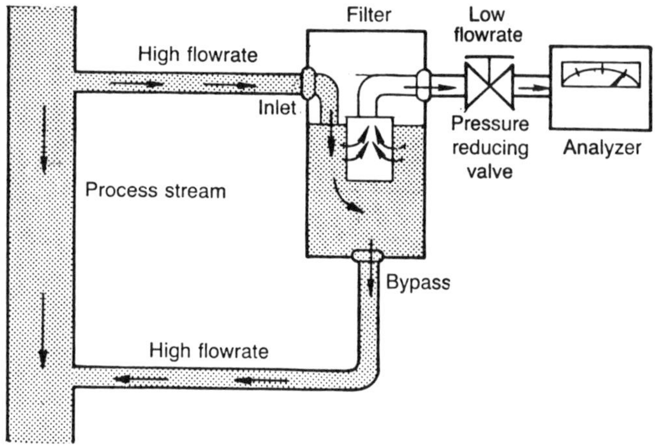

Figure 2. Slipstream or bypass filtration.

A slipstream filter requires inlet and outlet ports at opposite ends of the filter element to allow the high flow rate of the by-passed material to sweep the surface of the filter element and the filter reservoir, and a third port connected to the low flow rate line to the analyzer, which allows filtered samples to be withdrawn from the filter reservoir.

If bubble removal from a liquid is a requirement; this function may be combined with slipstream filtration, since the recommended flow direction for bubble removal is outside-to-inside, and the separated bubbles will be swept out of the housing by the bypass stream. In this case the liquid feed should enter at the bottom of the housing and the bypass liquid exit at the top of the housing.

A special problem arises in slipstream sampling if the filter is to coalesce and continuously drain suspended liquid from a gas stream or to coalesce liquid droplets from a liquid stream. As noted previously, the coalesced liquid is removed in the form of large drops from the downstream side of the filter. Therefore, the coalescing filter requires two outlet ports, one for the dry gas and one for the liquid drain. To combine coalescing and slipstream filtration, a filter housing would need four ports—two for inlet and bypass and two for filtered gas and coalesced liquid—which is not a practical design. Therefore, slipstreaming plus coalescing requires two stages of filtration (Fig. 3). The second (coalescing) stage must be located in the sample line to the analyzer, and should be as small as possible to minimize lag time. If the quantity of suspended liquid is not large, a miniature in-line disposable filter unit may be considered as a trap for the suspended liquid, to be replaced when saturated.

Stack Gas Sampling

When sampling hot, wet gas, such as stack gas, a filter capable of withstanding the gas temperature can be

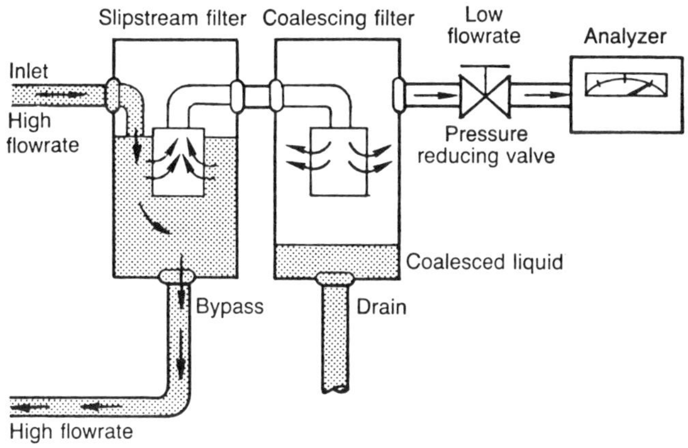

Figure 3. Slipstream plus coalescing filtration.

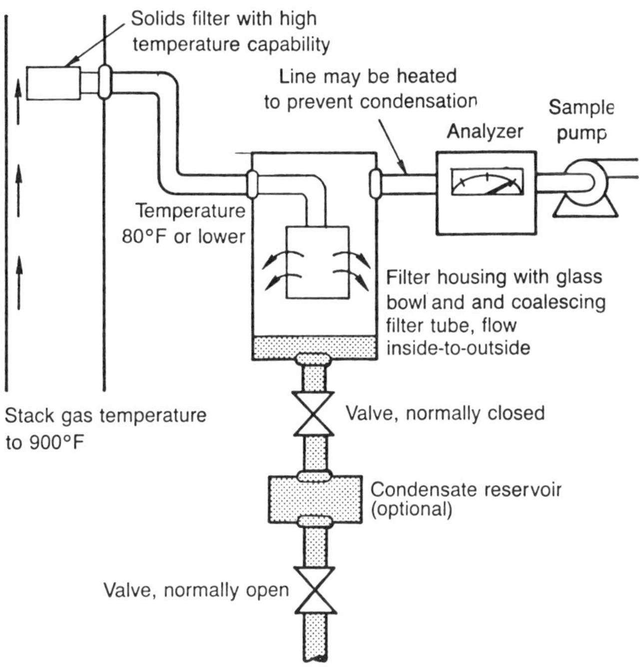

Figure 4. Stack gas sampling.

installed in the stack at the beginning of the sample line to prevent solids from entering the gas sample line. After the sample is cooled, a coalescing filter is used to remove suspended liquids before the sample goes to the analyzer (Fig. 4). Flow direction is inside-to-outside. Filter housings with Pyrex glass bowls are often used in this application to permit a visual check of the liquid level in the filter housing. Since there is often a considerable amount of liquid present at this point, steps must be taken to drain the housing to ensure that liquid does not build up and carry downstream to the analyzer.

The liquid coalescing filter should be located as close to the analyzer as possible to minimize the chance of condensation between the filter and the analyzer. Additional precautions that can be taken to avoid downstream condensation are to cool the sample below ambient temperature upstream of the coalescing filter, and to heat the line gently between the filter and the analyzer.

Sampling Liquid Effluent Streams

Liquid effluent analyzers usually deal with aqueous streams having a high solids content. In addition, the analyzers are often located in remote areas of the plant and are infrequently serviced. Therefore, the sample filter system must have long life between filter tube changes even in a high solids situation. The general recommendation for this requirement is a two-stage filter system, a 75-micron prefilter followed by a 25-micron final filter. The filters should be oversized as much as possible without causing excessive lag time. Plastic filter housings are usually a good choice.

Sampling High-Pressure Steam or Condensate

Measurements are often required in high pressure boiler systems of steam and condensate conductivity, specific ion concentrations and feedwater additive concentrations. In a continuous sampling system, the high pressure steam or condensate is cooled to below 100°F and then reduced to near atmospheric pressure for metering to the analyzers. Filtration is required upstream of the pressure reducing valves, to prevent pitting of the valve seats by suspended particles and to eliminate variations in flow rate to the analyzers.

A stainless steel filter housing with the appropriate pressure rating and 25-micron filter tube is recommended. Since the analyzer system is often located some distance from the sampling point, slipstream filtration is usually required. Figure 5 shows a sampling system in operation at a steam generating facility.

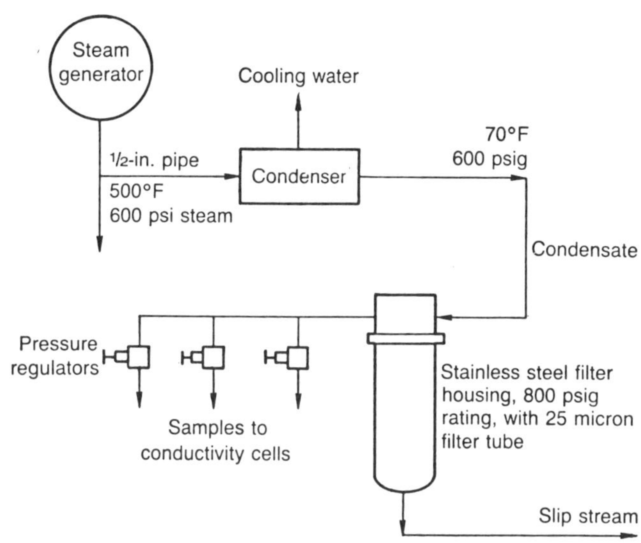

Figure 5. High pressure steam sampling.

18
Checklist for Orifice Flow Metering*

K. Arvid Johanson and R. Primrose

Mistakes and omissions are frequent in orifice flow metering system design, installation and operation. The real test of the orifice flow metering system is its operation under field conditions. The audit described in this article has been used successfully to obtain accuracies of 0.25 % material balance and of 0.43% average daily balance.

Precision in orifice flow metering systems has been the subject of many debates lately. The increasing use of computer control makes accurate flow and other measurements more important to obtain the benefits of better control. Suppliers and users must be prepared to defend their statement of the accuracy of their flow metering systems. Although the AGA Report No. 3[1] gives a design method for the orifice flow metering tube, and both AGA[2] and Dodge[3] give methods for calculating the error, no reference was available for insuring that all necessary points had been covered.

A program to upgrade metering would have to include adequate design, trained calibration personnel, precision test equipment and methods, control of data and revisions, and conducting an initial audit as part of the commissioning and a routine audit every two to four years. The following list serves well as an audit of the most common points that must be checked. Where a specific paragraph in the AGA Report No. 3 is referenced, the paragraph number is included at the end of the text question.

Orifice Plate

1. Does the roughness, flatness and cleanliness meet specifications? (2.2.1–2.2.5)
2. Is the orifice ID and bevel as measured within tolerances? (2.2.5)
3. Is the orifice plate OD correct for centering in the orifice flanges with a concentricity within 3%? (2.2.4)
4. Is the Beta within an acceptable range? (2.2.6)

Orifice Meter Tube

1. Is the purchased length at least thirty diameters upstream and ten diameters downstream of the orifice plate?
2. Does the meter tube meet the required smoothness and lack of irregularities such as scorings, pits, ridges, distortion and offsets?
3. Does the average inside diameter meet the specification? (2.3.4–2.3.5)
4. Is the orifice tap connection oriented to reduce the possibility of having two phases in leads (self-draining for gases, self-venting for liquids)?
5. Does the orientation of the orifice tap connection match the orifice drawing (loop sketch)?
6. Are the orifice tap connection centerlines exactly 1 in. from the edge of the orifice plate? (2.8.1)
7. If the orifice flange gaskets are thicker than 1/8-in., has an adjustment been made in tap location or in the raised face thickness?
8. Are the orifice tap connections the right size for the pipe and are they free of burrs on the inner wall of the pipe? (2.8.5)

Physical Data

1. Is composition of the stream known within ±0.2 percent by recent analysis?
2. Is composition constant or is it routinely or automatically analyzed?
3. Is actual stream data close (±5%) to data used in the orifice calculation?

* This chapter is based in part on the thesis "Improving the Orifice Flow Metering System for Ethylene to Inaccuracies Below One Percent" which was submitted to the University of Houston, Victoria, Texas, by Mr. Johanson on January 12, 1978.

The authors wish to thank the Union Carbide and University of Houston personnel who provided such excellent support on the original project and in reviewing the manuscript.

4. Is two phase flow possible within minimum and maximum process conditions?
5. Is the fluid density in a region of high sensitivity?
6. Are modes of operation possible that would over or under-range transmitters?
7. Is the fluid clean and not subject to flashing or condensation at all probable process and/or ambient temperature conditions?
8. Will the process fluid in the connection leads be of a single phase over the normal operating and normal ambient conditions?
9. Can there be an appreciable density difference in the two process leads due to process or ambient conditions?

Orifice Calculations

1. Is the orifice calculation made by a reputable method?
2. Is the flow calculated on a weight (mass) basis?
3. Is the density used in the calculation near the actual density?
4. Is the flow rate used in the calculation near the actual flow rate?

Instrumentation

1. Was the instrumentation purchased with minimum inaccuracy and with minimum temperature or other ambient effects?
2. Are full specifications and instruction manuals available for all instruments?
3. Are all special features such as selection for low temperature drift shown on instrumentation drawings?
4. Do the instrument hardware model numbers and ranges agree with the documentation (loop sketches)?
5. Is the damping in the differential pressure transmitter adjusted or fixed to reduce phase problems in compensation?
6. Were the instruments purchased and installed to be RFI and vibration insensitive?
7. Were the transmitters calibrated in place?
8. Were the mounting position effects on the transmitter calibrated out?
9. Was the calibration made through the computer for each transmitter?

Field Installation

1. Is the thermowell located in active flow, not in the downstream straight run, but less than 15 diameters from the orifice?
2. Is the line insulated between the orifice plate and the temperature or density element?
3. If in gas service, is the piping configuration installed to reduce the trapping of liquids?
4. If in liquid service, is the piping configuration installed to ensure that the pipe remains full of liquid?
5. Is the upstream straight run installed as shown on the drawings through the first two disturbances (fittings, valves, etc.)?
6. Does the upstream straight run meet the AGA specifications? (2.4.1-2.5.5)
7. Is the downstream straight run installed as shown on the drawings through the first disturbance?
8. Does the downstream straight run meet the AGA specifications? (2.4.1-2.5.5)
9. Are there any indications of pulsating or cyclic flow greater than 1 percent or with a period less than 10 times the sampling period?
10. Are there any indications of heavies, solids or bubbles in the process stream?
11. Is there evidence of corrosion or erosion?
12. Is there any evidence of polymerization or coating of the piping or the orifice plate?
13. Are the process connection leads short, clean, tight and properly oriented to be self-draining or self-venting?
14. Is the heat tracing, blowback, purge or self-purge in service?
15. Are pressure connections made to the downstream orifice tap connection (or upstream if so used in the flow calculation)?
16. Are the actual pressure and temperature measurements reasonably close to the data used in the orifice flow calculation?
17. Is the direction of flow through the primary element correct?
18. Are all pipe tie-ins found in physical inspections accounted for?

Computer Hardware

1. Is readout of each measurement variable available from -3 to $+103$ percent of span?
2. Are resolution and accuracy of the analog to digital converter provided at least two times minimum calibration accuracy, i.e., 0.05 percent for 0.1 percent calibration?
3. Is the speed of sampling ten times the maximum cyclic frequency in the process?
4. Is reduction in process and electrical noise documented and acceptable, i.e., noise does not increase the error or decrease the sampling speed?

5. Are differential input analog to digital converters used and is their common mode rejection acceptable?

Computer Programming

1. Is temperature input linearized in only one place, i.e., in computer or in instrument, but not in both?
2. If a thermocouple is used are the reference junction compensation method and temperature calculation acceptable?
3. Does the documentation include a simple flow sheet and details of the calculations of both density and flow?
4. Are alarm flags provided on the density calculation to ensure correction of bad process conditions?
5. Are alarm flags provided on out of range measurements?
6. Are multiple differential pressure cell switching routines based on new differential pressure, done at sampling speed rate, and correctly documented to provide minimum differential pressure error?
7. On multiple differential pressure installations, is the switching point set at 99.5 percent of the lower range?
8. Is the computer averaging routine acceptable even when handling outages?
9. Is the computer integration routine acceptable?
10. Are the conversion calculations acceptable?
11. If a contained product calculation is made, is it on an acceptable basis of current product compositions?
12. Is an alarm flag provided on data when the metering system is out of service *or* being calibrated?
13. Is the static pressure error minimized by zeroing at line pressure?
14. Are the measurement signals available directly as separately identifiable points in the engineering unit table?

Maintenance and Auditing

1. Is the history of instrument reliability maintained?
2. Do procedures exist to ensure calibration within the limits used in the error calculation?
3. Is adequately maintained test equipment used that can provide the required accuracy, i.e., 1/2 the acceptable inaccuracy (0.05 percent for 0.1 percent inaccuracy)?
4. Is documentation available on the complete system including all interfaces?
5. Are measurement variables within the stated range?

Production Records or Metericals Accounting

1. Is the accounting determination consistent with field calculations, i.e., avoid double corrections and double conversions?
2. Do the calculation constants used match those in the orifice calculation?
3. Are current readings consistent with past records?
4. Do these records verify design rangeability?
5. Are the correct compositions used for stream composition?
6. Is the computer integrated and averaged data used, i.e., charts if provided are for nonaccounting purposes?

References

1. American National Standard, "Orifice Metering of Natural Gas," (ANSI/API 2530) and American Gas Association (Report No. 3), 1978.
2. Ibid, p. 94.
3. Dodge, L. A., "Evaluating Flow Measurement Accuracy," *Automation,* May 1979, p. 67.

19

Flow Equation for Orifices*

K. R. Hall, P. T. Eubank, and J. C. Holste

Orifices have been used to measure flow rate for thousands of years, but economic pressures recently have dictated a careful evaluation of the procedure. The National Bureau of Standards currently has two major research programs (funded by API and GRI) underway with the goals of increasing the accuracy and range of orifice measurements.

The equations presented in this work represent a significant departure from those normally used. The new expressions are equally rigorous and possess many advantages compared with the conventional relationships.

DERIVATION

Assuming steady state, horizontal flow with no energy transferred as heat or work, the First Law of Thermodynamics becomes

$$h_1 - h_2 + (\dot{z}_2^2 - \dot{z}_1^2/2g_c) = 0 \tag{1}$$

where h is enthalpy per mass, $\dot{z}$ is velocity and g_c is the force/mass conversion factor. Subscript 1 denotes upstream and subscript 2 denotes downstream. The mass flow rate in steady state is

$$\dot{m} = \dot{z}A/v \tag{2}$$

where A is the flow area and v is volume per mass.

Rearranging Eq. 1 and substituting Eq. 2 into the result produces

$$\frac{\dot{m}^2}{2g_c}\left\{\left[\frac{v_1}{A_1}\right]^2 - \left[\frac{v_2}{A_2}\right]^2\right\} = h_2 - h_1 \tag{3}$$

which becomes, for circular cross sections, after inserting the equation of state

$$\dot{m} = \frac{\pi D_1^2 M P_1}{4 Z_1 R T_1}\left\{\frac{2g_c}{\psi}\left[h_2 - h_1\right]\right\}^{1/2} \tag{4}$$

where D is diameter, M is relative molecular mass, P is pressure, Z is compressibility factor, R is the gas constant, and T is temperature. The function ψ for compressible fluid flow is

$$\psi = 1 - \left[\frac{Z_2 T_2 P_1}{Z_1 T_1 P_2}\right]^2 \left[\frac{D_1}{D_2}\right]^4 \tag{5}$$

Completion of the derivation requires an expression for the enthalpy difference. The usual procedure uses

$$h_2 - h_1 = \int_{S_1}^{S_2} T ds + \int_{P_1}^{P_2} v dP \tag{6}$$

along with the definition

$${}_1f_2 = \int_{S_1}^{S_2} T ds - {}_1q_2 = \dot{z}^2\phi(\beta, N_R, \ldots) \tag{7}$$

where s is the entropy per mass, ${}_1q_2$ is the energy transferred as heat per mass, ${}_1f_2$ is "friction," ϕ is an empirical function of β ratio (d/D where d is the orifice diameter), Reynolds number, N_R, and possibly other factors. The primary disadvantage of this approach is the necessity of correlating ϕ (which gives rise to the discharge coefficient). Because ϕ reflects "friction," it is impossible to produce a rigorous correlation nor anything other than rough approximations.

An alternative, equally rigorous expression for the enthalpy difference is

* The following organizations provided financial support for this study: Gas Research Institute, Gas Processors Association, National Science Foundation, National Bureau of Standards, the donors of the Petroleum Research Fund (administered by the American Chemical Society), and the Texas Engineering Experiment Station.

Bill Lawler of Sun Gas and Ron Lindsay of Lone Star Gas provided the prime motivation to engage this project.

$$h_2 - h_1 = \frac{R}{M}\left\{ T_2 \int_0^{P_2} \left[\frac{\partial Z}{\partial 1/T}\right]_{P,x} \frac{dP}{P} - T_1 \int_0^{P_1} \left[\frac{\partial Z}{\partial 1/T}\right]_{P,x} \frac{dP}{P} + \int_{T_1}^{T_2} \frac{C_p^*}{R}\, dT \right\} \quad (8)$$

where x is composition and C_p^* is the ideal gas specific heat per mole. This expression requires correlations only for fluid properties that are relatively easy to develop. However, Eq. 8 does require temperature measurement, which for compressible fluids is significant. If the temperature did not change appreciably, the orifice would violate the energy balance and the square root term in Equation 4 would become imaginary. The new performance equation is then

$$\dot{m} = \frac{\pi D_1^2 P_1}{Z_1 T_1}\left\{\frac{g_c M}{8R\psi}\left[T_2 \int_0^{P_2} \left[\frac{\partial Z}{\partial 1/T}\right]_{P,x} \frac{dP}{P} - T_1 \int_0^{P_1} \left[\frac{\partial Z}{\partial 1/T}\right]_{P,x} \frac{dP}{P} + \int_{T_1}^{T_2} \frac{C_p^*}{R}\, dT \right]\right\}^{1/2} \quad (9)$$

For a given composition, it is possible to tabulate the pressure integrals in Eq. 9 (like F_{pv} in American Gas Association Report Number 3)

$$I_p(T,P) = T \int_0^{P} \left[\frac{\partial Z}{\partial 1/T}\right]_{P,x} \frac{dP}{P} \text{ for given composition} \quad (10)$$

It is also possible to tabulate the temperature integral as a function of the average temperature ($\overline{T} = [T_1 + T_2]/2$)

$$I_T(\overline{T}) = \int_{T_1}^{T_2} \frac{C_p^*}{R}\, dT \text{ for given composition} \quad (11)$$

Of course, it is preferable to use the equation of state and an expression for C_p^* and simply compute the integrals.

VOLUMETRIC FLOW RATE

Having the mass flow rate, it is a simple task to compute the volumetric flow rate at any condition. The procedure is to multiply the mass flow rate by the specific volume at the required conditions. This is true for ideal gas or real gas flow rates:

$$\dot{V} = \dot{m}v = \dot{m}ZRT/MP \quad (12)$$

$$\dot{V}_s = \dot{m}v_s = \dot{m}Z_sRT_s/MP_s \quad (13)$$

$$\dot{V}^* = \dot{m}v^* = \dot{m}RT/MP \quad (14)$$

$$\dot{V}_s^* = \dot{m}v_s^* = \dot{m}RT_s/MP_s \quad (15)$$

where $\dot{V}$ is volumetric flowrate, superscript * denotes ideal gas, subscript s denotes base conditions, Z is at T and P, Z_s is at T_s and P_s.

IMPLICATIONS

In Eq. 9, all frictional effects associated with flow through the orifice appear *only* in the observables: pressure drop and temperature change. The fluid properties, Z and C_p^*, are functions only of pressure, temperature and composition. Thus, it is not necessary to provide a correlation, such as the Buckingham or Stoltz equations, for frictional effects; they are observable. Nor is it necessary to provide an "expansion factor"; I_p and ψ correctly reflect compressible effects. Good correlations are currently available for Z and C_p^*, and others are becoming available. The accurate correlations do require a computer for utilization.

Another interesting observation is that damage to the orifice has no effect on Eq. 8 unless that damage changes D_1 or D_2. The pressure drop and temperature change adjust to reflect the damage. Thus, rounded, bent or scratched orifices would have no effect on calculation of $\dot{m}$.

In this analysis, the tap location has an obvious optimum: pipe taps (or at least similar taps). Pipe taps should be located at points where the fluid stream is attached to the wall. Thus, measuring pressure and temperature in these locations gives approximately the fluid values. In addition, D_1 equals D_2 for pipe taps thus damage to the orifice has no effect upon calculation of the flow rate. Flange taps have none of these advantages. They measure eddy pressures and temperatures (large fluctuations), $D_1 \neq D_2$ and damage to the orifice can have a detrimental effect.

SAMPLE PROBLEM

A fluid whose relative molecular mass is 17.38 flows in a 10-in. pipe through a 6-in. orifice. The upstream pressure and temperature are: 200 psig and 60°F. Using pipe taps, the pressure drop is 40-in. H_2O. Using a difference thermocouple, the temperature change is −0.134°F. The equation of state for the fluid is

$$Z = 1 - \{0.000156 + 0.3[1/T - 0.00192]\}P$$

The ideal gas specific heat is

$$C_p^* = 7.000 - T/1{,}000$$

Temperature units are °R in both equations; Pressure units are psia. Ambient conditions are 14.7 psia and 60°F with a g of 32.174ft/s^2. What is the flow rate in scfh with base conditions at 60°F and 14.73 psia?

$$\left[\frac{\partial Z}{\partial 1/T}\right]_{P,x} = -0.3P$$

$$T_2 \int_0^{P_2} \left[\frac{\partial Z}{\partial 1/T}\right]_{P,x} \frac{dP}{P} = -0.3P_2$$

$$T_1 \int_0^{P_1} \left[\frac{\partial Z}{\partial 1/T}\right]_{P,x} \frac{dP}{P} = -0.3P_1$$

$$\int_{T_1}^{T_2} \frac{C_P^*}{R}\, dT = \left[\frac{7.000}{1.986}\right][T_2 - T_1] - \frac{1}{1986}\left[\frac{T_2^2 - T_1^2}{2}\right]$$

$$I_{p2} - I_{p1} + I_T = 0.3\left\{\frac{[62.3664][32.174][40]}{1728[32.174]}\right\}$$

$$+ \frac{[7.000][-0.134]}{1.98586} + \frac{139.25}{3971.72} = -0.004178$$

$$\psi = 1 - \left\{\frac{[0.96642][519.536][214.7]}{[0.96623][519.67][213.26]}\right\}^2 \left[\frac{10}{10}\right]^4$$

$$= -0.01343$$

$$\dot{m} = \frac{\pi[100][214.7]}{[0.96623][519.67]} \left\{\frac{[32.174][17.38][-0.004178]}{8[1545.34][-0.01343]}\right\}^{1/2}$$

$$= 15.935 \text{ lbm/s} = 57{,}364 \text{ lbm/h}$$

$$\dot{V}_s = \frac{\dot{m} Z_s R T_s}{M P_s} = \frac{[57{,}364][0.9977][519.67][10.7315]}{[17.38][14.73]}$$

$$= 1{,}246{,}737 \text{ scfh}$$

20
Control Valves: A Review

M. Adams and D. Boyd

Picking a type and brand of control valve for a particular application used to be simple. Usually only one general type of valve was considered—the sliding-stem type. Each of a handful of major manufacturers typically offered one product suitable for the job, and the choice among them depended on obvious matters such as cost, delivery, vendor relationships and the user's own habitual preference.

Today, matters seem considerably more complicated—especially for engineers with limited experience or those who have not kept up with changes in the control valve industry. For all but the most unusual applications, more than a dozen leading manufacturers offer a bewildering assortment of sliding-stem valves in addition to ball and butterfly valves. Some are touted as "universal" valves for almost any size and service, while others are claimed to be optimum solutions for narrowly defined needs. Even the most knowledgeable user may wonder whether he is really getting the most for his money in the control valves he has been specifying.

Like most decisions, selection of a control valve involves a great deal of uncertainty. Consequently, the matter has to be simplified by establishing some ground rules. Presented here is an overview of the situation—in effect, a suggested set of simplifications—drawn from our regular contacts with users of control valves. The format for the discussion resembles a knowledgeable user's typical thought patterns: a categorization of available valve types and a checklist or set of criteria to be considered.

GENERAL CATEGORIES OF CONTROL VALVES

"Control valve" here means any power-operated valve, whether used for throttling or for on-off control. Varieties from which to select, as listed in Table 1, include sliding-stem valves (globe or angle configuration), and rotary valves—ball and butterfly. Other varieties such as motorized gate valves, louvers, pinch valves, plug valves, and self-operated regulators are not considered here. These major types, in turn, are further divided into a total of eight general subcategories according to relative performance and cost.

Table 1
Major Categories and Subcategories of Control Valves, with Typical General Characteristics

	Sliding stem valves			Ball valves		Butterfly valves		
	Regular	Bar-stock	Economy	Thru-bore	Partial	Swing-thru	Lined	Hi-perform.
Main characteristics	Heavy-duty General service	Machined from bar stock	Light-duty Inexpensive	On-off service	Characterized for throttling	No seal	Elastomer or TFE liner	Offset disk General service
Typical size range, inches	1/8	1/4 to 3	1/2 to 2	1 to 24	1 to 24	2 to 96	2 to 96	2 to 24
Typical standard body materials	Carbon steel Cast iron Stainless	Variety of alloys	Bronze (Cast iron) (Carbon steel)	Carbon steel (Stainless)	Carbon steel (Stainless)	Carbon steel Cast iron Stainless	Carbon steel Cast iron Stainless	Carbon steel Stainless
Typical standard end connections	ANSI Flange Welded Screwed	Flangeless Screwed	Screwed	Flangeless	Flangeless	Flangeless Lugged Welded	Flangeless Lugged	Flangeless Lugged
Typical pressure ratings	to ANSI 2500	to ANSI 600	ANSI 125	to ANSI 900	to ANSI 600	to ANSI 2500	to ANSI 300	to ANSI 600
Relative flow capacity	Moderate	Low	Moderate	High	High	High	High	High
Relative shutoff capability	Excellent	Excellent	Good	Excellent	Excellent	Poor	Good	Excellent

Despite variations found within each category—such as cage-guiding versus stem-guiding—all valves within a given subcategory may be considered very much alike in the early stages of the selection process. Selecting a valve involves narrowing down to one of these eight subcategories, and then comparing specific valves in that group from different manufacturers.

Sliding-stem Valves

The most versatile of control valves are the sliding-stems: globe, angle-body, and Y-body valves, particularly in the 1/2- to 8-in. range (Fig. 1). Within that span manufacturers offer twice as many size increments in sliding-stem valves as in rotary valves. Globe valves are available in cage-guided and post-guided designs, with flanged, flangeless screw-type or welded ends. Economical cast iron as well as carbon steel, stainless steel, and other high-performance body materials are available, as are pressure ratings to ANSI Class 2500. Their precise throttling capabilities, overall rangeability and general sturdiness can make sliding stem valves a bargain, despite their slight cost premium relative to rotary stem valves. The buyer gets a rugged, dependable unit intended for long, trouble-free service under the field conditions of piping stress, vibration, temperature changes and manhandling that control valves experience. And, in sizes through 3 in., incremental costs are low in contrast to the increments in benefits received.

Another subcategory of sliding-stem valves consists of those whose bodies are machined from metal bar stock (Fig. 2). Body sizes range from fractions of an inch up to three inches, flow capacities generally are lower than those of general purpose valves. End connections usually are flangeless (for mounting between piping flanges) or screwed. The main advantage of this type of valve is that far more materials are readily available in bar stock form than in cast form. Consequently, these valves are used where there are special corrosion considerations. However, their compactness and general high-quality construction make them attractive for flow rates below the range of the regular sliding-stem subcategory even when corrosion is not a consideration. Overall, they are an economical choice when they can be used.

The third subcategory, the lowest cost products among sliding-stem valves, are called "economy" bodies. These valves are often used for steam, air and water service, low-pressure and temperature applications that are not demanding. The sizes available range from 1/2 up to 2 in., and the body material is bronze, with screwed ends. Compared to regular sliding-stem valves, these units are very simple; their actuators are smaller; their pressure ratings are not as high; and they cost about half as much.

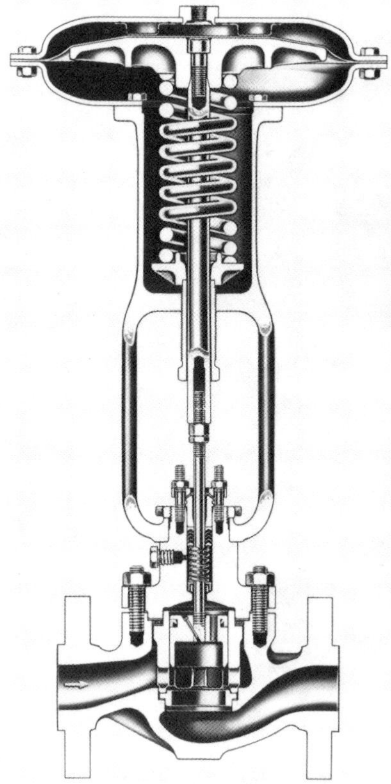

Figure 1. Typical "regular" sliding stem-valve.

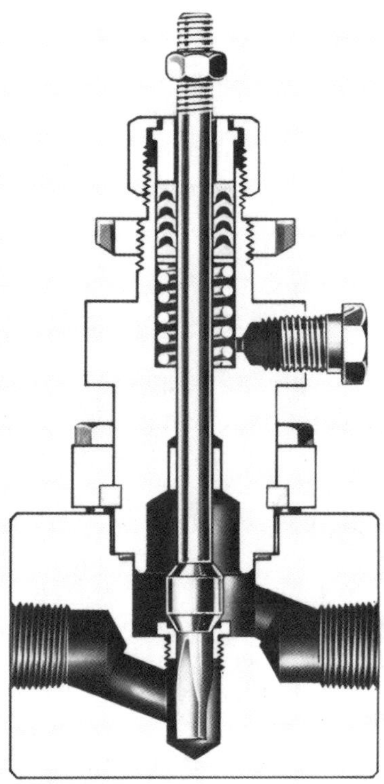

Figure 2. Typical "Bar-stock" sliding-stem valve.

Figure 3. Typical "through-bore" ball valve.

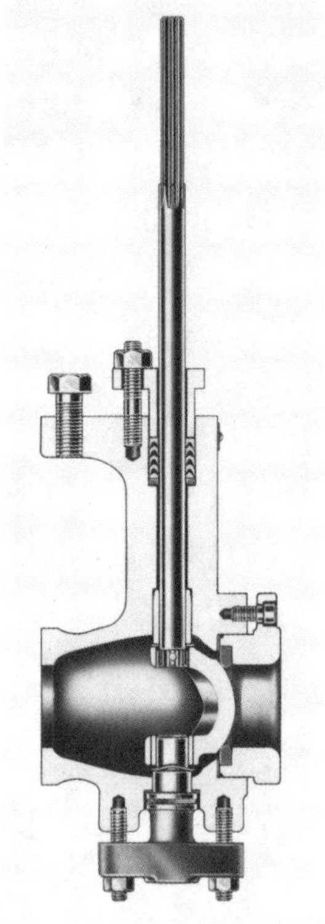

Figure 4. Typical "partial ball" valve.

Ball Valves

There are two subcategories of ball valves. The through-bore or full-ball type illustrated in Fig. 3 generally are used for on-off applications in sizes from 1/4 to 24 in. When fully open, they exhibit high flow capacity and low susceptibility to wear by erosive streams. Sluggish flow response to the first 20% of ball travel in either direction makes full-bore ball valves unsuitable for throttling applications. Newer designs in full-ball, reduced bore valves show faster response. Pressure ratings up to ANSI Class 900 are available, as are a variety of end connections and body materials.

The other kind of ball valve is the partial-ball variety, (Fig. 4). This subcategory is very much like the through-bore group except that the edge of the ball segment has a contoured notch shape for better throttling control and higher rangeability. Intended primarily for modulating service, not merely for on-off control, partial-ball valves are generally higher in overall performance than most full-ball varieties. They are engineered to eliminate lost motion, which is detrimental to performance. The use of metallic and fluoroplastic sealing elements allows wide temperature and fluid applicability. The inherent shearing action of the ball/ seal combination makes them ideal for control of fibrous slurry. Their straight-through flow design achieves high capacity close to that obtained in fullbore valves.

Butterfly Valves

Butterfly valves, in turn, are divided into three subcategories. The most rudimentary is the swing-through design. Rather like a stovepipe damper, but considerably more sophisticated, this kind of valve has no seals—the disk swings close to, but clear of, the body's inner wall. Such a valve is used for throttling applications that do not require shutoff tighter than about 1/2 to one percent of full flow. Sizes range from two to 96 inches. Body materials are cast iron, carbon steel, or stainless steel. Mounting is flangeless, lugged, or welded. Body pressure ratings up to ANSI Class 2500 are common, and wide temperature ranges are also available.

The next butterfly valve subcategory in order of cost is represented by the types with an elastomer or fluoropolymer (TFE) lining that contacts the disk for tight shutoff (Fig. 5). This subcategory is very much like the preceding swing-through group. However, the pressure and temperature ranges are considerably more restricted due to elastomeric materials, and welded ends are not available.

Finally, there are the high-performance butterfly valves such as the one in Fig. 6. These products are characterized by heavy shafts and discs, full rating bodies and sophisticated seals that allow shutoff to fairly high pressures. They begin absorbing energy from the process flow as soon as they begin rotating. Eccentric shaft mounting allows the disc to swing clear of the seal to minimize wear and seating torque requirements. The offset discs employed allow uninterrupted sealing, and a seal ring can be replaced without removing the disc. High-performance butterfly valves come in sizes ranging roughly from two to 36 in., with flangeless or lugged connections, carbon steel or stainless steel bodies, and pressure ratings up to ANSI Class 600. With their very

Figure 5. Typical "lined" butterfly valve.

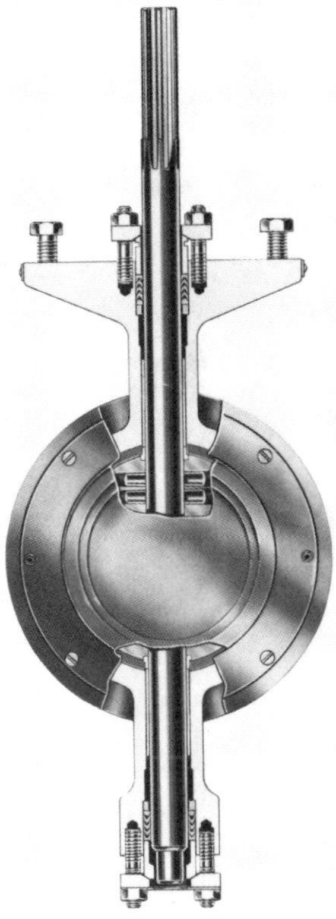

Figure 6. Typical "high-performance" butterfly valve.

tight shutoff, heavy-duty construction, and tight linkages, these valves are suitable for as many process applications as their sliding stem counterparts. Tight, metal-to-metal seals possible with advanced, eccentric disc design provide tight shutoff in high temperature applications that are too hot for elastomer-lined valves to handle.

This grouping of valves into a few major classes is a useful way to simplify the otherwise confusing variety of products on the market.

GENERAL SELECTION CRITERIA

Most of the considerations that guide the selection of valve type and brand are rather basic. However, there are some matters that may be overlooked by users whose familiarity is mainly limited to just one or a few valve types. Table 2 provides a checklist of important criteria to be discussed.

Pressure Ratings

Body pressure ratings ordinarily are considered according to ANSI pressure classes—the most common ones for steel and stainless steel being 150, 300, and 600.

Table 2
Suggested Check List of General Criteria for Selecting a Type and Brand of Control Valve
(No definite sequence is implied)

Body pressure rating
High and low temperature limits
Material compatibility and durability
Inherent flow characteristic and rangeability
Maximum pressure drop (shutoff and flowing)
Noise and cavitation
End connections
Shutoff leakage
Capacity versus cost
Nature of flowing media

(Source documents are ANSI Standards B16.34-1977, "Steel Valves," and ANSI B16.1-1975, "Cast Iron Pipe Flanges and Flanged Fittings.") For a given body material, each ANSI class corresponds to a prescribed profile of maximum pressures that decrease with temperature according to the strength of the material. Each material also has a minimum and maximum service temperature based on loss of ductility or loss of strength. For most applications, the required pressure rating is a foregone conclusion which does not often affect the selection of type and brand. Most valves are available in ANSI classes up to 600 or higher as standard items.

Temperature Considerations

Required temperature capabilities are usually also a foregone conclusion, but one which is more likely to narrow the valve selection. The considerations here include the strength or ductility of the body material as well as relative thermal expansion of various parts. Temperature limits may also be due to disintegration of soft parts at high temperatures or loss of resiliency at low temperatures. The soft materials under consideration include various elastomers, or rubbers, and TFE. They may be found in parts such as seat rings, seal or piston rings, packing, rotary shaft bearings and butterfly valve liners. Typical upper temperature limits for elastomers are in the 200°F range, and the limit for TFE is 350 to 450°F.

Temperature affects valve selection by excluding certain valves that do not have high- or low-temperature options—such as lined butterfly valves. It also may have some effect on the valve's performance. For instance, going from TFE to metal seals for high temperatures generally increases the shutoff leakage flow. Similarly, high-temperature metal bearing sleeves in rotary valves impose more friction load on the shaft than TFE bearings do, so that the shaft cannot withstand as high a pressure-drop load at shutoff.

Material Selection

The third criterion in Table 2—material compatibility and durability—is a somewhat more complex consideration. The issue here is corrosion by the process fluid and erosion by abrasive material, flashing, cavitation and—in the case of very high pressure drop at or near shutoff—by high-velocity fluid streams. The piping material usually dictates the body material. The trim materials in turn are usually a function of the body material, temperature range and erosive qualities of the fluid. When a body material other than the usual iron or carbon steel is desired at reasonable cost, a bar stock sliding-stem valve is strongly indicated for small sizes and a high-pressure butterfly valve for larger sizes.

Flow Characteristic

The next selection criterion—inherent flow characteristic—refers to the pattern in which the flow at constant pressure drop changes according to valve position. Typical characteristics are quick-opening, linear and equal-percentage. The choice of characteristic may have a strong influence on the stability or controllability of the process, since it represents the change of valve gain relative to travel. Many valves are carefully "characterized"—a desired flow characteristic has been established by such means as designing certain contours on a plug, cage, or ball element. Some valves are available in a variety of characteristics to suit the application, while others offer little or no choice. To quantitatively determine the best flow characteristic for a given application, a dynamic analysis of the control loop must be performed. In most cases, however, this is unnecessary; reference to established rules of thumb will suffice. For example, in a liquid level system where the valve experiences a constant pressure drop, a linear characteristic is recommended. See Table 3 for further recommendations.

Table 3
Liquid Level Systems

Control valve pressure drop	Best inherent characteristic
Constant ΔP	Linear
Decreasing ΔP with increasing load, ΔP at maximum load >20% of minimum load ΔP	Linear
Decreasing ΔP with increasing load, ΔP at maximum load <20% of minimum load ΔP	Equal-percentage
Increasing ΔP with increasing load, ΔP at maximum load <200% of minimum load ΔP	Linear
Increasing ΔP with increasing load, ΔP at maximum load >200% of minimum load ΔP	Quick opening

Rangeability

Part and parcel of a valve's flow characteristic is its rangeability, which is the ratio of its maximum and minimum controllable flow rates. Exceptionally wide rangeability may be desired for processes with drastic load swings, or in certain nonlinear processes such as titration. Generally speaking, rotary valves—especially partial ball valves—have greater rangeability than sliding-stem varieties.

Use of Positioners

To a certain extent, a valve with one inherent flow characteristic can be made to perform as though it had a different characteristic by using a nonlinear positioner-actuator combination; that is, by introducing variations in gain through those parts. The limitation of this approach lies in the positioner's frequency response and phase lag compared to the characteristic frequency of the process. Positioners are generally *detrimental* to good control for "fast processes such as liquid flow loops—aside from considerations of positioner cost and reliability.

Pressure Drop

Next in Table 2 is the maximum pressure drop the valve can tolerate at shutoff and when partly or fully open. Sliding-stem valves are generally superior in both regards because of the rugged, well-supported nature of their moving parts. Unlike most sliding-stem valves, many rotary valves are limited to pressure drops well below the body pressure rating—especially under flowing conditions, due to dynamic stresses imposed on the disk or ball segment by high-velocity flow.

Noise and Cavitation

These two considerations are often grouped together because they both usually accompany high pressure drops and are handled by special modifications of more-or-less standard valves. Cavitation is the noisy and potentially damaging implosion of bubbles formed when the pressure of a liquid momentarily dips below its vapor pressure through a constriction at high velocity. In controlling gases and vapors, noise results from the turbulence associated with high-velocity streams. When cavitation or noise is judged likely to be a problem, its severity must be predicted from the valve's specifications according to well-known techniques, and valves with better specifications must be sought if necessary. Cavitation-control and noise-control trims for various degrees of severity are widely available in regular sliding-stem valves—at a progressive penalty in terms of cost and flow capacity. Rotary valves are much more limited in this regard and are generally much more susceptible to cavitation and noise at a given pressure drop.

End Connections

At some point in the selection process, the valve's end connections must be considered. The question is simply whether the desired connection style is available in a valve being considered. In some situations, this matter can limit the selection rather narrowly. For instance, if a piping specification calls for welded connections only, the choice may be limited to general purpose sliding-stem valves. The few weld-end butterfly and ball valves that are available are rather expensive.

Shutoff Capability

Some consideration usually must be given to a valve's shutoff capability, which ordinarily is rated in terms of classes specified in ANSI Standard B16.104-1976. In actual service, shutoff leakage depends on many factors, including pressure drop, temperature, the condition of the sealing surfaces, and—very importantly for sliding-stem valves—on the force load on the seat. Since shutoff ratings are based on standard test conditions that may be very different from service conditions, service leakage cannot be predicted very well. However, the ANSI shutoff classes provide a good basis for comparisons among valves of similar configuration.

Valve users tend to overspecify on shutoff, incurring unnecessary cost. Actually, very few throttling valves really need to perform double duty as tight block valves. Shutoff should be considered in both directions—some rotary valves have poorer shutoff ratings in the reverse direction.

Cost Versus Capacity

Finally, the criterion of capacity or size can be an overriding constraint on selection, as it relates to cost. In some cases, capacity can dictate selection. For very large lines, sliding-stem valves are much more expensive than rotary types, if they are available at all. On the other hand, for very small flows, a suitable rotary valve may not be available. If the same valve is desired to handle a significantly larger flow at a future time, a sliding-stem valve with replaceable, restricted trim may be indicated. Rotaries generally have much higher maximum capacity than sliding-stem valves for a given body size. This fact makes rotaries attractive in applications where the pressure drop available for controlling by the valve is rather small. But it is of little or no advantage in high-pressure-drop applications such as pressure regulation or letdown.

At the risk of overgeneralizing, we may conceive of the relative prices for the major valve types over a range of flow capacity to be roughly as follows: For the very least demanding services, in which price is the dominant consideration, one might primarily consider economy sliding-stem valves for the smallest applications and butterfly valves for the largest, with ball valves in a narrow intermediate range. Leaving out the economy sliding-stems and the swing-through and lined butterflies due to service demands, we are left with the most typical spectrum of choices according to flow capacity: Bar-stock and regular sliding-stem valves for the lower ranges, ball valves for intermediate capacities, and high-performance butterfly valves for the very largest sizes. For sizes less than 3 in., general purpose sliding stem valves provide an exceptional value. For a minimal price premium over other types they offer unparalleled performance, flexibility and service life. In size 3 in. and larger, the premium for these devices over rotary products is substantial enough that a strong look at other types is warranted.

The obvious point here is this: different types of valves are appropriate to use in different size ranges, because they provide the most cost-effective solution in each given instance. If one sticks with the same type of valve over an extremely wide range of capacity, one sacrifices either performance at the low end or economy at the high end—or both. Putting it another way, if one standardizes on a valve type that is good for 95 percent of a wide range of capacities, it is probably more valve than is needed in 80 percent of the applications.

After going through all the other criteria for a given application, a specifier typically finds that nine out of ten times, he can use almost any type of valve. From there on, selection is a matter of price versus capacity as discussed here—coupled with the inevitable personal and institutional preferences. Since no single valve or valve-positioner-actuator package is cost-effective over the full range of capacities that are normally encountered, it is important to keep an open mind for alternative choices.

21

Metering Pumps: Types and Applications

J. P. Poynton

A metering pump is a reciprocating positive displacement type mechanism that offers inherent steady-state delivery accuracy and dynamic capacity adjustment. In its various configurations, it is capable of handling fluids ranging from abrasive slurries to corrosives and of generating high discharge pressure for injection applications. This chapter provides a state-of-the-art overview of metering pump configurations as well as the applications for each. It is intended to clear up much of the mystery that surrounds specification of this small but important family of process control elements.

Metering pumps contribute to the efficiency of many processes by automatically injecting a predetermined volume of fluid during a specified period of time. They are designed for extreme accuracy (±1 percent of setpoint is typical) and can be adjusted manually or automatically while in operation to vary the flow rate.

All metering pumps employ three basic mechanisms:

- A drive mechanism—A mechanical or hydraulic arrangement that permits capacity to be adjusted during operation
- A prime mover—A constant or variable speed electric motor or a pneumatic cylinder
- A liquid end or displacement chamber—A packed plunger, disc diaphragm or double diaphragm type configuration that contains all wetted parts and is tailored for a specific range of fluid types.

THE DRIVE MECHANISM

Metering pump operation depends upon the back-and-forth action of a plunger, which either displaces the process fluid directly or as a result of diaphragm flexure.

Capacity of a metering pump is a function of the cross-sectional area of the plunger, effective length of the plunger stroke and the stroking rate (speed). Because plunger area is constant for a prescribed pump, stroking length and pump speed are the variables available to adjust output capacity while the pump is operating. Speed- and/or stroke-length adjustments can be effected manually or automatically, in accordance with the process demand, to vary the volume of liquid delivered.

There are two main classifications for stroke-length adjustments. One varies the radius of eccentricity of the plunger drive mechanism, which is sometimes referred to as amplitude modulation. The second method, called lost motion, varies the portion of fixed crank travel that is transmitted for effective plunger displacement.

Amplitude Modulation

The slider crank mechanism shown in Fig. 1A is the forerunner of most modern drive designs. A turnbuckle-type adjustment permits stroke length to be altered by changing length of a pivot arm. Although many variations of this drive have evolved, such as the 3-dimensional or polar crank shown in Fig. 1B, all share the same basic operating principle and attach to the plunger. This drive generally is used in the higher-capacity pumps that operate under greater pressure.

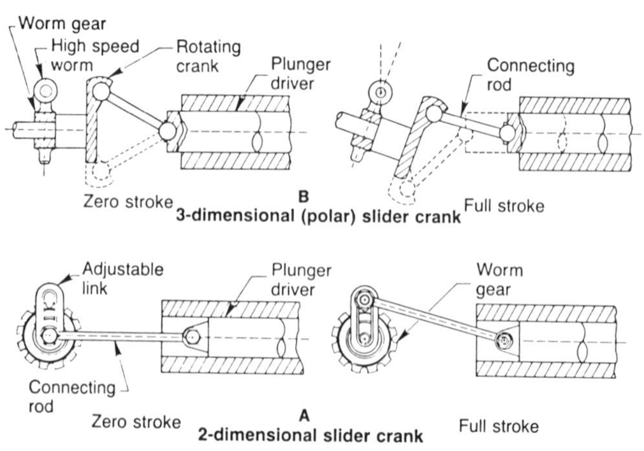

Figure 1. Examples of slider crank mechanisms.

Lost Motion Drives

For less demanding applications, an alternate method was developed to transfer power to the pumping element with equal accuracy. This method employs the lost-motion principle. As the name implies, lost-motion drives provide a method of altering the stroke length by somehow "losing" or not utilizing the drive thrust available over a full pumping cycle. Lost-motion drives include hydraulic and mechanical types.

Hydraulic lost motion (Fig. 2) entails a change in the effective rather than actual stroke length. Although the plunger continues to reciprocate the full length of the stroke at all times, some of the hydraulic fluid available to deflect the diaphragm escapes through a bypass valve and returns to the sump or hydraulic fluid reservoir. The bypass valve can be modulated externally while the pump is in operation. Although the lost-motion drive mechanism generally is less costly than the more complex slider crank, it should not be specified for capacities in excess of 100 gph to avoid excessive hydraulic shock.

A significant advantage of this drive is that the diaphragm is balanced hydraulically, improving its fatigue life and pressure-generating capability.

Mechanical lost motion differs in principle; the diaphragm is attached mechanically to the plunger or deflecting mechanism. The mechanism is driven forward by an offset cam, causing the diaphragm to deflect a predetermined amount. Flow is varied by limiting return of the mechanism (Fig. 3). Generally, a spring causes the driving mechanism to return to its original position after each cycle of the offset cam. If the plunger is allowed to return only half-way to the original position, the stroke and accompanying deflection will be decreased by 50 percent.

Mechanically actuated diaphragms are not used to generate pressures over 250 psi, or in applications where capacity requirements exceed 25 gph. Typical applications include water-treatment systems for cooling towers, boilers, water and sewage treatment plants as well as laboratory applications.

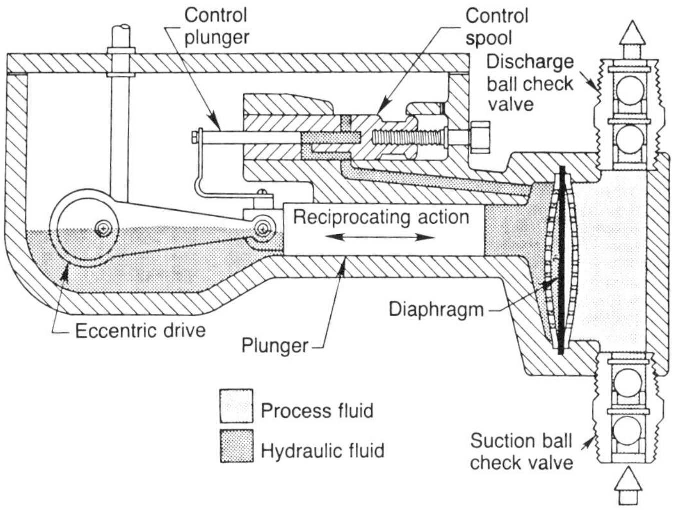

Figure 2. Hydraulic lost motion drive.

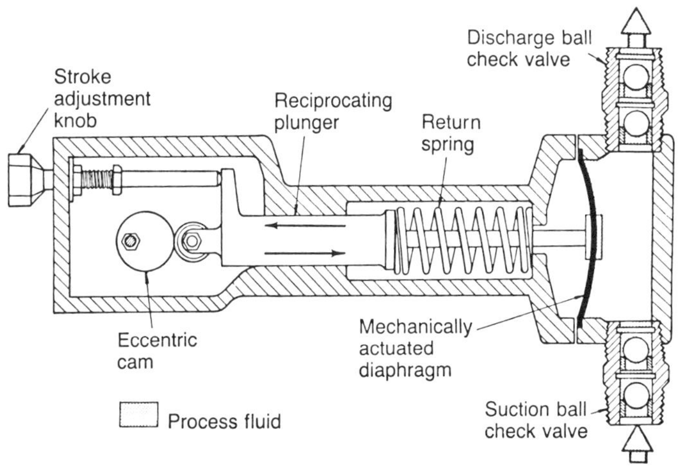

Figure 3. Eccentric cam drive (mechanical lost motion).

THE PRIME MOVER

The constant-speed AC electric motor is the most commonly used driver for metering pumps; the type of motor normally is specified by the purchaser and sized by the supplier to meet maximum operating conditions. Motors are either flange-mounted and coupled to the pump input shaft, or foot-mounted and supported on a bracket attached to the pump casing.

When specified, mechanical variable-speed drives are used with AC motors, and are either flange-mounted or supported on a common base with the pump drive. Drive speed is controlled by conical discs that are spring loaded to maintain belt tension. Belts are of the non-slip type to maintain constant speed on the output shaft. Adjustable frequency AC drives also may be employed.

Variable-speed DC motors are commonly used to vary pump output. Controllers employing silicon-controlled rectifier circuits control motor speed and often are paced automatically from a remote analog signal, as are adjustable-frequency AC drives.

Pneumatic drives use a reciprocating pneumatic cylinder to provide the necessary thrust differential between driving force and generated pressure. Pressure to the cylinder piston most often is provided by a four-way valve that alternately admits pressure to opposite sides of the cylinder piston. Actuation of the four-way valve may be accomplished either by a solenoid that changes the direction of the pressure porting or through an internal pressure porting arrangement that automatically cycles the cylinder piston. Pneumatic drives facilitate digital control because electrical on-off pulses to the solenoid are directly analogous to plunger reciprocation.

LIQUID END CONFIGURATIONS

Reciprocating metering pumps commonly utilize two types of liquid ends; packed plunger (or piston) and diaphragm. The packed-plunger liquid end contains a piston that directly contacts the pumped fluid. The diaphragm or bellows-type liquid end features some type of positive barrier arrangement (no dynamic seals) that prevents the pumped fluid from contacting the plunger.

Packed Plunger Liquid End

This is the basic design from which all types of liquid ends originate. Pumping is accomplished by one or more sealed plungers that draw in and expel fluid. One-way check valves on the inlet and outlet ends of the pump operate 180° out-of-phase to control filling of the displacement chamber during the suction (vacuum) stroke and to prevent backflow to the supply system during the discharge stroke (Fig. 4).

The plunger displaces a known volume of fluid during each cycle. Assuming it is fitted properly to the displacement chamber, this volume is equal to the volume of a cylinder having the same diameter and length corresponding to stroking distance. Actual capacity is likely to vary from the theoretical value for the following reasons:

- Fall-back loss or excessive time associated with check valve operation
- High discharge pressure (fluid compressibility)
- Entrapped gases
- Pump head elasticity
- Leakage losses through defective piping and seals.

The actual amount of liquid displaced by a properly operating metering pump should be about 95 percent of the theoretical capacity when the metered liquid is noncompressible. This actual value, known as volumetric efficiency, can be calculated as follows:

$$\text{Volumetric efficiency} = \frac{\text{Actual flow}}{\text{Theoretical delivery @ Maximum stroke} \times \text{Percent of stroke}/100}$$

The packed-plunger design has several advantages over other configurations. Among them are low initial

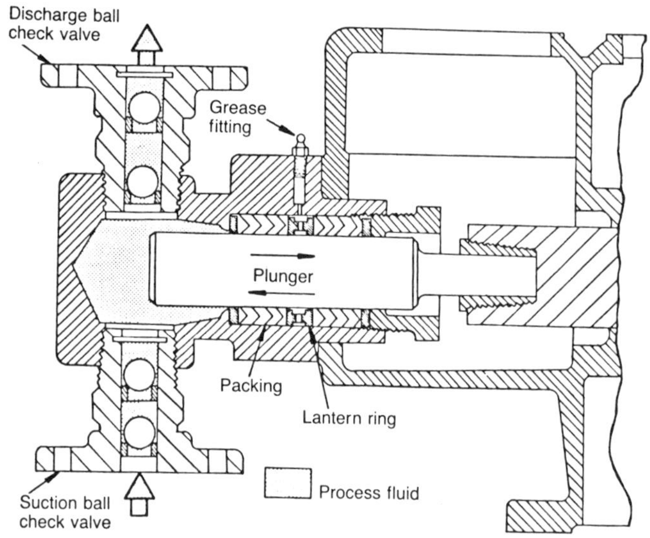

Figure 4. Packed plunger liquid end.

cost, simple design and the ability to develop high discharge pressures. In addition, the ability of the sealed piston to lift fluids to the pump inlet is limited only by the vapor pressure of the fluid. The packed-plunger pump, in its basic form, may be operated at pressures up to 7,500 psi.

The packed-plunger liquid end is a poor choice for certain applications, however, due to the design of the plunger mechanism. Packing is required to effect a tight seal between the plunger and plunger bore, and a small amount of controlled leakage of process fluid must be tolerated in most instances to cool the sliding interface. In addition, as the plunger reciprocates, the unavoidable friction that results wears the packing and increases leakage of process fluid past the plunger. Periodic adjustment is necessary since such wear decreases volumetric efficiency of the pump. Replacement eventually will be necessary as the packing progressively wears, further contributing to maintenance costs. Also, packing must be lubricated regularly.

In summary, packed-plunger pumps are employed in those applications requiring a low-cost pump, and where a certain amount of leakage and maintenance is permissible. Applications characteristically involve injection of nontoxic and noncorrosive chemicals in both batching and continuous operations.

Diaphragm Liquid Ends

Diaphragm metering pumps provide an effective solution to the leakage problem associated with packed-plunger pumps and thus are suited for handling corrosives. The plunger still travels through a bore, but instead of directly displacing the process fluid, the plunger actuates a diaphragm that acts as an interface between the plunger and the process fluid. On the opposite side of the diaphragm, a proportional amount of process fluid is drawn into the displacement chamber and discharged with each cycle of the plunger.

Diaphragms are actuated either mechanically (i.e., coupled directly to the plunger) or hydraulically. As mentioned previously, the hydraulically actuated diaphragm is inherently more suitable for continuous applications because the diaphragm is balanced between two fluids, thereby eliminating diaphragm fatigue problems and permitting higher discharge pressures.

The hydraulically actuated liquid end maintains accuracy as long as the proper volume of hydraulic fluid is maintained in the hydraulic chamber. A three-valve system is used in conjunction with amplitude-modulated drives to ensure that the hydraulic fluid is relatively free of air or entrained gases, and to protect against overpressure in both hydraulic and process fluids. Controlled porting of hydraulic fluid is used in hydraulic lost-motion drives to accomplish refill and gas-bleed action.

Check valves located on the suction and discharge sides of the pump admit and expel the process fluid in response to the negative and positive pressure exerted by the flexing diaphragm. Double check valves often are used to increase accuracy of the pump and provide redundant sealing action.

Disc Diaphragm Types

The disc diaphragm generally is made of TFE (tetrafluoroethylene), and flexes between two dish-shaped supporting (contour) plates with flow-through holes. The contour plates provide diaphragm containment and prevent rupturing under high pressure (Fig. 5). This design solves the problem of process fluid leaking past the plunger seals, and is very effective for pumping corrosive fluids such as acids. Diaphragm metering pumps mainly are used to handle liquids in applications where even minimal leakage is not acceptable at discharge pressures to 4,000 psi.

The disc diaphragm design can cause problems when viscous fluids and slurries are handled. Viscous fluids cause a very noticeable pressure drop across the small flow-through holes in the contour plates, often causing suction cavitation. The resultant random formation of gas bubbles in the fluid usually reduces the accuracy of the pump below acceptable levels. Also, particulate matter can settle out of slurries and build up between the diaphragm and contour plate.

Tubular Diaphragm Types

In the tubular diaphragm design, a plunger reciprocates as described previously, but a tube-shaped elastomeric diaphragm expands or contracts with the pressure provided by the hydraulic fluid. Constriction or expansion of the tube, combined with one-way action of suction and discharge check valves, creates a pulse of metered liquid through the pump. Cavitation problems are reduced since there is no contour plate on the process side of the tube and discharge pressures to 3,000 psi can be generated.

In the tubular diaphragm configuration (Fig. 6) the plunger acts upon hydraulic liquid located within the

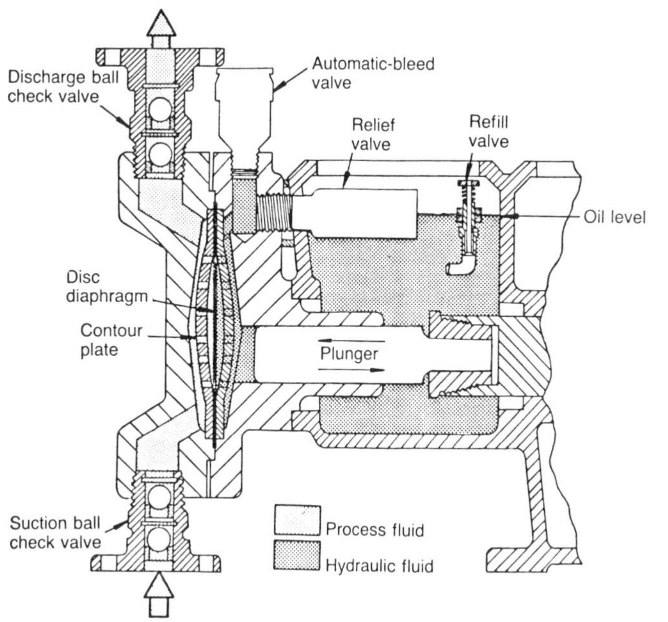

Figure 5. Disc diaphragm liquid end.

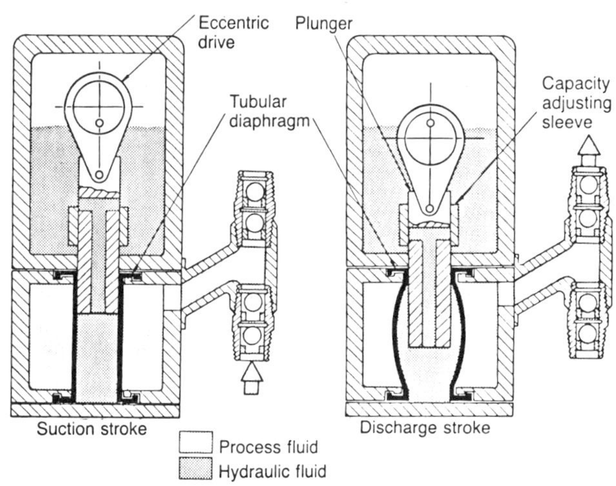

Figure 6. Single tubular diaphragm liquid end.

elastomeric tube. During the suction stroke, a vacuum is created within the tube when the plunger is withdrawn. This causes the tube to contract, drawing a predictable amount of process fluid through the suction check valve and into the displacement chamber located about the exterior of the diaphragm. The opposite occurs during the discharge stroke; pressurization of the hydraulic fluid within the tube causes it to expand and displace a proportional amount of process fluid through the discharge valve.

Other tubular diaphragm variations include the double tube (Fig. 7), and the combination disc/tube (Fig. 8). In these twin-diaphragm pumps, the plunger imparts hydraulic pressure to flex a primary diaphragm, which then flexes a secondary diaphragm via an intermediate liquid. The secondary diaphragm (tube) handles the process fluid while isolating it entirely from the pump head, precluding need for costly liquid end materials to resist corrosion by the process fluid. Further, the secondary diaphragm may be considered an extension of the suction and discharge piping, providing little resistance to flow. Such a pump ideally is suited to metering viscous liquids and abrasive slurries, and is not prone to sludge accumulation.

Twin-diaphragm pumps, in addition, offer the safety of a back-up diaphragm should the process diaphragm fail, and usually employ an inert hydraulic fluid between the diaphragms. This latter feature is useful when handling products that may react violently with hydraulic oils, or when contamination of the process chemical must be prevented in the event of secondary diaphragm failure. The intermediate chamber between the two diaphragms sometimes is equipped with a sight glass to allow monitoring. If the chamber is filled with a liquid of known pH reference color, change of color warns of diaphragm failure. Electrodes also may be provided in the intermediate chamber to monitor conductivity of the intermediate fluid. A change in conductivity can be detected by the electrodes to provide a warning of diaphragm failure.

Table 1 provides a listing of typical metering pump specifications.

METERING SYSTEMS

Metering pumps often are employed in open- or closed-loop systems. In a fairly stable process that does not dictate frequent capacity changes, a simple additive technique may be used whereby metered chemical flow rate is changed manually. Open-loop systems (Fig. 9) are characterized by the absence of feedback or corrective action by the controls; system operation is strictly proportional. Good sensitivity and a high degree of line-

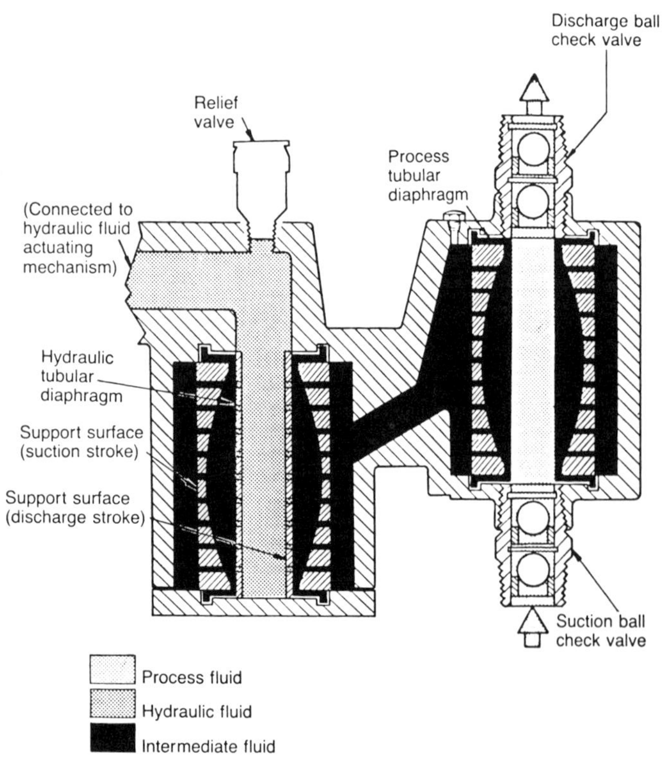

Figure 7. Double tubular diaphragm liquid end.

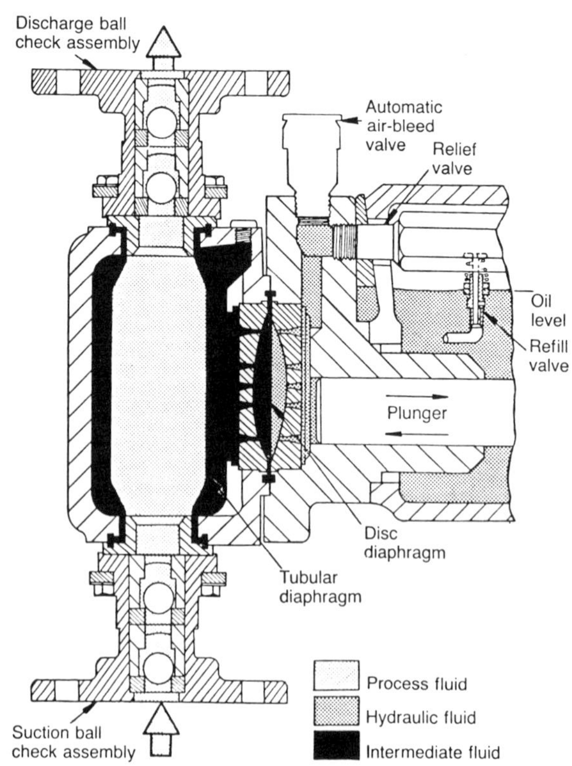

Figure 8. Disc/tubular diaphragm liquid end.

Table 1
Typical Metering Pump Specifications[1]

Pump type	Maximum pressure (psig)	Maximum capacity (gph)	Available materials of construction[2] Liquid end	Diaphragm
Packed plunger	7500	1500	A-K, Q	—
Mechanically actuated disc diaphragm	150	60	H through I	M-P
Hydraulically actuated disc diaphragm	4000	750	A through L	N
Tubular diaphragm	4000	300	A through K	M, O, P
Tubular/ disc diaphragm	1000	750	A through J	M, O, P
Double tubular diaphragm	300	250	A through K	M, O, P
Double disc diaphragm	4000	600	A through K	N, O, P

[1]Individual manufacturers may offer specialized designs that exceed pressure and capacity values shown, as well as other materials. Due to limited source selection, however, performance extremes and seldom-applied materials are not listed.

[2]Material list

A. Carbon steel
B. Grey iron
C. Ductile iron
D. 304 SS
E. 316 SS
F. Type 20 SS
G. Nickel-based alloys A, B and C
H. PVC
I. Polyethylene
J. Polypropylene
K. Acrylic
L. Polyvinylidene fluoride
M. FLU (fluorinated hydrocarbon) — viton
N. TFE (tetrafluoroethylene) — teflon
O. EPDM (ethylene propylene terpolymer)
P. CSM (chlorosulfonated polyethylene) — hypalon
Q. Ceramic

arity in the controls (including stroke control on the pump) are necessary features.

Open-loop systems require some way of adjusting the dosage to maintain the proper ratio between the measured process stream flow and the flow from the metering pump. As it usually is not necessary to attain zero-flow conditions on these systems, motor speed controls are practical. This frees manual stroke control for easy ratio (dosage) adjustment.

Closed-loop systems (Fig. 10) are characterized by feedback, or corrective action from the controller. Such controllers are quite sophisticated, and read the process variable after addition of the chemical reagent by the pump. Accuracy of the pump and its control (pneumatic or electric) is relatively important to maintaining the process variable close to the set point, or desired process conditions, without considerable modulation by the process controller.

In closed-loop systems, it is essential that the pump capacity be able to go to zero. It is more practical to attain zero output with stroke control than by varying motor speed, as many variable speed drives cannot be satisfactorily operated at low speed. Capacity controls in closed-loop systems usually are pneumatic or electric analog devices.

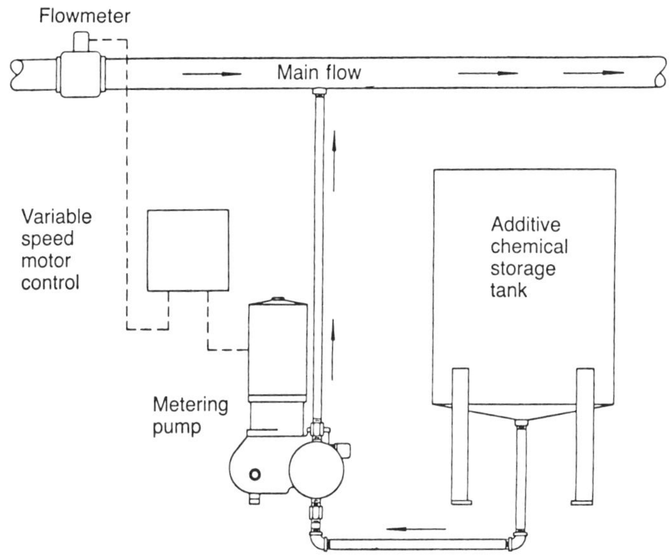

Figure 9. Typical open-loop control system with metering pump as the final control instrument.

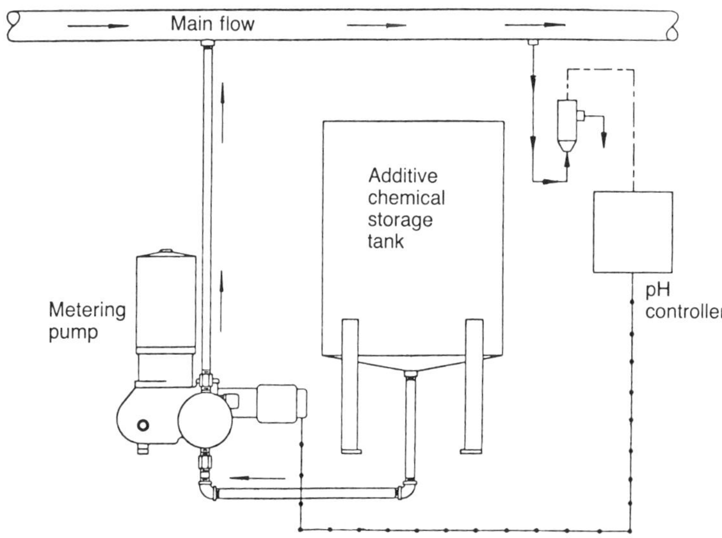

Figure 10. Typical closed-loop control system with metering pump as the final control instrument.

A metering pump can be used as a final control element for two process variable controllers. For example, when the main process liquid flow rate varies considerably, a closed-loop feedback system is not adequate for following large sweeps in flow rate as well as changes in the process variable. A combination of motor speed and capacity control can be used to achieve the desired range in this event, however. An open-loop proportional feed system cascaded to a closed-loop process variable control system, exemplifying combination control, is shown in Fig 11.

Turn-down ratios also should be considered in control selection. If a metering pump has a 10: 1 turn-down ratio, the output can be adjusted very accurately between 100 and 10 percent of rated flow. Below 10 percent, however, it is difficult to achieve accurate delivery without further refining the accuracy. Such action is accomplished by combining stroke control with motor speed control. With the motor running at 1/6 of rated speed, for example, and the capacity adjustment set at 10 percent, actual output would be .167 × 10 percent of total pump capacity.

Energy considerations also may enter into the selection process. Energy-efficient motors can be specified, and the sinusoidal output of metering pumps permits operation of more than one pump on the same drive without an appreciable increase in torque requirements. Figure 12 shows that torque requirements of simplex, duplex and triplex pumps are virtually identical when the pumps are phased properly.

In extremely hazardous locations, pneumatic drives and controls may be utilized. These direct-coupled air cylinder drives also provide control versatility by way of accessory electric modules such as a stroke pacer (for speed control), a stroke counter (for pumping a predetermined volume based on stroke count), or an analog-to-digital signal converter. Independent adjustment of plunger traverse speeds also is available on many pneumatic units for handling high-viscosity fluids and for improved suction lift.

PACKAGE CHEMICAL FEED SYSTEMS

Chemical feed systems are available as complete package units from many pump manufacturers. These

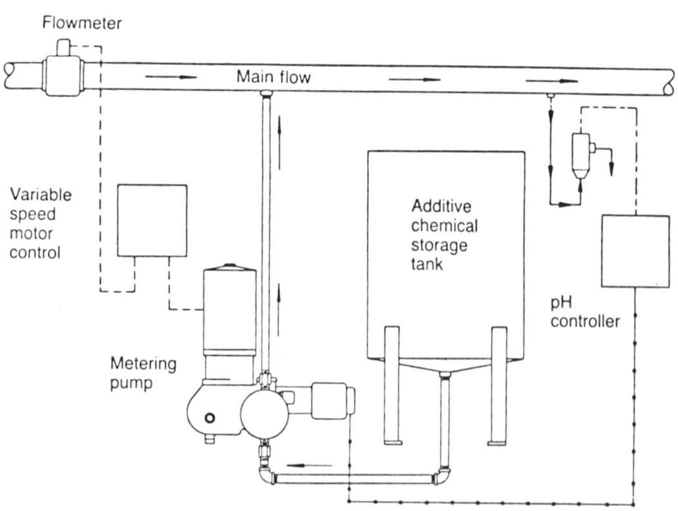

Figure 11. Combined open- and closed-loop control system with metering pump as the final control instrument.

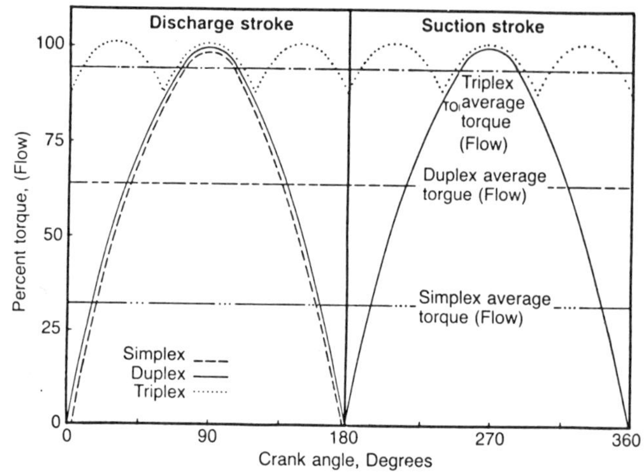

Figure 12. Peak torque requirements for simple, duplex and triplex metering pumps.

systems usually consist of a supply tank ranging in capacity from 50 to 1,500 gallons, an agitator, valves, piping and instrumentation, all mounted on a common frame or skid. The systems generally are provided ready for installation on the job site.

Chemical feed systems are used routinely for a wide variety of chemical processing applications and generally are produced as custom designed units. Water treatment feed systems usually are compact, and also may be specifically designed for each application.

ACCESSORIES AND CONTROLS

A wide variety of accessories may be used with metering pumps:

- Automatic controls to open valves and flush the pumping system
- Anti-siphon or back pressure valves, which prevent liquid from flowing through the pump without proper metering action
- Metering rate timers and counters to totalize pump pulses and to shut off flow when a specified limit has been reached
- pH analyzers/controllers, which frequently are used in closed-loop systems to adjust output of the metering pump
- Pulsation dampeners to overcome acceleration losses in the suction line and reduce pressure surging in the discharge line
- Safety relief valves, which prevent system overpressure
- Steam jackets to permit heating or maintain temperature of the process fluid
- Electric or pneumatic capacity controllers to permit automatic adjustment of capacity based on feedback from other process instrumentation.

22
Ultrasonic Flowmeters

E. A. DeVries

Ultrasonic flowmeters have been around since the 1960s and were used for oceanographic and military applications before they were introduced to the process industries. There are several different kinds of ultrasonic flowmeters, but they all work on the principle of frequency shift due to velocity. This chapter reviews our experiences, both good and bad, with both Doppler and transit time ultrasonic flowmeters.

The most common type of ultrasonic flowmeter is the Doppler meter. Its signal is reflected by particles, bubbles, or other discontinuities in the liquid, and the frequency is shifted by velocity of these discontinuities. This frequency shift renders a signal proportional to velocity.

Another common type of ultrasonic flowmeter is the "time-of-flight" or transit time meter. This meter uses two sensors that are lined up at an angle to the direction of flow, and that pulse alternately. A time-differential relationship proportional to the flow is calculated. In this case, the fluid must be free of entrainment or else scattering of the signal may induce an error.

There are other newer types of ultrasonic flowmeters, such as those that use combinations of these two principles, or ones that use the concept of multiple sonic waves in the pipe. However, we have not used these newer types so they will not be discussed.

DOPPLER METER LIMITATIONS AND STRENGTHS

Doppler meters have many limitations. First, they must have particles to work. The various Doppler manufacturers report that a minimum of 25 to 100 ppm of 30 to 100-micron-size particles are needed to obtain a good signal. Secondly, since these are velocity meters, a good flow profile is needed. This means that standard upstream and downstream straight runs, turbulent flow conditions, and a full pipe are all required. Another limitation is that Doppler meter signals do not penetrate air. This means that porous pipe materials such as concrete, concrete-lined, or many types of fiberglass cannot be used. Other limitations of Doppler meters include pipe temperature, pipe thickness and ambient extremes.

The principal strengths of Doppler flowmeters are the nonintrusive sensors and the relatively low price. This makes the Doppler meter attractive for measurement of certain nasty fluids, and for measurement of flows in large pipes.

APPLICATION EXPERIENCES

Our plant's first few experiences with Doppler flowmeters came in the mid-1970s with poor results. We first tried one on a polymer slurry that seemed to fit all the necessary application criteria. We were able to obtain an output signal, but we were unable to obtain any accuracy or repeatability, even with field calibration. Either our Reynolds numbers were marginal, or the particle size and density were varying. The result was that sometimes we had a good signal and sometimes we did not. We abandoned the Doppler meter for this application. We later installed a wedge element flowmeter, which has worked out well.

The second time we tried the Doppler meter was on the centrate out of a monomer slurry centrifuge. The centrate still contained about 1% particulate, and all conditions seemed to indicate that this would be an ideal application for a Doppler meter. However, we were again unable to get a repeatable signal, even with field calibration. After some lab tests, we verified that the particle size and density did vary greatly. We now have magnetic flowmeters in this service.

We now know that there is a relationship between the particle size and the minimum required particle concentrations. Figure 1 shows how the beam penetration changes with fluctuation in solids concentration. If the penetration depth is changed significantly, the Doppler meter will generally read incorrectly.

These phenomena have been a sizeable limitation in our Doppler meter installations. Now there are microprocessor-based Doppler meters that attempt to correct for this phenomena. However, we have not tried these newer types due to their relatively high cost.

Our first moderate success with Doppler meters came in the late 1970s on two wastewater flows. These meters were specified and installed by an environmental specialist who was familiar with the meter's limitations. The installations were difficult because of the type of fiberglass pipe that we used (Fig. 2). It was necessary to adjust both the sensor mounting, and the gain and frequency of the electronics, to overcome the trapped air bubbles in the wall of the fiberglass pipe. However, we did finally get a useable signal, and our field tests seemed to indicate that the accuracy of these meters was as good as that of head type flowmeters. Later however, we experienced some very heavy rains that caused the particle size and distribution in the wastewater to change significantly. Whenever this happens, it appears that we get large errors. However, we have been unable to perform any field accuracy checks to find out conclusively. On the whole, however, we have been happy with this particular application, and these Doppler meters are still in service. These were our first successful installations of ultrasonic flowmeters.

Our best success with Doppler meters was a recent application on chemical sludge flow at our effluent plant (Fig. 3). On startup we performed a field calibration test, and adjusted the transmitter slightly to get an accurate reading. We since have had one main electronic board failure on this transmitter, right after the warranty ran out. The service call for this was expensive—over half the price of a new flowmeter. However, on the whole we have been very pleased with this installation.

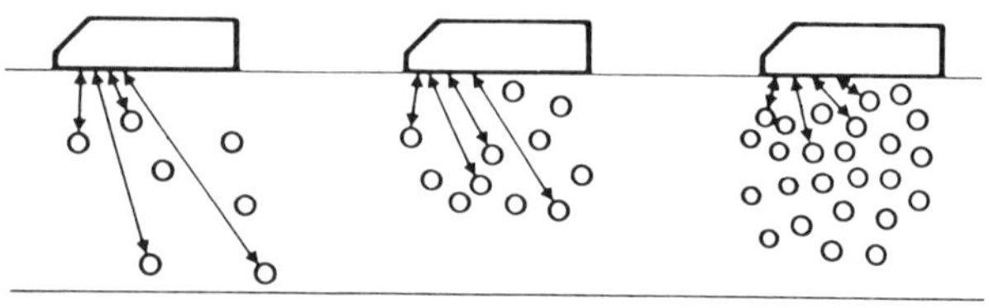

Figure 1. Beam penetration with changing solids concentration.

Figure 2. Clamp-on Doppler meter on a large fiberglass pipe for wastewater service.

Figure 3. Clamp-on Doppler meter on plant effluent sludge.

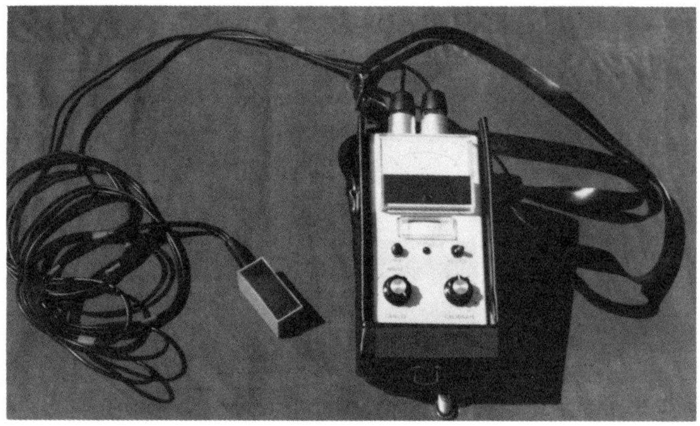

Figure 4. Hand-held Doppler meter for testing and calibration.

We are now trying a Doppler meter on a 50% caustic (sodium hydroxide) application. This has historically been a tough application for flowmeters. Thus far, everything looks favorable. However, we have not been able to perform any field calibration tests to verify the meter accuracy.

Figure 4 shows a hand-held portable Doppler meter that we use for testing and calibration. We are unable to use this where accuracy is important because its readings cannot be verified. However, it has been very useful for setting low flow switches and for testing parallel centrifugal pumps to verify that the flow is balanced between them. We have also used this meter for studies of cooling water distribution systems, but have not always been able to get a reading on the meter. Its beam is not

Figure 5. Time-of-flight meter being installed (top) on a molten sulfur line (much of the cabling is heat tracing) and final installation (bottom).

powerful enough. A great deal of particulate is needed to get a reading.

Time-of-flight type ultrasonic meters seem to be superior to the Doppler type in terms of accuracy and repeatability. They are also more expensive, and made for cleaner applications. We are using this type of meter for molten sulfur service because this was recommended to us by the Sulfur Institute. Figure 5 shows the installation. This meter has performed very satisfactorily in this service. The flow range was changed after startup, and the service cost was again relatively high. However, on the whole we have been pleased with this flowmeter.

In conclusion, ultrasonic flowmeters can be used for some tough applications, but they can also be extremely troublesome. All conditions must be right for them to work satisfactorily. There are some microprocessor-based ultrasonic meters available now that reportedly have overcome some of these limitations. However, we have avoided these new types due to their relative high cost. For tough applications, we prefer to use more tried and true flowmeters such as magnetic flowmeters, coreolis type mass flowmeters, or wedge element flowmeters. These are also relatively expensive, but our confidence in them is high. It seems to be easy to get good results from them. Most of the time it takes a good deal of effort to get acceptable results from ultrasonic flowmeters, particularly Doppler meters. However, there are a few applications that they are suited for, such as effluent sludge and molten sulfur.

Bibliography

Baird, J. D., "Recent Application of Unique Noninvasive Doppler Transducers for Flow Measurements," ISA Conference and Exhibit, 1982.

DiNapoli, L., "Applications and Principles of Doppler Flowmeter," ISA Conference and Exhibit, 1981.

Torrance, J. W., "Confusion and Delusion in Sonic Flowmetering," ISA Conference and Exhibit, 1973.

Waller, J. M., "Considerations in the Applications of Doppler Flowmeters," ISA Capital Cities Control Conference, 1980.

23
Controller Tuning

D. W. Juska

Quite often, controller tuning methods use information obtained by oscillating the control loop with a constant amplitude cycle under proportional-only (P) control. Simple calculations can then be made to arrive at settings for two-mode proportional and integral (PI) or three-mode proportional, integral, and derivative (PID) controllers. Calculations done in this manner are somewhat simpler and faster than those based on an open-loop process reaction curve. The problem, of course, is that very rarely can a loop be placed in a constant amplitude cycle.

This chapter describes a method by which information may be gathered identical to that obtained by cycling the loop. It uses only the open-loop process response curve. The advantage is simpler calculations related to these constant cycling methods.

A technique similar to the Ziegler-Nichols[1] process reaction curve method is used to obtain an approximation of the dead time, τ_{dt}, and capacity time constant, τ_1, assuming a self-regulating capacity.

PROCEDURE

1. Using the loop shown in Fig. 1, place the controller in manual, and step the controller output, Δm, by an amount sufficient to obtain a measurable steady-state change in process measurement, Δc. On a chart recorder, obtain the process reaction curve shown in Fig. 2. The process steady-state gain (final actuator, process, and transmitter), G_p, can now be calculated as: $G_p = \Delta c/\Delta m$ where Δc is the change in transmitter output in percent, and Δm is the change in controller output in percent.

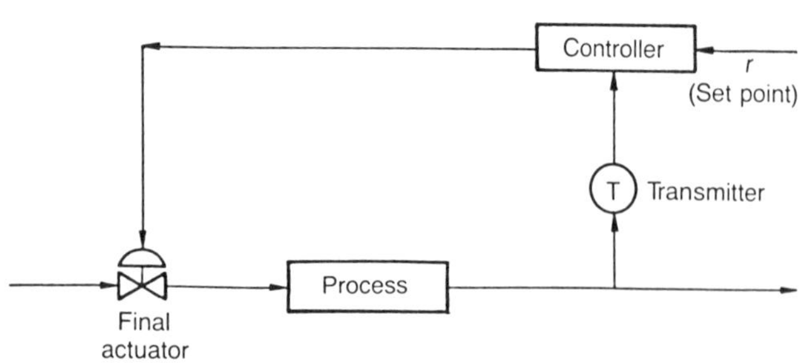

Figure 1. Loop block diagram.

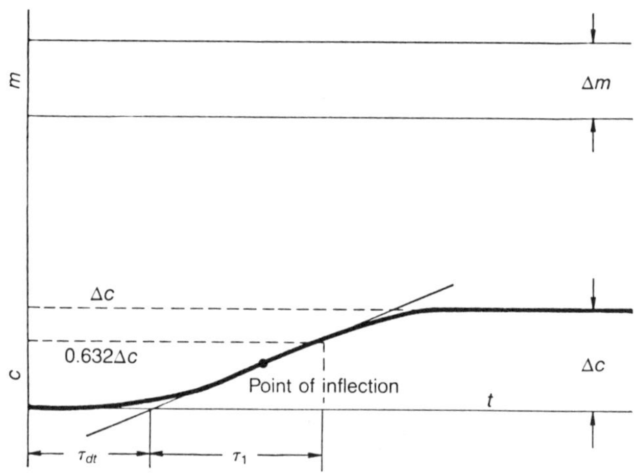

Figure 2. Process reaction curve.

2. Draw a line tangent to the point of inflection of the process response curve (Fig. 2). Note the point where this line crosses the X-axis. From this point to the point where the process response curve reaches 0.632 Δc is an approximation of the capacity time constant, τ_1. From the point in time when Δm was originally made to the point where the tangent line crosses the X-axis is an approximation of the dead time, τ_{dt}.

The rise of the process response curve away from the X-axis at the τ_{dt} point is because of multiple capacities in the loop that add both apparent dead time and capacitive lag. What, in fact, is happening is that the process is being modeled with a steady-state gain, G_p, a dead time, τ_{dt}, and a first-order lag with time constant, τ_1.

3. Use these values of τ_{dt} and τ_1 to arrive at an approximate dead-time-to-capacity ratio, τ_{dt}/τ_1, and enter Column 1 of Table 1 with this ratio.

Table 1
Relationship of τ_{dt}/τ_1 to Loop Natural Period and Controller Proportional Band

Column 1 τ_{dt}/τ_1	Column 2 τ_n/τ_{dt}	Column 3 $PB_u(\%)/G_p$
0	4	0
0.12	3.83	7
0.176	3.75	10
0.24	3.67	14
0.38	3.53	21
0.53	3.4	28
0.70	3.27	34
0.88	3.16	41
1.1	3.05	47
1.33	2.95	53
1.60	2.86	59
1.9	2.77	64
2.26	2.69	70
2.67	2.61	74
3.17	2.54	79
3.77	2.47	83
4.53	2.4	87
5.5	2.34	90
6.82	2.28	93
8.71	2.22	95
11.6	2.17	97
16.8	2.12	98.5
28.9	2.07	99.5
89.0	2.02	99.9

Columns 2 and 3 will give values of the natural period, τ_n, in terms of τ_{dt}, and the ultimate proportional band, PB_u, in terms of G_p, corresponding to this dead-time-to-capacity ratio. This information can now be used to calculate settings for one-, two-, or three-mode controllers to give a quarter amplitude damped response (see Appendix).

With this technique, it is not necessary to uniformly cycle a closed loop under P-only control to tune a PI or PID controller using the values of τ_n and PB_u. For example, use the process response curve shown in Fig. 2 to obtain: $\tau_{dt}/\tau_1 = 0.38$. Now enter the table in Column 1 at $\tau_{dt}/\tau_1 = 0.38$. This gives $\tau_n = 3.53\ \tau_{dt}$ and $PB_u = 21\ G_p$. Use the process response curve once more to obtain $G_p = \Delta c/\Delta m = 3.5$. Also from the process response curve, it is found that $\tau_{dt} = 10$ seconds so, $\tau_n = 3.53\ \tau_{dt} = 3.53\ (10)$ or $\tau_n = 35.3$ sec and $PB_u = 21\ G_p = 21\ (3.5) = 73.5\%$.

Suppose it becomes necessary to find tuning constants for a PID controller. Going to the Ziegler-Nichols constant cycling method given below to obtain the settings for a three mode PID controller for the process:

Controller	G	I(min/rep)	D(min)
P	0.5 G_c	—	—
PI	0.45 G_c	$\tau_n/1.2$	—
PID	0.6 G_c	$\tau_n/2.0$	$\tau_n/8.0$

Then,

$$G = 0.6\ G_c = 0.6(100/PB_u) = 0.6(100/73.5) = 0.82$$

or

$$PB = 100/G = 100/0.82 = 122.5\%$$

$$I = \tau_n/2.0 = 35.3/2.0 = 17.65 \text{ sec/rep}$$
$$= 0.3 \text{ min/rep}$$

$$D = \tau_n/8.0 = 35.3/8.0 = 4.4 \text{ sec} = 0.07 \text{ min}$$

Keep in mind that these controller settings are based upon information obtained at a certain loop operating point and therefore, subject to change with a load change on the process if appreciable capacity exists. Also, since ambiguities as large as ± 20 percent in the controller settings may exist in an analog controller, it may be necessary to slightly "trim" these adjustments for the optimum quarter amplitude damped response.

Finally, some rules of thumb may state that for $(\tau_{dt}/\tau_1) > 0.5$, derivative action may not be very helpful. When using the table and a very small value of derivative time is calculated for a large dead-time-to-capacity ratio, a two-mode PI controller rather than a three-mode PID controller may be more appropriate.

APPENDIX

Definition of Quarter Amplitude Damping

In Fig. 3, A is the amplitude of the initial excursion from the set point. τ_n is the period of constant amplitude oscillation of a control loop under proportional-only control. PB_u is the so-called "ultimate proportional band," or that value of the proportional band which

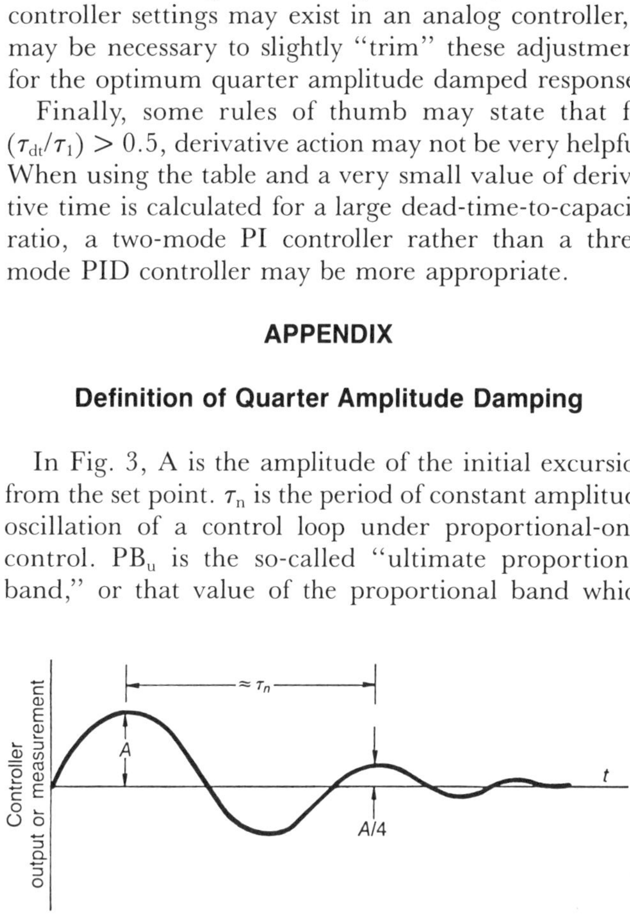

Figure 3

causes the control loop to uniformly cycle at τ_n. Proportional gain of a controller is $G_c = 100\%/PB$.

Sample Calculation for a Table Entry[2]

Given the following process model in Fig. 4 (including effects of the valve and transmitter):

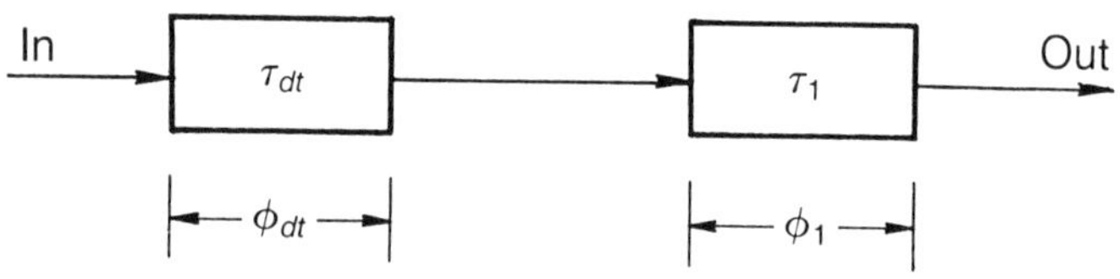

Figure 4

Where ϕ_{dt} = phase shift across dead time, τ_{dt}, and ϕ_1 = phase shift across capacity, τ_1.

1. Assume $\phi_1 = -78°$
2. Then at uniform oscillation, assuming $-180°$ phase shift across summer set to reverse acting:
   $\phi_1 + \phi_{dt} = -180°$
   $\phi_{dt} = -180° - \phi_1 = -180° - (-78°)$
   $\phi_{dt} = -102°$
3. Then: $\tau_n/\tau_{dt} = -360/\phi_{dt} = -360/-102 = 3.53$. This is Column 2 in the table.
4. $\tau_n/\tau_1 = -2\pi/\tan\phi_1 = -2\pi/\tan(-78°) = 1.33$
5. $\dfrac{\tau_n/\tau_1}{\tau_n/\tau_{dt}} = \dfrac{\tau_{dt}}{\tau_1} = 0.38$, this is Column 1 in the table.
6. Dynamic gain of the capacity τ_1 is given by:

$$G_1 = \left(1 + \left(2\pi \frac{\tau_1}{\tau_n}\right)^2\right)^{-1/2}$$

$$G_1 = \left(1 + \left(2\pi \frac{1}{1.33}\right)^2\right)^{-1/2} = 0.21$$

7. The loop gain
   $G_L = G_c G_{dt} G_1$; $G_p = \Delta c/\Delta m$
   where $G_{dt} = 1$*
   $G_L = (100/PB)\ G_p(0.21)$
   for uniform oscillation $G_L = 1$
   $1 = (100/PB)0.21G_p$
   $1 = 21G_p/PB$
   $PB/G_p = 21$

References

1. Ziegler-Nichols, "Optimum Settings for Automatic Controllers," Transactions of the ASME, November 1942.
2. Shinskey, F. G., *Process Control Systems,* 2nd Edition, McGraw-Hill (1979).

* Ibid., p.10

24

Specifying Control and Isolation Dampers

J. M. Rogers and T. F. Edgar

Dependable damper machinery is essential to the operation of a wide variety of equipment and processes throughout the hydrocarbon processing industry including power generation, process furnaces, heat and catalyst recovery systems and pollution control equipment. Air and feedstock preheaters, forced draft and induced draft fans, bypass ductwork and many other systems depend directly on the performance of damper machinery. Efficient dampers ensure maximum furnace and fuel efficiency, regulate air and gas flows to control processes under changing conditions, and isolate equipment for man-safety and maintenance to avoid total system shutdowns.

But careful design, construction, and installation standards for dampers have not been applied consistently in HPI facilities, resulting in a pattern of inefficient damper operation and failure throughout the industry. A study of some 250 damper installations at 18 HPI facilities throughout the U.S. has revealed numerous examples of misapplications, poor design and poor construction that have resulted in inefficiency and energy waste, costly shutdowns and unsafe working conditions.

Most of these failures are unnecessary. They can be avoided in new installations through a better understanding of damper functions, proper application of various damper types, careful design of individual dampers to meet specific system requirements and strict adherence to quality control standards in construction, testing and installation procedures. The purpose of this discussion is to inform the process engineer about these design considerations and hardware alternatives in the selection of dampers.

DAMPER TYPES AND HPI APPLICATIONS

There are two main types of dampers used in HPI facilities: slide gate (guillotine) dampers and multivane louver dampers. In addition, there are other less common applications that require specialized designs.

Slide gate dampers should be specified for most isolation applications where their excellent sealing capabilities, strength and low pressure drop (Fig. 1) can be most advantageous.

A standard slide gate damper is built around the "sliding blank" concept; a single solid blade which "slides" across a duct to present a solid wall of metal to the oncoming gas stream. The closed blade should extend six inches beyond the bottom edge of the duct and be surrounded by metallic seals at the top and bottom of the damper frame. When open, a slide gate damper presents a completely open duct to the gas stream, with no braces, seals, or any additional surfaces where fly ash, catalysts, or other particulates can collect.

Where man-safety requirements call for 100% isolation to allow duct entry for maintenance or repair, a zero-leakage slide gate damper (Fig. 2) includes metallic seals mounted on all four sides of the duct and adds a pressurized air seal around the closed blade. An external blower pressurizes the seal air chamber with clean purge air. Developing a pressure greater than that of the upstream gas, the blower ensures that only fresh air can leak into the duct.

Slide gate dampers are intended to be used in either the fully-open or fully-closed position, and the motion of the blade is relatively slow. They should be specified for isolation rather than control or modulation applications.

Slide gates are used in natural air draft doors, in flue gas desulfurization systems and in bypass ducts essential to the maintenance of other pollution control and heat recovery units.

Where there is more than one train of pollution control equipment, for example, downstream of a furnace or boiler, slide gate dampers are used to direct the gas flow toward or away from precipitators, baghouses, etc. This is often the case downstream of catalytic crackers, pyrolysis furnaces and other reformers where catalysts

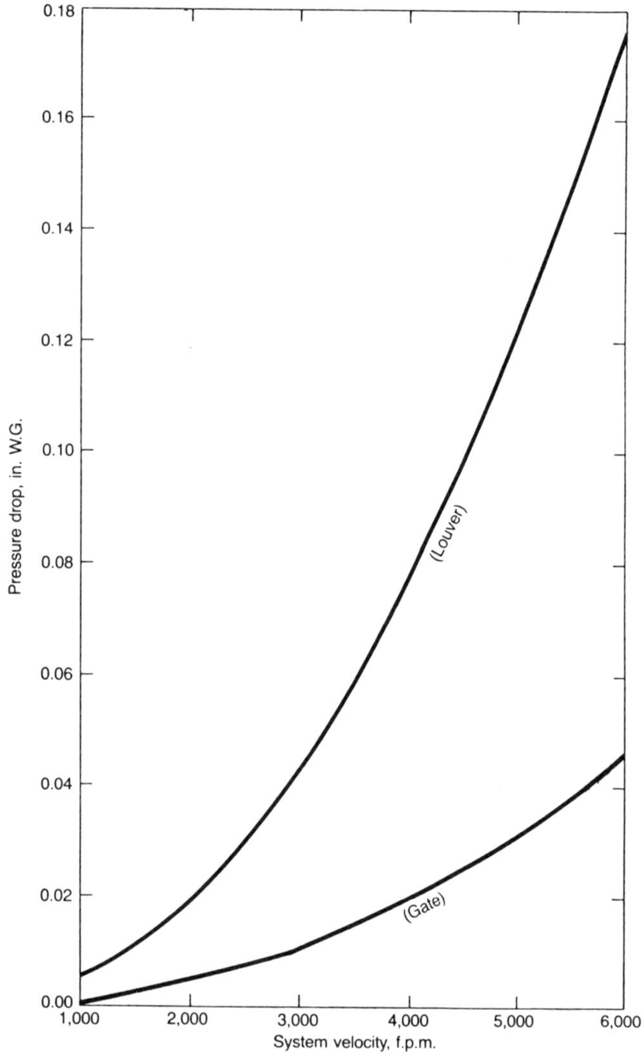

Figure 1. Total pressure drop vs system velocity at 300°F.

are used. The high temperatures found in these systems demand especially well designed dampers.

When more than one boiler or furnace exhausts into a common stack, slide gates can isolate an out-of-service unit without affecting the operation of the other(s). Standard (low leakage) slide gate dampers are satisfactory for natural draft stacks. The presence of an induced draft fan, however, might allow hot gases to be blown back into the equipment that is shut down. Zero-leak gates are then required.

Louver dampers are used to modulate the volume of gas flowing through ductwork or process equipment. They provide a wide range of adjustability as well as quick opening and closing capability. Actuator safety factors must be given special consideration because torque requirements are often higher than for slide gate dampers.

There are two basic types of multi-blade louver dampers: parallel-blade and opposed blade. Parallel-blade louver dampers (Fig. 3) include full metallic seals between blades, at the top and bottom of the duct, and at the side jambs. They provide tight shut off where a moderate amount of leakage (min. 1/2%) is acceptable and where their operating speed and compact design make them preferable to standard slide gates. Parallel-blade louver dampers are commonly used on the upstream side of induced draft fans, where the distorted

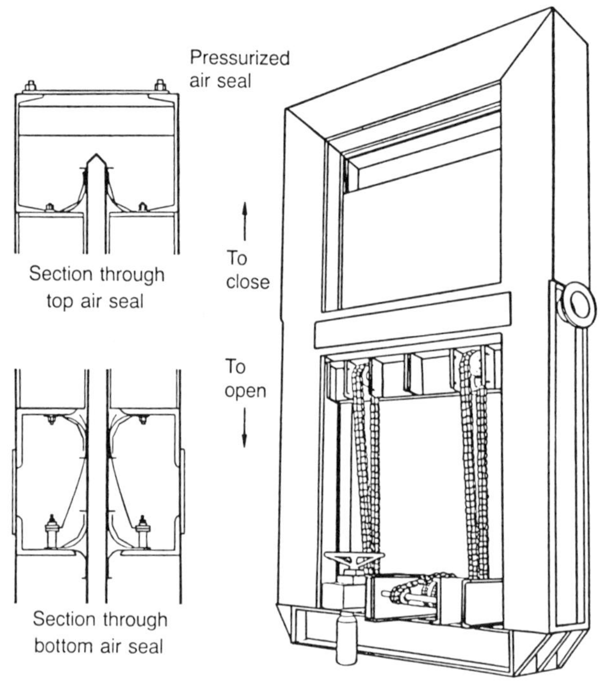

Figure 2. Zero leakage gate damper.

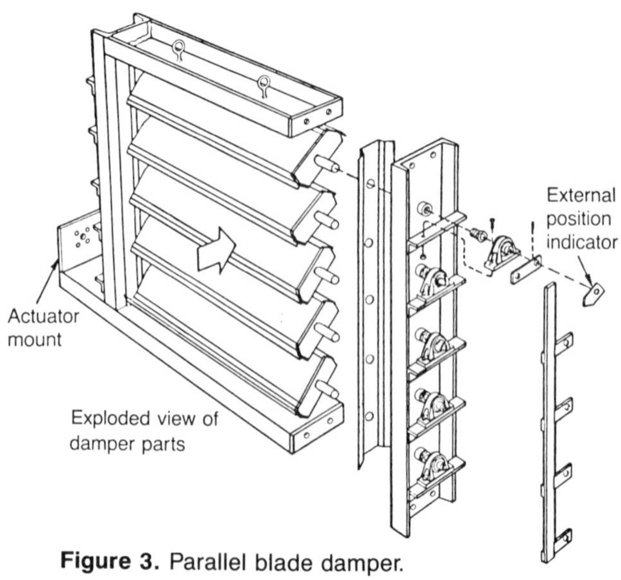

Figure 3. Parallel blade damper.

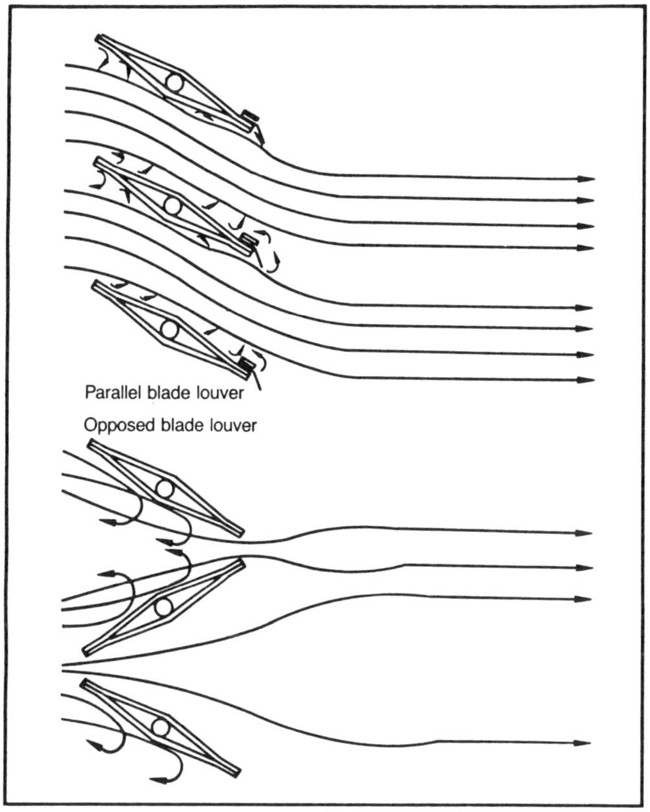

Figure 4. Air flow patterns for parallel and opposed blades.

(spinning) gas flow pattern created by the blades does not adversely affect fan efficiency (Fig. 4).

Parallel-blade louver dampers are also used with heat recovery units, such as air preheaters and waste heat boilers, when they are added to existing furnaces. They are typically installed in the stack directly above the furnace to force the hot gases into the heat recovery equipment. Quick opening ability is essential in this application to allow the furnace to stay on line in the event of an ID fan failure.

For modulation of gas flow exiting a forced draft fan, an opposed-blade louver damper can be adjusted to control both volume and static pressure throughout most of its open/close range while minimizing distortion of the system's gas flow pattern. Opposed-blade louvers, however, cannot generally provide as tight a seal in the closed position as parallel-blade louvers.

Modulating opposed-blade louver dampers are typically required to control and balance the varying drafts of two or more furnaces which exhaust into a common heat recovery device. They are also used to dissipate pressure downstream of fans. For example, a fan is often used to induce draft through an air preheater, forcing the heated air into the burner plenum of a furnace. Such furnaces usually operate in a balanced, or non-pressurized state, so fan outlet pressure must be dissipated.

Although multi-blade louver dampers are most often square or rectangular, they are commonly and effectively used with round ductwork. Transitions to and from square, rectangular and round configurations present no particular design problems.

Where quick closing speed is a consideration or where limited space around a duct makes a slide gate impractical, two louver dampers in series with a purge air blower between them can provide good isolation. A variation of this series concept has been applied in the design of gas turbine heat recovery systems for Alaska's North Slope installations. The louver on the upstream side of the damper series uses opposed blades and a separate actuator for efficient modulation. Parallel blades on the downstream side, plus the purge-air blower, provide tight shut off when necessary. In normal operation the downstream parallel blades are kept fully open to present minimum resistance in the gas stream while the upstream opposed blades are adjusted for modulation. In the event of boiler upset, both sets of louvers can be completely closed. The blower ensures tight shut off for safe entry into the duct.

Even in the fully-open position, louver damper blades and seals necessarily interfere with gas flow and cause static pressure drops that must be considered in fan specifications. These same obstructions also tend to accumulate particulates and are vulnerable to corrosives in the gas stream.

Special Purpose Dampers

Slide gate and multi-louver dampers are best suited for most isolation and control/modulation applications, but there are additional damper types which can be effective in specific applications.

Radial vane dampers are well suited for use at the inlets to centrifugal fans. The radial vane damper's design can boost fan efficiency because the vanes impart a whirling motion to the air as it enters the fan. When a radial vane damper is to be used with a high temperature or corrosive laden gas stream, it is essential to specify a "cantilever" vane design which keeps all support members and mounting bearings safely outside the gas stream (Fig. 5).

In small (less than 6 ft. diam.) ductwork, **single blade louver dampers** with seals can provide adequate isolation with leakage limited to the 1/2% range.

Wafer or butterfly dampers are widely used for stack control where leakage is not a critical factor and the wafer's simplicity makes it economically attractive. Bearing seizure is a common problem with wafer dampers, however, and the potential cost of repairing or replacing defective bearings should be carefully weighed against an initial savings afforded by low cost equip-

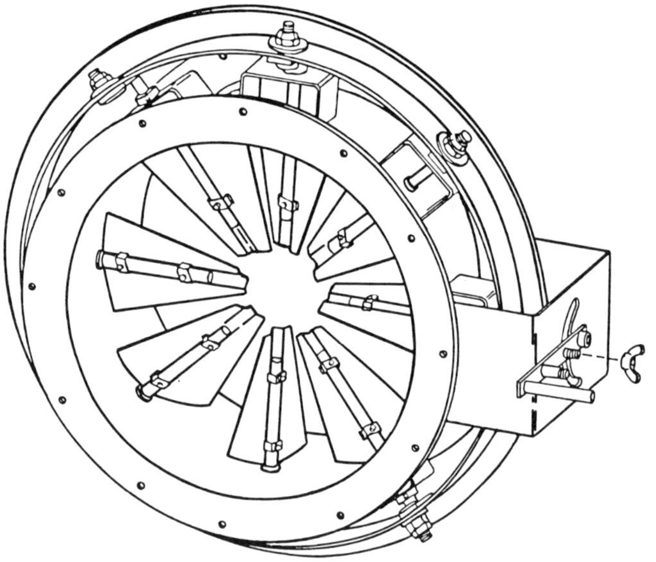

Figure 5. Radial fan dampers.

ment. Special bearings are often the most effective solution to this problem.

SPECIFYING AND DESIGNING INDIVIDUAL DAMPERS

Selecting the type of damper best suited to a particular application is only the first step in determining the design of the specific damper that will perform most efficiently in that application. Mechanical specification and design must include consideration of such factors as temperatures, system pressures, particulates, and corrosives carried in the gas stream. Each of these factors will affect the choice of materials; structure of the blades, frame and supports; seals; bearings; actuators; drives and couplings.

Temperature

Process temperatures of 1,500°F and higher make thermal expansion and contraction factors in damper design. The temperature on the upstream side of a tightly closed damper, for example, can be as much as 500°F higher than that on the downstream side causing the skins of a louver damper's blades to expand differently. Moreover, the blade temperature will often be different from that of the frame. Occasionally, refractory linings are applied to exposed damper parts. Insulating the exterior of the damper frame will help to minimize the effect of temperature differences by maintaining a more uniform thermal gradient throughout the entire damper structure.

Airfoil-shaped louver damper blades are best constructed of two steel sheets bolted at the mating edges and through the blade shaft. To keep the blades and shaft from shearing the bolts during expansion and contraction, the bolt holes must be slightly oversized.

Welding should be avoided whenever damper blades will be exposed to extreme temperatures. The rigid welds can stress relieve as the blades and shafts expand and contract leading to warping, binding and eventually to failure of the seal—and the damper.

Blade shafts must extend through the full length of louver blades, but cost and weight factors often make solid shafts impractical. This problem can be solved by using stub shafts at the blade ends, with a full length hollow shaft joining the ends. Varying rates of thermal expansion must again be considered.

Pressure

Operating pressures ranging from −30 to +50 in. W.G. are common in HPI damper applications. To withstand these pressures, a slide gate damper blade should be thick and solid. Thinner blades can buckle under high pressure surges near furnaces, and stress relief warping of thin blades exposed to extreme temperatures can result in damper failure from blade binding or loss of seal.

In louver dampers, all fasteners and linkages must be extremely tight fitting to resist the natural tendency of the blades to vibrate in a high velocity gas stream. Such vibration can destroy blade edge seals or crack blade shafts and bearings.

Particulates/Corrosives

The contaminated atmosphere in which many dampers must operate requires special consideration. Damper blades, seals and frames must be constructed or coated to withstand the corrosive properties of SO_x, NO_x, or any number of catalysts or chemicals present in various processes. Stack dampers, where temperatures are close to the dew point, and dampers around wet scrubbers are particularly susceptible to corrosive attack.

Particulates, including process stream catalysts and furnace carbon and fly ash, are also of particular concern. The interior of the damper frame must be free of all unnecessary obstructions including cross member bracing, internal struts, blade guides and (on louver dampers) angle and bar stops for blades. Such obstructions accelerate particulate buildup, inhibit gas flow and cause unnecessary pressure drops. The result can be damper malfunction and eventual failure.

Slide gate damper design must also determine whether the blade should enter the duct from the top, bottom or side. Bottom entry is most often preferred because the blade will naturally break up particulate buildup around the bottom seal, and the buildup will not be compacted in the damper frame. Where space limitations or other factors preclude bottom entry, side entry is the next best alternative. In certain systems, such as wet scrubbers, top entry protects the blade, drive and actuator members from the corrosive liquids that might drip down through the seals. The frame and seals on top entry gates can be protected against this corrosion through the use of special alloys. Table 1 lists some typical damper materials.

Table 1
Typical Damper Materials of Construction

	under 700°F	700°–1000°F	1000°F+
Frame	A-36	A-242 (Corten)	316, 316 L 317, 317 L, 317 LM Inconel 625, RA 330 TX Hastelloy G
Blades	A-36	A-588 (Corten)	316, 316 L 317, 317 L, 317 LM Inconel 625, RA 330 TX Hastelloy G
Seals	304 SS (Louvers) 301 SS Half Hard (Gates)	Inconel 625	Inconel 625

Blade Seals

Flexibility and resistance to corrosion and heat are important factors in the design of seals for both slide gate and louver dampers. Metallic seals are generally required; seals of nonmetallic composition materials often do not provide effective heat resistance. Inconel 625 and 317 stainless steel are the materials most often used.

Seals on slide gate dampers must prevent leakage into the frame and the atmosphere when the gate is open as well as closed. They must be mounted inside the frame where they can perform this open/closed sealing function without interfering with the gas flow or collecting contaminants.

Jamb seals between the blade ends and the frame on a louver damper must be flexible so that the fit is consistently tight without binding regardless of increasing or decreasing temperatures. Seals at the top and bottom of the damper should be bolted directly to the blades themselves.

Bearings

When mounted well away from the high temperatures of the gas stream, bearings on slide gate dampers present no particular problem. On louver dampers, however, the bearings are in direct contact with the hot blade shafts. The potential difficulties involved can be minimized by mounting the bearings in self-aligning pillow blocks that are bolted to metal steps on the exterior of the damper frame. To further protect the bearings from contaminants the shaft holes in the frame should be sealed with packing glands. Because of the effects of thermal expansion, the shafts should be locked to the bearings on the drive side of the frame only, allowing the shafts to expand and contract freely through the bearings on the opposite side.

Actuators

These must be specified to function under worst case conditions taking into account the system volume, pressure, velocity, temperature, contaminants and friction in the glands, bearings and linkage, as well as the weight of the damper components themselves. A drive safety (overdesign) factor of 3.0 will account for worst case conditions and for the component wear and particulate buildup that will gradually demand more of a damper's actuator over years of service. This same safety factor must also be applied to the total structure of the damper, including the linkages on both sides of louver dampers, to ensure against buckling from the excessive force applied by the actuator under extreme conditions.

Both electric and pneumatic actuators are well suited for HPI damper applications. For dampers that will be operated only occasionally, manual actuators are often the best choice.

Chain-and-sprocket drive systems generally provide the most efficient and reliable power transmission for slide gate dampers. The mechanical advantage that can be gained through the gearing allows for the use of smaller motors, and the sprockets and chains can operate effectively even when coated with gas stream particulates. A two-chain system with a 2.0 safety factor will allow each chain to support the entire weight of the blade if the other chain should fail (see Fig. 1).

Jackscrew systems, by way of contrast, are not effective where operating speed or long stroke (4 feet or more) is a factor. Jackscrews must move slowly since excess speed causes the mechanism to heat and seize. Moreover, jackscrew mechanisms require protective sleeves and fairly regular maintenance and lubricating to prevent failure due to corrosion from system contaminants.

A multi-blade louver damper's actuator should be coupled directly to the drive shaft whenever possible. This will eliminate problems that can arise from differences in expansion and contraction where the actuator is floor mounted and connected to the drive with a pipe or bar. Small thermal variations in such a system can result in widely varying differences in blade motion. A 1/8-inch difference in the length of the linkage, for example, can result in as much as a 1/2-inch difference in the motion of the top and bottom blades.

An additional factor that must be considered in the design of louver dampers is the running torque that develops as a result of the high and low pressure areas above and below the wing-like airfoil blades. The running torque on partially open blades in some systems may be stronger than the breakaway torque required to initially open the blades in the face of the system pressure.

OTHER CONSIDERATIONS

Any damper design proposal should include calculations pertaining to the design and choice of components and materials including:

1. Effects of seismic loading on each component of the assembly
2. The increase in system pressure drop caused by the open damper, based on readings taken 10 duct diameters upstream and downstream from the damper
3. Stress levels on major damper components under system design conditions
4. Fatigue life calculations for the entire damper and major components
5. Total leakage in acfm through the closed damper under design conditions
6. Blade deflection calculations at operating conditions (Blade deflection should not exceed the width of the duct divided by 360.)
7. Blade buckling calculations for gates
8. Torque calculations and related safety factors for louvers.

In addition, it is essential that the manufacturer be completely familiar with all design and engineering factors and requirements, and that the manufacturing process be subject to strict engineering controls. It is obvious that some dampers fail because of inadequate design or engineering. Less obvious, but just as important, are the problems created when separate companies perform the engineering and the manufacturing. The most effective way to avoid these problems is to specify that all design, engineering, construction and testing of damper machinery be performed at the manufacturer's own facilities. Final inspection and testing should be performed under strict purchaser supervision prior to shipment.

For efficient operation in extremes of temperature, pressure and contaminants, a damper must be structurally sound from the beginning. There are no specific codes for HPI damper design, but two general codes that can be applied are those found in the Steel Construction Manual of AISC and the Metals Handbook of ASTM and ASME.

The American Petroleum Institute has taken a first step toward ensuring proper application, design, construction and installation of damper machinery in the HPI with the proposed Recommended Practice (533) for Design of Air Preheat Systems. Further developments in standardizing damper codes will ensure more reliable installations. Meanwhile, refinery and process designers must continue to educate themselves—both to improve the efficiency and cost effectiveness of their dampers and to protect the personnel who must work with and around those dampers every day.

25
Three Ways to Calculate Control Scaling Factors

A. L. Nuaimi

Computational devices are used extensively in processing plant control systems. The accuracy of their scaling factors is essential for the controls to perform as intended. Three methods are described for the calculation of these factors.

Scaling factors are the constants of the equations of computational devices such as summers, multipliers, dividers, square rooters, etc. They can be pneumatic, electronic or software algorithms. Generally, they are employed whenever a control system is required to implement a mathematical process equation.

FUEL FLOWS EXAMPLE

As an example, an industrial boiler fires a combination of bunker C oil and gas as fuel. In the combustion control scheme a signal, F_t, is necessary to represent total fuel burned as measurement input to the total fuel, or Btu controller. The overall process equation is:

$$F_t = F_1 + F(F_g) \quad (1)$$

Where: F_t is total fuel flow in gpm, F_1 is fuel oil flow in gpm, F is a factor to convert fuel gas to fuel oil equivalent in g/Mscf, and F_g is fuel gas flow in Mscfm.

Instrument Diagram. The instrument loop diagram is shown in Fig. 1. A multiplier is used to express fuel gas flow in units of fuel oil equivalent, solving the process equation:

$$F_e = F(F_g) \quad (2)$$

The summer adds fuel oil to the fuel oil equivalent of fuel gas, F_e. The result is total fuel flow in fuel oil units, according to the equation:

$$F_t = F_1 + F_e \quad (3)$$

Multiplier and Summer. The multiplier and summer have, themselves, equations which depend on their make. Typical algorithms are:

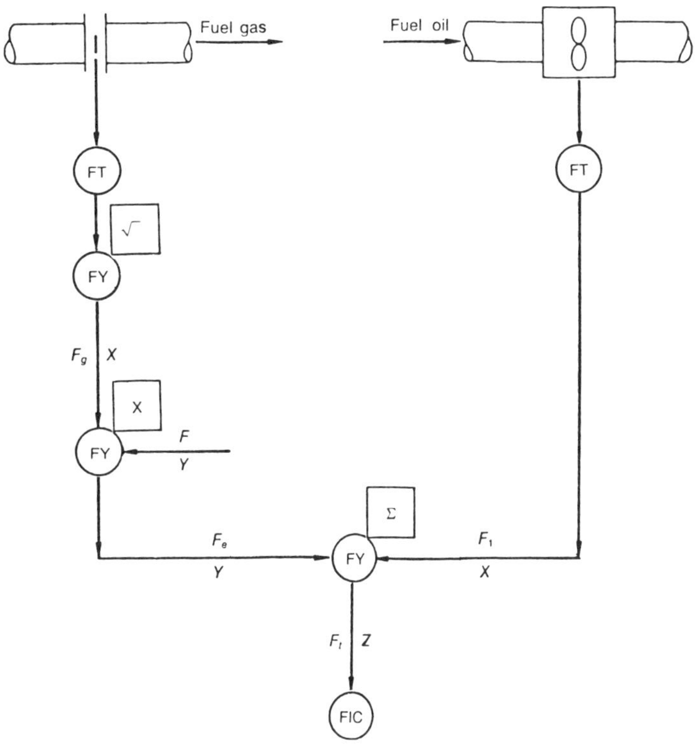

Figure 1. Total fuel flow scaling.

$$\text{Multiplier } Z = (A/100)XY + C$$

$$\text{Summer } Z = AX + BY + C$$

Where: X and Y are inputs ranged 0 to 100%; A and B are dimensionless constants; C is a constant, bias, valued between ±100%; and Z is the output ranged 0 to 100%.

To compute the scaling factors, A, B and C, we must first define the calibration of the signals involved.

Signal Calibration. Let the ranges and spans be as follows:

	Range	Engineering units	Span
F_g	0 to 12	Mscfm	12
F	0 8	g/Mscf	8
F_e	0 60	gpm	60
F_l	0 50	gpm	50
F_t	0 80	gpm	80

Scaling Factors. Three ways to evaluate the scaling factors will be demonstrated. All three produce identical values for the factors. The methods will be referred to as dimensionless equations, simultaneous equations and logical arithmetic, respectively.

Dimensionless Equations. Process equation (2) for the multiplier can be written as:

$$F_e\text{, gpm} = (F\text{, g/Mscf})(F_g\text{, Mscfm})$$

In dimensionless form this becomes:

$$1 = F(F_g)/F_e \quad (2a)$$

The multiplier equation, in dimensionless form, is:

$$1 = (AXY/100Z) + (C/Z) \quad (2b)$$

Equate the corresponding terms of (2a) and (2b), substitute the spans for the variables and solve for the constants.

$$A = 100ZFF_g/XYF_e$$
$$A = 100(100)(8)(12)/(100)(100)(60) = 1.6$$
$$CZ = 0,\ C = 0$$

The scaling factors for the multiplier are: A = 1.6 and C = 0%. The multiplier equation is Z = 0.016 XY

Simultaneous Equations. We will apply this method to the summer. Take the process equation (3) and calculate as many sets of values for the inputs and output as there are constants in the summer, in this case three.

F_l, gpm	F_e, gpm	F_t, gpm	X, %	Y, %	Z, %
0	0	0	0	0	0
20	60	80	40	100	100
50	30	80	100	50	100

With these values and percentages, form three summer equations and solve them simultaneously.

$$0 = (0)A + (0)B + C$$
$$100 = 40A + 100B + C$$
$$100 = 100A + 50B + C$$

The scaling factors are: A = 0.625, B = 0.75 and C = 0%. The summer equation becomes Z = 0.625 X + 0.75 Y

Logical Arithmetic. This is probably the most widely used method. It yields accurate results as long as the logic is right. Let us apply this approach to the summer:

Process equation: $F_t = F_l + F_e$
Summer equation: $Z = AX + BY + C$

By inspection we immediately see that C = 0. Make one variable, say, $F_e = 0$; this makes Y = 0. Let the second variable, F_l be maximum, i.e., 50 gpm. This makes X = 100%. With these conditions F_t will be = 50 gpm. This makes Z = (50/80)100% = 62.5%. Using the summer equation: 62.5% = A(100%); A = 0.625.

Now make the other variable, $F_l = 0$ and let F_e be maximum. Using arithmetic similar to the above, we find B = 0.75. These values for the scaling factors are the same as those obtained by the simultaneous equations method.

PROBLEM

If you feel in the calculational mood try the summer of Fig. 2 given that: The controlled temperature, T_c, of a processing column varies with the measured temperature and pressure according to the relationship:

$$T_c = aT + bP + c$$

where: a = 1.1
b = 0.5°F/psig
c = −15°F, T_c range 100 to 400°F, T range 100 to 360°F and P range 0 to 100 psig.

The scaling factors should come out as: A = 0.953, B = 0.167 and C = −1.667%.

Three ways to determine the scaling factors have been discussed. The practitioner may use the one he prefers and which suits the application best. The values from one method can also be checked against those from another.

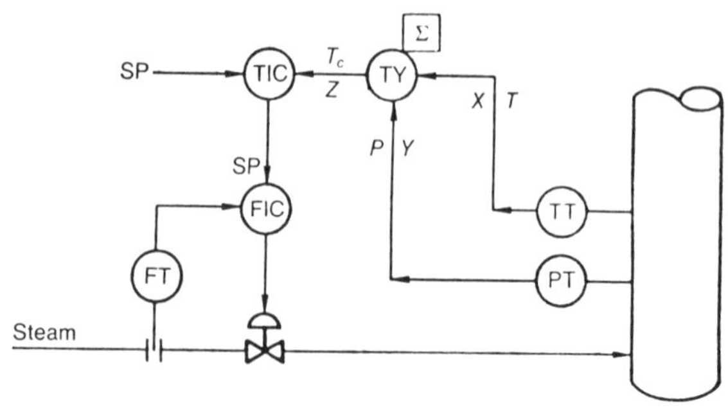

Figure 2. Temperature scaling.

26
Control System Fire Protection

G. K. Castle

There is a need to protect operating control systems, which include cables, pneumatic signal tubes, transmitters, remote operated valves and other control equipment. Goal: Provide enough time in case of emergency to shut off or divert the flow of hazardous fluids and gases in and around the affected unit. Another goal is to provide protection against loss or damage of the equipment itself.

Fire-related failures of unprotected control systems can occur within a few minutes of the accident. Usually 15 minutes or more of protection is needed to assure the control of critical operating equipment, stabilization of the situation and protection of the system.

It would be much too expensive to provide fire protection for every instrument or control system on a large operating unit. It is more realistic to protect parts of a control loop that are most critical to operations or require long replacement lead time or high cost outlays. Here, I will discuss the vulnerability to fire and appropriate methods of protecting such systems during a fire.

FIRE PROTECTION CRITERIA

To establish guidelines for protection, ask these questions:

1. Where is fire protection required? (Identify likely fire sources—location)
2. What are the boundaries of protection needed? (Identify the radius of vulnerability surrounding a possible fire source)
3. What control systems and components are critical within this area? (Identify components and loops)
4. How much protection is practical? (Set a time limit for fire exposure)
5. What fire protection methods and/or materials should be used? (Select systems which render the most economy, proper protection, and hinder operations least.)

TYPICAL CONTROL SYSTEMS

The variation of control systems involving different transmitters, control valves, transducers, lines, etc., is nearly infinite. Even though a control loop can be unique in several ways, almost every loop can be generalized into one of the following categories:[1]

1. A field-mounted output device (i.e., pressure, temperature, flow rate) and transmitter (sometimes with signal amplification) connected with a computer or analog controller either in a control center or mounted in the field. The signals provided may or may not cause action to be taken in another loop that has controller/computer connected input/output devices.
2. A field-mounted output device with an on-off control (i.e., a ball float level controller that turns a pump off or on at prescribed liquid levels).
3. A process-activated switch that causes an alarm, activates a shutdown system, is an input to a supervisory type control scheme or has some form of operator interface.
4. A remote-operated switch that starts/stops a pump, opens a valve or causes some type of process action to take place. These switches can be manually or automatically controlled and can be as simple as a push button at the end of a pair of wires or as complicated as a computer system transmitting various signals to operate address selected outputs.

COMPONENTS

In addition to the actual control loop, there are multi-pair transmission cables, pneumatic lines, power cables to motors, junction boxes, power supplies, valve activators and other auxiliary equipment and supports. These must be considered a part of the overall system when

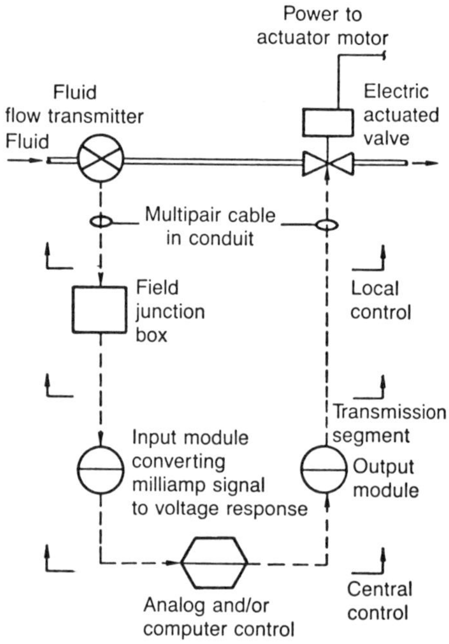

Figure 1. Typical control loop.

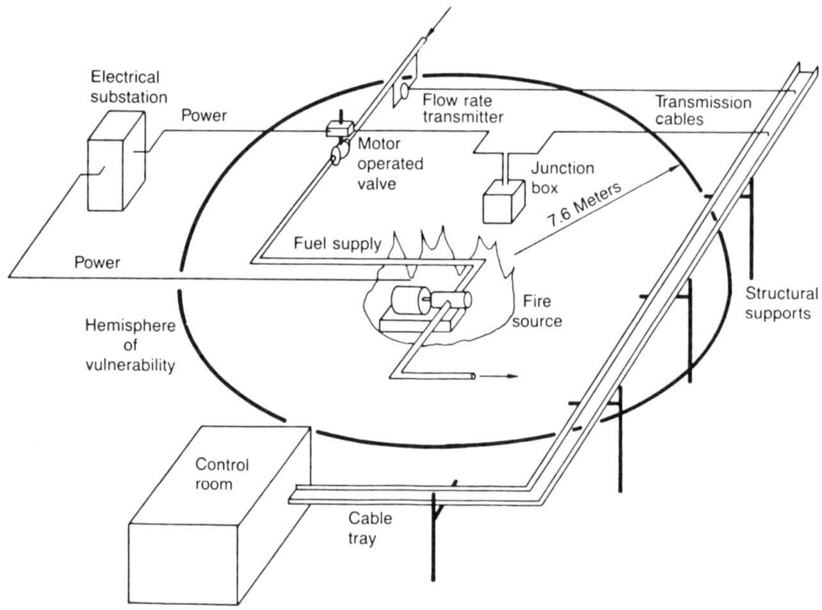

Figure 2. In-place field control system.

protection schemes are evaluated. Figure 1 shows a typical control system, identifying components. Basically this loop controls the operation of a valve that can limit the flow of fluids to perhaps a process tower or storage vessel. Next, we shall take this same loop and show the components as they are in an actual field installation, adding to the loop other components as previously mentioned that are just as vital to the overall operation of the system during a fire.

CRITICAL POINTS

Components of control systems are made by several different companies. Each has its own method of doing the same thing. These vendors may use widely varying materials to build the same basic device. All of these components and materials have various failure time-temperature limits. Protection against fire requires that each component or material be considered separately to ensure that protection is adequate, complete *system* protection is provided, and costs are kept to a minimum.

Where protection is sought mainly to preserve an expensive item of equipment or a long-lead-time component, it becomes simply a matter of isolating such items and providing fire protection that meets the failure criteria of the most vulnerable material or mechanism involved. Protection is localized. It is not required to consider or fit into a system. On the other hand, fire protection of a control loop including the controlled equipment (with the idea of preserving operational capability for 15 to 30 minutes) requires *complete system* protection to accomplish the goal.

VALVE CONTROL SYSTEM

Figure 2 shows a motor-operated valve control system. The system is a field application of the schematic loop in Fig. 1, but also shows cable trays, power sources and the general plan layout which might be found anywhere. Simply stated, the MOV functions to limit the flow of a hot flammable hydrocarbon. A pump moves the fluid. The pump is a fire source because of possible failure of its packings, seals or glands. All of the control system components required to permit operation of the MOV which fall within a given radius of the fire source (pump) are:

1. Valve actuator including motor and limit switches
2. Control lines (electrical) to actuator switches
3. Power cables to actuator motor
4. Junction box containing field termination of multipair transmission cables
5. Cable tray containing transmission cables leading to control room
6. Cable tray supports.

The radius chosen here is 7.6 meters. This would be a minimum distance considering that radiation from the flames of even a small fire on the order of 2 meters in radius can provide heat flux levels of 0.3 cal/cm^2-sec at this distance.

Note that the signal device and transmitter for flow rate are not included in the protection scheme. In this case, the idea is to preserve only the on/off (particularly off) capabilities of the MOV, not to monitor flow rates.

Assume that the possible fire source has been identified, a radius of vulnerability agreed on, and all critical control system components chosen within that radius. The next step is to select the proper fireproofing systems and/or materials.

VULNERABILITY

Most electronic instruments have discrete components that fail when they reach temperatures above 50°C. These devices will give erroneous readings above 50°C, with complete failure above 93°C. The "O" rings and diaphragm in pneumatic instruments are the weakest link. Most are made of materials that undergo serious thermal decomposition when reaching temperature levels of 200°C. Such decomposition, even locally, will permit gas leaks which cause erroneous signals after an exposure of only a few minutes. Pneumatic cylinders that provide the power to operate a valve can have the internal pressure changed during fire exposure by as much as 4 to 5 atmospheres. Such a change will almost certainly cause unwanted piston travel with attendant loss of control.

Transmission and power cables are manufactured with many types of insulation and wide ranges of temperature specifications. Normally, temperatures above 112°C will cause failure within a few hours. Temperatures in a fire will cause nearly instantaneous failure. Last, but not to be ignored, are structural supports for transmission cable trays and conduit. Steel will weaken upon exposure to heat, losing about 40 percent of its strength at 538°C. Under fire conditions, supports such as small angle braces or rods can reach this condition in less than a minute.[3]

Table 1 summarizes failure criteria for the control system components discussed above. With the exception of steel supports, all of the components discussed involve low failure temperature vulnerability.

Table 1
Approximate Failure Temperatures of Control System Components

Component - Location	Failure Temperature - ° C
Electronics	50-93° C
Transmitters	
Junction boxes	
Signal devices	
Alarms	
Computers	
Switches	
Motors	
Cables	112° C
Transmission	
Power	
Pneumatic devices	200° C
Air cylinders	
Seals	
Structural supports	538° C

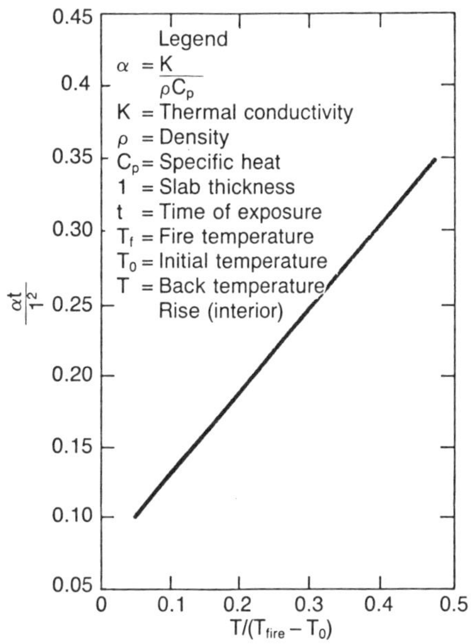

Figure 3. Backface temperature rise in slab exposed to free-burning hydrocarbon fire on front face.

METHODS OF FIRE PROTECTION

Fire environment in a free burning liquid or gaseous hydrocarbon fuel can produce temperature levels in excess of 1200°C and total heat flux levels of 2.65×10^5 W/M^2.[3] Because mixtures are not stoichiometric, the presence of carbon soot particles causes the fire to be nearly opaque, and 80 to 90 percent of the released energy is in the form of radiation. Depending on fire size and wind factors, normal average heat flux levels will be about 70 percent of the maximums given.

Thermal Analysis

The effect of such a thermal environment on protective materials can be assessed by using some simple thermal analysis. For example, consider an infinite slab (planar dimensions are an order of magnitude larger than thickness) with initial temperature, T_0, and the outer surface suddenly exposed to the fire temperature. The temperature rise of the rear face (interior of the protection system) can be quickly calculated using the dimensionless relationship shown in Fig. 3. If the criteria for protection is 15 minutes and the internal temperature is to be limited to 93°C, then Fig. 3 will give a value for the required dynamic thermal resistance ($\alpha t/1^2$) of 0.11 if the fire temperature is 982°C. Depending on the thermal properties of the protection system (materials) evaluated, this value corresponds to an insulation thickness of 20 to 50 cm considering today's refractory insulations.

Keep in mind, this analytical aid does not consider integrity of joints and seals that exist in any enclosure. Failure of these items will cause internal temperatures to rise more quickly. However, Fig. 3 provides a basis for checking systems to see if they are viable.

The failure and hence protection criteria for control components listed in Table 1 coupled with this type of thermal environment poses a fairly challenging problem even with today's sophisticated materials. Using copious thicknesses of high temperature insulation is not usually the best answer. Ideal fire protective systems must withstand weather, provide access to components for servicing/repair, be practical for field installation and be cost-effective.

Referring to Fig. 2, consider protection methods for each of the components within the radius of vulnerability. Starting with the valve actuator (MOV), two basic systems have been used to protect MOV's—flexible insulated bags and prefabricated box enclosures. The insulated bags are sewn to fit the particular operator in question. The bags have certain advantages in appearance and weight. However, keeping the bags sealed or closed tightly is a problem. And costs are high.

The prefabricated box enclosures are basically insulated boxes with variations on how they are installed and constructed. Some are quite versatile. They can be custom fabricated for any shape or purpose, provide access inside, allow extension of handwheels and clutches outside the enclosures and even indicators to show valve position for field viewing. When selecting such enclosures, talk to the manufacturer. Obtain *installed* costs. Find out what types of tests have been run and whether servicing of the MOV will be practical after the enclosure is installed. Some enclosure designs are fragile. They will not stand up to abuse.

Insulation

Leading from the valve actuator are two cables, one for control, the other for power. It is not unusual for these lines to be enclosed in conduit. From the actuator to the point where the power cable(s) leave the radius of vulnerability, protection is required. Similarly, the control transmission cable must be protected up to and including its journey through the cable tray.

The most economical and viable way of protecting single cables is insulation covered with stainless steel sheet and banded. Selection of the insulation should be based on lack of chloride content, minimum thickness required to limit cable temperatures to 112°C or less, ease of installation and high temperature capabilities. A tight seal must be provided between the actuator enclosure and the insulated conduit.

Various compounds have been used to spray or butter these individual cables. Although some protection is afforded, this method is not recommended, since thicknesses (and therefore expense) required for 15-minute fire resistance are prohibitive. Also weather resistance is a problem as well as cracking and achievement of uniform thicknesses during application.

Prefabricated Enclosures

The junction box and cable tray are links in the control system chain. They must be included also in the protection scheme, limiting their temperatures levels to 93°C and 112°C, respectively. Prefabricated enclosures are the best suited systems for this. Take care to ensure that the enclosure weight will not overstress the supports for the cable tray.

Insulation wrapping systems have been used for multi-conduit cable runs or cable trays. This method, while effective for fire protection, has disadvantages in cost, has no ability to handle tight and complicated areas, and provides no access into the cable tray without destruction of the protection system. As in the case of single cables, metal (stainless steel) outer coverings are needed.

It may be desirable to keep a pump running to escort as much fuel away from the fire as possible. Although this is not particularly the case in Fig. 2, since the remote actuated valve should stop fluid flow, there are many instances where this could be the case.

Shields; Sprinklers

Protection of power lines poses a special problem in fire protection. Since these lines are continuously carrying current of voltages of 220 VAC or more, insulation techniques will cause overheating due to the I^2R effect. The current-carrying conductors may be derated to allow for insulation. But this method is at best a last resort and almost always requires that larger cables be installed before fire protection measures are taken. The remaining choices for in situ fire protection are metal flame shields and sprinkler systems. Sometimes a combination of both may be necessary depending on how close the power cables are to the fire source.

Metal flame shields should be placed about 5 to 10 cm below the cables. They should extend outwards at least 1/3 meter on either side. The shields should also extend upwards around the cable to cut down side radiation effects. As in the case with sprayed or buttered mastic coatings, protection time is limited to somewhat less than 15 minutes. A double-walled shield with a 2-3-cm air gap between the metal surfaces will help considerably by reducing radiative heat transfer, but increases cost tremendously and still provides only limited protection. If possible, shields should be used in conjunction with sprinklers.

Table 2 lists items in the control system and methods of fire protection.

Table 2
Protective Equipment and Methods

Control system component	Recommended method of fire protection
Valve actuators (electric)	Prefabricated enclosures, insulated bags
Switches	Prefabricated enclosures
Junction boxes	Prefabricated enclosures
Controllers	Prefabricated enclosures
Alarms	Prefabricated enclosures
Computers	Prefabricated enclosures
Transmitters	Prefabricated enclosures
Signal cables (Single)	High temperature insulation with metal covers
Signal cables (Multigroups)	Prefabricated enclosures
Instrument cable trays	Prefabricated enclosures, Insulation wraps
Pneumatic actuators	Insulation wraps
Power cables	Metal shields, sprinklers
Structural supports	Weatherable mastic coatings

Completeness

If the fire-protective methods shown in Table 2 are used consistently with levels (time-temperature limits) of protection required, any control system can be kept operable during a fire. The key is *completeness* of protection, workmanship, and inspection.

CURRENT PROBLEMS

A problem that remains to be solved is a lack of meaningful fire-test data. Many fire-test facilities and methods are available but not many can truly represent a free-burning hydrocarbon fire. Manufacturers and suppliers of fire-protective materials and methods are as much a victim of poor and irrelevant testing as the end user. The best way to avoid the problem is to obtain actual performance data in full-scale outdoor hydrocarbon fire tests. Previous, current and future data of this type should be openly shared.

Power cables are the most difficult item to protect. Cable manufacturers, in harmony with inventive material suppliers, are making strides toward providing 15-minute protection within the cable itself. More work remains to be done.

Completeness of protection is seldom achieved in control systems unless the responsible designer or engineer *knows* control systems, their components and what is required to protect them. More education and in-plant seminars would be helpful.

References

1. From discussions with James Carter, Exxon Co., U.S.A., Baton Rouge, La.
2. Product Bulletin 1877, Samuel Moore and Co., 1972.
3. Castle, G. K., *Fire Protection of Structural Steel,* Loss Prevention, Vol. 8, 1974.
4. Carslaw, H. S., and Jaeger, J. C., *Conduction of Heat in Solids,* Oxford University Press, 1959.

Part 2
ADVANCED AND CONVENTIONAL CONTROL STRATEGIES

27
Dual Composition Column Control*

C. J. Ryskamp

Dual composition control, the closed loop control of the composition of both top and bottom products leaving a distillation column, has been implemented profitably on many columns. Dual composition control has also been accomplished on thermally coupled columns, where overhead vapor from one column is used as reboiler heat on a second column. This chapter presents new methods for understanding and implementing stable composition control.

Recent applications demonstrate that dual composition control yields substantial profit. Typically, 30 percent excess heat is expended to guarantee that compositions meet specifications. For columns which require 100 MMBtu per hour at $2 per MMBtu, the potential savings reach $500,000 per year in utility costs alone.

Among needs to be addressed when controlling both top and bottom product compositions are cost effectiveness, measurement integrity and control loop stability—especially with both loops closed. Also, the separation required for both products to meet specification must be attained either by increasing heat thruput or reducing feed rate.

Tight closed loop composition control can provide a higher average impurity level with less off-spec. product. If top and bottom products from a distillation column have the same value, controlling both compositions at the limiting specification will minimize energy required for separation. When the two products differ in value, and optimum compositions have been determined, dual composition control can assure maximum profit.

When the product from one column is the feed to a downstream column, controlling its composition reduces disturbances to the downstream column. This becomes especially important when a series of columns is used to provide separation of feed into multiple products. Without closed loop control, the operator must watch recorders and make adjustments in process variables. When process upsets occur and the operator is busy with other duties, he must also continue to monitor compositions and make even more adjustments. Closed loop composition control relieves him of this burden.

Cost of an analyzer and its proper installation is high, but the benefits of closed loop control are even higher. Management, engineers, operators and instrument men should understand the economic benefits of keeping the analyzers on closed loop control. Obviously, they must be confident that the measurement not only agrees with the real composition, but also will be dependable. By using limits on the output of the controller so that it has minimal effect on the process, the operator can gain confidence in the system without fear that it will cause a major upset. After his confidence is established, the limits can be widened gradually to allow the analyzer controller to make broader adjustments.

Because of long process lags, an analyzer control loop must respond slowly. Occasionally, operators make manual changes too fast for some columns. Stability of composition control loops presents a major problem, especially when both top and bottom compositions are on closed loop control.

UNDERSTANDING COMPOSITION CONTROL

Steady State Equilibrium

What constitutes steady state equilibrium—the condition when a plant is "lined-out and running smoothly" or "drawing perfect circles?" We tend to consider a plant lined-out when those variables measured and recorded are constant.

* The author is grateful to Arthur McCoy, Richard Rys and Noal McGee, all of The Foxboro Co., for their important contributions to the success of the control strategies.

With the development of automatic feedback control a half century ago, the concept of "lined-out" meant holding variables such as temperature, pressure and level constant, accepting the fact that the valve positions and flows had to change. Recently, we have learned how to measure compositions continuously on the process. The importance of putting these on closed-loop control and holding them constant is recognized today. Our concept of "lined-out" has changed to allow measurements like temperature and level to vary, provided compositions are held constant.

Today, and in the future, with increasing use of computers and heightened attention to efficiency, we will be maintaining calculated variables such as unit performance or profit at a desired level for overall plant optimization. Profit on one process unit may have to be altered because of its effect on other units. Closed-loop composition control is virtually a mandatory prerequisite to process optimization because it reduces the complexity of the process model used to maximize profit. Ultimately, those variables which are known to help yield more product or profit, or ensure safer plant operation, are the ones which must be "lined-out," while minor variables are changed as needed.

Upsets which move the process further away from the needed steady-state equilibrium take longer to recover and waste more energy. Better regulation can reduce the magnitude and duration of composition upsets and avoid waste of energy while assuring more valuable product.

Product Specifications

Most product specifications for distillation columns state the maximum allowable impurity level. A tight, or hard spec., is one where there is a high incentive to keep the product at or near the limit. A loose, or soft spec., allows more variation in product composition because change in composition minimally affects product values or utility cost. Generally, higher purity products will require tighter specs., not only because of the lower absolute concentration of impurities, but because of the higher utility cost involved in making the products. Increasing energy costs will require more products to meet tight specs.

To meet tight specs., process variables must be changed more frequently and in smaller amounts. Closed-loop composition control is often justified for products that must meet a tight spec. Since product composition can vary more with a loose spec., it may suffice to have the operator set the heat-to-mass thruput ratio for this product.

Composition Profiles

One way to understand the effects of control actions or changes in control strategy is to visualize the composition profile through the trays of a column. The center curve (Fig. 1) shows how liquid composition might appear on a typical binary column. The top to the bottom of the graph corresponds to the trays from the top to the bottom of a column, and the percentage of heavies is plotted from left to right, 0-100 percent, or percent lights from 100-0 percent.

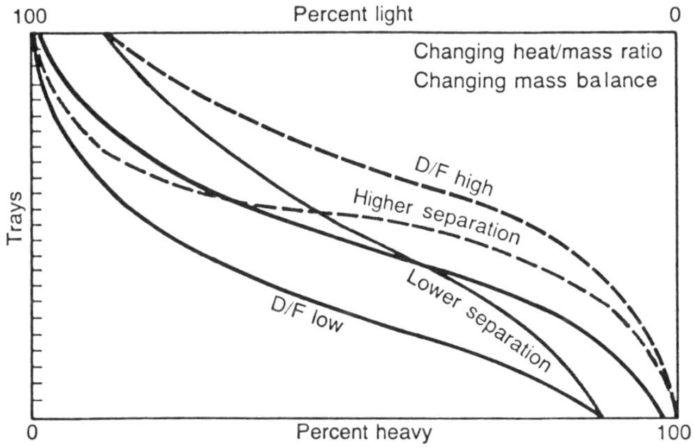

Figure 1. Composition profiles for distillation columns.

The curve approaches the left and right boundaries asymptotically as compositions become more pure. If the fraction of feed removed as top product (D/F) is increased, then the composition profile moves up as shown in the upper curve, with excessive amounts of heavy component in the top product. The bottom product is more pure, however. If the fraction of feed removed as top product is too small (D/F too low), then an excessive amount of lights will accumulate in the bottom product, while the top product will be overly pure. These curves are shown for concept only and do not account for any discontinuity or pinch at the feed tray.

If the ratio of heat thruput to mass thruput is held constant, the shape of the curve will remain relatively constant. However, if heat input and reflux are increased for constant feed and product flows, thereby increasing the heat to mass ratio, separation will be improved, and both products will become more pure (Fig. 1). The sharper curvature is sometimes called a sharper break in the composition profile. Conversely, if the heat-to-mass ratio is reduced, the curve will be extended and there may be excessive heavy components in the overhead and light components in the bottoms. Separation is poor in this condition.

With dual composition control, one composition control (AC) moves the profile up and down and the other

AC adjusts curvature (Fig. 1) to make two points on the curve meet specification. When required separation cannot be attained because of high feed rate and limited heat input, reflux or tray loading, then only one product composition can be controlled.

The composition profile also indicates the relative inventory of light and heavy components on the trays. The area above and to the right of the curve represents the light component inventory, and the area below and to the left of the curve represents the heavy component inventory. Total inventory on all the trays may remain relatively constant for small changes in operating variables. To move the curve up, as shown in Fig. 1, distillate flow would have to be higher than normal for a period of time to reduce the inventory of light components on the trays. Conversely, driving the composition profile down would require reducing distillate flow to accumulate light components on the trays. Bottoms product flow would have to move in the opposite direction during these changes to maintain overall mass balance. This accumulation of light components on the trays introduces an integrating lag into the process similar to the level in the tank.

When better separation is desired and the composition profile has a sharper break, inventory of both light and heavy components may remain the same, but a redistribution occurs with light components becoming more concentrated in the upper trays and heavy components becoming more concentrated in the lower trays. Therefore, a major change in mass balance would not be required on the column, but enough time must be allowed for the increased heat and reflux to redistribute components on the trays. This accumulation of higher purity products due to increased separation also introduces an integrating lag into the process response. Total tray inventory may increase slightly with an increase in liquid and vapor flows.

Energy must be expended to establish a composition profile, i.e., heat is introduced and removed from the column to separate the feed into the compositions present on the trays. Any abnormal mixing caused by upsets in column operation wastes some of the energy used to create the composition profile. This results in more energy consumption to force the profile back to normal. If control actions are taken to reduce disturbances to the composition profile and keep it at the desired steady-state equilibrium position, then energy waste due to mixing is similarly reduced.

Dynamics

Because of component separation in the distillation process, residence time of various components will differ. For a continuous stirred tank reactor, residence time distribution for all components is the same and, on an average, is the volume of the vessel divided by the flow rate through it. This residence time will be a first order lag if the vessel is perfectly mixed. When not perfectly mixed, it will be the sum of a first order lag, some higher lags and even some dead time. Likewise, the dead time for all components is the same through a pipe or on a conveyor belt. However, in a distillation column, residence time is a function of the inventory of a particular component divided by the flow rate of that component leaving the system.

Product compositions affect division of the inventory on the trays between light and heavy components. If both products are of the same purity, then the inventory will be split about 50-50. If the top product is of high purity and the bottom product is of low purity, then the inventory of lights in the trays will be greater than the inventory of heavies, as described previously. Conversely, if the bottom product is of high purity and the top product is of low purity, then the inventory of heavies on the trays will be greater than the inventory of light components. A smaller product flow or a larger inventory means a longer residence time, making control more difficult. Higher purity products also require higher reflux ratios, which mean a larger diameter tower and generally more trays. These combine to make liquid inventory larger and control more difficult. Also, the inventory in the overhead receiver, combined with the recycle path provided by the reflux stream, creates a dynamic response similar to a long time constant first order lag. The reboiler also introduces a lag.

With dual composition control, better overall response will be obtained if the composition controller for the component with shorter residence time is connected to adjust vapor flow rate through the trays. The controller for the component with longer residence time adjusts the liquid/vapor ratio. This will generate a relatively fast AC loop, which affects all trays concurrently (without creating waves and reflections in the composition profile), and a slow loop whose action gradually works overall tray liquid-vapor loading to provide the needed separation. Having one AC fast and the other slow reduces the interaction problem.

Derivative action is needed in the composition controller to compensate for exponential lags in the process. When a chromatograph is used as the onstream analyzer, two lags between the chromatograph and the controller can smooth steps in the composition signal. Time constants in these lags can be set for about one-half of the analysis time. This filters out high frequencies associated with a step change and allows a longtime constant derivative to be used in the controller without adverse effects from step changes (Fig. 2). Derivative times as

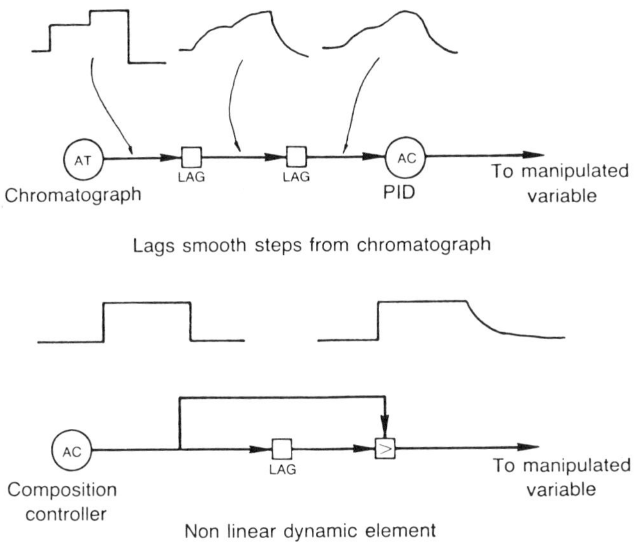

Figure 2. Devices improve chromatograph output.

long as 60 minutes have been used successfully with this method.

If there is concern that too much derivative action may make the product go off spec., a non-linear dynamic element may be added (Fig. 2). This consists of a lag and a signal selector on the output of the composition controller, which allows the variable to be manipulated quickly to yield a higher purity product, but slowly, depending on the lag time constant, in the direction which reduces product purity. For example, on top composition control, a high signal selector would be used for adjusting reflux or a low signal selector for distillate.

Continuous control with lags on measurement and derivative action in the composition controller has been at least as effective as sampled-data control. A distinct advantage is that standard functions may be used in either analog or digital systems. No special logic or programing is involved. Tuning is simple.

If heat input to the column is set independently for composition control, then heat removal in the condenser must be on pressure control. Conversely, if heat removal in the condenser is set independently for composition control, then heat input to the reboiler must be on pressure control. Changing vapor flow through the trays by either heat input or heat removal affects all trays almost simultaneously. Therefore, vapor flow is a proper variable to move the composition profile up or down. Columns with a large temperature difference from top to bottom will not follow this response because of the sensible heat of the liquid on the trays and the steel of the vessel and its internals.

Changing reflux always affects the top tray first and gradually works down the column. At least during the transient, it will change the curvature of the composition profile in the direction of the steady state objective. Therefore, reflux flow or reflux ratio is a proper variable for controlling separation on the column.

In conclusion, the product with the smaller component inventory or residence time on the trays will deliver faster response and is probably best for manipulating vapor flow through the column. Controls on the ends of the column should be structured to allow a change in mass balance. The product having the larger component inventory or residence time would be a slower composition control loop and should manipulate a heat/mass thruput ratio such as reflux ratio to control separation. These results generally agree with the conclusions from steady-state interaction analysis,[1] where it is recommended that the less pure product set mass balance and the high purity product set separation on the column.

Pressure Control

Ordinarily, tight pressure control is mandatory for good composition control. Disturbances in pressure will cause disturbances in the composition profile. This occurs because liquid on the trays of the distillation column exists at its bubble point. Any drop in pressure will cause the liquid to vaporize, resulting in an excessively high vapor flow rate. This extra vapor flow will cause heavy components to move up the column and may even cause column flooding. Pressure may be lowered slowly to increase relative volatilities and reduce utility costs required for given product compositions. The column can be operated even against the condenser constraint. However, pressure should be lowered slowly to allow the top composition controller to respond and keep composition on spec. A rapid increase in pressure is of less concern, provided there is no weeping of the trays, although it can reduce vapor flow enough to allow excessive lights in the column bottom.

Disadvantages are inherent to most methods of controlling pressure by manipulating condenser variables to affect heat removal. For example, adding and venting inert gas contaminates the product and causes product loss. Throttling cooling water can cause sediment and scale in the condenser tubes. On air fan condensers, throttling blade pitch or louvers has narrow rangeability and introduces maintenance problems. A hot vapor bypass works well only if the condenser is below the receiver, and even then response is slow when pressure is lowered, and pressure is aggravated by changes in receiver level. If a hot vapor bypass is used with a control valve on flow through the condenser, cost is much higher, not only because two controllers are required, but also because separate safety relief valves are required on the column and receiver.

Manipulating vent gas flow will readily control pressure when considerable amounts of noncondensibles exist in the feed. However, when overhead product is totally condensed, the overhead receiver can operate liquid full and the pressure controller output can be connected in place of the level controller output. Liquid is forced to back up into the condenser, where it inhibits heat transfer to control pressure. This method is always effective, is as fast as most other methods, has wide rangeability and does not require extra measurements or control valves.

An advantage of the liquid full receiver is that inventory of light components cannot shift between the receiver and the trays, causing unwanted upsets to the composition profile. Excessive subcooling of the reflux may be a disadvantage. Internal reflux flow may need to be calculated and controlled using temperature measurements and external reflux flow.

The pressure controller generally manipulates heat input or removal to maintain a heat balance. However, on a high pressure column with light components such as a deethanizer, the pressure controller should maintain a mass balance on the column. This is because the density difference between liquid and vapor is small, and removing more heat, for example, does not sufficiently reduce the volume of material within the system. Therefore, product flow should be manipulated directly by the pressure controller. With this type of column, it has been observed that a multi-stage reflux pump can continue to operate and pump the required reflux flow for hours at a time, even when the receiver is empty. Again, this is due to the small density difference between liquid and vapor. This is one case where reflux flow need not be reduced and have top product composition go off spec. because of insufficient cooling for a period of time.

When the pressure controller manipulates heat removal in the condenser, the pressure transmitter should be located at the top of the tower or on the overhead vapor line. When heat input is manipulated to control pressure, then the pressure transmitter should be located at the bottom of the column beneath the bottom tray. This choice of transmitter location tends to reduce tray loading when column pressure drop increases near the flood point. In the latter case, a high level alarm should be installed to alert the operator that the level may affect pressure measurement.

If pressure is allowed to change automatically, the temperature controllers should have either pressure compensation on the measurement, or should use a temperature differential between the control tray temperature and the top or bottom of the column where purity is near 100 percent. Alternately, a composition controller could be used to adjust the setpoint of the temperature controller as needed to compensate for pressure changes.

CONTROL STRATEGIES

The best control strategy for the top part of a distillation column is debatable. Generally, strategy cannot be determined without considering the controls on the bottom of the column, the feed, pressure control, as well as some process equipment details and operating conditions. For consideration here, feed flow and feed preheat are assumed to be independently set variables; bottom product flow is a dependent mass balance variable set by the bottom level controller, and heat input by the reboiler is the independent heat balance variable set by the bottom composition controller. This is a common arrangement of loops.

This assumption leaves two more manipulated variables, reflux (L) and distillate (D), and pressure which must be controlled (assuming the receiver runs full). One of the two, L or D, can be set independently to control top composition, but the other must be manipulated to control pressure. Its flow rate will depend on the heat balance. An increase in heat input will increase vapor flow to the condenser, heat removal and the liquid flow chosen as the dependent overhead variable, L or D.

Conventional Controls

In the conventional system, reflux (L) is set independently by the top composition control (Fig. 3). This leaves distillate flow on pressure control, dependent on the heat balance. Severe interaction will exist between the heat and mass control loops, because distillate flow, an independent mass balance variable, is a function of the heat balance. Therefore, disturbances in heat balance will cause disturbances in mass balance. The conventional system works well at low reflux ratios, e.g., $L/D < 1$.

With the conventional strategy, if more light components enter with the feed, and heat is increased manually or with a temperature or composition controller to keep excess lights out of the bottom, pressure will rise and increase distillate flow. Since reflux is constant until the top AC acts, heavies will increase in the overhead during the transient. Conversely, distillate purity will improve if lights in the feed drop. Heat input changes force the composition profile up or down because the mass balance is altered (Fig. 1).

Material Balance

In a material balance system (Fig. 4), distillate flow is set independently by the composition controller. Reflux is then manipulated by the pressure control and depends on the heat balance. This system greatly reduces interaction between the heat and mass control loops, and works well at high reflux ratios, e.g., $L/D > 5$.

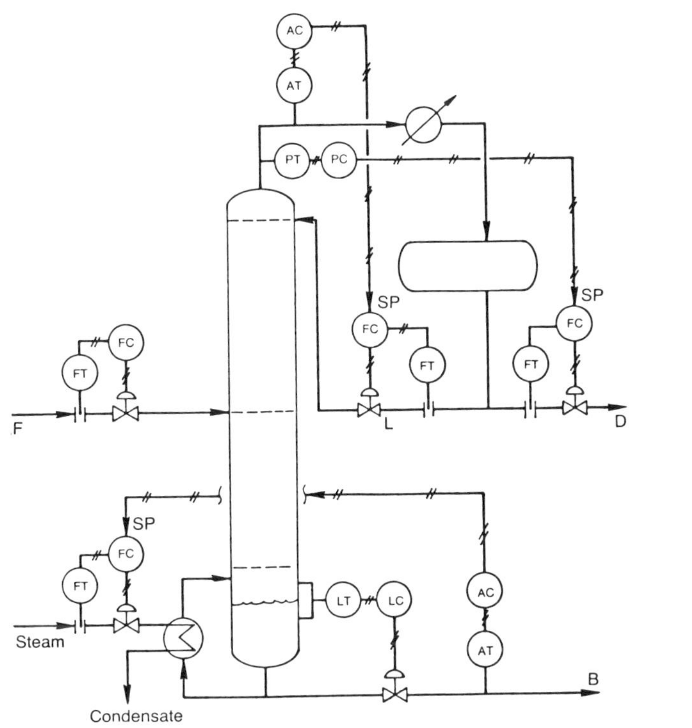

Figure 3. Conventional top composition control.

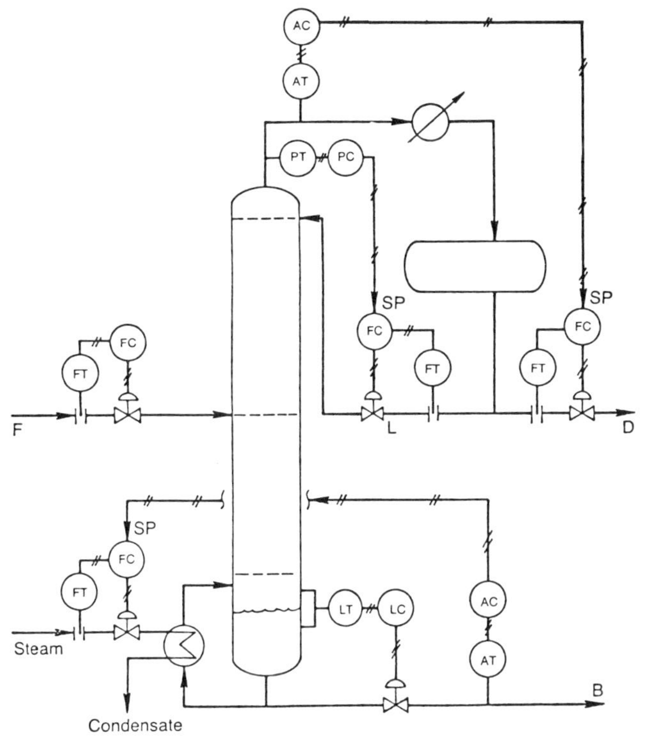

Figure 4. Material balance control.

Interaction will occur between top and bottom composition controls when both are in automatic. Relative gain theory[2] was developed as a mathematical tool to analyze this interaction. Chapter 10 of Shinskey[1] is devoted to the interaction and decoupling of control loops on distillation. Interaction between top and bottom ACs is greatly reduced with material balance control, especially at higher reflux ratios. However, the bottom AC cannot function with the top AC in manual. Control structures should be avoided where composition control on one end of the column requires that temperature or composition control on the other end be in automatic. These can too easily be left in manual, and the closed loop response needed from the other controller will not occur. Pressure and level controls are normally left in automatic and can be assumed to be active whenever composition control is used.

With the material balance strategy, if lights in the feed drop and distillate flow is constant, the distillate will contain more heavies, even when heat input is reduced. This is objectionable but not unstable. If more light components enter with the feed, and a bottom temperature or composition controller increases heat input, the system easily becomes unstable if the top AC is in manual, and winds up until a limit is reached. The small increase in feed light components which can be tolerated is limited to the small amount of heavy components originally leaving in the distillate stream. Beyond this, more lights in the feed will remain in the process because of the fixed distillate flow. As lights accumulate on the trays and reach the bottom of the column, more heat will be introduced to avoid excessive lights in the bottom product. The additional heat causes reflux to increase, but does not remove lights from the column. A higher purity distillate spec. worsens the problem. The same phenomena can occur, but may not be recognized, when the upper TC or AC is very slow to respond and the bottom TC or AC responds much faster. It appears as more severe interaction between the top and bottom control loops.

Figure 5 shows a feedforward material balance system with feed flow as the measured load variable. Feedforward helps maintain a constant lights inventory, even with the top AC in manual, but does not correct for feed composition changes. With additional cost and complexity, a feed composition measurement could be included in the feedforward system. However, interaction still exists when a bottom AC adjusts heat input. Changing heat input affects the composition profile as shown in Fig. 1.

NEW CONTROL STRATEGY

A new system which combines reflux and distillate controls has been found to work well at moderate reflux ratios ($1 < L/D < 5$), and at a high reflux ratio ($L/D = 15$). Also, it will probably work, slightly modified, at reflux ratios below 1, but this technique has not been

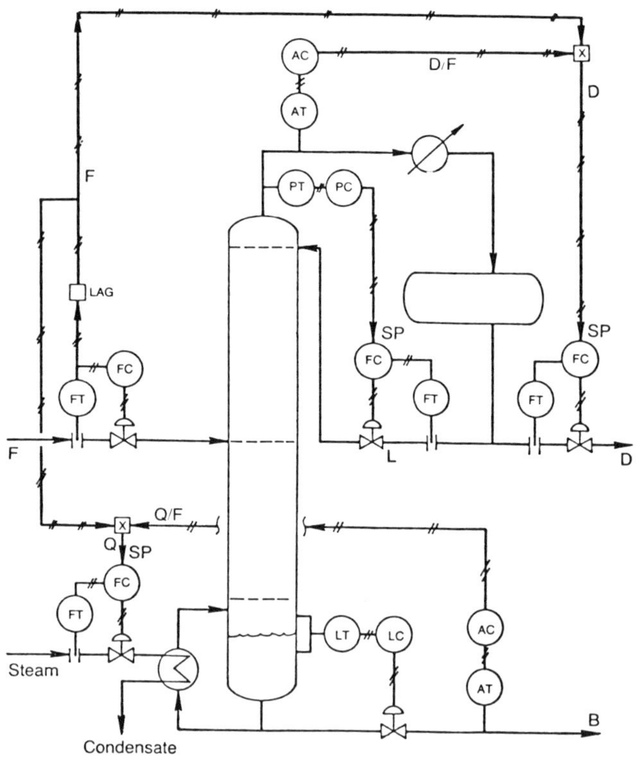

Figure 5. Material balance with feed forward.

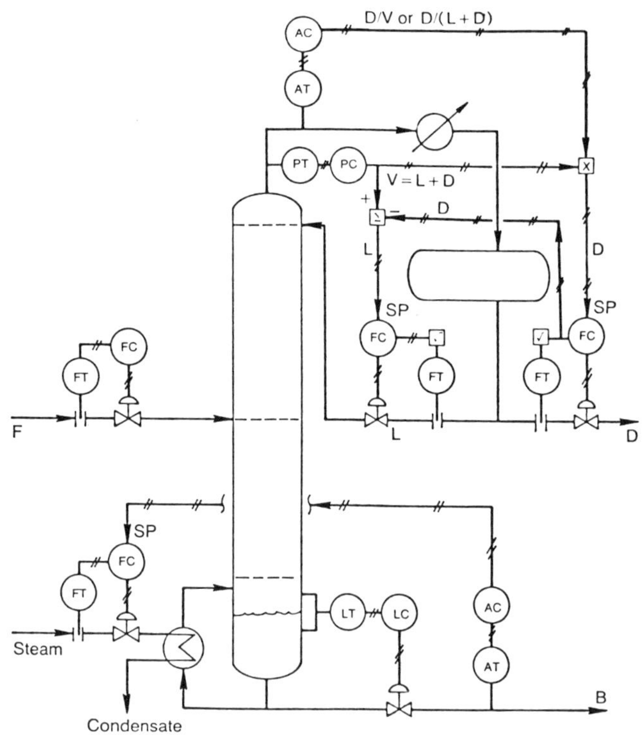

Figure 6. Combined reflux and distillate control.

tried and, indeed, may not yield a benefit because of the short residence time of light components on the trays. In the combined system (Fig. 6), the independent variable set by the composition controller is the reflux ratio, or the ratio of distillate to total overhead flow D/(L + D). The dependent variable is total overhead flow (L + D), which is set by the pressure control. Overhead vapor flow (V) equals L + D in the steady state.

Implementation

The combined control strategy can be implemented quite easily using one multiplying function and one summing function. Since both reflux and distillate flows must be linear, taking the square root of these signals is required. The output of the pressure controller corresponds to total column overhead flow (L + D), and is connected to both the multiplier and the summer. The other signal into the multiplier is the fraction of total overhead which is to be withdrawn as distillate product, D/(L + D). The output of the multiplier is the distillate flow signal "D."

Distillate flow measurement goes to the distillate flow controller, and also goes to the summing function as a negative input where it is subtracted from total overhead flow to obtain the reflux flow signal "L." Signal "L" is passed to the setpoint of the reflux flow controller which maintains the corresponding reflux flow.

Operation

When pressure rises, the pressure controller increases both reflux and distillate in the proportion fixed by the signal D/(L + D). Conversely, when pressure drops, flows will be lowered in the same proportion. Having a constant reflux ratio helps to keep distillate composition constant. Signal D/(L + D) may be set either by the operator or adjusted automatically by a tray temperature or top composition controller.

In any of the three previously described strategies, if the overhead receiver does not run liquid full, the level controller output can be connected in place of the pressure controller output. The level controller then manipulates the dependent heat balance variable, which is the distillate flow in the conventional system, the reflux flow in the material balance system, and the sum L + D in the new system.

Benefits of Combined L + D Control

This simple addition to the controls for the top of a distillation column produces many benefits. Its primary advantage is its ability to hold the curvature of the composition profile constant in the upper section of the column. Stated more technically, it holds a constant slope on the rectifying operating line on a McCabe-Thiele diagram, which is a graphical method of making distillation calculations. This results in distillate composition which is more nearly constant than the conventional or material balance solutions when changes occur in heat input and overhead vapor flow rate. Therefore, inherent decoupling of the top AC and heat input exists. Decoupling of the pressure controller and top composition controller is explicit in the design.

With the combined control strategy (Fig. 6), if more light components enter with the feed, and heat input is raised to keep them out of the bottom, both reflux and distillate will be increased proportionally. This removes the additional light components from the column. Moreover, keeping reflux ratio constant tends to keep distillate composition constant, significantly reducing the need for a top composition controller, or reducing its burden when used. Predominantly, the composition profile varies in the lower section of the column with changes in heat input.

If temperature or composition controls are used on both the top and bottom of the column, the new control strategy provides partial decoupling by reducing interaction between these loops. On an isobutane-normal butane splitter with the new control strategy on top, the bottom AC setting heat input exhibited a period of 1 to 1½ hours, whereas, with the material balance strategy on top, the bottom AC period was extended to as long as 6 to 12 hours. In addition, if the top composition controller must be put in manual to service the analyzer, the bottom composition controller may be left in automatic. (Again, with the material balance scheme, Fig. 4 or 5, the bottom AC must be put in manual whenever the top AC is put in manual.)

The combined strategy provides another benefit when components increase in the feed that are lighter than the light key, e.g., ethane in a depropanizer feed. These components are more difficult to condense, and the problem is heightened when the receiver runs liquid full with the pressure controller directly setting the dependent heat balance flow. In the conventional strategy, increasing pressure (or level) increases distillate flow, removing the lighter than light components faster and reducing upset. However, an increase in heavies in the distillate may result since reflux is not increased. If the material balance strategy is used, reflux is increased but distillate flow is not. Therefore, the lighter than light components accumulate, compounding the problem. Again, the new strategy represents a compromise. Distillate flow is increased, helping to remove the lighter than light key components, and reflux is also increased, keeping the heavy components down in the bottom of the column. This is a major benefit when it reduces propagation of an upset down through a sequence of columns.

Another method was implemented early in 1979 when a feedforward material balance system was modified. The output of the pressure controller was connected not only to set reflux, but also as a multiplier on distillate flow with the feed signal in place of the AC output signal. The pressure controller could then raise reflux and distillate flows simultaneously. As an expedient move, the AC output was connected back to the same multiplier/divider card as a divisor. This made it possible to put the AC on closed loop control without more hardware, and some of the benefits previously described were recognized. As an unexpected benefit, the nonlinear action of feedback trim in the denominator compensated somewhat for nonlinear process response. However, the multiplier-divider was difficult to scale because current values of feed, reflux and distillate had to be used, as well as transmitter spans, and the AC rangeability was narrow. The multiplier had to be rescaled for large changes in feed composition or reflux ratio.

The new combined reflux and distillate control is extremely easy to scale and is based only on transmitter spans. No operating data or distillation calculations are required. When feedforward is used, the feed flow only has to set heat input (Fig. 7). As soon as vapor generation rate is changed, pressure is affected and distillate flow will be adjusted the correct amount in the proper direction to maintain mass balance at the new feed rate. This same action would have been obtained with a feed to distillate feedforward signal.

Figure 7 also shows heat transfer rate control on a hot oil reboiler. Heat transfer rate (Q) is calculated by multiplying hot oil flow by its temperature drop, and controlled by adjusting the flow controller setpoint. Maintaining a constant Q/F tends to sustain a constant bottom composition and minimizes control action required by the bottom composition controller, which retains bottom composition closer to spec.

Composition Measurement can have a significant impact on the success of closed loop control. One analyzer should be used for the top composition and another entirely separate analyzer should be used for the bottom or any other stream, thereby avoiding leaks in a sampling system which could cause errors in composition going to the analyzer. Also, a single stream analyzer provides faster response. Span of the measurement should be adjusted so that the desired composition is somewhere near midscale. At 5 or 10 percent of scale, control action is severely restricted with measurement below setpoint.

When an intermediate stream composition is controlled, the ratio of the light key to the heavy key can be

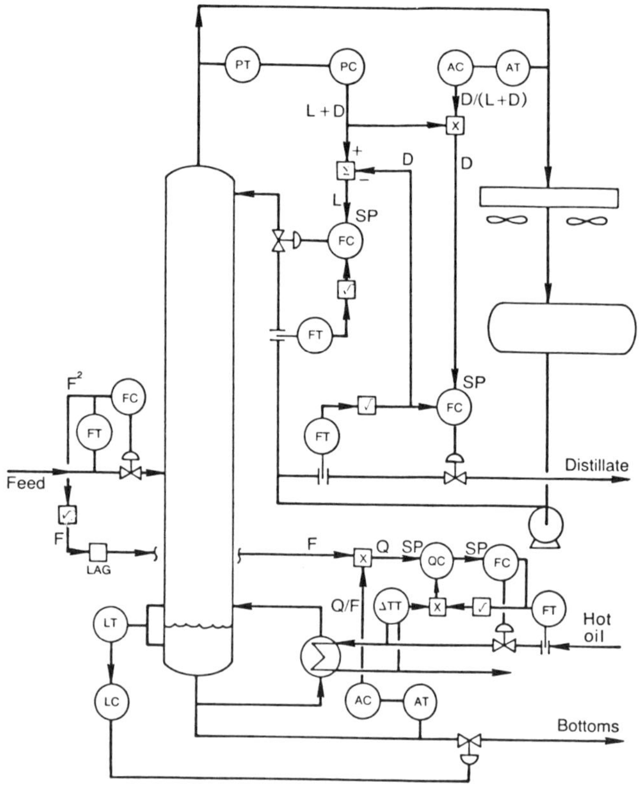

Figure 7. Combined *L* and *D* with feedforward.

calculated and used as the measurement to the controller. When this stream is further processed and the final product leaves the plant, the light key to heavy key ratio will remain the same. Therefore, the light impurity in the final product will be constant. This technique avoids process time lags between the point in the process where the composition is created and the point where the product finally leaves. This technique also avoids some analyzer drift problems such as variable amplifier gain or variable sample size on a chromatograph.

When a composition loop is unstable, the chart record normally shows sharp peaks separated by wide valleys because of the nonlinear response of the process. For example, composition of the top product from the distillation column changes logarithmically as light components accumulate on the trays. The most direct solution to this problem is to control the log of the measured concentration instead of concentration directly. This would require that the setpoint be entered as the log of the specification. The midscale of the composition controller should be the specification concentration and the maximum measurement should be a factor of 5 times the spec. The minimum measurement should be about 1/5 of the spec. providing a maximum 25 times the minimum. The widest practical span is the maximum factor of 10 above spec. and the minimum 1/10th of spec., which would make the top of the span 100 times more concentrated than the bottom. Alternately, process nonlinearity could be corrected by letting the measurement to the controller automatically adjust controller gain. However, tuning is more difficult. One other method is to use a divider and a summer computing elements. This is implemented easily with analog hardware.[3]

Overrides and Limiters

Setpoint limits on reflux flow and heat input are needed to prevent tray flooding from excessively high flow rates or losing the seal and allowing weeping or back-trapping to occur at low flow rates. When a composition controller is first put in service, the output can be limited to a very narrow span so that the operator need not be concerned that the composition control loop will upset the process. After the controller has been in service for a while and tuned, the limits can be made wider to allow more control action. This is one way to help the operator accept closed loop composition control as he gains confidence in the whole system.

Override controls are used on the overhead receiver level to reduce reflux flow and/or distillate product flow, preventing the pumps from losing suction. On columns with a relatively small bottoms product flow, a low level override on the bottom may be needed to reduce heat input. It must be remembered that these are rather drastic control actions and should be used only to protect equipment because they will surely cause composition upsets.

When a temperature controller responds too slowly to be used in the composition control loop, or if the temperature measurement is too uncertain in its correlation with composition, the temperature controller still may be used as an override to protect the process on analyzer failure. For example, temperature in the bottom of a depropanizer that contains butanes and gasoline will not correspond to propane concentration or propane to isobutane ratio. However, if the bottoms analyzer is setting heat input to the column, the temperature controller can be used as an override to raise heat input when temperature falls significantly below normal. This could occur if the process underwent a major upset and the analyzer could not respond quickly, or if the measurement to the analyzer controller failed. Overrides also can be used to cut heat input when column differential pressure goes too high to avoid tray flooding.

External reset feedback to a controller may be used to prevent reset windup on the master controller when the slave is limited and the master no longer can affect the process.[4] External reset feedback is available in both analog and digital systems.

Overrides and limiters should be applied only where problems are known to exist with the column. Their use makes control more complex and troubleshooting more difficult. Lack of understanding of the overrides or their improper implementation can be a deterrent to successful closed loop composition control.

THERMALLY COUPLED COLUMNS CONTROL SOLUTION

Increasing cost of energy relative to materials and labor has forced the use of process designs which reduce energy consumption, even though they may introduce other problems. One such design uses the overhead vapor from one distillation column as reboiler heat on another. The common reboiler-condenser allows both columns to operate with a utility cost about equal to the utility cost of only one. However, this design creates control problems because of the process interaction between the columns.

Interaction between the two columns is created by the common reboiler-condenser (Fig. 8). To eliminate this interaction, a control scheme has been created which maintains a constant heat transfer rate in the reboiler-condenser. This is achieved by holding a constant flow rate of liquid leaving the condensing tubes side of the exchanger. In the steady state, vapor flow into the tubes and heat transfer rate also will be constant. This is similar to a steam reboiler where the leaving condensate is on flow control. The tubes fill with liquid until the remaining surface area condenses vapor at a rate equal to the liquid removal rate. Heat transfer rate through the common reboiler-condenser determines vapor loading on the first column. The operator sets this loading as the value of reflux plus distillate (L + D) or as overhead vapor flow (V).

The L + D signal, vapor overhead, goes to a multiplier and a summer. The output of the top composition controller (AC) is the fraction of overhead vapor to be withdrawn as distillate product, D/(L + D). The balance of the response is the same as the combined reflux and distillate control described earlier. When distillate is adjusted to obtain the needed composition, reflux is changed an equal amount in the opposite direction, keeping total flow (L + D) constant. This constant heat transfer rate inhibits interaction between columns.

Since heat removal is fixed, heat input to the first column must be on pressure control to generate vapors as fast as they are being condensed, while maintaining the heat balance. On the second column, heat input on a

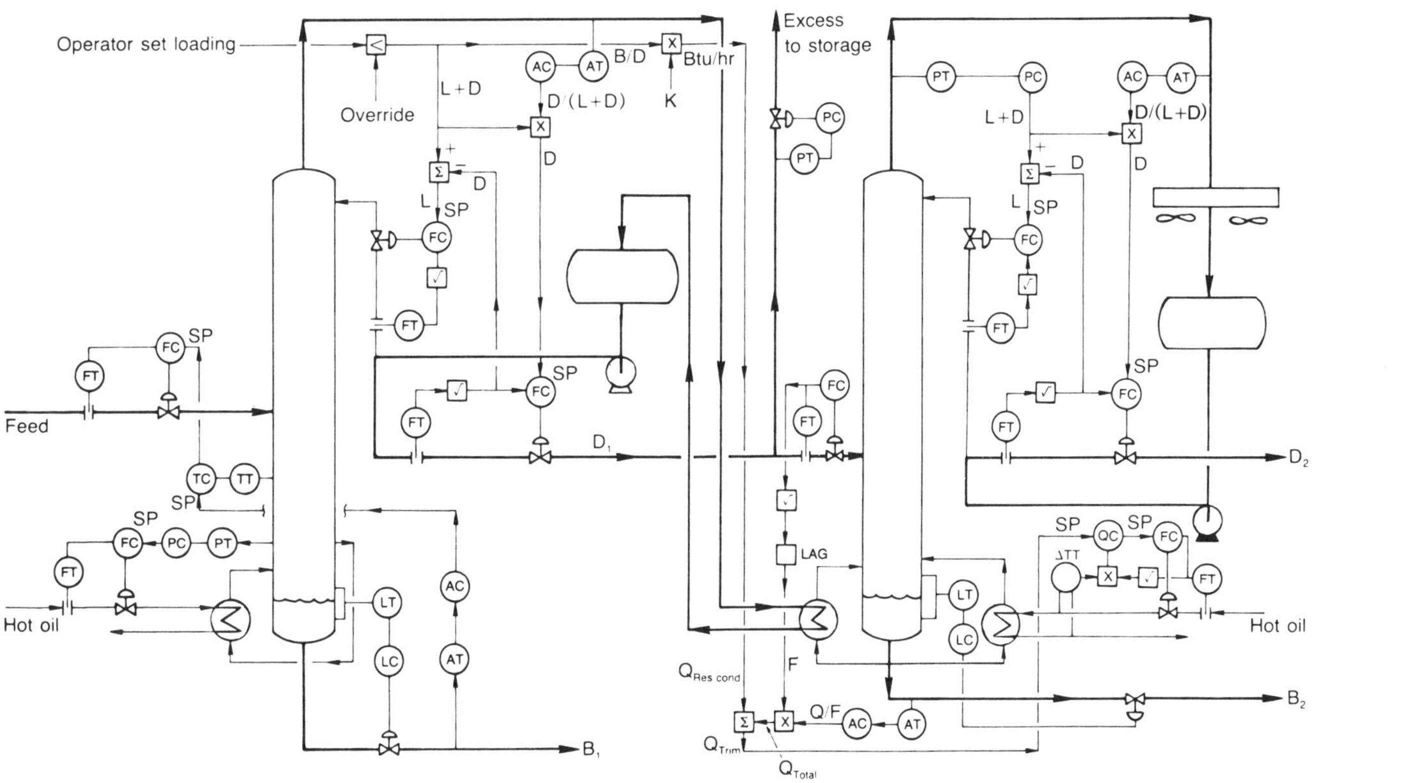

Figure 8. Dual composition control on thermally coupled columns.

trim reboiler is adjusted for bottom composition control. Its pressure controller must manipulate heat removal.

Distillate flow (D) on the first column is an independent mass balance variable because L + D is a constant and the signal D/(L + D) is adjusted for top composition control. Thus, from the mass balance equation (F = D + B), it is obvious that either feed (F) or bottom product flow (B) can be set independently, and the other must be on bottom level control to maintain the mass balance. The bottom AC can set the one chosen as the independent mass balance variable to maintain bottom composition. Bottom level is the more critical loop because it can severely upset heat input. Therefore, LC setting bottom flow was chosen because it avoids dead time in the level loop, if level were to set feed rate. This leaves feed rate to be set from the bottom composition control. More feed forces the profile down increasing lights in the bottom, and less feed lets the profile rise increasing bottom purity.

Because there was a large temperature rise from the feed tray to the bottom, and most of the feed is liquid, a tray temperature controller near the bottom sets feed rate. The bottom AC sets tray temperature. The TC provides fast response and substitute control when the bottom analyzer is out of service. If feed rate was determined by an upstream process, the bottom AC would have to manipulate the vapor flow (L + D), but slowly because of its effect on heat input to the second column.

Combined reflux and distillate control is also used on the second column with feedforward on heat input and dual composition control. The only added feature is that heat transfer in the reboiler-condenser is subtracted from total heat required and the difference sets heat transfer in the trim reboiler.

This control strategy, with an October 1979 startup, was the first case in which all four compositions, the top and bottom of both columns, were held successfully on closed loop control. The customer was impressed and satisfied with the results of this control strategy, which still performs well after six months in service.

References

1. Shinskey, F. G., *Distillation Control for Productivity and Energy Conservation,* McGraw-Hill, New York, 1977, pp. 306–307.
2. Bristol, E. H., "On a New Measure of Interaction for Multivariable Process Control," *IEEE Transactions on Automatic Control,* January 1966.
3. Shinskey, F. G., op. cit., pp. 258–259.
4. Shinskey, F. G., op. cit., p. 262.

28

Improve Reliability of Advanced Tower Control

R. J. DiBiano

Excellent control results have been obtained by using internal liquid and vapor rates as inferential parameters for tower control. The inferred internal reflux ratio is used as an intermediate controller in a double cascade compositional control scheme. This arrangement also offers a convenient way to approach flooding and capacity constraints, so that the required separation is obtained at maximum distillation tower thruput.

Distillation tower control with these inferential parameters does not provide composition control directly. It is more closely related to separation efficiency in the tower. Of course, good separation implies good composition when the rest of the unit is operating as designed. Therefore, it does offer convenient backup control when chromatographs are serviced or not functioning properly.

Several of the chronic problems associated with distillation towers can be handled with advanced control better than with traditional temperature and flow control. Some of these problems are:

- Holding product composition close to specifications
- Compensating for effects of weather on air-finned condensers
- Inverse process response to control actions
- Changing feed compositions
- Changing feed flowrates
- Changing liquid levels and rates in towers and receivers

Inferential control solves the chronic problems of distillation tower control mentioned above *more reliably* than direct composition control. This kind of inferential control has been installed at several locations, in some cases on towers which have liquid side draws. The internal reflux ratio was calculated for the tray below the side draw, so that constant separation was obtained in the main part of the tower which was below that point.

The liquid reflux rate was inferred from the external reflux rate plus the condensation from the cold reflux minus the liquid side draw. The vapor rate was taken as the total liquid flow from the accumulator drum plus the condensation. If the lag between the calculated vapor rate and actual tower conditions caused any problems, the accumulator drum and tower levels were used to compute internal flowrates more precisely.

Since the basis for this control are process flows and temperatures, response to changes in operating conditions is very rapid. It is possible to maintain high purity overhead product through normal changes in thruput and feed composition. In one case, feed/forward control from feed thruput and composition was used to control heat input to the reboiler while this inferential control was used to control reflux. Operator attention is required only when attempts are made to operate the units beyond physical constraints of the hardware.

With all of the pomp for using advanced control to reduce energy consumption coming from management, it is easy to lose sight of other incentives for improved control, such as improved product split in separation processes. While chemical applications offer a clearer incentive for improving product splits due to large value differential in product streams, such incentives also exist in refining, though they are harder to identify.

The literature is filled with process control ideas for distillation control based on both laboratory[1-4] and industrial[5-8] experience. In 2-product distillation, after objectives and constraints have been resolved, the problem essentially reduces to maintaining two interacting manipulated variables at setpoints. Probably the most difficult variables to select, but also the most rewarding ones to control, are composition control on both overhead and bottoms. Dual composition control has been accomplished successfully in the laboratory[2] without difficulty, even without considering the coupled nature of the problem. For commercial applications, Jafarey and

McAvoy[4] have indicated the difficulties that could be expected in attempting dual composition control. Jafarey concluded that certain systems; i.e., difficult separations, are not amenable to some types of dual composition control. Burford and others,[7] also indicate that they have had difficulties in maintaining a commercial system on dual composition control. Castellano[8] claims to have put an unnamed azeotropic system on dual composition control, but gives no dynamic composition control data to enable the reader to judge his success. In summary, dual composition control could be highly profitable in some applications, but reliability is usually a problem.

CONTROL PROBLEMS

Dynamic problems with distillation towers have always plagued certain petrochemical and refining operations. For petrochemical operations, the problem of holding a product composition on specification during an externally generated upset is a lucrative goal easily recognized even by the casual observer. With relatively unsophisticated but advanced feedback control splits may be manipulated readily to hold one product composition, so long as separation is adequate. Since separation can be related to energy input, minimizing energy input while satisfying both product composition constraints becomes a second objective for distillation tower control. Directly satisfying this second objective is no longer so easy. By operating close to two composition constraints rather than one, interaction becomes a problem. Chemicals distillation towers often have large reflux ratios and many trays due to purity requirements and difficulty in separating components with close boiling points. Purity requirements can be so stringent that components must be measured in the ppm range to control them. While it may be possible to design a process chromatograph to measure in these ranges, too much must be compromised in the way of delays, peak separation and maintenance requirements.

Interaction between streams under composition control for difficult separations is extremely severe. In separating two close boiling hydrocarbons, control of the heavy key in the overhead by reflux is very sensitive with constant heat to the reboiler. However, once the bottoms product is placed on composition control via heat input to the reboiler, and overhead composition is returned to control by reflux, the process gain increases for the loop making overhead composition considerably less sensitive to reflux. This is because heat input to the reboiler now will react to added reflux in keeping light key in the bottoms constant. Hence, some way around complete dependence on feedback control with its inherent delays is desirable.

For refining applications, reflux ratios are generally small, separation much easier and specifications are on physical properties such as endpoint, flashpoint, overpoint, metals, vapor pressure and viscosity rather than purity. Therefore, it would seem that refiners would not have the same problems. Distillation units, like crude units and FCC main fractionators, in fact, have lower reflux ratios which facilitate less interaction between streams on composition control. Unfortunately, this advantage is often offset by the added complexity of several sidestreams, all of which interact, and the inability to relate physical properties to composition.

INFERENTIAL SOLUTIONS

In general, inferential control on composition for one of the product streams from the distillation tower with feedback control on the other works well. Also, intermediate variables such as inferred tower internal flows and temperatures are used as an intermediate cascade for the stream on feedback control (Fig. 1). Controlling one of the compositions in open loop while closing the loop on the other avoids the loop interaction problem. The disadvantage is the loop chosen for feed/forward control will be held to its setpoint only as well as your model. Often, however, the constraints and requirements on the distillation tower products may be fulfilled by maintaining tight composition control on one of the product streams and loose control on the other.

Figure 2 shows a chemicals distillation tower on this type of control. It splits products A and B in the side draw from product C in the bottom. Light ends which escape upstream removal are drawn off by a small drag stream from the overhead in a so-called "pasturization

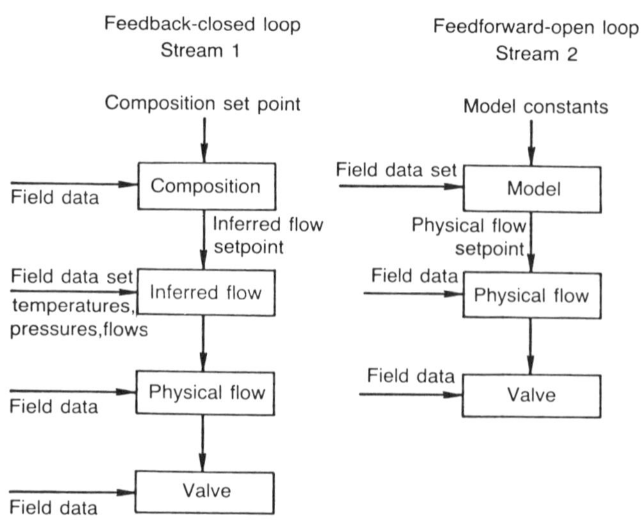

Figure 1. Block diagrams of open and closed control loops for the product streams.

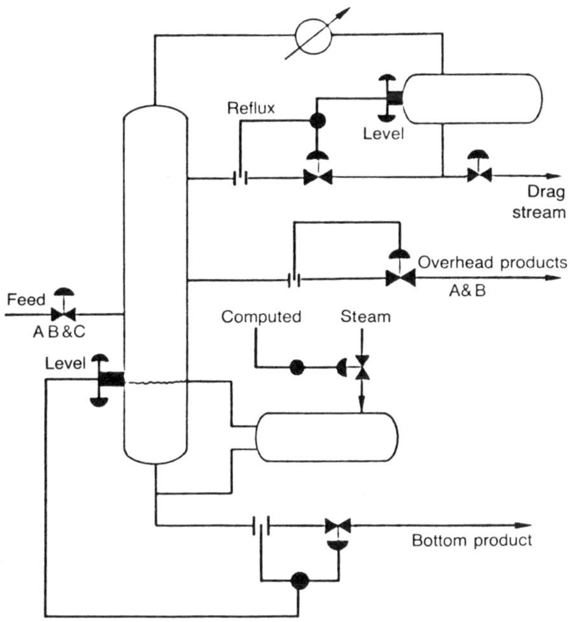

Figure 2. Chemicals distillation tower on inferential control.

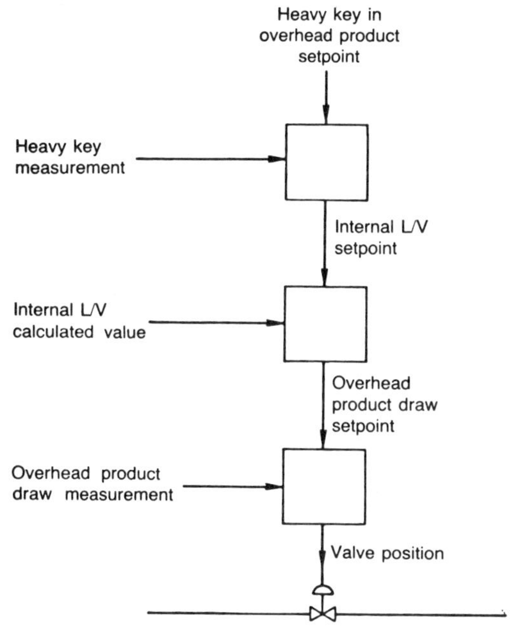

Figure 3. Block diagram of control on the tower overhead.

section." The drag stream is too small to control the level in the overhead accumulator effectively, and using the side draw to control level (a second possibility) presents inverse response problems. When the overhead accumulator level rises, the side draw should increase. Initially, this has the effect of cutting internal reflux which heats the tower and causes more overhead vapor and therefore increases the level of the accumulator still further. Controlling the level by means of reflux, a third possibility, is not satisfactory either. Any change on the condensation rate (airfinned condenser) of the overhead vapors will have a pronounced effect on tower operation since the reflux will change dramatically.

A satisfactory solution to the level control problem is controlling the overhead accumulator level with both the external reflux and the side draw to keep the internal reflux virtually constant. Ultimately, the internal reflux is not kept strictly constant in the tower for two reasons. First, it is ratioed to another flow quantity, usually the feed rate or, as we have done, the internal vapor rate. It is this dimensionless quantity that is held virtually constant. Secondly, as shown in Fig. 3, this inferred parameter is used as input to the secondary controller in cascade, much as flow is traditionally used as input to a secondary controller in cascade with a primary level controller. The slave or secondary controller holds the flow setpoint constant only within its dynamic time frame, taking resets from the primary level controller, or as in our case, the primary controller for composition.

Figure 4 shows control on the bottoms of the tower which consists of a thermosyphon steam reboiler. Feed

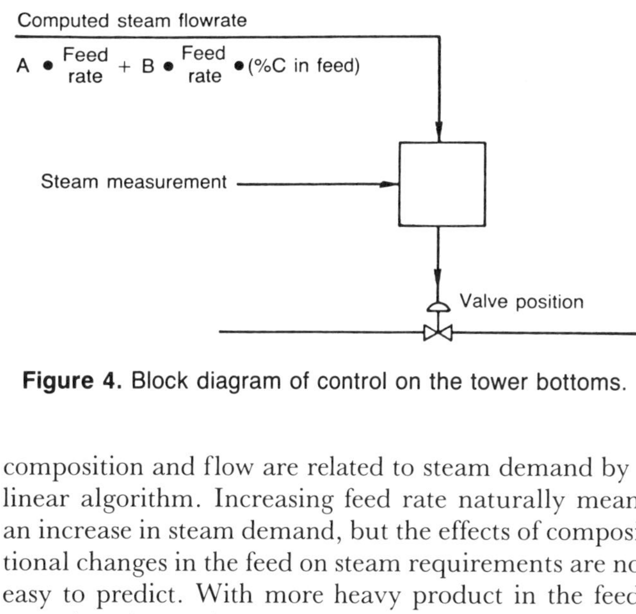

Figure 4. Block diagram of control on the tower bottoms.

composition and flow are related to steam demand by a linear algorithm. Increasing feed rate naturally means an increase in steam demand, but the effects of compositional changes in the feed on steam requirements are not easy to predict. With more heavy product in the feed, more heat is required. Apparently, this is because of the heat balance around the tower. With more heavy product leaving the heat balance envelope at a higher temperature than the light product, it is clear that more heat will be required.

The major problem with the control system occurred because of a misapplication of the intended control strategy. After control system installation, operations required constraints on all flows controlled by chromatograph analysis in case a chromatograph malfunction occurs and is not picked up by error checking. One of the constrained flows is steam flow to the reboilers.

When feed rate to the complex is cut back, the feed/forward is programmed to cut back the steam flowrate appropriately. However, the steam flowrate will hit the low flow constraint and be maintained higher than it ought to be. The operator is now faced with the decision of allowing the steam to cut back beyond the flow constraint and setting up a new one, or changing the constants in the steam flow control algorithm. When the operator elects to change the constants for the steam control algorithm, some undesirable consequences result.

While light key stays out of the bottom with the added steam flow, heavy key at the top will increase suddenly. Feedback control on overhead composition will eventually bring the composition back to setpoint but this will take time. Bottom product is put into a tank and sold directly. High purity requirements on this product make the operator reluctant to risk contamination of the product tank. Overhead product goes into a reactor. Composition of this stream is important to operation of that reactor. A certain amount of heavy key is required to maintain proper reactor temperature and optimize system operation. Excursions from proper overhead composition cause the reactor temperature to go out of control. To further aggravate the situation, the composition controller on the overhead product has overtuned from lowering reset time. This was done to shorten excursion time from setpoint during these upsets. The excursion time has been shortened, but at the expense of having the overhead composition cycle during normal operation.

While the operators will change the steam control algorithm constants to keep the steam from cutting back below the flow constraint, they seldom change the algorithm back because extra heat in the reboiler assures that the operator will not have to worry about the product stream compositions going off specification. They have that assurance, however, only at the expense of using extra energy.

Another inferential control application is shown in Fig. 5. This is a reformate splitter in which many hundreds of substances of all types are present; i.e., paraffins, olefins, naphthenes and aromatics. They are present in varying proportions depending on, among other things, severity of the upstream reactor and the type of crude processed. The heavy key in the light reformate is the C_9 aromatic which will most contaminate the C_8 aromatic product downstream. There is no further provision to remove C_9 aromatics. The light reformate goes through an extraction unit only to recover aromatics from other species. The light key on the bottom is the C_8 aromatic. As there are several other components with boiling points between the C_8 and C_9 aromatics in varying proportions, temperature alone is not a useful parameter in controlling either overheads or bottoms.

Overhead product is used to control level, and external reflux is used to control internal reflux ratio. There is no inverse response problem as in the previous side draw example and overhead product flow is large enough to comfortably control level. Also, overhead product goes to a storage tank so wide fluctuations in the flowrate will not disturb the process. External reflux was chosen in this example to control internal reflux ratio since control moves for this inferred parameter are generally milder than control moves made to control level.

On the bottom of the tower, the light key component is a very valuable product and market conditions still dictate that the best possible recovery of this product is worth the extra energy consumption. Hence, a control strategy other than dual composition control which keeps light key out of the bottom can be rationalized as being as good or better than closed-loop feedback composition control. It is desirable to keep the light key to a minimum and not to control it. Level is cascaded to bottoms product flow while bottoms temperature is cascaded to heat input as shown in Fig. 6. This control differs from the classical oil-fired reboiler control in which outlet temperature is controlling heat input. The advantages of this control are first, bottoms temperature, though admittedly a poor measure of light key in the bottoms, gives a better measure than heater outlet temperature. The second reason is more complicated. The bottoms product exchanges heat with the feed to the stabilizer upstream. An upset in bottoms product flow is

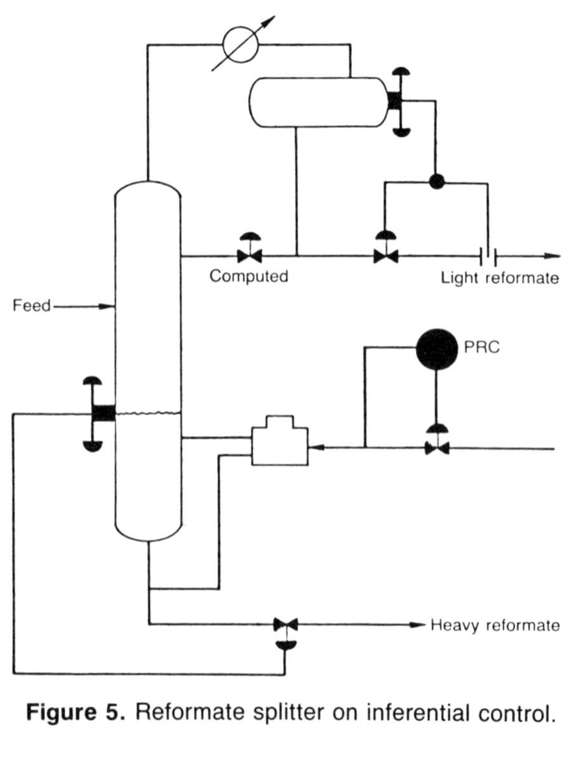

Figure 5. Reformate splitter on inferential control.

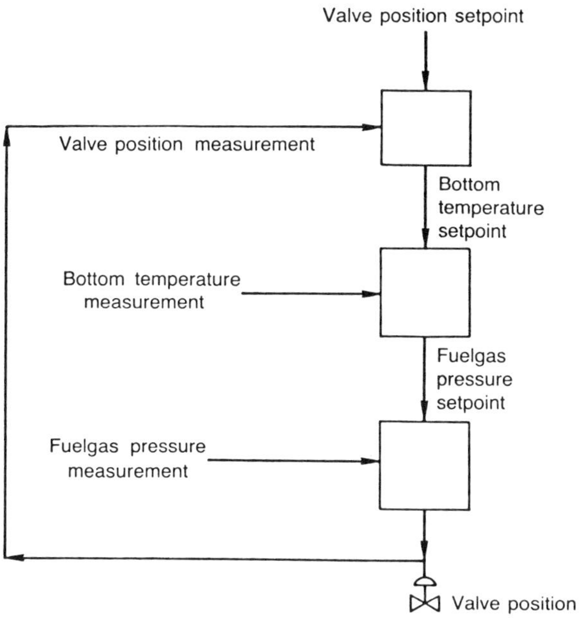

Figure 6. Block diagram of control on reformate splitter bottoms.

reflected in an upset to the stabilizer which again upsets the reformate splitter and results in positive feedback to the original upset in the reformate splitter bottoms. Also, the reboiler alone does not have the capacity to control bottoms level, but by tuning the level to bottoms flow cascade loop to react slowly and having the reboiler on a high-gain bottoms temperature control, the problem of positive feedback was solved. This resulted because high-level upsets in reformate splitter bottoms result usually in a low bottoms temperature. Hence, the reboiler works with the bottoms product flow to bring the level back to setpoint with a minimum of bottoms flow upset.

Above the temperature controller is a primary controller which uses the fuel gas valve position as a measurement and maintains a setpoint near, but not at, full open. Thus, the reboiler stays in control yet it remains near its full-capacity constraint.

References

1. Schwanke, C. O., "Experimental Multivariable Control of a Pilot-Plant-Scale Distillation Column," University of Texas, Austin, Dissertation, 1976.
2. Bilek, R. and Wood, R. K., "Multivariable Frequench Domain Controller Design for a Binary Distillation Column," 86th AIChE Meeting, Houston, 1979.
3. Tyres, B. D., "Multivariable Control System Designed for an Industrial Distillation Column," *Ind. Eng. Chem. Proc. Des. Devel.,* Vol. 18, Nov. 1, pp. 177-82, 1979.
4. Jafarey, A. and McAvoy, T. J., "Degeneracy of Decoupling in Distillation Columns," *Ind. Eng. Chem. Proc. Des. Devel.,* Vol. 17, Nov. 4, 1978.
5. Cutter, C. R. and Ramaker, B. L., "Dynamic Matrix Control—A Computer Control Algorithm," 86th AIChE Meeting, Houston, 1979.
6. Seeman, R. C. and Nigenfeld, A. E., "Digital Distillation Tower Control Offers Flexibility," *Oil and Gas Journal,* Vol. 73, Nov. 43, pp. 57-60, 1975.
7. Burford, B. N. et al., "Computers Conserve Energy in NGL Fractionators," 86th AIChE Meeting, Houston, 1979.
8. Castellano, E. N., et al., "Digital Control of a Distillation System," *Chem. Eng. Prog.,* Vol. 74, Nov. 4, pp. 56-60, 1978.

29

Using Relative Gains to Design Dual Composition Controls*

L. M. Gordon

The need to minimize energy used in distillation has recently focused increased attention on the problems of controlling composition simultaneously at both ends of a column. First among these is the problem of interaction between the two composition control loops. Although decoupling control systems have often been viewed as the solution to this problem, these schemes have met with limited success. Their application may be viewed as treating the symptom rather than the disease, since the degree of interaction depends upon the variables chosen by the instrument engineer to close the composition loops. The concept of relative gains has been established to describe the amount of interaction in multivariable control schemes. However, practical techniques for evaluating the gains for dual composition control systems have not been available. This chapter presents simple algebraic expressions that quantify the relative gains in binary separations for each of the control system structures available to the designer, so that interaction can be minimized in design. With further attention to dynamics, the problem of interaction often can be essentially prevented.

ECONOMIC INCENTIVES

There are often strong economic incentives for operating a column closer to both specifications. Figures quoted for savings vary widely in the literature. Luyben (1975) states that energy savings often may not be worth the increased cost and complexity of instrumentation. Griffin (1981) states that a well-designed control system can cut energy costs by as much as 20 percent, typically with five to 10 percent increased throughout. Ryskamp[5] states that, typically, 30 percent excess heat is expended to guarantee that compositions meet specifications. Jafarey and McAvoy[1] describe a deisobutanizer using 37 percent excess energy, costing $327,000/yr due to overrefluxing. Each case must be evaluated individually. Potential for cost reduction will depend on production rate, relative value of products and fuel, cost of cooling, availability of waste fuels, cost of installing and maintaining analyzers, and other operating costs, such as fouling or material-handling problems.

One common mistake is failure to consider what will be done with the energy that might be saved. No gain is achieved by reducing steam consumption in an individual unit only to have it vented to atmosphere by header pressure controls. To be meaningful, reductions in energy used in unit operations must be reflected back into the utilities area of the plant to reduce fuel consumption.

Where economic analysis has provided the incentive to install a system for dual composition control, the designer runs head-on into the problem of designing the control system structure. Specifically, this means deciding how to pair the controlled and manipulated variables. In a typical column, there are as many as five variables to be controlled, including top and bottom composition, bottom and accumulator level, and column pressure. Historically, a like number of variables has been used for manipulation, including distillate and bottom flow, reflux flow, heat input, and heat removal. This five by five matrix can be paired in 120 unique ways. Even after eliminating the obviously unworkable pairings, such as control of accumulator level by manipulating bottoms flow, the designer is left with more than two dozen feasible schemes.

Choosing among this many possibilities has always been the stumbling block in designing control systems for dual composition control. Faced with this many options, designers have tended to return to the traditional

* The author expresses his appreciation to F. Greg Shinskey and Carroll Ryskamp of The Foxboro Company for their comments and suggestions during the preparation of this article.

reflux and boilup controls shown in Fig. 1. However, with distillation columns as varied as their products and feeds, this scheme has yielded widely varying results. And where controls are not successful and responsive, they are simply placed in manual and ignored by the operator.

Classical decoupling schemes[3] have not provided a solution to the problem of interaction because of practical problems encountered on real columns.[7] In computers, initialization routines may prevent placing classical decoupling schemes on-line without upsetting product compositions. Decoupling systems that include overrides can drive to saturation when constraints are encountered. Most seriously, decouplers applied to systems with negative interaction (defined later) may have very little tolerance for errors in decoupler gains. For this group, which always includes the interaction encountered in reflux and boilup controls, small errors can transform a system that provides complete decoupling into one that provides no control at all. Since proper decoupler gains depend on the process gains, which inevitably change with variations in feed rate, product specifications, and column characteristics, these systems require constant attention and adjustment beyond the ordinary capability of plant operating personnel.

Difficulties associated with the application of decoupling systems have prompted a reexamination of interaction itself. The problem may be postulated in two ways:

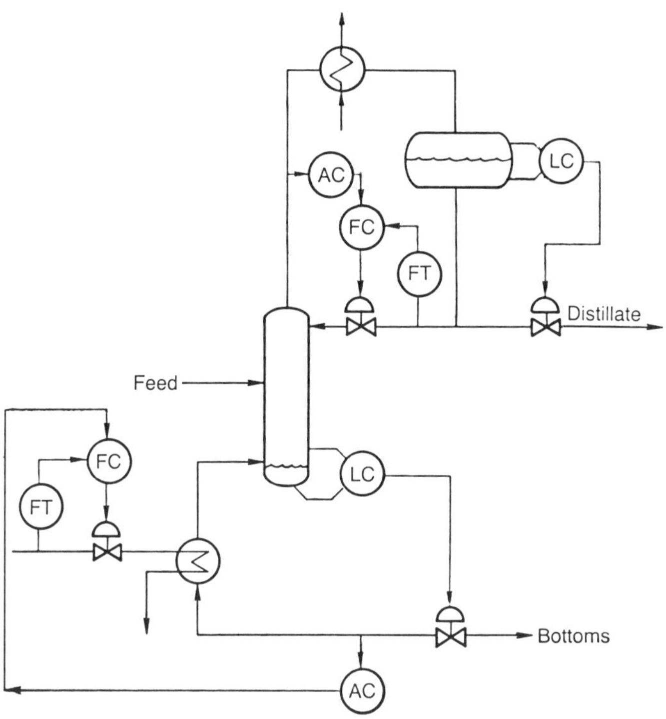

Figure 1. Traditional reflux and boilup controls.

- For a given column, is the interaction equally strong in each of the control structures available to the designer?
- For a given control structure, will the interaction be equally strong in every column where it is applied?

The stumbling block of loop assignment may, in this way, be converted into a stepping stone by providing the opportunity to use a control structure that will exhibit minimum interaction in any particular application.

Shinskey and Ryskamp have given consistent guidelines for assigning loops to minimize loop interaction. Thus, Shinskey[2] suggests that the controller assigned to the less pure product should manipulate the mass balance while the controller assigned to the more pure product should manipulate separation (defined later). Ryskamp[5] suggests that the controller for the component with the shorter residence time should adjust vapor flow while the controller for the component with the longer residence time should adjust the liquid/vapor ratio. An analytical approach has been refined that verifies these guidelines for most columns and identifies those columns where exception should be taken.

SEPARATION

Any analytical approach to minimizing interaction requires mathematical models of the column. Designers are thoroughly familiar with overall and component mass balances across the column. The relation between the energy balance and pressure control is also well established. A definition of how completely the mixed feed is split into its components is equally fundamental, and may be expressed in terms of the product compositions, as in Shinskey[6]

$$S = y(1 - x)/(x(1 - y)) \qquad (1)$$

where S = separation index
y = mole fraction of light component in distillate
x = mole fraction of light component in bottoms

Thus, as distillate and bottoms products become more pure, y approaches 1.0 and x approaches 0.0, causing S to approach ∞. Conversely, if there is no separation, both y and x equal z, the mole fraction of the light component in the feed, causing S to approach 1.0.

This definition, however, measures the degree of separation that *has been achieved* in a column. Predicting the separation that *will be realized* requires a different equation that models the relation between separation and the design and operating parameters for the column. Shinskey[2] has proposed a model that relates separation to the ratio of boilup to feed rate, given the number of trays, the tray efficiency, and the relative volatility. Recently, a more accurate model has been proposed by Jafarey, Douglas, and McAvoy.[1] For saturated feed,

$$S = \left[\frac{\alpha}{\sqrt{1 + \frac{D}{Lz}}}\right]^{nE} \tag{2}$$

where α = relative volatility (assumed constant across the column)
D = distillate product flow rate
L = reflux flow rate
n = the number of actual trays
E = tray efficiency (assumed constant across the column)

This model incorporates the influence of energy per unit thruput in terms of the reflux ratio, L/D, and also includes the effect of changes in feed composition on separation. By this model, separation may be added to the set of variables available for controlling composition, through manipulation of the reflux ratio. Alternatively, other combinations whose values follow from the reflux ratio may be manipulated, including L/V and D/V, where V is the rate of overhead vapor flow.

The definition of separation given in Eq. 1 is a single relationship between two variables, x and y. Therefore, specifying the separation does not fix product compositions; rather, it specifies a set of related compositions. Likewise, the material balance may be expressed in terms of the product compositions. Overall,

$$F = D + B$$

where F = feed rate to the column
B = bottom product flow rate

and, for the light component,

$$Fz = Dy + Bx$$

then:

$$D/F = (z - x)/(W - x) \tag{3}$$

For a given feed composition, z, this is also a single relationship in two variables, and specifying the fractional split of the feed between top and bottom products only defines a related set of product compositions.

The relation among the material balance, separation and product compositions can be visualized on a diagram proposed by Shinskey (Fig. 2), which plots impurity in the bottom product, x, against impurity in the top product, 1 − y. Against these axes, feed composition is defined by the coordinates x = z and 1 − y = 1 − z, and lines of constant fractional split plot as straight lines radiating from these coordinates. Lines of constant separation plot hyperbolically against these axes. Taken together, a given separation and fractional split define an intersection that fixes the compositions of both products.

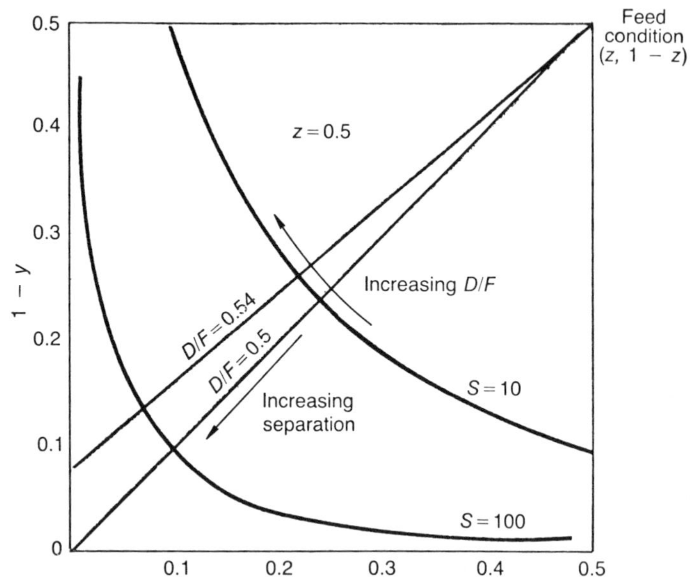

Figure 2. Relation among product purities, separation and the material balance.

RELATIVE GAIN

The concept of relative gain provides a measure of interaction that may be expected in closing a particular loop. First proposed by Bristol[4] and developed by Shinskey,[2,6] the concept may be used to analyze possible control scheme configurations for minimum interaction. Given a control scheme by which a set of variables, C, is controlled by manipulating a set of variables, M, the relative gain, λ_{ij}, for the loop that controls c_i by manipulating m_j is defined as:

$$\lambda_{ij} \equiv \frac{\left.\frac{\partial c_i}{\partial m_j}\right|_m}{\left.\frac{\partial c_i}{\partial m_j}\right|_c} \tag{4}$$

The numerator, or open-loop gain with all other loops open, indicates the sensitivity of c_i to m_j when all other manipulated variables are held constant (i.e., when all other controllers are in manual). The denominator, or open-loop gain with all other loops closed, indicates the same sensitivity when all the other controlled variables are held constant (i.e., when all other controllers are in automatic). Since interaction is created by reciprocal disturbances among control loops in automatic, the ratio of these two gains measures the effect of the interaction with other loops on the feasibility of controlling c_i by manipulating m_j. For $\lambda_{ij} = 1.0$, the influence of m_j over c_i is not changed when the other loops are closed, indicating that this loop does not interact with the others. As the value of λ_{ij} departs from 1.0, it indicates an increasing degree of interaction. Values greater than 1.0 indicate that the interaction reduces the sensivity of c_i to m_j, which degrades performance when all control loops are closed. Values of λ_{ij} that approach zero indicate that most of the influence of m_j over c_i is exerted through interaction, by way of the other closed loops, causing poor performance when the other loops are opened. Negative values of λ_{ij} indicate that the interaction reverses the direction in which c_i is driven by changes in m_j, causing the stability of this loop to be conditional on the auto/manual status of other loops.

Relative gains for all possible pairings may be gathered together in a relative gain array (RGA), defined as:

$$\Lambda = \begin{array}{c|ccc} & m_1 & \cdots & m_n \\ \hline c_1 & \lambda_{11} & \cdots & \lambda_{1n} \\ \vdots & \vdots & \lambda_{ij} & \vdots \\ c_n & \lambda_{n1} & \cdots & \lambda_{nn} \end{array}$$

Bristol has shown that the sum of the numbers in every row and column must equal 1.0. Thus, arrays with values of λ greater than 1.0 must also contain negative values of λ. An array that includes negative values is said to exhibit negative interaction. Conversely, an array in which all relative gains are positive, and therefore less than or equal to 1.0, is said to exhibit positive interaction. In the case of two by two arrays with positive interaction, as one value of λ approaches 0.0, the other must approach 1.0, and values of λ that approach 0.5 indicate maximum positive interaction. In the case of matrices with negative interaction, values of λ that approach $\pm\infty$ indicate maximum interaction.

RELATIVE GAIN AND DUAL COMPOSITION CONTROL

The set of possible controlled and manipulated variables for a distillation column forms a five-by-six matrix,

$$\Lambda = \begin{array}{c|cccccc} & D & B & S & V & L & Q \\ \hline y & \lambda_{yD} & & & & & \\ x & & & & & & \\ l_a & & & & \text{etc.} & & \\ l_b & & & & & & \\ p & & & & & & \end{array}$$

where: Q = heat removal rate
l_a = accumulator level
l_b = bottom level
p = column pressure

However, evaluating the complete set of relative gains suggested by this array gives a misleading representation of interaction in dual composition control. The definition of relative gain given by Eq. 4 requires evaluation of each open-loop gain with *all* other loops open. Since separation actually results from the ratio of internal vapor and liquid rates, if boilup and reflux are held constant, variations in D or B will have no effect on composition (only on the liquid levels) so that the relative gains for controlling compositions with these variables will be zero, indicating that these assignments will not control. Yet, these assignments do control, through the action of the level and pressure control loops that reflect these changes into reflux or boilup, if distillate or bottoms flow is used to control composition.

To evaluate relative gains properly for composition control, the overall matrix must be reduced to two by two subsets that set x and y against possible pairs of manipulated variables, under the assumptions that the remaining variables are used to control both levels and column pressure, and that these loops are closed and holding their measurements at set point. Since these loops are an order of magnitude faster than the composition loops, their ability to achieve this is a reasonable assumption. The open-loop gains (numerators) for the pairs in these matrices are then evaluated with only the other composition controller in manual. For example, in the subset that assigns D and S to control y and x,

$$\Lambda_{DS} = \begin{array}{c|cc} & D & S \\ \hline y & \lambda_{yD}(\Lambda_{DS}) & \\ x & & \end{array}$$

the relative gain for control of top composition by manipulation of distillate flow is given by:

$$\lambda_{yD}(\Lambda_{DS}) = \frac{\left.\dfrac{\partial y}{\partial D}\right|_{S,\, l_a,\, l_b,\, p}}{\left.\dfrac{\partial y}{\partial D}\right|_{x,\, l_a,\, l_b,\, p}}$$

In this expression, identification of the subset is essential. The relative gain λ_{yD} appears in more than one subset, and the value of $\partial y/\partial D$ will depend on which other manipulated variable is held constant.

Determining the values of these relative gains is not the formidable task it seems at first. Since the overall mass balance requires equal and opposite changes in D and B,

$$dD = -dB$$

which is achieved by assigning at least one of these variables for level control to close the overall mass balance. Since this substitution may be made in both the numerator and denominator of the relative gain, the signs cancel and the value of the relative gain will not be changed. For example, (dropping the notation for the always-constant levels and pressure),

$$\lambda_{yD}(\Lambda_{DS}) = \frac{\left.\frac{\partial y}{\partial D}\right|_S}{\left.\frac{\partial y}{\partial D}\right|_x} = \frac{-\left.\frac{\partial y}{\partial B}\right|_S}{-\left.\frac{\partial y}{\partial B}\right|_x} = \lambda_{yB}(\Lambda_{BS})$$

Next, eliminating heat removal, Q, as a candidate for composition control reduces the problem to six unique two by two composition subsets,

$$\Lambda_{DS} = \begin{array}{c|cc} & D & S \\ \hline y & & \\ x & & \end{array} \qquad \Lambda_{DL} = \begin{array}{c|cc} & D & L \\ \hline y & & \\ x & & \end{array} \qquad \Lambda_{DV} = \begin{array}{c|cc} & D & V \\ \hline y & & \\ x & & \end{array}$$

$$\Lambda_{SV} = \begin{array}{c|cc} & S & V \\ \hline y & & \\ x & & \end{array} \qquad \Lambda_{SL} = \begin{array}{c|cc} & S & L \\ \hline y & & \\ x & & \end{array} \qquad \Lambda_{LV} = \begin{array}{c|cc} & L & V \\ \hline y & & \\ x & & \end{array}$$

which contain 24 relative gains. However, under the assumption that both levels and pressure are invariable, the requirement that the sum of the elements in each row and column must be 1.0, reduces the problem to determining just six unique values. For example,

$$\Lambda_{DS} = \begin{array}{c|cc} & D & S \\ \hline y & \lambda_{yD}(\Lambda_{DS}) & 1 - \lambda_{yD}(\Lambda_{DS}) \\ x & 1 - \lambda_{yD}(\Lambda_{DS}) & \lambda_{yD}(\Lambda_{DS}) \end{array}$$

Finally, collecting the six unique relative gains together yields the following set of definitions:

$$\lambda_{yD}(\Lambda_{DS}) = \frac{\left.\frac{\partial y}{\partial D}\right|_S}{\left.\frac{\partial y}{\partial D}\right|_x} \qquad \lambda_{yD}(\Lambda_{DL}) = \frac{\left.\frac{\partial y}{\partial D}\right|_L}{\left.\frac{\partial y}{\partial D}\right|_x}$$

$$\lambda_{yD}(\Lambda_{DV}) = \frac{\left.\frac{\partial y}{\partial D}\right|_V}{\left.\frac{\partial y}{\partial D}\right|_x} \qquad \lambda_{yS}(\Lambda_{SV}) = \frac{\left.\frac{\partial y}{\partial S}\right|_V}{\left.\frac{\partial y}{\partial S}\right|_x}$$

$$\lambda_{yS}(\Lambda_{SL}) = \frac{\left.\frac{\partial y}{\partial S}\right|_L}{\left.\frac{\partial y}{\partial S}\right|_x} \qquad \lambda_{yL}(\Lambda_{LV}) = \frac{\left.\frac{\partial y}{\partial L}\right|_V}{\left.\frac{\partial y}{\partial L}\right|_x}$$

which may be derived by differentiating the equations that model the column:

$$S \equiv y(1 - x)/(x(1 - y)) \tag{1}$$

$$S = \left[\frac{\alpha}{\sqrt{1 + \frac{D}{Lz}}}\right]^{nE} \tag{2}$$

$$D/F = (z - x)/(y - x) \tag{3}$$

plus a mass balance across the condenser,

$$V = L + D \tag{5}$$

Results

The results of these derivations may be collected into a set of factors that quantify the interaction of each possible dual composition control scheme.

$$\lambda_{yD}(\Lambda_{DS}) = \frac{1}{1 + \frac{(y - z)x(1 - x)}{(z - x)y(1 - y)}} \equiv \lambda \tag{6}$$

$$\lambda_{yD}(\Lambda_{DL}) = \lambda + (1 - \lambda)\epsilon \tag{7}$$

where:

$$\epsilon = \frac{nEy(1 - y)}{2((zL/D) + 1)(y - x)} \tag{8}$$

also:

$$\lambda_{yD}(\Lambda_{DV}) = \lambda + (1 - \lambda)\epsilon(1 + (D/L)) \tag{9}$$

$$\lambda_{yS}(\Lambda_{SV}) = 1 - \lambda + \frac{\lambda}{\epsilon(1 + (D/L))} \quad (10)$$

$$\lambda_{yS}(\Lambda_{SL}) = 1 - \lambda + (\lambda/\epsilon) \equiv \sigma \quad (11)$$

$$\lambda_{yL}(\Lambda_{LV}) = 1 - \sigma + \sigma(1 + (L/D))(1 - \epsilon) \quad (12)$$

EXAMPLE

Consider a column having 100 theoretical trays receiving a feed of 30 percent lights and separating it into a top product of 99 percent purity and a bottom product 95 percent pure, using a reflux ratio of 12. Then,

$y = 0.99$ $\quad nE = 100$
$x = 0.05$ $\quad L/D = 12$
$z = 0.3$

From these values, the matrices may be completed as:

$$\Lambda_{DS} = \begin{array}{c|cc} & D & S \\ \hline y & 0.07 & 0.93 \\ x & 0.93 & 0.07 \end{array} \qquad \Lambda_{DL} = \begin{array}{c|cc} & D & L \\ \hline y & 0.176 & 0.824 \\ x & 0.824 & 0.176 \end{array}$$

$$\Lambda_{DV} = \begin{array}{c|cc} & D & V \\ \hline y & 0.186 & 0.814 \\ x & 0.814 & 0.186 \end{array} \qquad \Lambda_{SV} = \begin{array}{c|cc} & S & V \\ \hline y & 1.496 & -0.496 \\ x & -0.496 & 1.496 \end{array}$$

$$\Lambda_{SL} = \begin{array}{c|cc} & S & L \\ \hline y & 1.543 & -0.543 \\ x & -0.543 & 1.543 \end{array} \qquad \Lambda_{LV} = \begin{array}{c|cc} & L & V \\ \hline y & 17.221 & -16.221 \\ x & -16.221 & 17.221 \end{array}$$

For this column, applying traditional reflux and boilup controls, (Λ_{LV}) will lead to strong negative interaction between the composition loops. Opposite, but equally strong positive interaction will exist for the material-balance scheme (Λ_{DV}), which assigns distillate to control top composition, and boilup to control bottom composition. (Using B instead of D to control bottom composition suggests an attractive pairing. However, this subset, (Λ_{BV}), must be rejected because no appropriate variable would be available for control of column base level.) For both of these schemes, although the individual loops may work well, performance will be very poor when both loops are closed. The subset Λ_{SL} contains a likely pairing for top composition. However, the assignment of reflux for controlling bottom composition may show poor dynamic performance due to liquid holdup with the column, unless accumulator level or column pressure is controlled by manipulating boilup. The first two subsets, Λ_{DS} and Λ_{DL}, are also material-balance schemes. Between the two, the subset Λ_{DS} shows excellent pairings, after assigning B in place of D. However, this scheme is subject to a subtle dynamic problem described by Shinskey[6] for the control scheme designed around the Λ_{DV} subset.

Open-loop response of the bottom composition controller in such a scheme includes three separate phases. In the first, increasing boilup reduces the impurity in the bottom product, x. At the same time, with the top composition controller in manual, the pressure controller carries the increased heat load by raising reflux to lower the level in the flooded condenser. Later, this increased reflux reaches the bottom of the column and the response enters the second phase, during which the impurity in the bottom product reverses and begins to increase. Its final value may be predicted from Fig. 2. With constant distillate flow, increasing reflux improves separation along a line of constant D/F, making both products more pure. Thus, x will be lower and y will be higher than their original values.

If the top composition controller is in automatic, the response then enters the third phase. Since D is overpure, the top composition controller will increase distillate flow. To hold pressure, the pressure controller simultaneously enforces an equal reduction in reflux, which eventually reaches the bottom of the column and causes x to reverse a second time. Its final value may be determined from the relative gain, which is the ratio of its net change at the end of the second phase to its net change at the end of the third phase. For the example given above, $\lambda_{xV}(\Lambda_{DV}) = 0.186$. Thus, the final change in x will be 5.4 times its change at the end of the second phase. This double reversal effectively extends the dead time in this response to the beginning of the second reversal, causing an excessive period of oscillation in this loop.

The same line of reasoning may be applied to the Λ_{BS} subset. The only difference is that the bottoms composition controller manipulates boilup indirectly, through the bottoms level controller. This case, $\lambda_{xB}(\Lambda_{BS}) = 0.93$, and the second reversal is not so drastic. Nevertheless, the period of this loop will still be excessive.

The subset Λ_{SV} represents the best choice for dual composition control on this column (Fig. 3). This is the scheme proposed by Ryskamp.[5] The top composition controller manipulates separation by driving the distillate-to-vapor ratio, D/V, which is uniquely related to the reflux ratio by a mass balance across the accumulator, i.e,

$$V = L + D$$

from which

$D/L = (D/V)/(1 - (D/V))$

Bottom composition is controlled by manipulating boilup in cascade, which leaves bottoms level to be controlled by manipulating bottoms flow. Operating the receiver liquid full eliminates accumulator level as a controlled variable and allows pressure control by manipulating the total of reflux plus distillate flow.

The dynamics of this scheme will be superior to those of the material-balance subsets for two reasons. First, the feedforward path from D to L avoids any delay from the pressure controller. Second, since $\lambda_{yS}(\Lambda_{SV}) > 1.0$, the second reversal does not occur, and the bottoms composition loop will have a much shorter period of oscillation.

A closer look at Eq. (6) through (12) shows how design of the control system is influenced by factors that vary from application to application. The fundamental significance of the separation and material balance is embodied to the factor λ, which appears in all other relative gains. Depending only on feed and product specifications, this factor is independent of column design and varies from 0.0 to 1.0. Where bottom product is very pure and/or the feed is primarily lights, this factor will approach 1.0. Where top product is very pure and/or the feed is primarily heavies, this factor will approach 0.0.

In the first case, this subset will suggest a control scheme with little interaction. In the second case, substituting B for D will provide attractive pairings. However, since these relative gains are less than 1.0, the double reversal in the bottom composition loops may make the dynamics of this loop unattractive. As the separation becomes more symmetrical, i.e., as top and bottom compositions approach equal purity and the feed approaches an even mix of lights and heavies, λ will approach 0.5, indicating that these material-balance schemes will be highly interactive.

The influence of column design resides in the value of ϵ, which depends on the number and efficiency of the trays and on the reflux ratio. High purity products and high reflux ratios reduce the value of ϵ. As ϵ approaches 0.0, $\lambda_{yD}(\Lambda_{DL})$ and $\lambda_{yD}(\Lambda_{DV})$ approach λ from above, making column design less significant in determining control system design. Conversely, as ϵ approaches 1.0, both $\lambda_{yD}(\Lambda_{DL})$ and $\lambda_{yS}(\Lambda_{SL})$ approach 1.0, with $\lambda_{yD}(\Lambda_{DL})$ approaching from below and $\lambda_{yS}(\Lambda_{SL})$ approaching from above. In this case, both of these subsets offer attractive pairings that require control of bottom composition by manipulating reflux. The dynamics of these arrangements will almost certainly be unacceptable unless accumulator level or column pressure is controlled by manipulating boilup. Given this assignment, the bottom composition loop of $\lambda_{yD}(\Lambda_{DL})$ will still be subject to the double reversal described earlier.

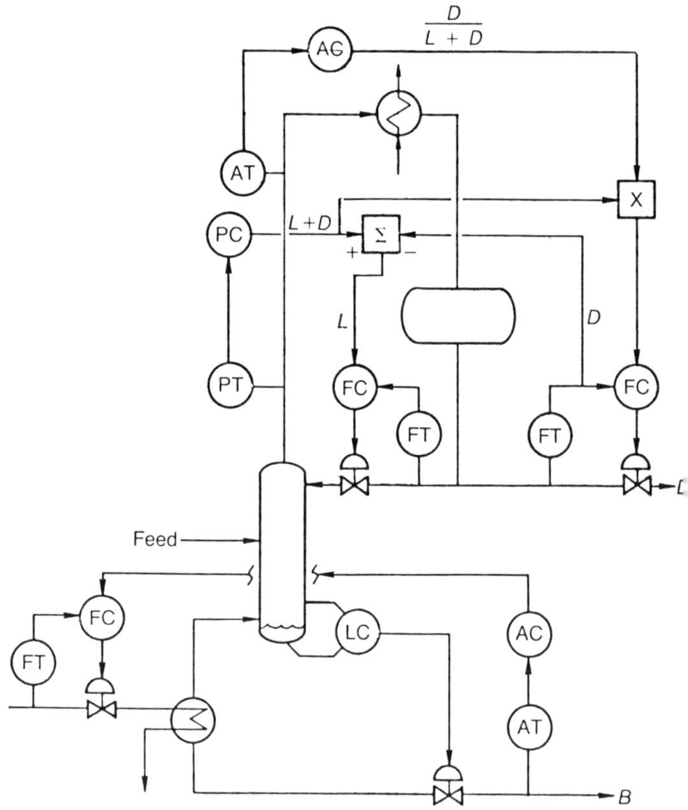

Figure 3. Reflux ratio control scheme for the Λ_{SV} subset.

Reflux ratio also influences the difference between $\lambda_{yD}(\Lambda_{DL})$ and $\lambda_{yD}(\Lambda_{DV})$. At high reflux ratios, these relative gains approach equality with $\lambda_{yD}(\Lambda_{DL}) < \lambda_{yD}(\Lambda_{DV})$. Similarly, the factors $\lambda_{yS}(\Lambda_{SL})$ and $\lambda_{yS}(\Lambda_{SV})$ approach equality at high reflux ratios, with $\lambda_{yS}(\Lambda_{SV}) < \lambda_{yS}(\Lambda_{SL})$. Finally, large reflux ratios make a strong contribution to $\lambda_{yL}(\Lambda_{LV})$, raising this factor the the large positive numbers that indicate strong interaction and predict poor performance for traditional controls. This is consistent with the observation that traditional controls have been most successful on columns with low reflux ratios.

CONCLUSIONS

Dynamics must be considered in applying the results of the relative gain analysis. In general, top composition control by manipulating a variable at the bottom of the column may be acceptable because of the rapid response of vapor flow to changes in heat input. However, response of bottom composition to manipulation of variables at the top of the column is usually not feasible unless accumulator level or column pressure is controlled by manipulating boilup. Given that assignment, the period of the bottom composition loop in systems designed around relative gains slightly greater than 1.0 will be shorter than the period of the same loop in systems designed around relative gains slightly less than 1.0. This is because the double reversal, which adds extra dead time into the response, is not present for relative gains greater than 1.0.

A final consideration is that the designer should avoid control systems in which response of either composition loop is inhibited when the other loop is out of service because of analyzer failure or maintenance, or because the operator has placed the controller in manual.

References

1. Douglas, J. M., Jafarey, A., and McAvoy, T. J., "Short-Cut Techniques for Distillation Column Design and Control," *I&EC Process Design and Development,* Vol. 18, pp. 198–202, April 1979.
2. Shinskey, F. G., *Distillation Control for Productivity and Energy Conservation,* McGraw-Hill Book Co., New York, p. 307, 1978.
3. Luyben, W. L., "Distillation Decoupling," *AIChE Journal,* Vol. 16, No. 2, pp. 198–203, March 1970.
4. Bristol, E. H., "On a New Measure of Interaction for Multivariable Process Control," *IEEE Transactions on Automatic Control,* 1966, pp. 133–134.
5. Ryskamp, C. J., "New Control Strategy Improves Dual Composition Control," *Hydrocarbon Processing,* June 1980.
6. Shinskey, F. G., "Predicting Distillation Column Response Using Relative Gains," *Hydrocarbon Processing,* May 1981.
7. Shinskey, F. G., "The Stability of Interacting Loops With and Without Decoupling," presented at the IFAC Symposium on Multivariable Control, Fredericton, NB, July 4-8, 1977.

30
Predict Distillation Column Response Using Relative Gains

F. G. Shinskey

A typical distillation column requires control of four or five variables through the manipulation of a like number of valves. A principal factor influencing column performance is the effectiveness of its control-system structure. The structure can be as simple as four or five single loops with no interconnections. However, there are many choices to be made, even regarding single-loop combinations: the number of different configurations varies factorially with the number of controlled variables. Thus, a four-loop system could be configured 24 ways, and a five-loop system 120 different ways. Performance of three control configurations will be analyzed, using relative-gains, as an aid to control system selection.

Before the best control configuration selection can be made, some information must be available concerning expected responsiveness of the column. Such responsiveness is difficult to predict, and consequently many columns are controlled less effectively than is necessary. Field reconfiguration is also common—and even expected—in large projects where substantial incentives exist for minimizing operating costs. Configuration has even become a controversial issue between those who promote distillate-boilup manipulation and those who prefer reflux-boilup manipulation for control of product compositions. Further confusion results from the fact that, while the former approach is more effective with high-reflux-ratio columns, the latter seems to be better with those requiring a low reflux-ratio.

Between the two extremes, neither configuration may be as effective as desired, and another approach is needed, combining the features of both. One such arrangement, employed successfully by C. J. Ryskamp,[1] features manipulation of distillate-to-boilup ratio and boilup rate for controlling overhead and bottoms compositions.

This article will attempt to predict performance of all three of these configurations relative to each other, so that their merits may be compared to facilitate selection. Used for evaluation is the relative-gain analysis method developed by the author[2] and applied specifically to distillation columns.[3]

RELATIVE-GAIN ANALYSIS

Any proposed control loop can be evaluated for its open-loop gain; that is, response of the selected controlled variable c_i to the selected manipulated variable m_j. If other control loops are involved in the same process, observed open-loop gain may be affected by the action of the other controllers. Relative gain is the ratio of the open-loop gain of the observed loop while other controllers are in manual, to the same gain evaluated with the other controllers in automatic. Mathematically, it may be stated as the ratio of the open-loop gain in terms of the manipulated variables, to the open-loop gain in terms of the controlled variables:

$$\lambda_{ij} \equiv \frac{\left.\dfrac{\partial c_i}{\partial m_j}\right|_m}{\left.\dfrac{\partial c_i}{\partial m_j}\right|_c} \qquad (1)$$

If this ratio, i.e., the relative gain, is unity, no interaction is evident because the influence of the other controllers on loop gain is nil.

A relative gain approaching zero indicates that closure of the other loops destroys the effectiveness of the observed loop. Values much greater than unity indicate increasing interaction and also loss of effectiveness. Negative values show that closure of the other loops reverses the gain of the observed loop, so that its stability is conditioned on their performance.

Sets and Subsets

Values of relative gain λ_{ij} may be tabulated in an array Λ for a distillation column:

		Manipulated variables D	L	V	B
Controlled variables	y	λ_{yD}	λ_{yL}		
	x	λ_{xD}	etc.		
	b				
	p				

where D = distillate flow
L = reflux flow
V = vapor boilup
B = bottom-product flow
y = mol fraction light component in distillate
x = mol fraction light component in bottom
b = liquid level in column base
p = column pressure

One property of the relative-gain array is that the sum of all the numbers in each column and in each row is unity. It is thus possible to fill in all entries without evaluating all relative gains individually—only $(n - 1)^2$ terms must be calculated to fill an n^2 array.

However, it is much more convenient to break the 4 by 4 or 5 by 5 array into significant 2 by 2 subsets, and then compare the terms in each subset against the others. This is possible in distillation-column control because composition variables (x and y) respond at least an order of magnitude more slowly than inventory variables (liquid levels and pressure). Thus they are scarcely capable of upsetting the faster loops, although they may strongly interact with each other.

The strategy used to predict response here is, then, evaluation of 2 by 2 subsets containing compositions x and y. The pair of manipulated variables selected to control x and y will differ for each case examined, and consequently the relative gains will also differ. Consider the case where distillate flow D and boilup V are chosen to control x and y. (This presumes base level is controlled satisfactorily by bottom-product flow, and pressure by reflux.) This 2 by 2 subset then appears as $\Lambda_{D,V}$:

$$\Lambda_{D,V} = \begin{array}{c|cc} & D & V \\ y & \lambda_{yD} & 1 - \lambda_{yD} \\ x & 1 - \lambda_{yD} & \lambda_{yD} \end{array} \qquad (2)$$

Because the relative gains of the 2 by 2 subset are complementary, only one term—here, λ_{yD}—needs to be evaluated.

BOTTOM COMPOSITION RESPONSE

It commonly has been observed that response of bottom composition to changes in boilup seems to depend greatly on configuration of the remaining loops on the column. To determine the reasons behind this behavior, the relative gain of this single loop, i.e., λ_{xV}, will be evaluated for different choices of manipulated variable m used to control y:

$$\lambda_{xV} \equiv \frac{\left.\dfrac{\partial x}{\partial V}\right|_m}{\left.\dfrac{\partial x}{\partial V}\right|_y} \qquad (3)$$

It is assumed that the other two manipulated variables are under control of base level and pressure, in evaluating both the numerator and denominator of Equation (3).

Relative Gains for Pairs of Single Loops

The configuration shown in Fig. 1, wherein distillate flow is manipulated for composition control, is a form of material-balance control; the column's external balance is manipulated directly to control distillate composition. Its 2 by 2 subset was given in Equation (2). Relative gain $\lambda_{xV} = \lambda_{yD}$, whose derivation appears in Reference 3, for a binary separation:

$$\lambda_{xV}(\Lambda_{D,V}) = \frac{1}{1 + \dfrac{(y - z)x(1 - x)}{(z - x)y(1 - y)}} \qquad (4)$$

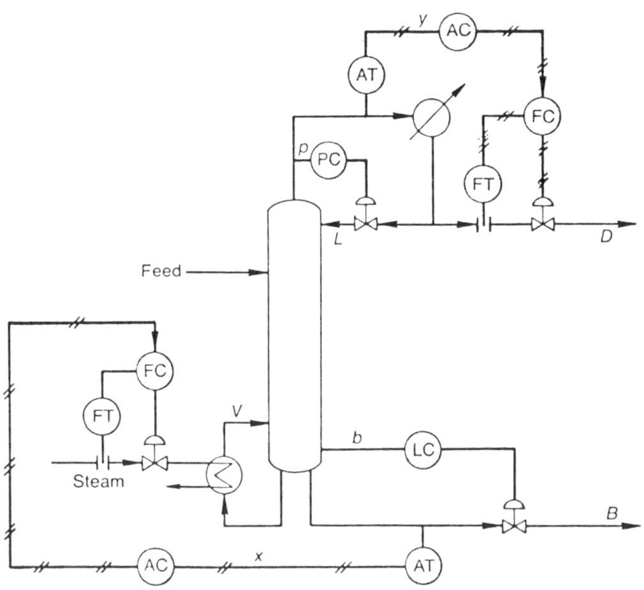

Figure 1. Typical material balance system where distillate and boilup are manipulated to control compositions.

where x, y, and z represent mol fractions of the light component in bottom product, distillate and feed, at design considerations. Fortunately, this relationship is completely independent of such things as stream size, number of trays, relative volatility, and other such factors, so that it can be calculated reliably from a minimum amount of information.

(Because the balance of this article is dedicated to evaluating λ_{xV} for various subsets ($\Lambda_{m,V}$) the representation $\lambda_{xV}(\Lambda_{m,V})$ will be shortened to simply Λ_m.)

Relative gain Λ_D always lies between 0 and 1, principally determined by the ratio x/(1 − y). If the top product is purer than the bottom product, x>(1 − y), and Λ_D will approach zero. If the bottom product is purer, λ_D will approach unity. When compositions are nearly equal and feed is about a 50-50 mixture, Λ_D will be about 0.5.

Relative gains for manipulation of reflux and boilup to control composition (Fig. 2) are always much greater than unity. The derivation described in Reference 3 yields:

$$\Lambda_L = \Lambda_D \left[1 + \frac{\beta(y - x)^2}{(z - x)y(1 - y)}\right] \qquad (5)$$

where β is a function of the number of trays and their efficiency, and relative volatility of the components. It can be calculated from the boilup-to-feed ratio and compositions of the products:

$$\beta = \frac{V/F}{\ln\left[\frac{y(1 - x)}{x(1 - y)}\right]} \qquad (6)$$

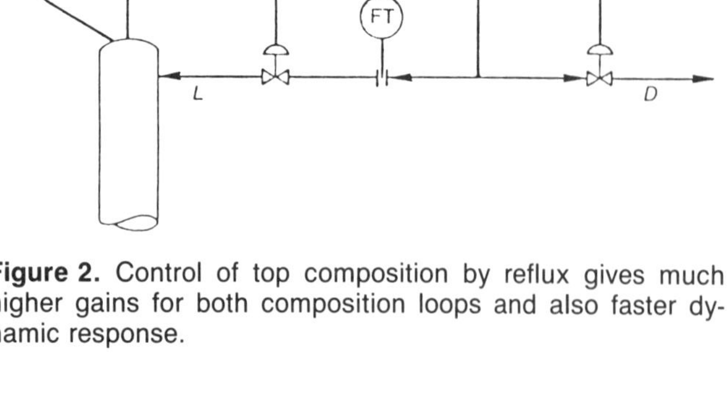

Figure 2. Control of top composition by reflux gives much higher gains for both composition loops and also faster dynamic response.

Typically, β varies in the range of 0.1 to 0.5. While the model of Equation (6) is not a sufficiently exact basis for column design, it describes adequately operation about a particular point, and therefore is suitable for calculating relative gains.

Dynamic Responses

Relative gain is useful in predicting dynamic response in that it is a numerical ratio of the steady-state changes in bottom composition before and after action is taken by the other controllers. A change in boilup initially disturbs all controlled variables, but configuration of the other loops determines what happens next. Consider reflux control of top composition (Fig. 2). With the top-composition controller in manual, a change in boilup will cause only a single response in bottom composition, because the pressure and base-level controllers affect only the discharges from the column.

With the top-composition controller in automatic, it will react to the upset caused by the change in boilup and raise reflux in the same direction. Ultimately, top composition will return to set point, and bottom composition will stabilize near its original value, having changed only $1/\Lambda_L$ as much as it did with reflux constant. Expected shapes of response curves for both conditions of the reflux loop appear in Fig. 3.

The example used to generate Fig. 3 is a column having the following characteristics: x = 0.05, y = 0.95, z = 0.50, V/F = 1.5, L/D = 2.0. Solution to Equation (4) gives Λ_D = 0.50, and solution to Equations (5) and (6) yields Λ_L = 5.3. Consequently, the steady-state change in x for constant reflux is 5.3 times as great as that for controlled top composition.

When the loops are arranged as in Fig. 1, a change in boilup will cause an identical change in reflux through the action of the pressure controller. As in the previous example, the effect of the reflux change on x is delayed by the time required for the reflux to reach the bottom of the column. However, the pressure loop tends to be faster than the top-composition loop, and therefore, return to steady-state may be somewhat faster. Fig. 4 compares response under constant distillate flow to that achieved under constant reflux flow.

Also shown in Fig. 4 is the effect of controlling y by manipulating D. The top-composition controller reacts after the pressure controller has increased reflux, and its action causes reflux to decrease somewhat. As a result, bottom composition shows three distinct responses occurring serially. Composition x falls, rises, and falls again, before reaching a final steady state. This double reversal actually creates 180 degrees of phase shift. In effect, bottom composition response is delayed by the

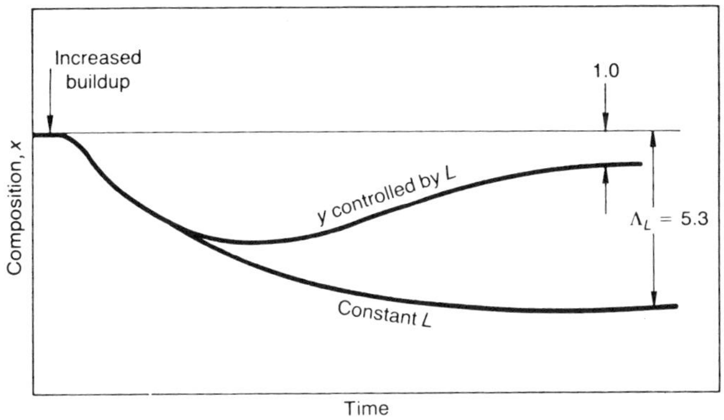

Figure 3. With constant reflux, *x* responds to boilup unimodally; however, the action of the top composition controller reduces its steady-state gain.

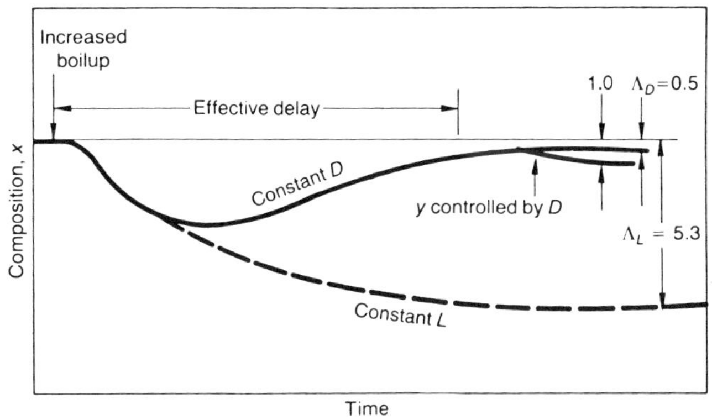

Figure 4. Action of the pressure controller forces reflux to increase to the point where the bottom composition change is reduced beyond what is required for top composition control.

time between change in boilup and second reversal in the curve. (See Reference 2, pp. 236, 237, for a comparison between inverse response and dead time.) Closed-loop simulations have revealed that the natural period of the loop containing a response curve with a double reversal is approximately four times the effective delay shown in Fig. 4, making it essentially the same as dead time. In effect, dynamic response of the bottom-composition loop is conditioned on dynamic performance of the top-composition loop. If the top-composition loop is in manual, bottom-composition control may be lost altogether.

The degree to which these adverse effects are realized is directly related to the value of Λ_D. In a column where the bottom product is much purer than the distillate, Λ_D approaches unity, and the effect is minimal. However, the period of the bottom-composition loop still appears to be excessive. A case in point is a styrene-ethylbenzene column where y is about 0.90 and x is approximately 0.005. Relative gain Λ_D is 0.95 or more, so little interaction exists. Yet the period of the bottom-composition loop has been observed to be about 2.5 hours compared to about 40 minutes for the top-composition loop.

Decoupling

To reduce interaction naturally occurring in these single-loop configurations, a decoupler may be added. In its simplest form, the unilateral decoupler adjusts distillate or reflux flow as a function of manipulations in boilup, as shown in Fig. 5. (A bilateral decoupler would also adjust boilup as a function of reflux or distillate. However, this second loop complicates the system, and presents the opportunity for closed-loop instability.[2]) The coefficient for decoupling reflux from boilup must be slightly less than 1.0, since a value of 1.0 would result in constant D. In fact, there is a linear relationship between relative gain of the decoupled process and the de-

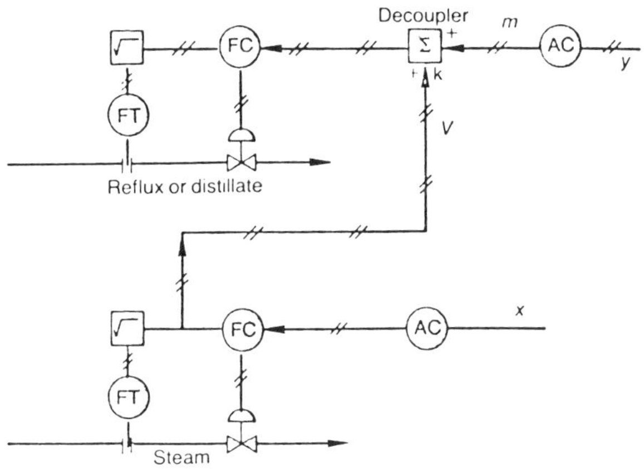

Figure 5. Decoupler automatically adjusts *D* or *L* in proportion to changes.

coupling coefficient. Considering that, in the example of Fig. 3, $\lambda_{xV} = 5.3$ when $\Delta L/\Delta V = 0$, and 0.5 when $\Delta L/\Delta V = 1.0$, we have:

$$\lambda_{xV} = 5.3 - 4.8\Delta L/\Delta V$$

or in general,

$$\lambda_{xV} = \Lambda_L - (\Lambda_L - \Lambda_D)\frac{\Delta L}{\Delta V} \tag{7}$$

By setting Equation (7) to unity, the decoupling coefficient kL for reflux can be found:

$$k_L = \frac{\Delta L}{\Delta V} = \frac{\Lambda_L - 1}{\Lambda_L - \Lambda_D} \tag{8}$$

For the example of Fig. 3, $k_L = 0.896$.

One of the difficulties of decoupling is sensitivity to errors. Variations in $\Delta L/\Delta V$ or the relative gain terms of

Equation (7) will cause variations in λ_{xV}, causing it to change from its desired value of unity. Because relative gains are functions of compositions, they will vary with compositions. Also L and V are usually the largest flow rates in the column, and are subject to variations in heat balance, principally heat losses, subcooling of reflux, and changes in feed enthalpy. Therefore, significant variations are to be expected in performance of the reflux decoupler. An increase in $\Delta L/\Delta V$ from 0.896 to only 1.0 will reduce λ_{xV} from 1.0 to 0.5, in the example given.

The distillate decoupling coefficient happens to be the complement of the reflux coefficient:

$$k_D = \frac{\Delta D}{\Delta V} = \frac{1 - \Lambda_D}{\Lambda_L - \Lambda_D} \quad (9)$$

Its value in the example cited is 0.104. Relative gain for the column with distillate decoupling is:

$$\lambda_{xV} = (\Lambda_L - \Lambda_D)\frac{\Delta D}{\Delta V} + \Lambda_D \quad (10)$$

In the material-balance system, distillate flow is not affected by variations in heat balance. Furthermore, percentage errors in $\Delta D/\Delta V$ will have much less effect on λ_{xV} than the same percentage errors in $\Delta L/\Delta V$, because it is so much smaller.

While decoupling under material-balance control has been effective, it does not appear to improve dynamic response of the bottom-composition loop. Like the styrene-ethylbenzene column which did not need decoupling, the bottom-composition loop tends to be much slower than the top-composition loop.

DISTILLATE-TO-VAPOR RATIO CONTROL

An alternate to the single-loop configurations and their decoupled forms is the system of Fig. 6, in which distillate-to-vapor ratio is manipulated for top-composition control. This is the system described by Ryskamp.[1] Because D = V − L, fixing D/V fixes L/D and L/V as well.

This system responds to changes in boilup as follows: increasing boilup raises pressure, which the pressure controller counters by increasing its output, which represents total overhead vapor, V = L + D. This signal is multiplied by the output of the composition controller D/V to set distillate flow D. It also enters a subtractor where measured distillate flow D is subtracted to produce the set point for reflux flow L. Consequently, the pressure controller raises both L and D in a ratio de-

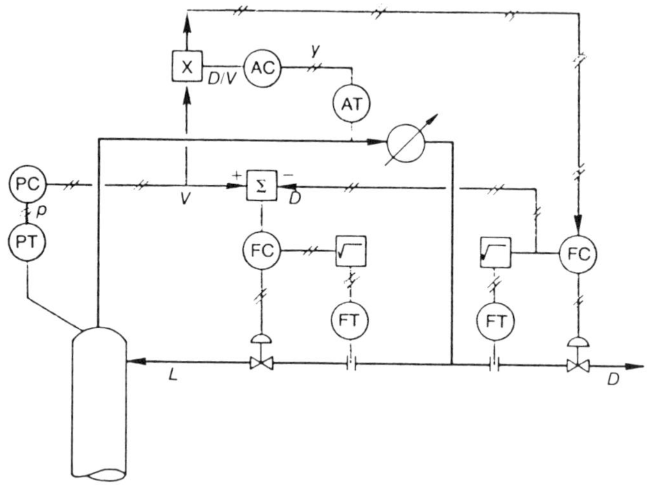

Figure 6. Pressure controller manipulates both reflux and distillate in a ratio adjusted by the top composition controller.

termined by the output of the top-composition controller.

On a change in top composition, the controller adjusts D/V and hence D, whose new value causes a recalculation of L in the subtractor. In this way, reflux responds immediately to the composition controller, without upset to column pressure and resulting delay of the pressure loop.

Relative Gain

Manipulation of D/V to control top composition gives a bottom-composition response between that of constant L and constant D. Derivation of relative gain yields:

$$\Lambda_{D,V} = \Lambda_D\left\{1 + \frac{(y - x)}{y(1 - y)\ln[y(1 - x)/x(1 - y)]}\right\} \quad (11)$$

While its form is similar to Λ_L, its values tend to be much lower, typically in the range of 1.5 to 3.5. For the numerical example used thus far, $\Lambda_{D/V} = 2.1$.

A second method for finding $\Lambda_{D/V}$ is to use the linear relationships developed in Equation (7) or (10). In the example, L/V is 1/1.5, which then would be the value of $\Delta L/\Delta V$ in Equation (7). Solution for λ_{xV} gives the same 2.1.

Dynamic Response

Because relative gain of the distillate-to-vapor ratio system is always greater than unity, steady-state is approached from the same direction as when reflux and boilup are used to control compositions. Thus reversal in the step response characteristic of distillate-boilup configuration is avoided. Bottom-composition response then is as fast as under conditions of constant reflux. Figure 7 compares step responses of bottom composition to boilup under constant reflux, constant D/V, and manipulation of D/V to control top composition.

If it were desirable to decouple the system having D/V manipulation, the decoupling coefficient would be calculated as before:

$$k_{D/V} = \frac{\Delta(D/V)}{\Delta V} = \frac{\Lambda_{D/V} - 1}{\Lambda_L - \Lambda_D} \quad (12)$$

for the example being used here, $k_{D/V} = 0.229$.

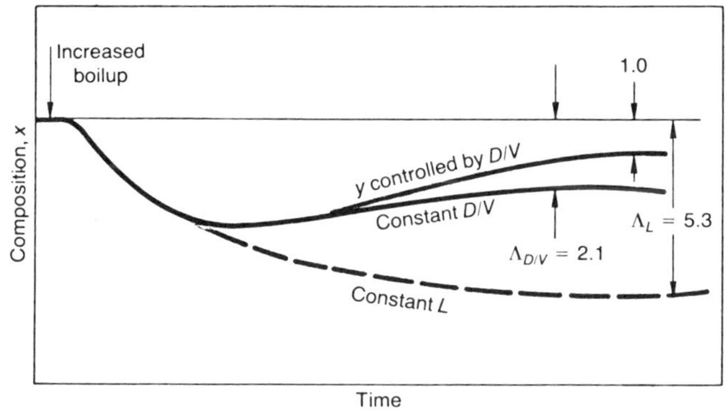

Figure 7. Manipulating *D/V* for top composition control preserves the fast dynamic response of the bottom composition loop while reducing interaction and sensitivity to disturbance in heat balance.

Disturbance Propagation

Magnitude of the decoupling coefficient indicates relative effort required by the top-composition controller to counteract changes in the heat balance, whether they are undesired disturbances, or induced to adjust bottom composition. In essence, k is the amount the top manipulated variable must change to restore top composition to set point, per-unit boilup change ΔV. A unit disturbance in V would require reflux to change 0.896 units, distillate 0.104 units, and the D/V ratio 0.229 units, for the example under discussion. Thus, material-balance control is the least upset by heat-balance disturbances, only 11.6 percent as sensitive as reflux manipulation. By contrast, manipulation of D/V is 2.2 times as sensitive as material-balance control, but only 25.6 percent as sensitive as reflux manipulation.

CONCLUSIONS

Relative-gain determination is seen as an extremely useful tool to evaluate expected performance of various control-loop configurations on virtually any distillation column. Moreover, it has confirmed the effectiveness of Ryskamp's system structure which manipulates distillate-to-vapor ratio for top-composition control. Because relative-gain values for this configuration seem to fall in a fairly narrow and desirable range, his system may well be applicable to most distillation columns, thereby simplifying the designer's task as well as training required by plant operators.

References

1. Ryskamp, C. J., New Strategy Improves Dual Composition Control, *Hydrocarbon Processing,* June 1980.
2. Shinskey, F. G., *Process-Control Systems,* 2nd Edition, McGraw-Hill Book Co., New York, 1979, Chapter 8.
3. Shinskey, F. G., *Distillation Control,* McGraw-Hill Book Co., New York, 1977, Chapter 10.

31

Distillation Unit Computer Control*

R. Rinne, H. Sunell, P. R. Latour, and K. K. Paynter

A computer-based control system was implemented in early 1981 for an 850 tph (140 Mbpd) atmospheric crude distillation unit of Neste Oy at Porvoo Refinery, Finland.

The objectives were to maximize the yields of on specification distillates, increase crude capacity, reduce fuel consumption, and enhance the operators' tasks.

Nonlinear, adaptive, feedforward, interacting, and constraint control techniques were employed to control preheat, furnaces, tbp (true boiling point) cutpoints, pumparounds, overflash, temperatures, pressure, steam and flooding.

Improvements in distillate yields reduced long residue yield more than 2.2 percent crude. Excess long residue exceeding 2 percent of crude charge which formerly bypassed the loaded vacuum unit to storage was eliminated. Heat loss to cooling water was reduced 6 MW, representing 6 percent of unit energy consumption. Naphtha yield was increased 0.4 percent on crude and quality variations were cut in half. Crude switch transient periods were cut in half.

Reasons for operators' good acceptance are discussed. System availability and operating experience are shown to be excellent.

INTRODUCTION

Neste Oy's larger refinery at Porvoo, Finland, 40 km east of Helsinki, can process 1600 tph (265 Mbpd) crude oil. Neste studied the feasibility of computer control and decided the best initial application would be on the largest and most efficient atmospheric crude distillation unit. It can process 850 tph (140 Mbpd).

In September 1979, Neste Oy selected a petroleum fractionation optimizing control technology,[1,2] integrated into a basic process control computer system, to be implemented on this atmospheric crude distillation tower. The package was customized for this unit based on the functional specification prepared by Neste engineers.

The system was installed in the central control room and interfaced to existing electronic analog instruments. A few additional temperature and flow signals were needed, and analyzers for gas oil cloud point, heavy gas oil color and naphtha endpoint were purchased.

The detailed applications engineering was completed with a program specification report by December 1979 and programming began in January 1980. The applications software was demonstrated in Houston in April 1980 on a separate computer and integrated into the system in Helsinki in June 1980. Engineer and operator training with graphics was finished at the factory by September 1980 and the computer was installed during a normal crude unit shutdown in October 1980.

Control commissioning and initial tuning were finished by mid December 1980. Final tuning and testing was done from mid January to mid February 1981, when the project was completed successfully, within budget and ahead of schedule. The project principles and activities followed those described elsewhere.[3,4,5]

This chapter includes a process description and reports on each of the main control objectives and the computer control functions employed to achieve each objective. The results to date are given with comments

* Originally presented at the National Petroleum Refiners Assoc. Computer Conference, Houston, Texas, Nov. 1-4, 1981.

Mr. Matti Kolari, Neste instrument engineer and Mr. Juha Jakkula, Neste process engineer, made important contributions to the project specifications. Mr. Perti Pekkala of Nokia Electronics directed the computer software and hardware integration and installation. Mr. Carl Wetzel of SETPOINT, INC. did the applications programming and coordinated project activities with Pekkala and Rinne. Nokia Electronics, Helsinki, was prime contractor and computer system supplier.

on operator acceptance and refinery operating experience.

PROCESS DESCRIPTION

The crude unit processes a wide variety of crudes with rates from 700 tph to 850 tph (90–140 Mbpd) depending on crude quality. A mixture of two to four crude types in the feed is normal.

The crude oil is heated in two parallel furnaces. Tower heat is removed in two circulating reflux systems. Upper circulating reflux transfers heat to the deisopentanizer reboiler, light ends tower reboilers, boiler feed water and cooling water. The lower circulating reflux gives heat to a debutanizer, depentanizer and crude charge.

The products are naphtha, kerosine, diesel oil, gas oil, heavy gas oil and long residue. The naphtha goes to hydrotreating and distillations. Its endpoint is limited by the catalytic reformer. The kerosine is hydrotreated o jet fuel or stored for heavy fuel oil blending. The diesel and gas oil are hydrotreated for final products with cloud point as a quality specification. The heavy gas oil is FCCU feed. The long residue goes to vacuum units, with vacuum gas oil to FCCU. Vacuum unit capacity limits crude unit charge, and they normally are run at maximum throughput.

Significant crude switches only occurred about twice a month and unit operation normally had been rather steady.

CONTROL OBJECTIVES AND FUNCTIONS

There are five main objectives for this computer system: stabilize atmospheric residue flow, maximize valuable distillates yield, maximize energy efficiency, maintain product qualities within specifications and enhance the operators' tasks. A set of comprehensive and complementary control functions meet these objectives in a coordinated way.

The technology is based upon rigorous mass and energy balance calculations for tower internal liquid and vapor rates. These are repeated automatically every two minutes from on line data. The resulting information is used to derive performance indicators for separation (internal reflux ratios), boiling ranges (tbp cutpoints) and constraints (tray loads, pumparound limits). Since the correlations do not require any stream analyzers or crude charge information, the computer can deduce current information about energy balance, side draw quality and available capacity, in real time for fast feedback/interacting control from existing instrumentation signals. This effectively uses the crude column itself as a crude analyzer.

Atmospheric Residue

This crude unit is the most efficient of the Neste units and should be loaded maximally at the expense of the less efficient units. One of the bottlenecks to total crude throughput is the downstream vacuum unit. Rather than fix crude charge rate and take varying atmospheric (long) residue flows caused by 1) different crude types and 2) atmospheric bottoms level control to long residue, the control objective was to stabilize the long residue flow to allow the operator to set it to load maximally the vacuum unit, without sending excess long residue to storage.

The computer control strategy is shown in Fig. 1. The bottoms long residue flow was put on flow control, set by the operator through the computer. Both crude charge rate controllers are adjusted to maintain bottoms level, with feedforward adjustments from long residue flow adjustments.

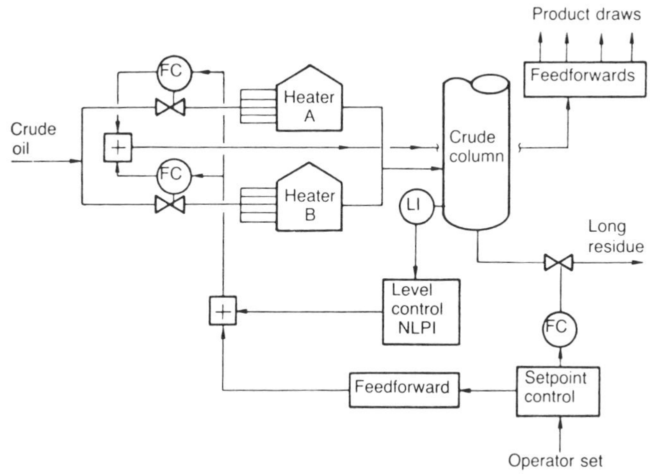

Figure 1. Atmospheric crude unit bottoms level control strategy.

A nonlinear proportional-integral (NLPI) algorithm for level control makes small, slow crude rate changes, allowing the bottom level to swing on either side of the desired target within a wide acceptable range. By minimizing crude rate changes, disturbances to the crude heaters and tower were minimized. This maintains operating efficiency while keeping the bottoms level within safe limits. Feedforward control minimizes level disturbances when the operator changes the long residue flow rate. Feedforward controllers to the draws were synchronized dynamically to minimize upsets from crude changes to the quality controls.

Switches between different crude tanks reach the unit in 30 to 40 minutes. These are detected first at the heaters and bottoms level. A switch from light North Sea crude (26 percent long residue yield) to heavier mid east crude (40 percent long residue) will show a rapid rise in level calling for a decrease in crude flow. Complex crude switch programs are not required.

Distillate Yields

Another control objective was to maximize the yields of valuable distillates by increasing vaporization at the flash zone and by minimizing overflash.

Heater pass outlet temperature balance controls, a column minimum pressure constraint controller and a maximum HGO (heavy gas oil) draw constraint controller increase the flash zone crude vaporization and decrease overflash. Heater pass outlet temperature balance control is shown in Fig. 2. Minimum pressure and maximum HGO constraint controls are shown in Fig. 3.

The pass outlet temperature balance controls manipulate the heater pass valve positions to equalize the pass outlet temperatures, thereby allowing a greater combined outlet temperature without coking or exceeding tube limits.

The minimum pressure controller manipulates the overhead accumulator pressure controller to minimize column pressure subject to column limits including tray flooding, off gas flow, HGO color, and pumparound duty requirements. Pressure corrected temperatures for tbp cutpoints eliminate quality variations.

HGO off color can be caused by either excess vapor velocity lifting entrainment to the HGO tray (pressure is too low) or insufficient HGO wash to the flash zone (overflash is too low). Color followed overflash quite well on this unit.

The maximum HGO constraint controller manipulates the HGO draw flow controller to maximize HGO draw subject to the following limits: HGO color, overflash, HGO stripper level and HGO stripper level valve position.

Energy Conservation

Energy is conserved by increasing heater efficiency and maximizing heat recovery to crude and other process streams.

The controls balance heater efficiency, maximize column pumparound duty to crude and other process streams, and minimize heat transferred to cooling water. The heater efficiency balance control loop is in Fig. 2. The control strategies for maximizing heat recovery in the upper and lower circulating refluxes are in Figs. 4 and 5.

The TIC VP (valve position) controller (NLPI algorithm) in the UCR (upper circulating reflux) keeps the column top temperature controller valve position in control range by making small, slow changes to the cooling water flow. The minimum UCR flow constraint controller manipulates the heat exchanger bypass flow to minimize UCR flow subject to limits including UCR DPC (differential pressure control) valve position and UCR return temperature.

The deisopentanizer reflux-to-feed overriding constraint controller manipulates the UCR flow through the DIC5 (deisopentanizer) reboiler to minimize heat loss to cooling water in the UCR.

The minimum external reflux constraint controller minimizes external reflux flow (and thereby heat lost in

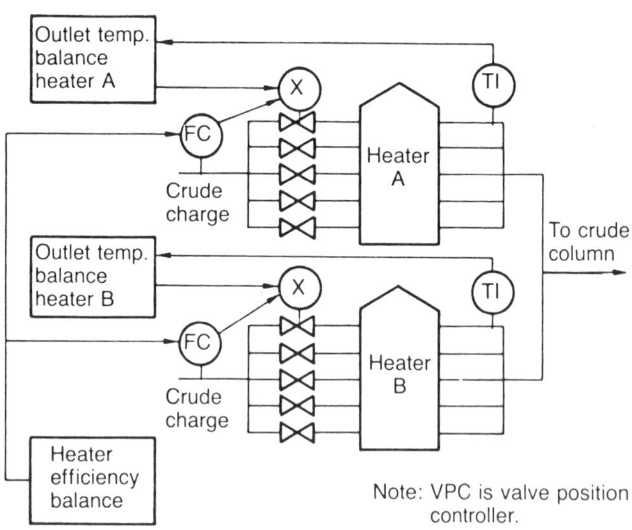

Figure 2. Heater pass outlet temperature balance controls.

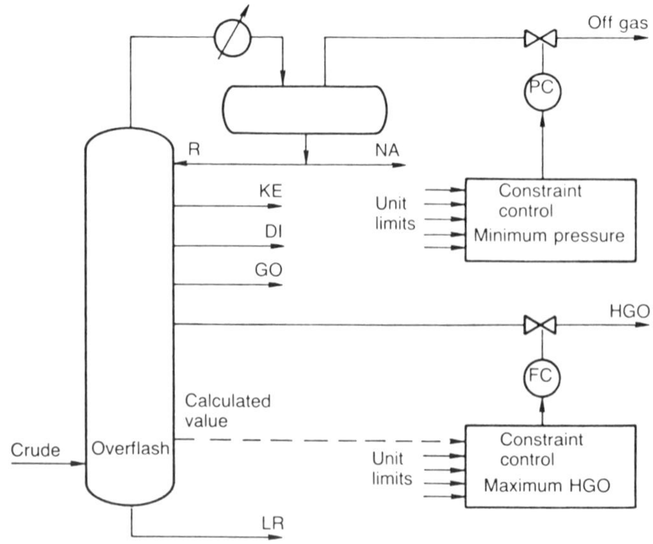

Figure 3. Overflash and column pressure controls.

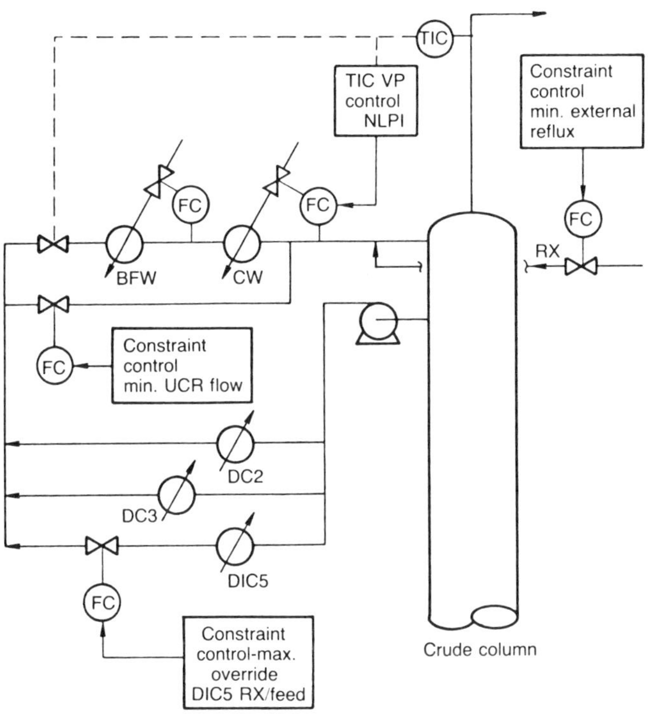

Figure 4. Upper circulating reflux controls.

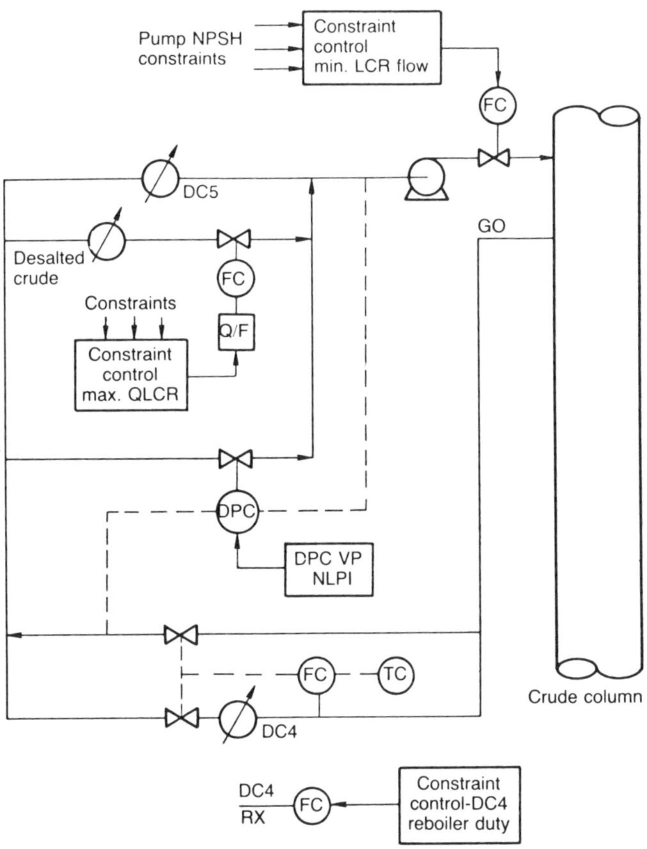

Figure 5. Lower circulating reflux controls.

the overhead condenser) subject to top temperature controllability limits.

In the lower circulating reflux, the maximum QLCR constraint controller manipulates the LGR (lower circulating reflux) duty-to-feed controller to maximize the duty to the desalted crude heat exchanger subject to limits, including DI/GO (diesel/gas oil) cutpoint controllability, DI yield, crude heater inlet temperature, column jet flood, heat exchanger approach to max duty and flashing of LCR pump suction.

Since LCR pump suction flashing often limits this duty and crude unit thruput, a correlation for pump NPSH (net positive suction head) was developed to predict available heat removal duty before pump cavitation. The correlation accounts for temperature, pressure, hydraulics and gas oil vapor pressure/flash point at standard conditions. The latter depends upon the DI/GO tbp cutpoint.

The DC4 (debutanizer) reboiler duty overriding constraint controller maximizes the LCR duty by manipulating debutanizer reflux flow when the crude heat exchanger reaches its maximum duty limit. The DPC VP controller (NLPI algorithm) manipulates the LCR DPC setpoint to maintain control of the LCR differential pressure controller valve position.

The LCR heat exchanger arrangement was revamped during the October 1980 shutdown. The analog controls were designed in 1979 for the new arrangement and implemented in November 1980.

Product Quality

Maintaining product specifications during crude rate and composition changes and changes in long residue flow is an important objective.

The quality specifications are naphtha endpoint, kero endpoint, diesel cloud point, gas oil cloud point and heavy gas oil color/metals.

The primary quality controls are the true boiling point cutpoint controls. They provide fast feedback and interacting adjustments to the draw flows and LCR duty to hold the boiling ranges at setpoints. Multivariable techniques are required.

Setpoints are either operator set or reset by outer cascade loops from available quality analyzers. These are shown in Fig. 6.

The calculated cutpoints do not require analyzers, and they account for pressure changes, stripping steam, internal tray conditions, non-condensables and heat duties.

The DI/GO cutpoint is more difficult to estimate than the others because this section has a large circulating reflux and only three trays. Since they are predominantly heat transfer trays rather than fractionation trays, and

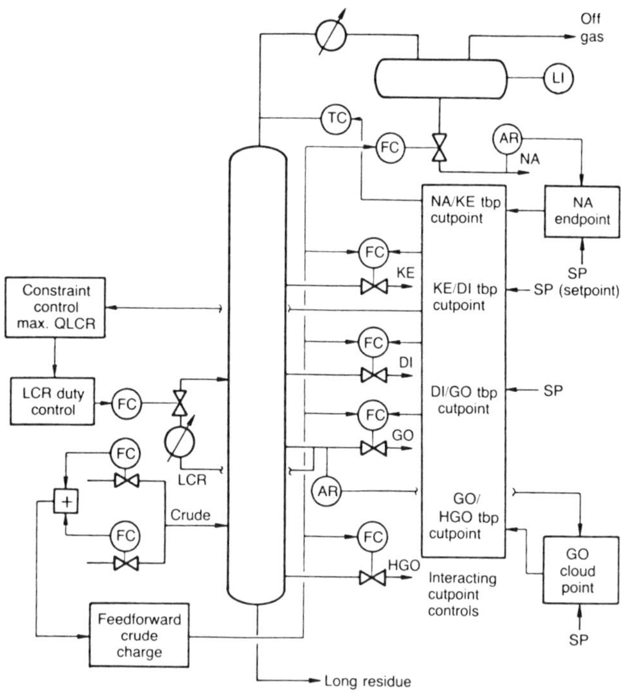

Figure 6. Product quality controls.

since thermal equilibrium is not established between leaving vapor and liquid, the cutpoint calculation is more complex. Diesel draw tray is total draw off without returning diesel as reflux to gas oil.

The dynamic effect of LCR duty on GO/HGO cutpoint and gas oil cloud point was significant.

Adaptive gain techniques are employed widely to keep loops in tune with the process.

Controllers based on naphtha endpoint and gas oil cloud point analyzers trim the top and bottom cutpoint controllers to enhance control of the naphtha and gas oil product specifications.

Feedforward controls minimize disturbances to the quality loops during changes in crude charge rate.

OPERATOR TASKS

Process computer systems should make the operators' job easier, more interesting and less routine. This ensures his understanding, acceptance and lasting use of the system.

To assist the operator in minimizing disturbances to other units, NLPI controllers were used for the overhead accumulator and diesel stripper level controls. They minimize swings in feed to the gas plant and hydrotreating units. Stripping steam ratio controllers follow changes in product flows to hold adequate separation with reduced steam consumption.

RESULTS

The computer eliminated hydroskimming, increased distillate yields, improved energy efficiency and held product quality.

End of Hydroskimming

Vacuum unit feed was to be stabilized to reach maximum thruput. At the same time crude feed was varying in quality.

Previously, crude column level was controlled by adjusting a slip stream flow bypassing the vacuum unit to storage. For proper level control this flow had to vary over a broad range. At its minimum it was 1 to 2 percent of crude rate.

With the new level control scheme in the computer, there is no longer any need to bypass the vacuum unit, and thus the uneconomical hydroskimming refining was eliminated.

Increase of Distillate Yields

With manual pass temperature balancing of heaters the temperature difference between the hottest and coldest tube was typically 5 to 7°C. Due to this, average outlet temperatures exceeding 370°C could not be maintained without the risk of excessive coking.

Computer control brought the temperature deviations down to 2°C which allowed us to increase the setpoint of outlet temperature controllers 3°C. We believe that further temperature increases will be possible after the present tube coking rate is determined.

So far this control function alone has improved distillate yield by 0.5 percent of total feed rate.

Conventional tower pressure control maintains top pressure at a constant design value. At an early phase of the project, it became obvious that our column could be operated at a substantially lower pressure.

By manual adjustments we reduced the pressure by 0.5 bars. Another 0.5 bar average reduction was realized by computer control. Increased distillate yields benefit was 1.2 percent of total feed in each case.

Lack of information on the amount of actual overflash resulted in noticeable loss of HGO to long residue.

On-line calculation of overflash made it possible to control the amount of overflash to optimum which increased HGO yield by 0.5 percent of crude.

Energy Efficiency

Problems in one air preheater have prevented the use of heater efficiency balancing control.

Average heat losses to cooling water have been reduced by 6 MW. This is about 6 percent of the total energy consumption of the unit. About 80 percent of this extra heat recovery has been used for crude preheating, the rest for better separation in fractionating columns. As a result, no extra fuel was required for increasing heater outlet temperatures.

Product Qualities

Although the final specification of naphtha product is defined as ASTM endpoint, our normal practice is to analyze 95 percent point of laboratory gas chromatograph distillation. According to our experience, a good correlation exists between these two methods.

Computer control reduced the standard deviation of measured naphtha endpoint from 3.3°C to 1.8°C. The data are based on regular laboratory analyses and the results obtained during the acceptance test run, respectively.

This allows us to increase the naphtha endpoint operating target by 1.5°C without any additional heavy ends in the reformer feed. Therefore, 0.4 percent more of the crude can be recovered to increase naphtha yield.

Better control of the diesel cut results in standard deviation of diesel cloud point of 0.8°C, which is half of that experienced earlier.

The standard deviation of gas oil cloud point was reduced from 2.3°C to 0.9°C.

In spite of disturbances caused by optimizing controls (crude rate, pressure, maximum LCR), variations in diesel and gas oil compositions are maintained much closer to their targets compared to operation with constant separation.

These results are summarized in Table 1.

Flow meter accuracy has been excellent, providing hourly average material balance closure less than 1 percent of crude. Instrumentation performance has provided hourly average heat balance closure of less than 5 percent of heater fired duty. This is not required for good control.

Table 1
Summary of Results

	Before	After
1. Long residue bypass, %	>1-2	0
2. Heater temp, °C	370	373+
Δ distillates, %	0	>+0.5
3. Δ tower press, bar	0	−0.5
Δ distillates, %	0	+1.2
4. ΔHGO from overflash, %	0	+0.5
5. Δfuel consumption, MW	0	−6
%	0	−6
6. Naptha EP σ, °C	3.3	1.8
ΔEP, °C	0	+1.5
Δnap. yield, %	0	+0.4
7. Diesel cloud σ, °C	1.6	0.8
8. GO cloud σ, °C	2.3	0.9

OPERATORS' ACCEPTANCE

The operators accepted the system from the beginning, and this attitude has not changed during the first year of operation. This is not to say the operators gave no criticism. In fact, comments like, "I liked the old way better" guided us in pinpointing the problematic loops that required closer examination in design or tuning. On the other hand, engineers never had to ask questions like, "Who took those controls off?" The most probable reasons for this positive attitude are discussed below. Some of them may be unique for this application but most of them can be adopted for any similar computer project.

The basis for the system design was the production strategy itself. In this way, the computer only solves problems that operating people have in their work. Therefore, operators have found it to be a valuable tool, not an inevitable burden.

At the risk of losing some theoretically calculated benefits, complex solutions purposely were avoided to assure high availability of the system. All the control functions are configured by using simple modules which are connected to each other in hierarchic order like in normal cascade control.

A rule without exception has been that the operators alone are allowed to take controls on or off and make changes in setpoints, constraint limits, etc.

Every effort was made to provide the operators with an easy to use computer interface. Current value report displays were designed in cooperation with the operators (Figs. 7 and 8). Interactive control diagram displays were built so that the hierarchic level of control being used and the limiting constraints can be seen (Figs. 9 and 10).

The system was marketed to the operating staff throughout the project life. Project engineers spent a lot of their time in the control room explaining the new control ideas and how these would solve current operating problems.

The classroom training of operators and shift foremen took place a few weeks before commissioning. The main objective of this training was to give full understanding of the purpose and functioning of all controls. The most valuable training session was, however, the commissioning period.

Commissioning was carried out by teamwork between the vendor, project engineers and operating staff. This work required activity from all parties. Thorough knowledge of the system enabled us to proceed with

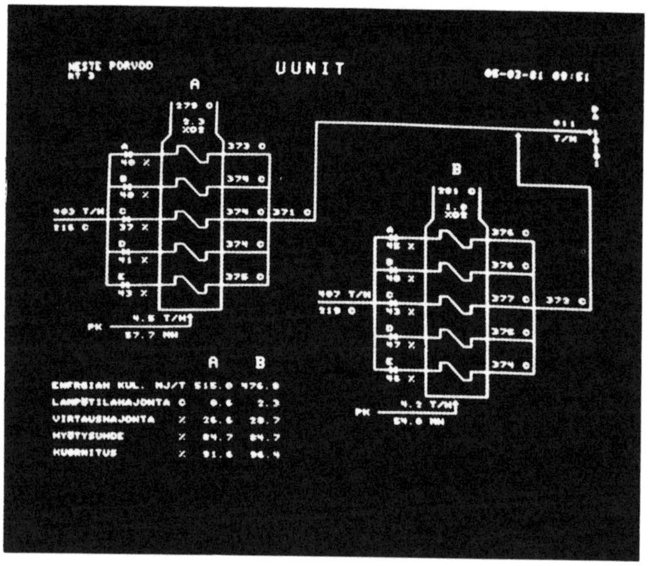

Figure 7. Heater current value display.

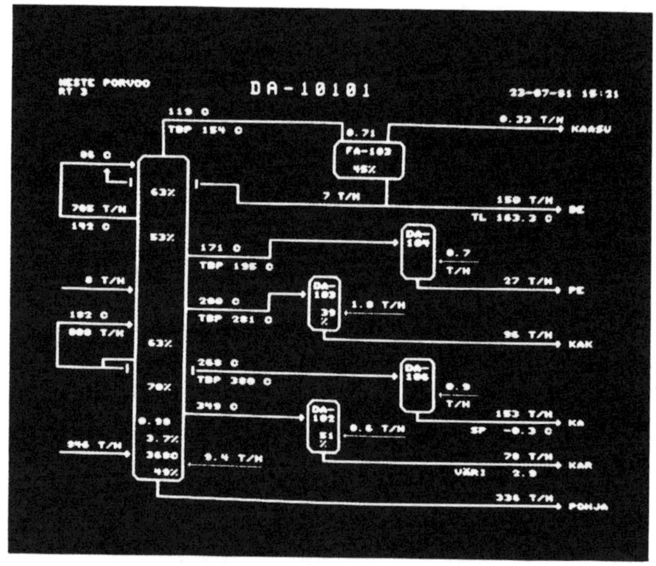

Figure 8. Crude tower current value display.

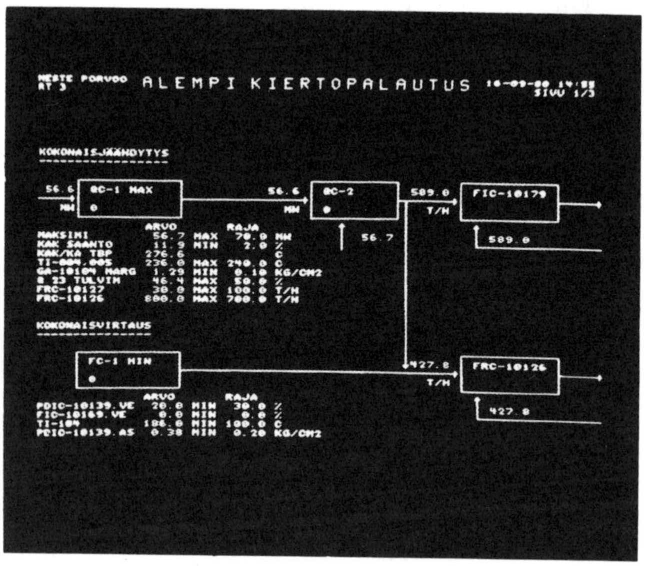

Figure 9. Control hierarchy display-max. LCR.

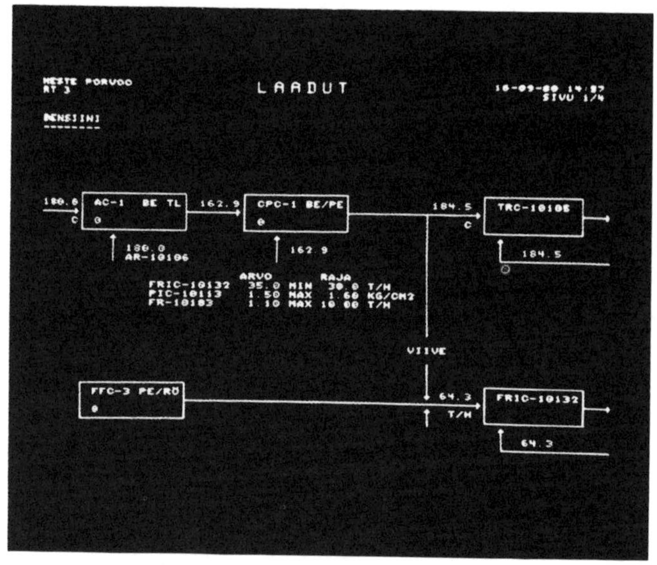

Figure 10. Control hierarchy display-naphtha cutpoint, endpoint.

commissioning in spite of problems that arose since they could be traced and solved immediately on site. The progressive, short commissioning period also strengthened the operators' confidence in the system.

SYSTEM AVAILABILITY AND OPERATING EXPERIENCE

The availability of the computer control system has been practically 100 percent. The only exception is heater efficiency balancing control due to process problems explained previously.

Naturally, if the computer fails, the computer controls are lost temporarily. As an example, statistics of a selected month are shown in Table 2. The computer crashed once for two hours because of loss of cooling in the computer room. The average down time of individual control loops was four hours, i.e., most of the loops were back in closed loop control within two hours after computer restart. Cloud point control (AC-2) was off over a weekend due to sampling problems. The rather low figures for VPC-4, CPC-1, AC-1 and CPC-2 were caused by troubles in the pressure control of the deisopentanizer.

Table 2
Availability of Computer System and Individual Control Loops Computer Availability: 99.7%

Month: May 1981

Loop ID	Function	Availability % Computer run time
LC-1	Crude column bottom level	99.7
EDC-1	Heater efficiency balance	0.0
TDC-2	Heater A pass outlet temp. balance	99.9
TDC-3	Heater B pass outlet temp. balance	99.8
VPC-3	TIC valve position	99.8
FC2	Min. UCR flow	99.7
RRC-1	DIC5 RX/F	99.4
VPC-4	Min. CW valve	85.8
QC-2	LCR Q/F	99.3
QC-1	Max. QLCR	99.3
FC-1	Min. LCR flow	99.1
QC-3	DC4 reboiler duty	99.6
VPC-1	LCR DPC valve position	99.6
VPC-2	QDC4 override control	99.8
TDC-1	Suction temp. tracking	99.8
PDC-1	Suction press tracking	99.7
CPC-1	NA/KE cutpoint	98.8
AC-1	Naphtha endpoint	85.8
CPC-2	KE/DI cutpoint	98.5
CPC-3	DI/GO cutpoint	99.2
CPC-4	GO/HGO cutpoint	99.7
AC-2	GO cloud point	92.8
LC-2	Overhead accumulator level	99.7
LC-3	Diesel stripper level	99.5
SRC-1	KE stripper stm./KE	99.9
SRC-2	DI stripper stm./DI	99.9
SRC-3	GO stripper stm./GO	99.9
SRC-4	HGO stripper stm./HGO	99.8
SRC-5	Bottoms stm./LR	99.8
FC-4	Max. HGO	99.7
PC-1	Min. pressure	99.9
FC-5	Min. external reflux	99.6

Knowing that computers can and do fail, we were concerned about continuity of plant operation on such occasions. The two computer crashes we have experienced so far have taught us that transfer from computer to analog control does not cause any upset in the process. This was made possible by careful analysis of the behavior of every individual controller before and after computer failures.

Although we have not performed any particular tests on reliability of tower calculations, we have found the results consistent with laboratory and analyzer data. Neither have we noticed any invalidity in calculations during changes of operating modes or normal process upsets. The only questionable part of the program was the flooding calculation giving low values for tray loadings. The reason was an error in conversion of Anglo-Saxon units to metric.

One of the major reasons for selecting a system utilizing tbp cutpoint calculations was its minimal dependence on process analyzers. After seven months of operation, we have learned that all three analyzers perform well and are useful for different reasons.

The tbp cutpoint of naphtha correlates closely with the endpoint analyzer. Even if the latter suffers from remarkably long process lag, it provides a useful feedback to compensate for any errors in tbp calculations due to instrument drifts.

A similar good correlation was found between gas oil cloud point and tbp cutpoint (see Fig. 11). The cloud point, however, is affected by crude quality, which can be seen in Fig. 12. The computer raised the setpoint for tbp cutpoint by 4°C to keep the cloud point at − 1.5°C.

To our knowledge, the color analyzers used in refineries until this project were of the alarm type only. The continuous analyzer we installed provided valuable information about the process, and together with the overflash calculation provide a powerful tool for maximizing HGO draw off rate. Figure 13 shows clearly how an insufficient amount of overflash lets HGO go black. The color analyzer allows the operator to adjust the limit for minimum overflash for optimum color of HGO.

The number and severity of crude switches have increased because of the many new crude oils processed today. Due to the new control system, smooth operation has been maintained for the transient periods. The transient periods have been cut in half.

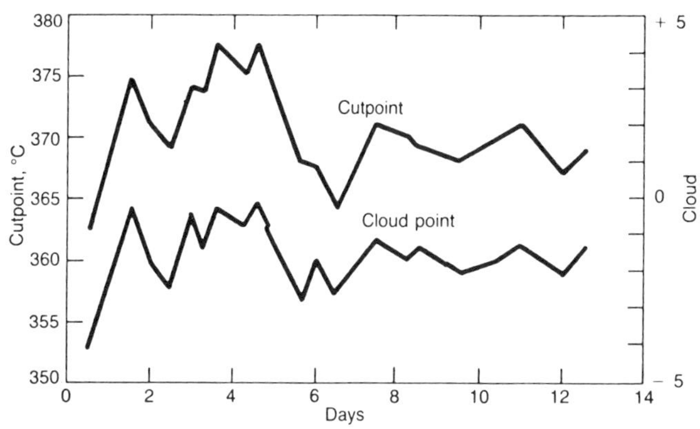

Figure 11. Cloud point and tbp cutpoint of gasoil without computer control.

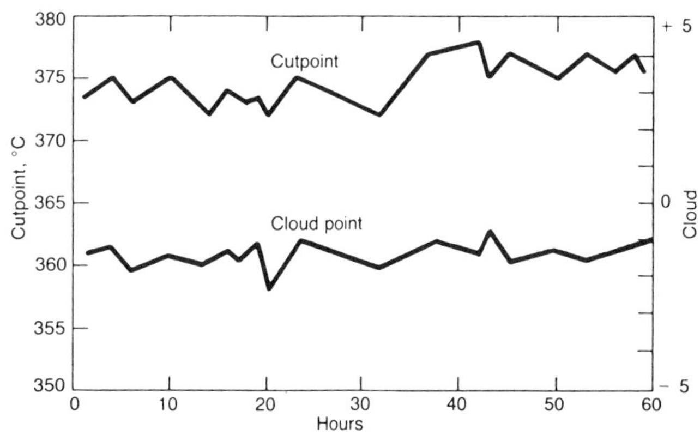

Figure 12. Cloud point and tbp cutpoint of gasoil with computer control.

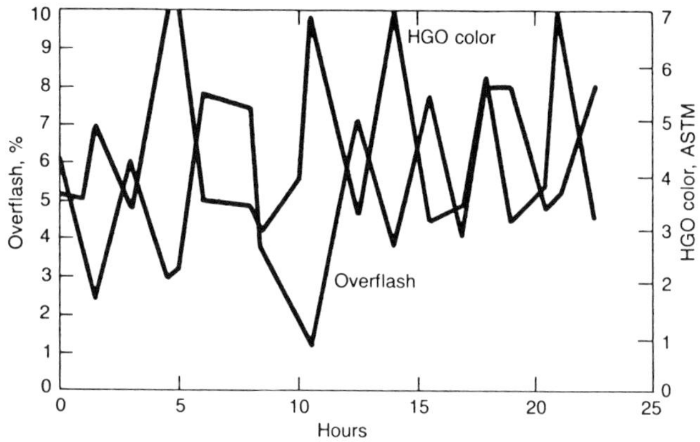

Figure 13. Correlation between HGO color and overflash.

Although instruments play an important role in operating reliability of a system like ours, we have not seen any sign or extra work load on the instrument maintenance department, except the analyzer group.

Yet, more attention is required from both instrument and operations staff when carrying out normal maintenance routines. A transmitter that normally affects only one controller may now cause troubles in several parts of the process via the interacting control loops. Good examples are transmitters connected to tower calculations or feedforward controls. After some harmful upsets, we were forced to write instructions on how to proceed when zeroing or repairing each transmitter. The instructions tell the technician which controls are affected and should be put on manual and which computer inputs should be replaced by manual entries.

The project was justified by economic benefits and each control function has a certain role in reaching the final result. We believe that during the training and commissioning period we succeeded in transferring the economic way of thinking to the operators whose former concern mainly was to keep the flows and levels at their setpoints. The system helps the operators understand interactions between unit material and energy flows.

The financial success of this project (payout less than one year) has given confidence to management to such an extent that another crude unit project is in the design phase. The control philosophy will be identical, only the split of responsibility between the computer and the instrumentation will be different: most stabilizing and even some optimizing controls will be handled by a digital distributed control system.

References

1. Van Horn, L. D., "Crude Unit Computer Control—How Good Is It?," *Hydrocarbon Processing,* V59, n4, April 1980, p. 145.
2. Latour, P. R., "Online Computer Optimization 1: What It Is and Where To Do It," *Hydrocarbon Processing,* V58, n6, June 1979, p. 73.
3. Latour, P. R., "Online Computer Optimization 2: Benefits and Implementation," *Hydrocarbon Processing,* V58, n7, July 1979, p. 219.
4. Van Horn, L. D., "Consulting Services, Interface Sources, and Projects," *Canadian Controls and Instruments,* Vol. 17, n6, June 1978.
5. Latour, P. R., "Requirements for Successful Closed-Loop Optimization of Petroleum Refining Processes," IFAC Conference Proceedings, *Digital Computer Applications to Process Control,* Dusseldorf, FRG, Pergamon Press, October 1980, p. 11.

32

Group Methods for Advanced Column Control

D. E. Haskins, F. Tolfo, and L. Chauvin

Group method calculations are suited for distillation column advanced controls in real time. Five calculation methods will be discussed and compared for accuracy by computing the internal reflux for several columns and operating conditions. Practical design criteria are defined and illustrated by comparing two distillation column reflux control schemes. The resulting conclusions include the best location and preferred order of group method calculations.

The most rigorous methods of calculations for distillation columns are the stage-to-stage and the successive approximation calculations. These techniques require a minimum number of simplifying assumptions and are most rigorous because they involve material and energy balances for each stage in the column. However, for control calculations these rigorous methods are not attractive because they consume large amounts of computer resources and may be considered too rigorous for control problems.

Group methods of calculations for distillation columns avoid the tedious stage-to-stage or successive approximation calculations. These techniques relate the number of stages to the composition at the top and the bottom of the column and do not provide individual stage information. Because group methods are simpler than stage-to-stage or successive approximation methods, they are more suitable to online computer calculations. Group methods are considered accurate enough for control purposes.

GROUP METHODS CONSIDERED

Group methods can be subdivided into two further categories. The first category consists of shortcut design methods like the Gilliland[1] or Erbar-Maddox[2] techniques. The second category consists of non-design estimation techniques that can be adapted to control calculations, such as the binary distillation Fenske-Gilliland equations,[3] the Shinskey separation factor approach,[4] or the Smith-Brinkley method.[13]

Gilliland Shortcut Design Method

Gilliland proposed his shortcut design method in 1940.[1] Since then it has continued to be the most widely accepted shortcut technique. This method is based on a graphical curve that is the result of over 50 binary and multicomponent stage-to-stage calculations. The Gilliland correlation relates two variables, ϕ and ω

$$\phi = (N - N_{MIN})/(N + 1) \quad (1)$$

where N = number of theoretical stages
N_{MIN} = minimum number of theoretical stages for the separation at total reflux

$$\omega = \frac{(L/D) - (L/D)_{MIN}}{(L/D) + 1} \quad (2)$$

where (L/D) = design reflux ratio
$(L/D)_{MIN}$ = minimum reflux for the separation with infinite stages

The use of this correlation requires estimates for N_{MIN} and $(L/D)_{MIN}$. The estimate for N_{MIN} is usually done by using the Fenske equation, sometimes referred to as the Fenske-Underwood equation because it was proposed independently by Fenske[5] and Underwood.[6]

$$N_{MIN} = \frac{\ln\left[\left(\frac{y_l}{y_h}\right)\Big/\left(\frac{x_l}{x_h}\right)\right]}{\ln\alpha} \quad (3)$$

where y_l = molar fraction light key in distillate
y_h = molar fraction heavy key in distillate
x_l = molar fraction light key in bottoms
x_h = molar fraction heavy key in bottoms
α = relative volatility of light key to heavy key

The estimate for $(L/D)_{MIN}$ has been proposed by several authors including Gilliland. The most convenient and most often used method is the Underwood equations.[7]

$$\left(\frac{L}{D}\right)_{MIN} + 1 = \sum_{i=1}^{R} \frac{\alpha_i y_i}{\alpha_i - \theta} \qquad (4)$$

$$1 - q = \sum_{i=1}^{R} \frac{\alpha_i z_i}{\alpha_i - \theta} \qquad (5)$$

where a_i = relative volatilities of the R components
y_i = distillate fractions of the R components
z_i = feed fractions of the R components
θ = roots of the equations (there are R roots)
q = feed conditions (equal to 1 for feed at boiling point, equal to 0 for feed at dew point, or between 0 and 1)

The Underwood equations are solved iteratively to find the root, θ, that lies between the relative volatilities of the key components.

Erbar-Maddox Short-Cut Design Method

The Erbar-Maddox[2] shortcut design method is similar to the Gilliland method. They both are graphical and involve N_{MIN} and $(L/D)_{MIN}$. The same correlations for N_{MIN} and $(L/D)_{MIN}$ that are used with the Gilliland correlation could be used with that of Erbar-Maddox. The Erbar-Maddox correlation is also based on stage-to-stage calculations, and it is claimed by its originators to be better than Gilliland's correlation. Nevertheless, it has been less popular than Gilliland's method. One reason for the avoidance of the Erbar-Maddox method is the requirement for a combined mass and energy balance to provide an estimate for the top plate L/D value needed for the correlation.

Binary Fenske-Gilliland Method

One of the more complicated parts of using the two methods above is the iterative solution of the Underwood Eqs. 4 and 5. The need for the Underwood equations can be eliminated if one can assume that the feed composition and quality (fraction vaporized) are known and that the distillation is pseudo-binary.[3] Pseudo-binary means that other components in a multicomponent feed mixture do not distribute between column overhead and bottoms. With these assumptions, the minimum reflux for a variable quality, pseudo-binary feed can be expressed as:

$$\left(\frac{L}{D}\right)_{MIN} = \frac{\left(\frac{y_l}{z_l}\right)[\alpha - q(\alpha - 1)] - \left(\frac{y_h}{z_h}\right)[1 + q(\alpha - 1)] + (\alpha - 1)(q - 1)}{\alpha - 1} \qquad (6)$$

The equation for $(L/D)_{MIN}$ and the Fenske Eq. 3 for N_{MIN} can be combined with the Gilliland correlation to provide a single relationship. The Gilliland correlation is notoriously difficult to fit with a single equation. A recent article[9] shows that a number of equations are adequate but none is a superior fit. A curve fit by General Electric[8] is as good as any and is probably the simplest.

$$e^{(1-\phi)} = 2.73 - 1.27(1 - \omega) \qquad (7)$$

The work by General Electric[8] also made a further simplifying assumption for Eq. 1.

$$1 - \phi = \frac{N_{MIN} + 1}{N + 1} \qquad (8)$$

Assume that N_{MIN} and N are both much greater than 1.

$$1 - \phi \simeq \frac{N_{MIN}}{N} \qquad (9)$$

Combining Eqs. 2, 3, 6, 7, and 9 results in the following equation:

$$\frac{1}{N \ln \alpha}\left[\frac{y_l/y_h}{x_l/x_h}\right] = 2.73 - \frac{1.27\,D}{V(\alpha - 1)}\left\{\frac{y_l}{z_l}[\alpha - q(\alpha - 1)] - \frac{y_h}{z_h}[1 + q(\alpha - 1)] + q(\alpha - 1)\right\} \qquad (10)$$

Thus a number of assumptions have reduced the method of Gilliland to a single equation. However, note that the equation is transcendental, i.e., variables appear on both sides of the equation in a logarithmic manner that precludes a direct solution. Therefore, a computer is needed to do the iterative calculations.

Shinskey Separation Factor Method

This method is attributed to Shinskey because it is discussed in two of his books.[4,10] The concept of separation factor was not invented by Shinskey and is discussed in other references, for example King.[11] The separation factor appears in the Fenske Eq. 3 and, therefore, is the final equation of the binary Fenske-Gilliland method discussed previously.

$$S = \frac{y_l/y_h}{x_l/x_h} \tag{11}$$

Shinskey's contribution was to note that:

$$S = f(V, \alpha, n, E, F, z) \tag{12}$$

where V = overhead vapor rate (L + D)
α = relative volatility
n = number of actual trays
E = tray efficiency
F = feed rate
z = feed composition

To try to identify the form of that functionality, Shinskey[10] used a computer program based on the following assumptions:

- binary distillation
- constant molal overflow
- constant relative volatility in column
- constant tray efficiency in column
- material balance only.

Shinskey found that defining two functional groups, X and Y, reduced the functionality of Eq. 12 to a single curve of Y versus X.

$$Y = \frac{\ln S}{nE\log\alpha} = \frac{N_{MIN}}{N} \tag{13}$$

$$X = \frac{V\alpha^{1.68}}{F(nE)^{0.32}} \tag{14}$$

The X-Y curve, like the Gilliland correlation curve, is not easily fit by one equation. Shinskey notes that most columns operate in the linear portion of the curve so that in most cases

$$\frac{V}{F} = \frac{\ln S}{0.286(nE)^{0.68}\alpha^{1.68}\log\alpha} \tag{15}$$

Because of the assumptions made in the development of this equation, the final Eq. 15 is not recommended for column design purposes.[10] However, the important concept for control calculations is the form of the final equation. Eq. 15 suggests that it would be possible to say that:

$$\frac{V}{F} = \beta\log S \tag{16}$$

where β is a function at least of n, E, and α and most probably the feed composition, z.

In fact, Shinskey has suggested[4] that Eq. 16 could be very useful, and calls β the "column characterization factor." This factor is an obvious choice for computer identification and model updating.

Eq. 16 is much simpler than Eq. 10 for the binary Fenske-Gilliland method. Both equations involve the same variables but it is not readily apparent that the functionality is similar. A linearized form of Eq. 16 has been used in control work.[12]

$$S = a + b(V/F) + c(V/F)^2 + \ldots.. \tag{17}$$

Smith Brinkley Short Cut Design Method

Smith and Brinkley[13] presented a general equation for the mole fraction of a component i leaving in the bottom product:

$$\frac{(1 + S_{n,i}^{N-M}) + R(1 - S_{n,i})}{(1 + S_{n,i}^{N-M}) + R(1 - S_{n,i}) + h_i S_{n,i}^{N-M}(1 - S_{m,i}^{M+1})} \tag{18}$$

In Eq. 18, $f_i = (BX_B/FX_F)_i$ is the fraction of i going to the bottom product. S is an equilibrium factor for component i defined for the upper (n) and lower (m) sections of the column:

$$S_{n,i} = K_i V/L, \quad S_{m,i} = K_i' V'/L' \tag{19}$$

K_i = y_i/x_i is the equilibrium ratio of the vapor to liquid molar composition
V, L = molar flow rates (supposed constant) of vapor and liquid in the upper and lower (′) parts of the column
N, M = number of stages in the entire column and in the lower section, respectively
R = external reflux ratio = L_{N+1}/D
h_i = quantity dependent on whether the feed is mostly liquid or vapor

If the feed is mostly liquid,

$$h_i = \frac{K'_i}{K_i}\frac{L}{L'}\left(\frac{1 - S_n}{1 - S_m}\right)_i \tag{20}$$

If the feed is mostly vapor,

$$h_i = \frac{L}{L'}\left(\frac{1 - S_n}{1 - S_m}\right)_i \tag{21}$$

K_i values are estimated at some appropriate average temperature, which could be the arithmetic average between feed tray (M + 1) and top tray (N) for the upper section and reboiler (1) and feed tray (M + 1) for the lower section. When computing the minimum reflux for a given separation between light and heavy key, the equation can be rewritten more conveniently:

$$(L/D)_{MIN} = \frac{f_i h_i S_{n,i}^{N-M}(1 - S_{m,i}^{M+1}) - (1 + S_{n,i}^{N-M})(1 - f_i)}{(1 - S_{n,i}) - (1 - f_i)} \tag{22}$$

When used for reflux prediction on a specific column, the model requires some adjustment. This is usually introduced via the tray efficiency whereby the effective number of trays is computed from the actual number.

$$N_{EFF} = \epsilon N$$

COMPARISON BETWEEN METHODS

Group method calculations are used for distillation control and for economic optimization of operations.

A "good" method should satisfy the following criteria:

- It must be accurate; (L/D) predictions are used in the determination of the economic optimum in the check of approaching process constraints and in the calculation of feedforward signals.

  The quality of the model prediction has a direct impact on the reliability of the benefits estimated from optimization. When approaching constraints, inaccurate predictions result in larger safety margins and reduced benefits.
- It must be simple; complex models requiring many inputs are difficult to fit with live process data. Inaccuracies from process measurements are all combined in the parametric identification, and the model loses its predictive accuracy.
- It should not take up too much computer resources, and therefore iterative methods should be avoided.

Based on these considerations, we have selected three group methods: Smith-Brinkley, Fenske-Gilliland and Shinskey, and compared them for three different light ends fractionation columns: a de-ethanizer, a de-propanizer and a de-butanizer.

First, the parameters were calculated so that the model would predict the process design performance. Then, the models were used to calculate the reflux necessary to achieve the design minimum and maximum separations. In total, we have nine points of comparison between the three models (three for each column). The results are summarized in Table 1.

Table 1
Method Comparison Results

	Composition (Mole %)		Reflux rates (Mole % of feed rate)		
	Light key in bottom X_L	Heavy key in top Y_h	Smith Brinkley	Fenske Gilliland	Shinskey
De-ethanizer					
Design values	1.25	3.49	59.19	59.15	59.16
Maximum purity	0.96	2.90	67.43	66.43	66.47
Minimum purity	1.54	4.08	53.28	53.55	53.23
De-propanizer					
Design values	1.26	1.45	110.08	109.31	109.49
Maximum purity	1.30	1.50	109.39	108.22	108.30
Minimum purity	1.22	1.40	111.41	111.01	111.23
De-butanizer					
Design values	0.1	1.45	110.96	105.33	108.51
Maximum purity	0.05	1.50	109.24	103.78	107.79
Minimum purity	0.10	1.40	118.59	111.80	111.69

It is worth noting that for the columns for our example, the Smith-Brinkley model only works when the reflux is calculated for the percent heavy key component in the top. When the reflux is calculated on the basis of the percent light key in the bottoms, the predictions are unrealistically large.

From the comparison it is apparent that the reflux rates predicted for the three models are remarkably close and within the accuracy of flow measurements. Therefore, based on this numerical comparison of accuracy, any one of the three models could be used for advanced control and online optimization.

Considering then the other two criteria (simplicity and ease of calculations), the model of Shinskey seems more attractive. It is easily used for the online parametric identification of the model through the simple series expansion given in Eq. 17.

PRACTICAL DESIGN CRITERIA FOR ADVANCED CONTROLS

The design of advanced controls has to be based on sound theories. Various group method calculations for distillation columns have been discussed in the previous sections. There are also other practical design criteria for advanced controls:

- Simplicity of controls
- Optimal use of system resources
- Hierarchical control structure
- Flexibility

One of the necessary requirements for a control system design is simplicity. Simple controls are cheaper to install and maintain. The training problems for technical people and operators will be much fewer with simple controls. In general, controls that are not understood will not be used. Therefore, the simplest controls that accomplish the tasks are the best.

The second criterion for advanced control system design is the optimal use of the system resources. This can mean the proper choice of functionality between hardware and software. It can mean the proper choice of control structure so that the system is not overloaded. In computer control systems, advanced controls should be designed so that computer memory or, in the case of real-time systems, computer processing time is minimized. There is an obvious relationship between simplicity and optimal resource allocation. However, for a given complexity of controls there are still different amounts of resource allocation based on different designs.

A third criterion for advanced control system design is the need for hierarchical control structures. This need is predicated on two practical facts. First, the structures of distributed control algorithms lead naturally to control hierarchies. Second, a belief in the accuracy of Murphy's law creates a need for hierarchies of control. The different levels in control hierarchies should be designed with principles of modularity and invulnerability. Modularity means that controls should be designed so that functions are separated and that only a portion of the controls can be turned off when things go wrong. Invulnerability means that a problem in one part of the controls does not result in all of the controls being incapacitated. This principle results in the practice of designing the least reliable controls at the highest control level in the hierarchy.

The last criterion is that of flexibility. Control systems are dynamic in that they tend to be constantly modified. Furthermore, the designer should be aware that, although the original design should be simple, there may have to be field changes. The hierarchy structure should allow changes to be made without making major revisions to the remainder of the control system.

Figure 1 shows two hierarchy structures. From the preceding discussions of the practical design criteria, it should be obvious that control structure B is preferred to control structure A. The insertion of a distillation model just above a final control element causes too many potential problems.

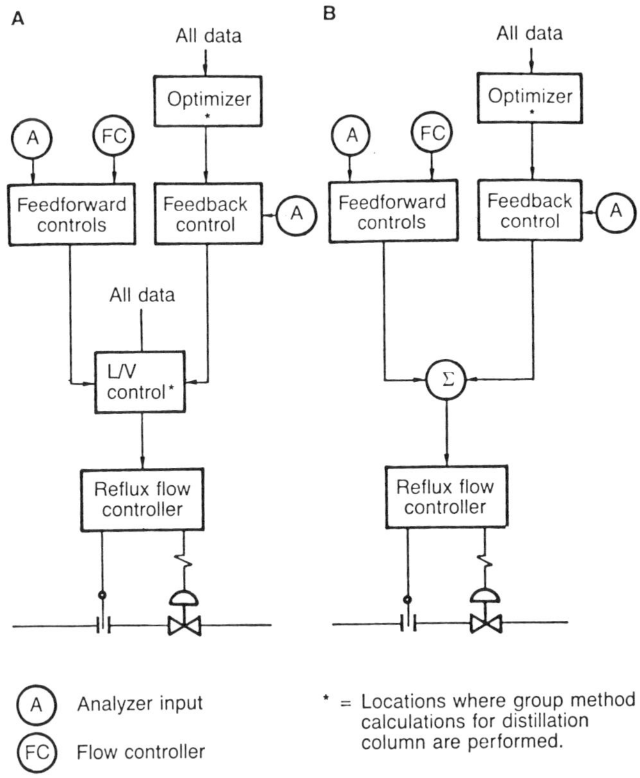

Figure 1. Hierarchy structures.

Once a hierarchy structure like scheme B is selected, the question then becomes one of choosing the distillation model for the highest level optimizer. The same practical design criteria should be considered in the model selection process. Of the methods discussed here, the Shinskey method is superior, based on simplicity, computational efficiency, and invulnerability. The superiority of the separation factor approach has also been noted in an earlier paper[12] which presented data demonstrating that a simple linear separation factor model was adequate. The binary Fenske-Gilliland method suffers from computational problems and a dependency on feed information, which may or may not be available. Other methods add complexity and computation requirements. The major problem is to find the simplest model that is adequate for the control tasks.

Another reason for starting with a simple model is that the optimizer may prove to be unnecessary. Therefore, it is best to minimize design efforts at the start of a project. For example, a little experience with the distillation columns and the cost objective function may show that the column always has an optimum at the same constraint. In this case, an optimizer is not needed and the task becomes one of designing a constraint pusher.

Figure 2 illustrates the control scheme for hierarchy structure B.

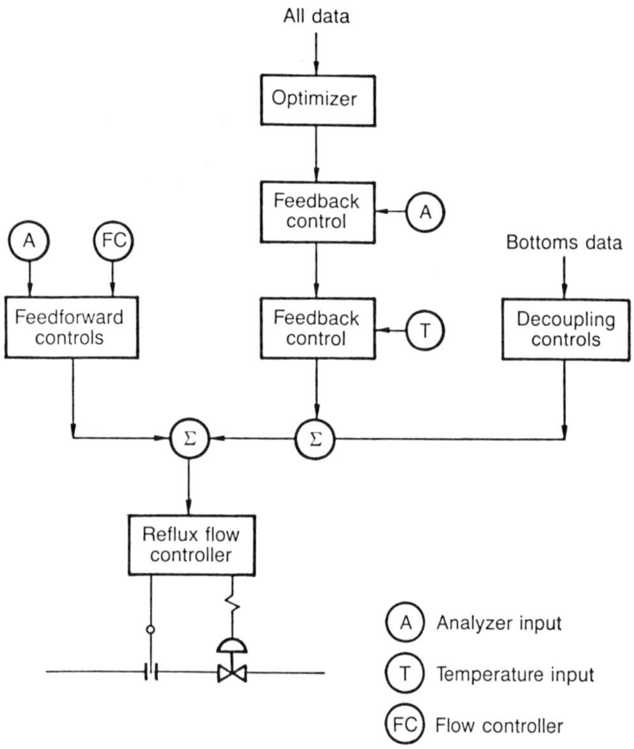

Figure 2. Control hierarchy B expanded.

CONCLUSIONS

Because of the importance of the practical criteria for control system design, control hierarchies should be arranged like scheme B, Fig. 1. The most important result is that distillation column model calculations, even the relatively simple group methods, should be done at the highest levels in the hierarchies.

Also, because of the practical criteria for control system design, the order of preference for the column model calculations should start with the most simple and proceed, only if necessary, to the most complex. The order of model preference is then:

1. Shinskey separation factor method
2. Binary Fenske-Gilliland method
3. Gilliland short-cut design method (using Underwood equations)
4. Erbar-Maddox short-cut design method

If none of those calculation methods are good enough for control, the only resource is stage-to-stage calculations or successive approximations. The justification for this step would have to be well defined because of the excessive requirements for computer resources.

References

1. Gilliland, E. R., *Ind. Eng. Chem.,* 32, 1220, 1940.
2. Erbar, J. H. and R. N. Maddox, *Petrol. Refiner,* 40 (5): 183, 1961.
3. King, C. J., *Separation Processes,* pp. 445–448, McGraw Hill Book Company, 1971.
4. Shinskey, F. G., *Energy Conservation Through Control,* pp. 255–258, Academic Press, 1978.
5. Fenske, M. R., *Ind. Eng. Chem.,* 24, 482, 1932.
6. Underwood, A. J. V., *Trans. Inst. Chem. Engr.,* 10, 112, 1932.
7. Underwood, A. J. V., *Chem. Eng. Progr.,* 44, 603, 1948.
8. *Process Analysis Seminar,* General Electric Company, Volume E2, October 1967.
9. Chang, Huan-Yang, "Gilliland Plot in One Equation," *Hydrocarbon Processing,* p. 146, October 1981.
10. Shinskey, F. G., *Distillation Control,* pp. 43–46, McGraw Hill Book Company, 1977.
11. King, C. J. Ibid, pp. 25–51.
12. Van der Grinten, P. M. E. M. and Buys, J. J. H., "Construction, Up-dating, and Evaluation of Statistical Models for Optimising Control of Distillation Columns," *I. Chem. E. Symposium Series,* 32, pp. 6:13–6:21, 1969.
13. Smith, B. D. and Brinkley, *Am. Inst. Chem. Engr. I,* 6, 446, 1960.

33
Indirect Dual Composition Control of a Deethanizer

P. A. S. Ralston, P. B. Deshpande, P. Hanik, and S. Bender

An indirect dual composition control scheme for a high-purity, industrial deethanizer is presented. A feed-forward algorithm varies the reflux flow setpoint to minimize energy demand and a cascade feedback loop inferentially controls bottoms composition. The feed-forward (supervisory) algorithm responds to changes in feed rate and/or feed composition. Changes in feed composition are measured in terms of the total C_2 to total C_3 mole ratio. The algorithm is based on actual data, and is a correlation between the reflux ratio and feed composition. The controller equations for all other tower loops are also solved by the computer. After three months of normal operation, results indicate that tops and bottoms target compositions are maintained without "overpurification" with the scheme. Also, both reflux and steam flow are reduced an average of 10 percent as compared with previous column operation under similar conditions.

INTRODUCTION

This work is the result of a collaboration between the University of Louisville and Northern Petrochemical Co. in Morris, Ill. The deethanizer studied is one column in the six column fractionation section of an ethylene plant. The deethanizer (Fig. 1) has a feed stream composed of small amounts of methane, along with acetylene, ethylene, ethane, propylene, propane, 1,3-butadiene, 1-butene, n-butane and n-pentane. Deethanizers function to remove ethane and lighter components from the feed; thus the desired separation is between ethane and propylene. The tower is 109 feet tall, 8 feet in diameter in the enriching section and 10.6 feet in diameter in the stripping section. It contains 48, two-pass, split-flow valve trays. The distillate is withdrawn as vapor from the top of the reflux accumulator. Bottom product is withdrawn beneath the last tray, and the remaining liquid circulated through the tube side of the thermosiphon reboiler. To produce polymer-grade ethylene, the propylene content in the distillate must not exceed 1,200 ppm. To produce polymer-grade propylene, the ethane content in the bottoms product must not exceed 200 ppm.

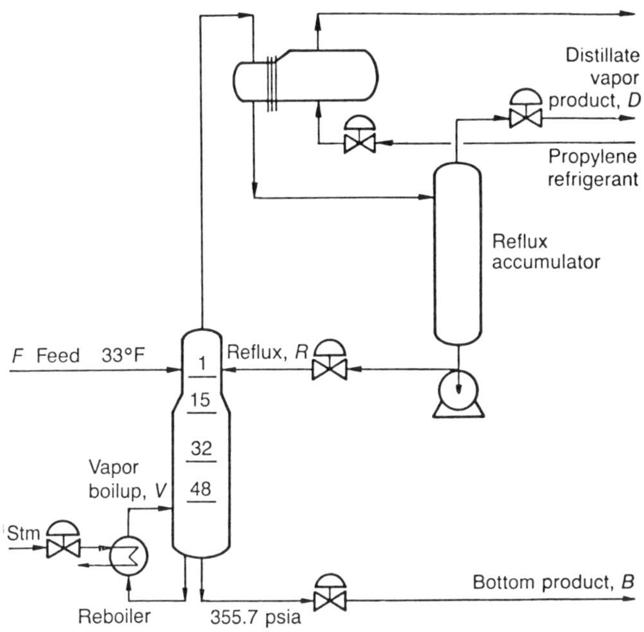

Figure 1. Deethanizer.

CONTROL SCHEME

Originally, the column operated under a single composition control scheme (Fig. 2 without the highlighted control computer). This scheme was first implemented via conventional board mounted controllers. In the middle 1970s, minicomputers were installed for the ethylene plant. Direct digital and supervisory control options were provided. The original panelboard instrumentation was retained for 100 percent redundancy. The computer uses the same input field measurements as the

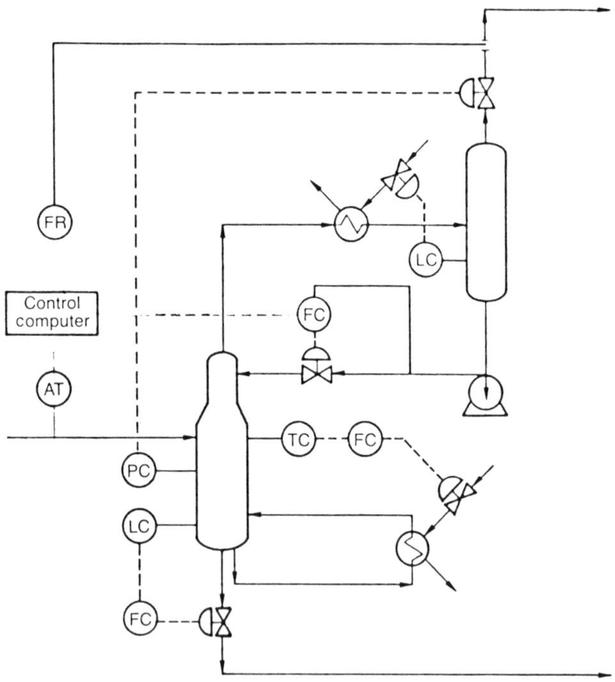

Figure 2. Original and revised (control computer addition) control schemes.

Table 1
Steady State Relative Gains

		2% Increase			2% Decrease	
(DV)		D	V		D	V
Λ_{DV} =	y	.998	.002	y	.037	.963
	x	.002	.998	x	.963	.037
(RB)		R	B		R	B
Λ_{RB} =	y	.977	.022	y	.01	.99
	x	.022	.977	x	.99	.01
(RV)		R	V		R	V
Λ_{RV} =	y	$-3.95E^{-04}$	1.000395	y	1.00022	$-2.2000E^{-04}$
	x	1.000395	$-3.95E^{-04}$	x	$-2.2000E^{-04}$	1.00022

		0.2% Increase			0.2% Decrease	
(DV)		D	V		D	V
Λ_{DV} =	y	.985	.015	y	.051	.949
	x	.015	.985	x	.949	.051
(RB)		R	B		R	B
Λ_{RB} =	y	.96	.04	y	.125	.875
	x	.04	.96	x	.875	.125
(RV)		R	V		R	V
Λ_{RV} =	y	−.005	1.005	y	1.003	−.003
	x	1.005	−.005	x	−.003	1.003

		0.05% Increase			0.05% Decrease	
(DV)		D	V		D	V
Λ_{DV} =	y	.967	.033	y	−.016	1.016
	x	.033	.967	x	1.016	−.016
(RB)		R	B		R	B
Λ_{RB} =	y	.887	.113	y	2.512	−1.512
	x	.113	.887	x	−1.512	2.512
(RV)		R	V		R	V
Λ_{RV} =	y	−.03	1.03	y	1.028	−.028
	x	1.03	−.03	x	−.028	1.028

analog controllers. The original scheme was easily implemented on the computer. With the original scheme, operators adjusted the reflux setpoint based on a twice daily distillate composition analysis. Because of the infrequent composition data, the operators often "overfluxed" the tower yielding products purer than specifications require and wasting energy at the condenser and reboiler. Consequently, it was desired to improve the column operation with an energy saving supervisory control scheme for the deethanizer.

Since composition at both ends was important, a detailed relative gain analysis was undertaken with the aid of a commercial distillation design package to assess the potential for dual composition control.[1] The results of both a steady state and dynamic relative gain analysis are inconclusive due to inconsistencies of the results. Calculation of the RGA for a positive change in manipulated variable suggests a scheme exactly opposite to that suggested from the RGA for a negative change in manipulated variable. Some authors contend that these results suggest single composition control.[1,2,3] Other research on binary systems suggests that such ambiguous results are due to system nonlinearities.[4,5] In fact, McAvoy[5] has stated (in reference to binary systems), "If process gains and relative gains change with the step in manipulation of variables, this indicates either that the convergence criterion is too loose or that too large a step has been made." However, for the deethanizer, reduction of perturbation size had no effect on the RGA results (Table 1). The simulation package was one of the best commercially available. The convergence criterion was as tight as economically and realistically feasible. The RGA failed to yield definitive information for all three dual composition schemes attempted. Dynamic RGA analysis provided little information due to inadequate dynamic models of complex columns. However, indirect dual composition control is possible using a single point scheme in combination with a feedforward loop to minimize energy usage.

Jafarey and McAvoy[6], Shinskey[7], and Luyben[8] have worked with energy minimization in columns with reported success. Of specific interest is the work of Jafarey and McAvoy. Their indirect dual composition control scheme consists of a feedforward/feedback loop to minimize energy consumption. The feedforward loop varies the boilup to feed ratio (V/F) in response to changes in feed composition according to an analytical model.

A variation of this concept is employed in the development of the supervisory control scheme for the deethanizer (Fig. 2 with control computer addition). The scheme consists of a digital feedforward control algorithm that varies the reflux flow setpoint to minimize energy usage and a cascade feedback temperature loop inferentially controls bottoms composition.

The feedforward algorithm must respond to variations in feed rate and/or feed composition caused by transient process fluctuations. Most of these are due to change in the number and/or configuration of cracking furnaces. The algorithm developed is a correlation between reflux ratio and feed composition.

Correlation Development

A correlation is developed in terms of external reflux ratio, R/D; therefore, the reflux setpoint will be changed by multiplying the reflux ratio by distillate flow. For energy minimization, changes in feed rate are handled by ratioing reflux to feed rate, while controlling bottom product composition.[8] A feed rate disturbance, after an initial transient period, causes the distillate flow to change in proportion to the feed rate change. Since reflux flow is set by multiplication of reflux ratio by distillate flow, reflux will also change in proportion to the feed rate change. Steady state simulation results confirm this (Fig. 3). This figure was generated by varying the feed rate while maintaining distillate and bottoms composition specifications. Hence, the reflux ratio may remain constant for feed rate changes. Figure 4 shows the variation in reflux ratio with feed rate for four different furnace configurations. The four different furnace configurations correspond to four different feed compositions. Each data point represents an average of four daily readings. Also, the reflux flow was set by operators; therefore, it could vary for a similar feed rate and feed composition depending on which operator was in charge of the column. In view of this, Fig. 4 tends to verify that reflux ratio is independent of feed rate changes at a given feed composition. Therefore, a feedforward algorithm based on a correlation between reflux ratio and feed composition will respond correctly to feed rate changes.

The feed composition changes to the deethanizer are primarily changes in the total C_2 to total C_3 mole ratio. Ethylene and ethane comprise the total C_2 content, and propylene and propane comprise the total C_3 content. The mole ratio C_2/C_3 is used as the parameter to indicate feed composition. A correlation between reflux ratio and the C_2/C_3 ratio was developed. The correlation is based on plant data (over the complete rage of furnace

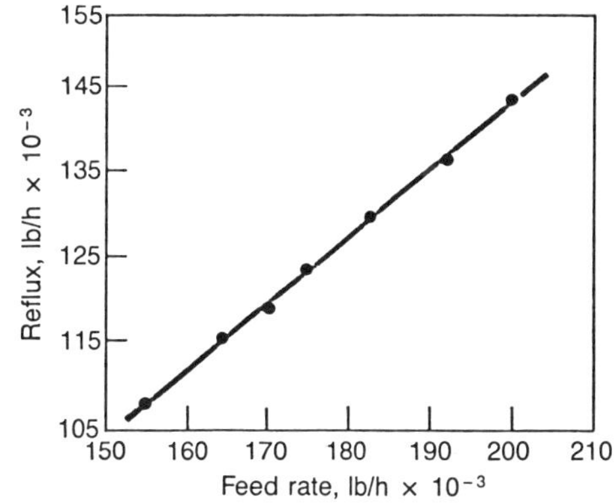

Figure 3. Reflux flow vs. feed rate.

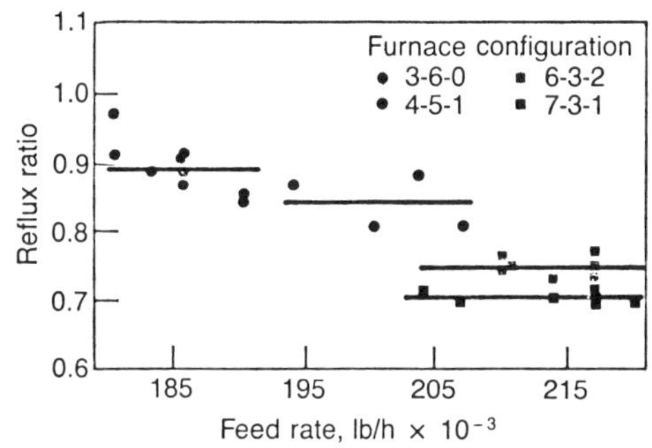

Figure 4. Reflux ratio vs. feed rate.

configurations and conversion rates), and verified by computer simulation. For these data, the C_2/C_3 varies from 3 to 10 as the reflux ratio varies from 1.0 to 0.7. A parabolic fit made to the data results in the following equation:

$$R/D = 0.00547(C_2/C_3)^2 - 0.1006(C_2/C_3) + 1.1855 \quad (1)$$

Modifications were made and dynamic parameters determined online by trial and error. The response of reflux flow to feed composition changes may be modeled approximately as first order plus deadtime. Note that the control scheme shown in Fig. 2 is not quite an indirect dual composition control scheme since the inferential composition control loop has not been provided with any feedforward adjustment.

Software Development

The supervisory strategy consists of a feedforward algorithm to change the reflux flow setpoint in response to feed rate and feed composition changes. The reflux setpoint is updated in response to feed rate and feed composition changes. The program calculates the appropriate reflux ratio from the correlation, multiplies the reflux ratio by the distillate flow value, and outputs the resulting reflux flow setpoint to the controller. All other tower loops are on DDC. The feedback controller equations are solved on the computer, but the setpoint for each of these loops remains unchanged.

Observation of the deethanizer distillate flow over time shows that the flow fluctuates significantly. Therefore, an average value that dampens these fluctuations is used (Fig. 5). It is obtained from a filtering equation with an experimentally determined filter factor.

The code was written according to the logic flow shown in Table 2. Once compiled, the program was linked to all existing special routines and control utility programs.

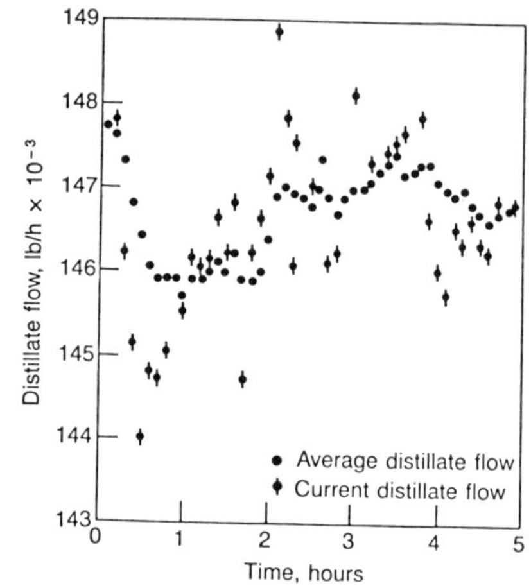

Figure 5. Average and current distillate flow.

IMPLEMENTATION AND RESULTS

The deethanizer supervisory program has been online for six months at this writing. During this time, feed rates varied plus or minus one to two percent about 195,000 lb/hr, with furnace configuration constant at eight ethane furnaces and two propane furnaces. (Feed composition changes due to changing furnaces were incurred.) Initial testing verified that the delay and lag time constants selected are reasonable. A program cycle time of six minutes provides ample response. Gains in the correlation were modified slightly.

To assess the usefulness of the supervisory program, a comparison is made between the new scheme (Fig. 2 with control computer added) and the single point DDC scheme which held reflux constant (Fig. 2 without the control computer).

The results are shown in Figs. 6 through 8. Figure 6 shows that the modified feedforward scheme has achieved a reduction in reflux flow required to hold product compositions within specifications. Figure 7 shows a similar reduction in reboiler steam flow. Figure 8 shows an improvement in the impurity level in the bottoms product as measured by an online analyzer. No analyzer exists for the distillate composition. However, twice daily laboratory analyses show a constant propylene content of 1,200 ppm with the new scheme as compared to 500 ppm with the original scheme.

Even though results are presented after an initial testing period, it is conclusive that the new scheme works. Specifications are maintained on both ends. However, maximum benefit would be obtained by adding feedforward control to the temperature controller. Since it is difficult to mentally track the column behavior, operators may unnecessarily raise the temperature controller setpoint in response to the larger bottoms ethane content. This counteracts benefits or the supervisory algorithm, sending more propylene overhead, and using unnecessary steam. Nonetheless, economic benefits exist without this addition.

It is good practice to rigorously evaluate economic performance after one year of operation. In this case, a

Table 2
Logic Flow in Supervisory Program

Step	Description
1	The feed composition is measured and stored in terms of mole percent C_2s and C_3s.
2	The ratio, C_2/C_3, is calculated and checked against minimum and maximum acceptable ratios. If out of bounds, the last acceptable value is used.
3	Feed composition is delayed and lagged.
4	An average distillate flow is calculated.
5	Status of the reflux controller is checked. If not on DDC, error is printed and program controls transferred to end of code.
6	Predicted reflux ratio is calculated from the empirical correlation.
7	Actual reflux ratio is calculated by dividing current reflux flow by average distillate flow.
8	Status of supervisory switch is checked. If in manual, go to end of program. If 1st pass in automatic, a bias is calculated for bumpless transfer and counter for RAMP function is initialized. If in automatic, continue.
9	The actual reflux ratio and predicted reflux ratio are compared. If their difference is larger than an acceptable value, the actual is retained for outputting to controller.
10	The stepchange in reflux ratio is changed to a ramp function (due to difference in predicted and actual).
11	The new setpoint for the reflux controller is calculated from the product of the ramped reflux ratio and the average distillate flow.
12	This calculated setpoint is checked against minimum and maximum acceptable limits. If value is greater that the upper limit, the maximum value is used; if it is less than the lower limit, the minimum is used.
13	The setpoint value is output to the reflux flow controller.
14	Bias is cleared, and present feed composition is set equal to last feed composition.
15	Continue, return to step 1.

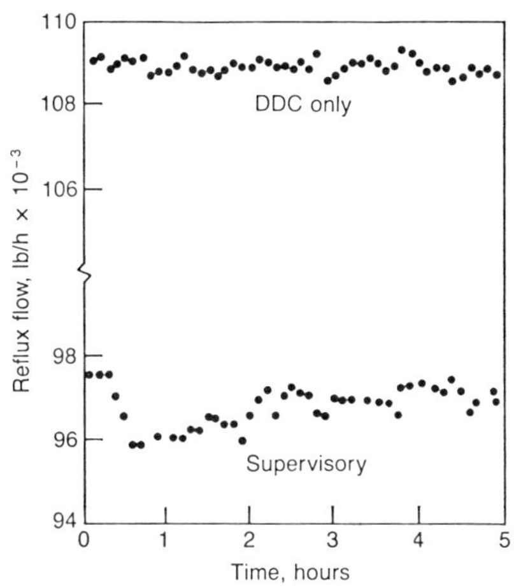

Figure 6. Reflux flow for supervisory and DDC control.

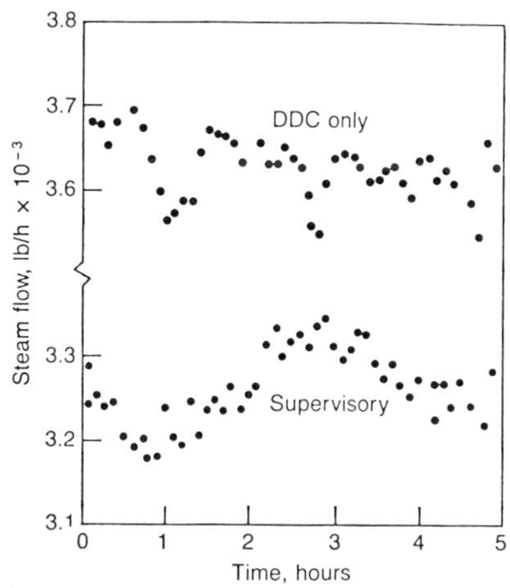

Figure 7. Steam flow for supervisory and DDC control.

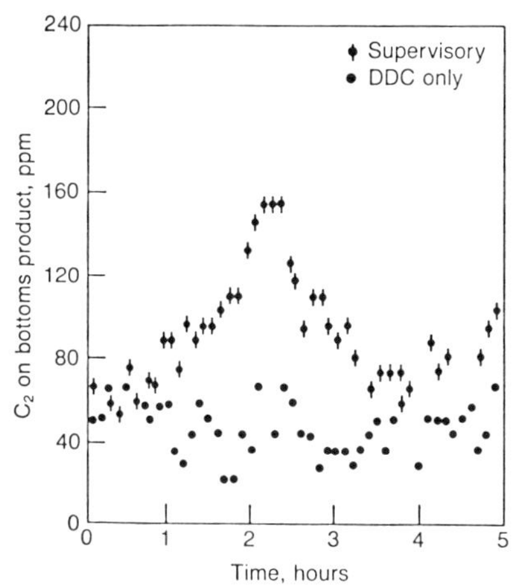

Figure 8. Product impurity for supervisory and DDC control.

projected yearly benefit is based on initial results. A savings of \$240,000 per year is estimated based on steam savings, propylene refrigerant savings and accounting for the loss of 700 ppm propylene. The material goes on to the C_2 splitter and is recycled back to the furnaces and finally used as fuel gas. Also, 20 days per year downtime is assumed. The value used for liquid refrigerant is calculated based on an energy quantity cost and not quality. It is also calculated from the cost of 600 psi steam. Quality is assessed according to thermodynamic availability, the ability to convert energy to work. For 600 psi steam, less than 40 percent of the heat energy can be extracted as work, whereas close to 80 percent of the heat energy in liquid propylene can be extracted as work. If this is considered, an even higher savings is realized.

CONCLUSIONS AND SIGNIFICANCE

1. The results of the steady state and dynamic relative gain analysis did not provide definitive information for dual composition control design in this case.
2. It is possible to achieve dual composition control for the deethanizer. It is accomplished indirectly using a feedback loop to control bottoms composition in combination with a feedforward supervisory algorithm to minimize energy usage.
3. The computer program that executes the supervisory algorithm performs well, and results indicate great potential for utility savings.
4. It is recommended that to obtain maximum benefit from the scheme, feedforward control should be provided for the bottoms temperature (inferential composition) control loop. Specifically, a feedforward algorithm should calculate the tray temperature controller setpoint based on the feed rate and feed composition.

References

1. Stark, P. A., Deshpande, P. B., and Bender, S. K., "Relative Gain Analysis for an Industrial Deethanizer," *Proceedings of the American Control Conference,* San Francisco, Calif., Vol. 3, p. 1,139, 1983.
2. Ralston, P. A. S., "Advanced Control of a Commercial Distillation System," PhD Dissertation, University of Louisville, 1983.
3. Thurston, C. W., "Computer-aided Design of Distillation Column Controls, Part 1," *Hydrocarbon Processing,* Vol. 60, No. 7, p. 125, July 1981.
4. Kim, Y. S., and McAvoy, T. S., "Computing the Relative Gains for Pressure and Composition Control of a Single Distillation Tower," *Proceedings of Automatic Control Conference,* Arlington, Va., Vol. 1, pp. 74–80, 1982.
5. McAvoy, T. S., "Interaction Analysis," ISA Monograph, Chapter 2, 1983.
6. Jafarey, A., and McAvoy, T. J., "Steady State Feedforward Control Algorithm for Reducing Energy Costs in Distillation," *ISA Transactions,* 19 (4), 89 (1980).
7. Shinksy, F. G., *Distillation Control for Productivity and Energy Conservation,* McGraw Hill, Inc., 1977, pp. 43–45.
8. Luyben, W. L., "Steady-State Energy Conservation Aspects of Distillation Column Control Design," *Industrial and Engineering Chemistry Fundamentals,* 14 (4), 321 (1975).

34

Predicting Tray Temperatures

V. L. Rice

An analysis of distillation columns often requires data for individual tray temperatures, either specific ones or the entire profile. A common approach to obtain this temperature information is through use of a rigorous tray-by-tray distillation simulation, usually with a main-frame computer system. Unfortunately, this rigorous approach is either impractical or just too much trouble in many cases. For example, an on-line optimizing control scheme rarely has enough space (computer memory) or real time for a rigorous calculation of distillation column tray temperatures.

A shorter method is presented in this chapter that predicts the tray temperatures of simple distillation columns. Following the theoretical discussion of the method, some examples of its use are presented.

BASIS OF THE METHOD

The approach taken to obtain the tray temperature relationships is the same one used by Smith, et al, to develop a shortcut rating of simple distillation columns. The basic assumptions are as follows:

- Constant vapor and liquid flow rates in specified sections (either rectifying or stripping).
- Constant relative volatilities in the same sections.

With these assumptions and the symbols identified in Fig. 1, the following equation is derived:

$$f_i = \frac{(1 - S_{ni}^{N-M}) + R_{pi}(1 - S_{ni})}{(1 - S_{ni}^{N-M}) + R_{pi}(1 - S_{ni}) + h_i S_{ni}^{N-M}(1 - S_{mi}^{M+1})} \quad (1)$$

where $f_i = Bx_{bi}/Fx_{fi}$
N = total number of theoretical trays
M = number of theoretical trays below feed tray
$S_{ni} = KV/L$
$S_{mi} = K'V'/L'$

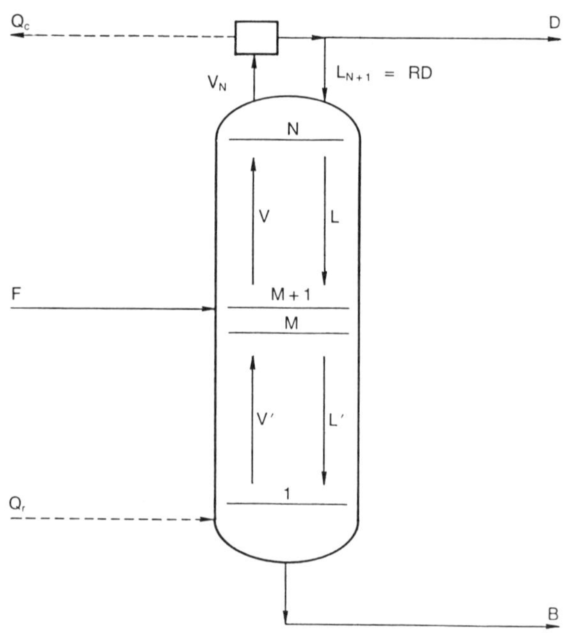

Figure 1. Column configuration helps define symbols.

$R_{pi} = R/K_{(N+1,i)}$ for a partial condenser
$R_{pi} = R$ for a total condenser

$$h_i = \frac{K_{(m+1,i)}}{K_i}\left[\frac{L(1 - S_{ni})}{L'(1 - S_{mi})}\right]$$

This method requires specifications for N, M, vapor-liquid traffic in both column sections and product rates.

A combination of the last two specifications can be given in several ways. The column parameter inputs for the examples that follow are N, M, F, D/F, R, and q (the mole fraction of feed that is liquid). These inputs will fix all of the necessary specifications. With these inputs, both product flowrates and compositions are calculated.

Now, based on the same previous assumptions, two equations are developed to determine the liquid composition on any tray. The derivation is similar to the one used for Eq. 1. Although the details of the derivation are not shown here, the final equations are:

For stripping section,

$$x_{mi} = Bx_{bi}\left[\frac{S_{mi}^{m} - S_{mi}}{S_{mi} - 1} + 1\right] \tag{2}$$

For rectifying section,

$$x_{ni} = \frac{x_{di}}{K_i}\left[\frac{L/(K_i - L)(S_{ni}^{N} - S_{ni}^{n})}{V(S_{ni}^{N} - S_{ni}^{N-1})} + 1\right] \tag{3}$$

where x'_{ni}, x'_{mi} = unnormalized mole fractions
n, m = individual tray numbers in the rectifying and stripping sections, respectively

Notice these equations yield mole fractions that must be normalized to give an approximate material balance on any given tray. As long as the assumption of constant vapor-liquid flows is not grossly violated, this normalization will be valid.

Since these equations give the liquid composition on a tray, a bubble-point calculation will yield the tray temperature.

To solve the equations for composition and bubble point, the following information is needed:

- Section terminal composition, i.e., distillate composition for rectifying section trays and bottoms composition for stripping section trays.
- Section vapor and liquid flow rates.
- Component K-values at some average temperature for the section.
- Theoretical number of trays in the column.

The first two items would normally be available from column operating data (for a given service) or supplied by a rating calculation for a new service. The last two items require some special consideration. In the convergence of the Smith-Brinkley shortcut method,[1] a temperature parameter, T_p, is calculated. The average temperature used for K-values should be the linear average of the section terminal temperature and T_p. The fact that T_p depends on the value of N, the theoretical trays, further complicates the matter. If this tray temperature method is used in conjunction with a Smith-Brinkley rating calculation, T_p and N (and M) will be available.

If a new column configuration is being considered, the rating can be done by a Smith-Brinkley calculation using the input parameters discussed before. Then, all the information required for the tray temperature algorithm will be available. If tray temperatures are desired for an already existing column service (i.e., fixed product compositions), the following procedure is recommended:

Step 1. Use the Smith-Brinkley method to determine the total theoretical trays required to give the specified product compositions. This will be a trial-and-error procedure.
Step 2. Calculate the desired tray temperatures using the values of N, and T_p determined in the first step.

Reasonable approximations can be made for K-values and theoretical trays. The feed tray temperature is usually a good approximation to the value of T_p. Any of several methods could be used to approximate the required number of theoretical trays such as: accepted tray efficiencies for a particular column service, a Fenske-type calculation, etc. Results obtained using these approximations should be checked carefully and compared with what would be reasonable.

Many applications of this shortcut method are feasible. The major advantage offered by this method is that a small amount of computer memory is needed for what is usually a memory intensive calculation. Also, the speed of the calculation is orders of magnitude faster than the rigorous methods usually employed. A later example will show these advantages can be achieved with only a very small loss of accuracy.

SMALL COMPUTER APPLICATIONS

The design or rating of a distillation column is normally performed with a commercial (or in-house) steady-state process simulator on a main-frame computer. These simulators yield the composition, temperature, pressure, and flow rate of each stream associated with the distillation column. Depending on which type of calculation is performed (design or rating) other information yielded by these simulators are: trays required, column tray temperature and composition profiles, vapor-liquid rate profiles, and reboiler and condenser duties.

The utility of the method described herein is that a complete preliminary design or rating of a simple distillation column can be accomplished on a hand calculator with its inherent advantages over a main frame system. In our case, we used a calculator with 2.2 KBYTES of memory. Depending on the required accuracy this method yields the final results needed for many different kinds of studies.

Table 1
High Purity Butane/Pentane Splitter

Feed		Configuration and operating data
Comp.	**Mol frac.**	N = 18, M = 12.
i-C_4	0.0042	Overhead pressure = 109 psia
n-C_4	0.2238	Bottoms pressure = 114 psia
i-C_5	0.3997	Feed rate = 6,500 lb-mol/h
n-C_5	0.3615	D/F = 0.238
n-C_6	0.0106	R = 6.6624
n-C_7	0.0002	Feed q = 1.0

Table 2
Rating Results Compared

	Distillate		Bottoms	
Comp.	**Shortcut**	**Rigorous**	**Shortcut**	**Rigorous**
i-C_4	0.0176	0.0176	0.0000	0.0000
n-C_4	0.8968	0.8948	0.0135	0.0142
i-C_5	0.0692	0.0690	0.5029	0.5030
n-C_5	0.0164	0.0187	0.4692	0.4686
n-C_6	0.0000	0.0000	0.0139	0.0139
n-C_7	0.0000	0.0000	0.0002	0.0003
Total flow, lb-mol/h	1,547	1,547	4,953	4,953
Temp., °F	158.1	159.6	224.5	226.1

Tray temperature profile:

	Temperature, °F	
Tray No.	**Shortcut**	**Rigorous**
18 (top tray)	158.1	159.6
17	164.1	165.4.
16	171.9	173.0
15	180.7	181.8
14	189.7	190.2
13	198.6	197.3
12	202.1	199.7
11	205.4	202.4
10	208.4	205.3
9	211.0	208.2
8	213.4	211.1
7	215.5	213.9
6	217.3	216.4
5	218.9	218.7
4	220.3	220.7
3	221.6	222.5
2	222.9	224.2
1 (Reboiler)	224.1	226.1

Secondly, it more accurate results are required, the results from this method could be used to reduce the time spent on a main frame system. If case studies are required, these can be performed using the calculator system to eliminate any obviously infeasible cases, thus reducing the cases run on the main frame. Also, the calculator method will generate product compositions and temperature profiles to use as initial estimates for a rigorous tray-by-tray method; thus, reducing the computer time needed for convergence.

An example of the use of this calculator system is the rating of a column with specifications given in Table 1. The column was also analyzed using a rigorous commercial process simulator[3] for comparison. The K-value algorithm is the one previously developed by the author.[4] The results are compared in Table 2. As can be seen the shortcut procedure, results are very close to those obtained by the more rigorous approach.

PROCESS CONTROL APPLICATIONS

Real-time simulation for use in process control applications require speed and usually a minimum amount of computer memory. The shortcut method herein can be adapted for this sort of simulation. For example, the method was applied to one section of a natural gas plant. The section involved two distillation columns having tray temperature control loops. By adding second-order lags to steady-state calculations based on the Smith-Brinkley and tray temperature algorithm, the output behaved very much like the ones from the actual plant. Thus, the simulation was accurate enough to allow operators to practice operating the plant under normal conditions, upset conditions, or startup/shutdown conditions.

References

1. Smith, B. D. and Brinkley, W. K., "General Short-Cut Equation for Equilibrium Stage Processes," *A.I.Ch.E.J.*, Vol. 6, No. 3 (1960), pp. 446–450.
2. Fleischer, M. T. and Prett, D. M., "Simplified Techniques for Simulation of Complex Distillation Columns," A.I.Ch.E., 88th National Meeting, Philadelphia, June 1980.
3. Erbar, J. H., "MAXI*SIM," Super Consultants, Inc.
4. Rice, V. L., "Program Performs Vapor-Liquid Equilibrium Calculations," *Chemical Engineering*, Vol. 89, No. 13 (June 28, 1982), pp. 77–86.

35
Simple Optimization for Dual Composition Control

L. M. Gordon

Optimization has become the ultimate objective for designers of distillation control systems. This is as it should be, since optimization promises to adjust the operation of distillation columns for maximum profit, even as the cost of energy and the values of product streams continually vary.

In pursuit of this elusive objective, much effort has been focused on developing elaborate computer models of various separations to aid in determining optimum set points. Many of the models are capable of excellent results. However, they are also quite expensive, especially if they are developed by outside sources. Beyond their initial cost, they are difficult and expensive to calibrate and maintain, their assumptions and limitations may not be well-understood by operating personnel, and plant conditions inevitably change. Consequently, all too often, they lose their accuracy and fall into disfavor and disuse.

This chapter describes an algorithm that uses a simple algebraic model to predict optimum composition set points for dual composition control of two-product columns.

Although this technique will not equal the accuracy of more elaborate models, it can reliably predict whether an optimum lies within the range of allowable compositions, and whether it is sharp or broad. Where an optimum does exist, rough economic data can give a first approximation of the economic incentive for applying more powerful analysis techniques. An optimum that lies beyond the allowable operating region indicates the need for constraint controls that can maintain the separation at its operating limits.

MODELING THE COLUMN

The starting point for any effort in model-based optimization must be the development of the process model. In this analysis, the model evolves from material and energy balances, definition and derivation.

Material Balances

Figure 1 shows a simplified schematic of a two-product column. Feed to the column is assumed to be some mixture of liquid and vapor. The products from the column are saturated or subcooled liquids. Material balances across the column lead directly to a very fundamental relationship among the composition of the feed and the products. For this column,

$$F = D + B \tag{1}$$

where F = feed rate in moles/time
D = overhead product rate in moles/time
B = bottoms product rate in moles/time

For the light key,

$$Fz = Dy + Bx \tag{2}$$

where z = mole fraction of light key in the feed
y = mole fraction of light key in the distillate
x = mole fraction of light key in the bottoms.

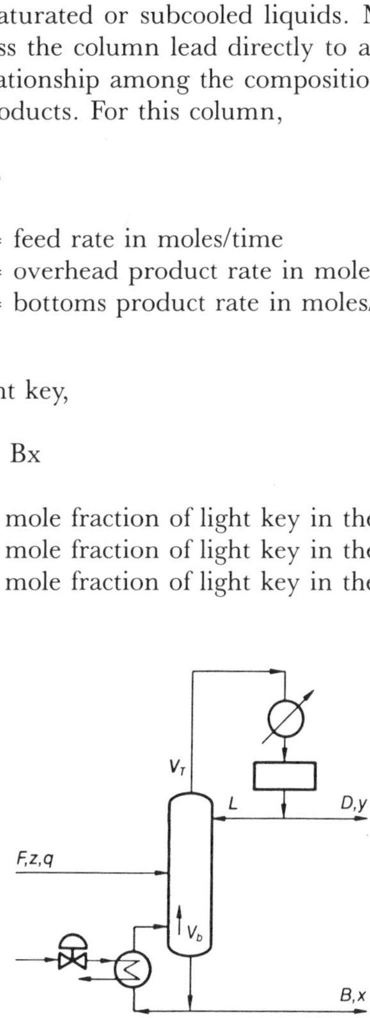

Figure 1. Major streams of a two-product column.

Combining Eqs. 1 and 2 yields:

$$D/F = (z - x)/(y - x) \quad (3)$$

where z and F are given. Through manipulation of the set point of a distillate flow controller, or of the valve directly, the control system will fix the value of D/F. The result is a single relationship between two unknowns, the product compositions.

The implication of this equation is that specification of the material balance across the column forces the two product compositions into a set of possible combinations that satisfies Eq. 3.

For the two-product column shown in Fig. 1, all of the overhead vapor flow is condensed and divided between reflux and overhead product. Then,

$$V_t = L + D$$

where V_t = flow rate of the overhead vapor, moles/time
L = flow rate of the reflux, moles/time

By rearranging,

$$V_t/D = L/D + 1 \quad (4)$$

Also, the vapor component in the feed is added to the vapor generated in the reboiler,

$$V_t = V_b + Fq$$

where V_b = flow rate of vapor from the first tray, moles/time
q = fraction of the vapor in the feed

assuming constant molar overflow across the column. The value of q normally falls between 0.0 and 1.0, with 0.0 indicating a saturated liquid and 1.0 indicating a saturated vapor. Values outside of this range are also possible, indicating a subcooled or a superheated feed.

Combining this last relation with Eqs. 3 and 4 yields:

$$V_b/F = (L/D + 1)((z - x)/(y - x)) - q \quad (5)$$

Separation

The purpose of a distillation column is to separate a mixed feed into relatively pure product streams. An index of separation, which describes the purity of the products has been defined, following the Fenske equation, by Shinskey.[1]

$$S = y(1 - x)/x(1 - y) \quad (6)$$

where S = separation index.

Thus, as each of the product streams approaches purity, S approaches ∞. Conversely, as separation is reduced each of the product compositions approaches the feed composition and S approaches 1.0.

This definition of separation, however, only measures the purification that has been achieved in a column. To predict the separation that will be realized requires a model that relates separation to the design and operating parameters of the column. Such a model has been derived by Jafarey, Douglas, and McAvoy,[2] from Smoker's equation, which provides an exact analytical solution for binary distillations when the relative volatility is constant. By this model,

$$S = \left[\alpha \sqrt{1 - \frac{1 - q + L/D}{(1 + L/D)(1 - q + (L/D)(z + z_{LL}))}}\right]^{nE} \quad (1)$$

where α = relative volatility (assumed constant across the column)
n = number of actual trays in the column
E = tray efficiency (assumed constant across the column)
z_{LL} = mole fraction of the lighter-than-light key in the feed

Taking the number of trays, the relative volatility, and the feed composition and percent vaporization as given for a particular distillation column, separation thus increases with the reflux ratio for the column.

For a given purification, i.e., a given separation, Eq. 6 provides a second equation in the two unknown compositions. Figure 2 presents a graphic representation of Eqs. 3 and 6. The axes measure the impurity in the two

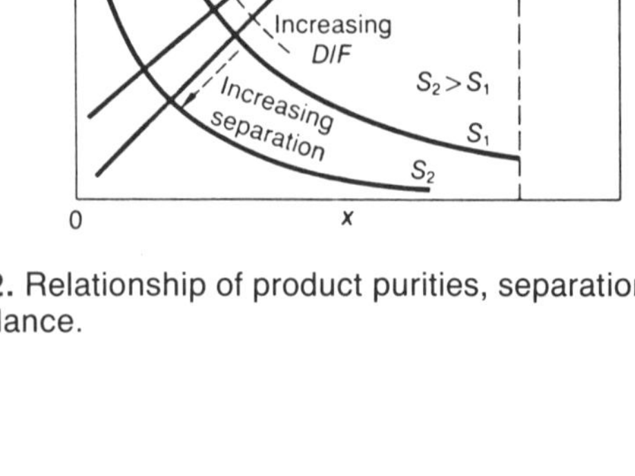

Figure 2. Relationship of product purities, separation and material balance.

product streams; thus the origin represents a condition of perfectly pure products. Similarly, a condition in which no purification takes place is represented by the coordinates x = z, 1 − y = 1 − z. Each point within this coordinate space represents a possible combination of product compositions. Lines of constant D/F plot as diagonals against these coordinates, each one radiating from the coordinates representing the feed condition, and becoming more horizontal as D/F increases. Lines of constant separation plot as hyperbolas against these coordinates.

Each line of constant separation may be thought of as representing a given amount of purification work. At the extremities of the lines, this work is concentrated on purifying one of the products at the expense of the other. Conversely, in the middle of a separation curve, the work is evenly allocated and the products are of roughly equal purity. Because higher values of separation represent higher purity products, these hyperbolas move closer to the origin at the higher reflux ratios which yield higher separation.

For a given column and its operating parameters, the intersection of the material balance and separation curves is the graphic representation of the simultaneous solution of Eqs. 3 and 6 which fixes the composition of both product streams. For example, consider the separation of a 50-50 mix of isobutanes and normal butanes in a 55-tray column with a reflux ratio of 6.4. For these keys, the relative volatility is approximately 1.35. Further, assume tray efficiency to be just under 78 percent and the feed to be 10 percent vapor. Then,

$z = 0.5$	$q = 0.10$	$n = 55$
$E = 77.8$	$L/D = 6.4$	$\alpha = 1.35$

and, from Eq. 7,

$S = 1{,}065$

Then, if half of the feed is taken as overhead product, D/F = 0.5 and the simultaneous solution yields:

$y = 0.97 \qquad x = 0.03$

Thus, each of the two-product streams will contain 3% impurity.

SEEKING THE OPTIMUM SET POINTS

The algorithm for determining the optimum set points reverses these equations, determining product values and energy costs for assumed product compositions and the given operating parameters.

First, consider the more common, simpler, case in which the values of the product streams are unrelated to their composition, as long as they do not exceed impurity limits. In this case, it is not necessary to examine the entire coordinate space; the optimum set points must lie along one of its boundaries.

Operating Regions

As shown in Fig. 2, the intersection of the material balance and separation lines defines the current operating point for a distillation column. Changes in the mass and energy flows within the column will cause movement to a new operating point, in a direction that depends on the nature of the change.

For example, increasing energy input at a constant D/F generates additional vapor overhead which is all returned as reflux, increasing the reflux ratio and, consequently, the separation. This moves the operating point toward the origin along the line of constant D/F, and forces improved purity in both product streams, without changing the volume of either one. If the value of the products is unrelated to their purity, such a change would increase the cost of making the separation without increasing the value of the products. The net result would be a decrease in the operating profit for the column.

Since the costs of operating a distillation column are related to the energy thruput, it is useful to identify the effect of changing the operation of the column along a line of constant vapor rate. Figure 3 shows the relation of the lines of constant V/F to the lines of constant S. If V/F remains constant as D/F increases, reflux must be reduced, causing a reduction in the reflux ratio and in the separation index. The line of constant V/F must

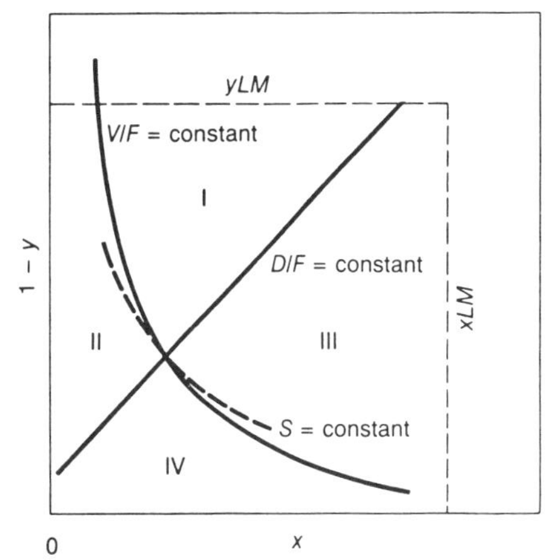

Figure 3. Coordinate space divided into four distinct regions.

therefore fall inside the line of constant S, as D/F increases.

Drawing more distillate product without increasing the energy input will, therefore, increase the volume of the top product at the expense of lowering its purity. Simultaneously, the volume of the bottoms product would be reduced and its purity would be increased. Depending on the relative value of the products, this change might either increase or decrease the total value of the products, while the cost of the energy thruput remained constant.

Consider a column producing products that are both purer than the specifications will allow. The constant V/F and the constant D/F curves which pass through the current operating point divide the coordinate space into four distinct regions, bounded by the axes and the maximum allowable impurity in each product. These regions are labeled I, II, III, and IV in Fig. 3.

Movement from the current operating point into each one of these regions causes predictable changes in the net return for the column, depending on the relative value of the products. For example, in the usual case where the distillate product is more valuable than the bottoms product, movement into Region I will result in increased total product value because of the higher fraction of feed going to the more valuable product, even as the cost of operating the column is reduced because of the lower energy input. Overall return is thus increased, if the regulatory control system is capable of controlling the product compositions closer to the allowable impurity limits without exceeding them.

The following tables summarize the possible combinations.

Case A: Distillate more valuable than bottoms

Region:	I	II	III	IV
Product value	Higher	Higher	Lower	Lower
Energy required	Lower	Higher	Lower	Higher

Case B: Bottoms more valuable than distillate

Region:	I	II	III	IV
Product value	Lower	Lower	Higher	Higher
Energy required	Lower	Higher	Lower	Higher

Inspection of these possibilities shows that in the first case the operating point should be moved into Region I. However, at each new operating point, the same analysis will apply, with the result that the operating point should be moved until it encounters the limit on the maximum impurity in the distillate product. By the same reasoning, in Case B, the operating point should be moved until it reaches the limit on the maximum allowable impurity in the bottoms product.

Where the value of the product is not constant, but increases with its purity, the problem is more complex. In this case, the location of the optimum will not necessarily be found along one of the impurity limits, as there may be some benefit to increasing energy thruput to make the products more pure, even if the material balance remains constant.

The Optimization Algorithm

The exact location of the optimum within the allowable composition limits depends on the relative value of the products and the cost of energy. Specific examples will be given below. In general, if the product values are fixed and nearly equal, the optimum will lie near the intersection of the impurity limits, where the cost of making the separation is lowest.

Where there is a significant differential in fixed product value, or if the product value increases with purity, there will be some benefit in shifting the operating point to produce more of the more valuable product, or to improve purity, until the cost of the additional energy required overtakes the increased value of the products. Where energy is "free" because it is recovered from waste fuels or by heat integration with other units, the optimum will lie at one of the edges of the coordinate space. Such a column should be driven to the verge of flooding, or to the limit of available heat transfer.

The algorithm for finding the optimum set points calculates the value added per unit of feed using the reasoning and the model equations presented previously, along with information on the physical properties of the keys, the values of products and the cost of energy.

The optimum product compositions are those values for which the difference between the net value of the products and the cost of making the separation is the highest. Where the value of the products is constant, one of the optimum values is known, and it is only necessary to examine this difference as the other composition varies along this impurity limit. Where the values of the products are not fixed, it becomes necessary to examine this difference throughout the entire coordinate space bounded by the impurity limits and the axes in Fig. 3.

At each pair of assumed compositions, Eq. 3 fixes the material balance and allows a calculation of the net value of the products per unit of feed. Next, Eqs. 6 and 7 allow calculation of the necessary reflux ratio, and Eq. 5 may be used to calculate the moles of vapor generated per mole of feed. This may be combined with the cost of energy to calculate the value added per unit of feed.

TYPICAL RESULTS

To illustrate the results of this algorithm, consider the butane splitter described earlier. Assume the specifica-

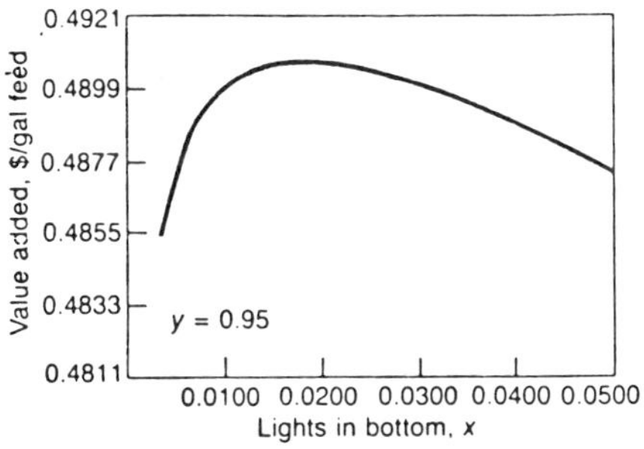

Figure 4. For constant value products, the optimum lies along the impurity limit for the more valuable product.

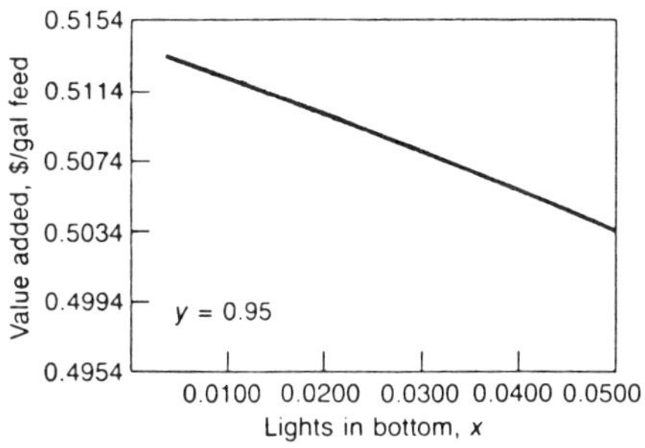

Figure 5. When energy is "free," purity of the less valuable product should be maximized.

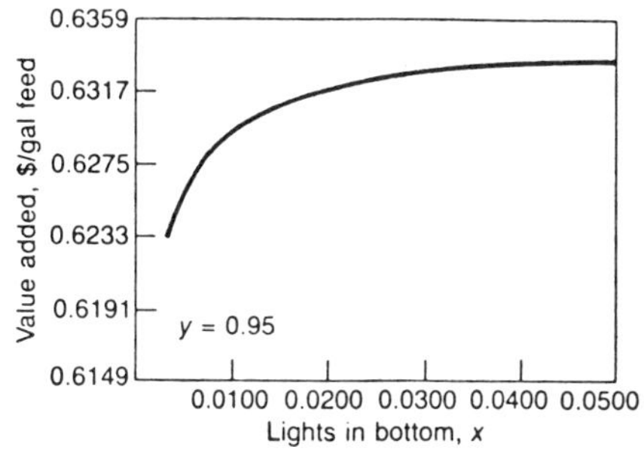

Figure 6. With products of similar value, both should be controlled at their impurity limits.

tions allow as much as 5% impurity in the product streams. Let the distillate product stream be valued at $0.70/gal, the bottoms stream valued at $0.30/gal, and the cost of energy to be $6.50 per million Btus. Since the distillate is more valuable, the column should be operated to keep the normal butane content of the isobutane product at the upper impurity limit, 5 percent, while the composition of the bottoms product is adjusted to maximize the operating profit of the column. Figure 4 presents the results of the objective surface calculation.

For this set of data, the figure shows an optimum range for the set point of the bottoms composition controller. In this case, it centers around 2 percent light key in the bottoms product. As is typically true, the variation in the value added is a small fraction of its absolute value.

However, although the variation with the bottoms composition appears small, over time, the economic incentive for operating at an optimum can become quite significant. In this case, the estimated value added at the current operating point is $0.485/gal of feed. At the optimum, the estimated value added is $0.491/gal of feed. Assuming a feed rate of 10,000 b/d,

10,000 b/d(42 gal/b)
350 d/yr(0.006 $/gal) = 880,000 $/yr

Figure 5 presents the results of the analysis for this column, under the assumption that the energy input is "free." As expected, the profit/gal of feed steadily increases as the material balance shifts toward the top of the column and the purity of the bottoms product increases. For these data, the estimated value added at the current operating point is $0.504/gal of feed. If the distillate is held at its impurity limit, and the bottoms impurity can be reduced to 1 percent, the estimated value added increases to $0.512/gal of feed. For one year,

10,000 b/d(42 gal/b) × 350 d/yr(0.008 $/gal)
= 1,180,000 $/yr

Figure 6 presents the results of the algorithm for a case where the cost differentials are not so large. Distillate is valued at $0.70/gal, and bottoms at $0.60/gal. The cost of energy is set at $7.00 per million Btu.

As expected, there is no incentive to overpurify the products, or to shift production into either one of the products. For these data, the estimated value at the operating point is $0.63/gal of feed. If the composition of both products can be held at their impurity limit, the estimated value added increases to $0.634/gal of feed. For one year,

10,000 b/d(42 gal/b)350 d/yr(0.004 $/gal)
= 590,000 $/yr

When both compositions are varied, the value added becomes a three-dimensional surface over the composition plane. Figure 7 shows the results of the analysis for the butane splitter, under the assumption that the values of both products are linear functions of their impurity. The surface shows a maximum for variations in both variables. The highest value added falls at $x = 0.013$, $y = 0.026$; the top plane of the bounded space rests on this optimum.

Figure 8 shows the results of the calculations for the case where energy is "free" and the product values increase with purity. As expected, both products should be made as pure as possible, up to the operating limits of the column. The value-added surface appears linear because the assumed value function is linear; nonlinear and stepwise functions will product similar variations in the shape of the surface.

LIMITATIONS

The results of this algorithm are subject to the assumptions used in the calculations. First of all, the algorithm uses an algebraic model of separation (Eq. 7), rather than tray-to-tray calculations. Reference 2 discusses the accuracy limitations of this model.

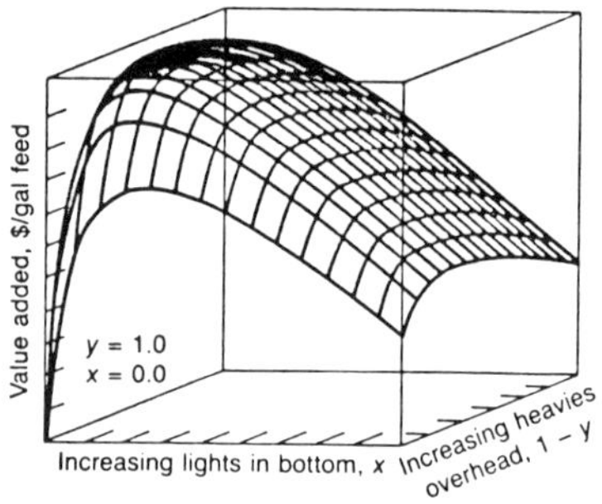

Figure 7. When product value increases with purity, the optimum set points may not fall on either impurity limit.

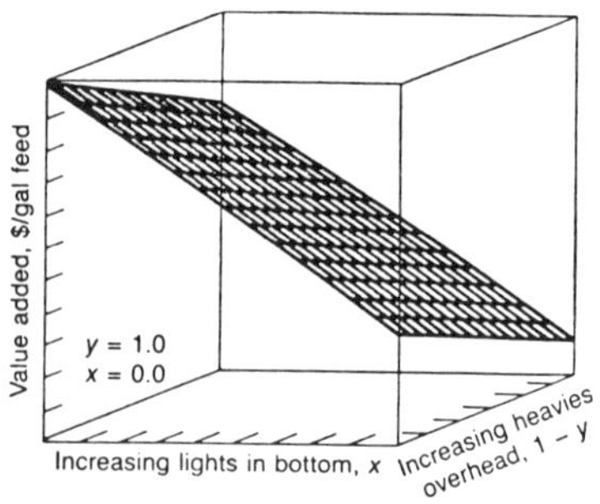

Figure 8. If product values increase with purity and energy is "free," both product purities should be maximized.

In particular this model does not explicitly consider distortions in the composition profile because of substantial reflux subcooling or incorrect location of the feed tray. For new columns, the feed tray is usually correct, and there are often several alternate locations. However, over time, feed and product compositions often change, leading to a mismatch between the composition of the feed and the inventory on the feed tray. This reduces the efficiency of the trays in the immediate vicinity of the feed tray. In the algorithm, these distortions appear as extreme values for the calculated efficiency or relative volatility, signaling the need to review current operations for the column.

The model is not limited to binary feeds. Off-key components may be present in the feed, so long as they do not contribute a significant fraction of the inventory on the trays within the column. Typically, the fractional inventory for the off-key components is much less than their fraction in the feed, unless their molecular weights are close to those of the light and heavy keys. For example, the inventory of noncondensibles in the feed is typically limited to the receiver and perhaps the top two or three trays of the column.

A second limitation of the algorithm resides in the estimation of the physical properties of the product streams at the assumed compositions. The calculations estimate these properties as linear combinations of those of the light and heavy keys. This assumption will be acceptable as long as the column is separating homologous components where polarity of the molecules does not significantly affect the relative volatility, and any differences in the size, structure or polarity of the molecules allow the assumption that volume is conserved.

CONCLUSIONS

Although the algorithm presented is quite simple, it provides several significant insights into the optimization of two-product columns.

First, the variation in the value added through optimization is relatively small, compared with its absolute value. Consequently, calculating the precise value of the optimum set points is of secondary importance to achieving the good regulatory control that allows enforcement of the chosen set point values.

Where more than one source of energy is available, and the sources are valued differently, it becomes essential to design controls that are capable of properly sequencing these sources. Where energy is recovered from heat integration with other units, it becomes essential to design controls that can minimize interaction between these units. This will stabilize the operation of both units, so that each can be held closer to optimum operation.

Second, in many cases, the optimum operation of the column is predictable without calculation. In cases where there is no significant difference between fixed values of the two products, optimum operation will be achieved by controlling the composition of both products at their impurity limits. In cases where the energy input is "free" or limited in amount, designing constraint controls that can maintain the column at its physical or operating constraints becomes the key to maximum profit.

Finally, where improved regulatory controls and any constraints imposed by the market or downstream units allow even modest improvements in the choice of the operating point of the column, the potential accumulated economic benefit can become quite large. Almost invariably, this incentive is many times greater than the cost of the instrumentation required.

References

1. Shinskey, F. G., *Distillation Control for Productivity and Energy Conservation,* McGraw-Hill Book Co., New York, 1978.
2. Douglas, J. M., Jafarey, A., and McAvoy, T. J., "Short-Cut Techniques for Distillation Column Design and Control," *I&EC Process Design and Development,* Vol. 18, April 1979.

36

Improve FCC Profitability With Better Control*

J. C. Brice and K. V. Krikorian

The FCC is a complex process both from a planning and control standpoint. This is primarily because the basic process variables and hence control loops are very interacting. Operators are typically concerned with keeping the unit stable rather than in forms of optimization which push the unit closer to the limits. These very difficulties and incentives make the FCCU an ideal application for closed-loop computer control. The financial benefits a well designed computer control system can bring will be typically worth 5 percent to 10 percent of the process revenue.

HIERARCHY OF FCC PLANNING AND CONTROL

Before embarking on the detailed justification of computer process control systems it is useful to review the typical hierarchy of FCCU operations planning and control. The starting point for most integrated oil companies is the market place. The availability of crude oils and demands for products together with market prices are the inputs to the supply and logistics departments of most refiners. This information is used to develop an optimum operating plan for each of the oil companies refining centers and also a set of boundary economics and volume limits.

This information can be used locally within the refinery to input to the offline FCCU monitoring and optimization system. This system checks that the operating data received from the unit indicates satisfactory performance and quickly identifies any developing problems affecting yields. The optimization system has the further objective of identifying the most attractive operating strategy based on the latest information on unit limitations, feedstock availability and quality, etc. The information provided by this system will provide specific operating targets and objectives for the online computer control system.

* Based on material originally presented at Katalistics 3rd Annual Fluid Cat Cracking Symposium, Amsterdam, May 26, 27, 1982.

NATURE OF THE FLUID CAT CRACKING PROCESS

The reactor regenerator and main fractionator present a complex interactive control problem. The important operating variables are shown in Table 1. In a modern riser cracker the feed is injected into a hot stream of regenerated catalyst, and the reaction occurs over a short contact time in the riser before the catalyst and reacted products are separated in the reactor. The degree of conversion and coke yield are a complex function of the reactor temperature, catalyst-to-oil ratio, riser residence time and the condition (activity, carbon level, metals content) of the regenerated catalyst. After separation, the spent catalyst is stripped and regenerated by burning off the carbon before circulating back to the feed injector. If temperatures in the reactor and regenerator are to be stable, the two vessels must heat-balance. This occurs when the heat removed in the reactor

Table 1
Unit Operating Variables

Typically not at limit values	Range	Remarks
(1) Reactor temperature	480–550°C	Set for conversion
(2) Feed quality	—	—
(3) Recycle—ratio total/fresh feed	1.0–1.3	Improves distillate yield
(4) Feed preheat temp	120–425°C	Depends on heat balance
(5) Catalyst-to-oil ratio	3–9	Set by cat circulation
(6) Carbon on regen catalyst	0.05–0.3wt%	—
(7) Steam to feed and spent cat stripping	—	—
(8) Regenerator temperature	650–750°C	Limit from metallurgy
(9) CO vol % flue gas	ppm-6%	—
Often at limit values		
(1) Feed rate	—	—
(2) Combustion air rate	—	—
(3) Cat gas compressor loading	—	—

to vaporize and crack the feed balances the heat released by combustion of coke in the regenerator. Similarly, the two vessels must carbon-balance with the carbon laid down on the catalyst burned off in the regenerator to as low a level as possible.

Summarizing the effect of operating conditions on conversion yields in the reactor these can be grouped into two simple terms. One is the reactor temperature, the second can be called the "catalytic severity," which is the combined effect of catalyst-to-oil ratio, catalyst contact time and catalyst activity. The unit operator can measure reactor temperature, but he has no direct measure of the vital catalytic severity.

Table 2
Examples of Complex Interactions Between Process Variables

Assume an increase in regen, cat slide valve opening

Base condition—HTR, full CO combustion, excess oxygen in flue gas (oxidizing conditions)

Response under open loop conditions	**Speed**
1. Catalyst circulation rate (CCR) increases	Fast
2. Reactor temperature increases (RT)	Fast
3. Coke yield and conversion increase RT and CCR	Fast
4. Flue gas O_2 drops	Moderate
5. Small regen T adjustment to new balance	Slow
6. Recycle quality changes, result of (2) on fractionator	Slow
7. Feed quality change (6) increases coke yield	Slow
8. Regen temperature increases due to (7)	Slow
(etc.)	

The unit will slowly cycle to a new equilibrium condition

An insight into the interactive nature of the process can be gained from the response under loop conditions to a change in position of the regenerated catalyst slide valve as shown in Table 2 with a unit running at HTR (high temperature regeneration) condition with full CO combustion and excess oxygen in flue gas. The direct result is an increase in catalyst circulation and an increase in reactor temperature causing increased feed conversion and coke yield. This is followed in the regenerator by combustion of additional coke, a drop in oxygen content of the flue gas and an approach to a new equilibrium temperature in the regenerator that will gradually feed back, affecting reactor temperature. The change in reactor temperature will have fed into the downstream fractionator causing a change in the quality of the heavy cycle oil that will affect total feed quality, assuming a recycle operation. The net effect will be oscillations in all the major variables as the unit slowly settles down to a new equilibrium. The dynamics of the unit are characterized by some variables that respond very rapidly and some that respond very slowly, causing feedback to the system.

Table 3
Response of System Changes with Partial CO Combustion in Regenerator

Assume same regenerator slide valve increase

Base condition—HTR, partial CO combustion—no excess oxygen

Response under open loop conditions	**Speed**
1. Catalyst circulation rate increases (CCR)	Fast
2. Reactor temperature increases (RT)	Fast
3. Conversion and coke yield increases	Fast
4. CO/CO_2 ratio increases to burn coke	Moderate
5. Regen temperature drops due to CO/CO_2 and CCR	Slow
6. Reactor temperature falls (RT) due to (5)	Slow
7. Conversion and coke yield fall due to (6)	Slow
8. Recycle quality changes result of (2) and (6) on fractionator	Slow
9. Lower regen T and CO increase carbon on regen catalyst affects yields increases coke	Slow
(etc.)	

The unit will cycle to a new equilibrium condition

Note that a control scheme selected for previous oxidizing conditions may not be viable under this set of starting conditions

An alternative starting point is shown in Table 3 in which the unit is operating in the regenerator at HTR conditions oxygen-starved so that the CO level rises. The same change in catalyst circulation will have a more profound effect in this case because the additional coke formed causes an increase in the CO/CO_2 ratio, which reduces the heat release in the regenerator. This together with the circulation rate increase causes the regenerator temperature to fall, which will cause regeneration efficiency to fall and coke level to increase on the catalyst.

In a typical conventional control scheme the reactor temperature is controlled by catalyst circulation rate, which is a very fast loop, and stabilizes the operation of the downstream fractionator and gas concentration unit. Under HTR conditions and with excess oxygen available, carbon balance will be achieved automatically.

Note that this control scheme for reactor temperature is not suitable when the unit is in a partial CO combustion mode at HTR conditions. The heat available to the reactor front the regenerator is reduced by increasing cat circulation rate since the increased coke yield increases CO/CO_2 ratio, which decreases heat available from the regenerator. An alternative control scheme is generally selected, such as feed preheat to control reactor temperature.

Conventional analog control systems do not cope well with this complex situation, and it is no wonder that the unit is often operated away from the optimum conditions. The operator is not enthusiastic about approaching limiting operations and finding himself in a corner.

OBJECTIVES OF AN IDEAL FCCU CONTROL SYSTEM

The objectives of the control system are summarized in Table 4. These are first to ensure safe operations with an effective, reliable emergency system. The essential factors are described in Table 5. Second, the control objectives are ideally suited to the application of a computer. The qualities of the computer control systems that make them so valuable are *flexibility* to test different control configurations quickly and safely and to arrive at the best regulatory system. Also, the computers allow *multivariable control techniques* where several manipulated variables can be used to achieve stable non-interactive control of each variable. The interacting nature of the control system effectively counteracts the interactions of the process. Finally, computer control systems are *"intelligent"* and can be used to maximize objectives such as conversion up to the physical unit limitations.

Table 4
Objectives of an Ideal FCCU Control Scheme

Function	Objective
(1) Safety	Reverse flow protection
(2) Heat balance	Constant reactor T, regen T within limits
(3) Carbon balance	Clean catalyst regeneration improves yield
(4) Conversion	Achieve most profitable level depending on economics and unit constraints
(5) Interactions	Overcome process variable interactions achieve smooth control
(6) Constraints	Operate up to the physical limits restricting further profitability

Table 5
Emergency System

Basic functions include:

(1) Reverse flow pretection
   Triggers + Low slide valve ΔP signal
   + Low temp in regen cat standpipe
   + High temp in spent cat standpipe
   The ΔP plus the corresponding temperature signal to be satisfied before emergency system comes into operation
   Actions + Slide valves shut
   + Feed diverted to main column
   + Feed injection steam cut in
(2) Reactor high temperature protection caused by loss of feed
   Action is to close regen cat slide valve since reactor is not designed for regeneration conditions
(3) Combustion air blower failure
   Action + Feed diverted to main column
   + Feed injection steam cut in
   + Steam cut in into blower discharge line protects regenerator and blower (from catalyst)

DEVELOPMENT OF OPTIMUM CONTROL CONFIGURATION

Because of the interactions of the process many combinations of controlled and manipulated variables can be considered. The flexibility of the software of a computer control system allows various configurations to be tested in developing an optimum scheme. A typical scheme could end up with reactor temperature controlled by feed preheat and/or recycle, CO content controlled by circulation rate, regenerator temperature could be floating or modified by the rate of catalyst addition (i.e., catalyst activity). Once the basic system is installed and operating, the advanced functions such as multivariable controls, constraint controls and feed-forward controls can be added.

To develop these systems knowledge of the dynamics and interactions of the process must be obtained. In an existing plant this knowledge can be gained by parametric tests manipulating one variable at a time while holding all others steady and recording all of the time-varied responses. For a new plant, a model of the process can be used to give steady-state interaction factors. Normally there is little dynamic data available unless an expensive dynamic engineering simulation is constructed. In the absence of quantitative data, good control judgement and process experience will result in a workable solution.

MULTIVARIABLE CONTROLS

Multivariable controls or interacting controls are control systems that interact in themselves to minimize or eliminate interactions in the process. In its simplest application, instead of one valve being manipulated by one variable, two or more valves will be adjusted that will also be manipulated by at least one other variable.

A typical system is shown in Fig. 1 for carbon and conversion controls. An objective of the system is to prevent the carbon balance control mechanism, which adjusts the regen cat slide valve, from affecting reactor temperature. Since the cat circulation rate will affect reactor temperature, the multivariable controller must make an adjustment to the feed preheat and/or recycle rate at the correct time so as to eliminate any change in reactor temperature. The dynamic operators F1 and F2 will be derived from an initial control engineering study or parametric tests.

These systems must be digital and microprocessor or computer based. Analog controls do not have the functionality, repeatability or long time constants required. The effect of the multivariable control is to stop any continuous cycling of the unit and to minimize the effect of disturbances on the plant. Furthermore, the transition from one operating mode to another can be made smoother, faster and without loss of conversion or feed rate.

LIMIT CONTROL—A BENEFIT OF A PROCESS COMPUTER

Because of the interacting nature of the FCC process and the difficulties controlling it with conventional analog instrumentation, the plant normally operates at a "safe" distance away from perceived limit values. Often this value is not reevaluated when conditions change so it becomes a worst-case limit value.

In its simplest form, a limit control system will drive a variable to a maximum value until a limiting value of another variable is reached. A more complex scheme is shown in Fig. 2 that is designed to maximize feed rate as constrained by one of the major equipment limits shown. Assuming naphtha is cut at the minimum cut point on the main fractionator, the debutanizer tower may be limited in capacity below the feed. The computer can estimate the loading from the mass balances of the tower and key temperatures. This loading can be compared with a maximum value above which flooding is known to occur. Similarly, the air blower may be limited in capacity by the availability of steam to the turbine driver, and the gas compressor by an overload power limit on the motor driver. The preheat furnace may be limited on heat fired, which can be calculated by the computer based on feed rate and coil inlet and outlet temperatures. The limit control system can scan each factor, estimate the margin to the limit value and translate that margin into the equivalent margin on feed rate. With no limiting values indicated, the system would gradually increase feed rate. Once one of the variables reached a limit value the system would hold feed rate constant. This system is very valuable for situations where due to changes in feed quality, for example, the

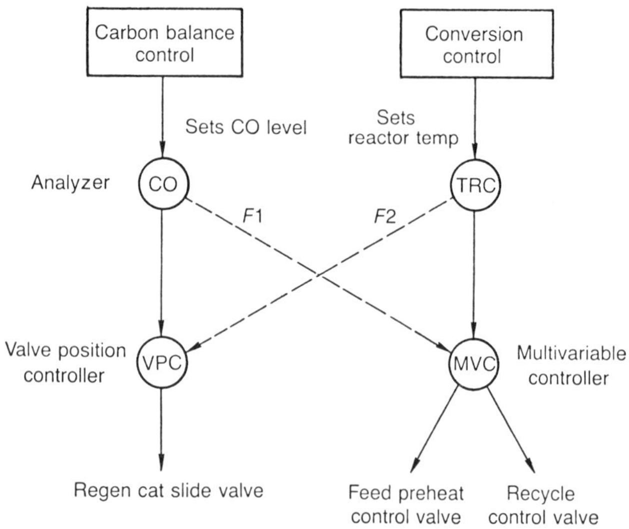

Figure 1. Multivariable controls.

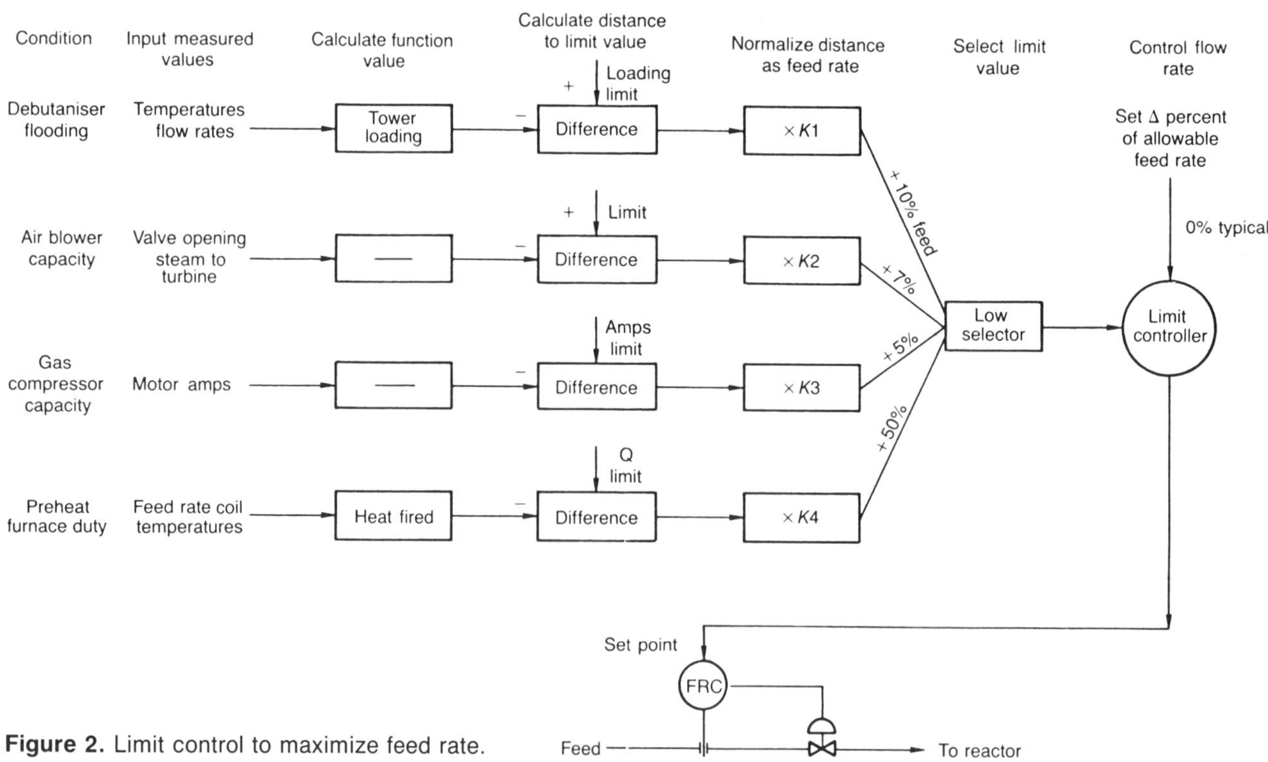

Figure 2. Limit control to maximize feed rate.

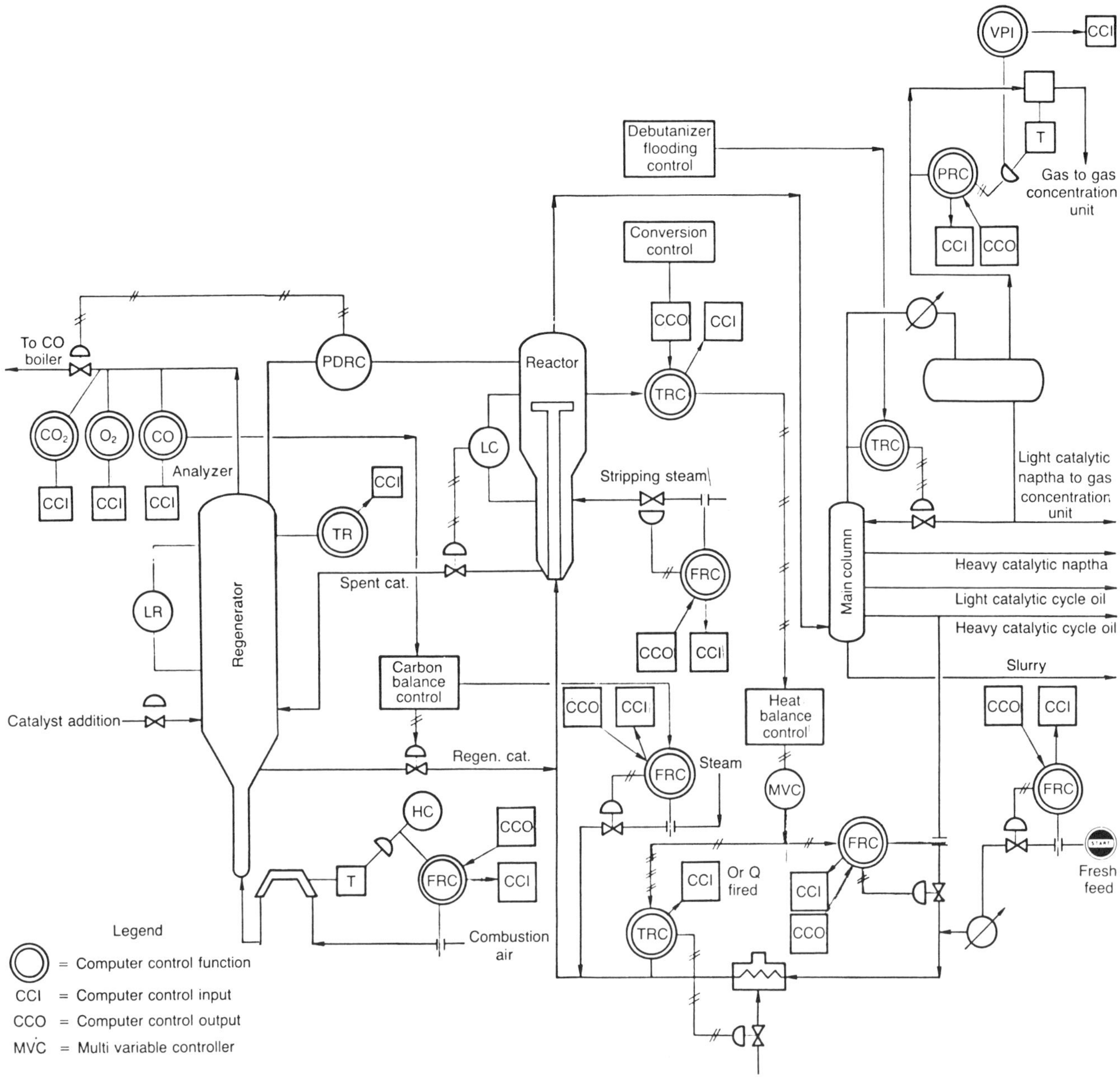

Figure 3. Comprehensive computer control system.

limiting factor changes. The system automatically recognizes the new limit and adjusts feed accordingly.

Several such strategies could be developed to cover situations such as feed rate fixed with conversion to be maximized. The operators need simply decide which major operating mode they are in and then commission the appropriate system. These systems must be designed with many startup and shutdown checks to make operation easy and safe. They may also be designed to check the status of all potential limit variables, advise on the planned course of action and ask for confirmation to proceed.

It is in this area that the interaction with the offline unit monitoring and optimization system is of value. Often if one limit is reached but other constraints are loose, a more profitable operation can be reached by adjusting operations so that more than one limit is reached. For example, if the unit has been operated at maximum feed rate at high severity and reaches the limit on gas compressor capacity, a further improvement in profitability may be realized by reducing severity to go to a distillate-type operation, at which point the feed rate can be further increased. The unit may eventually reach a point where air rate and gas compressor both limit. This type of evaluation of alternative operating strategies needs to be carried out offline with the selection of the constraint strategy keyed in by the operator.

Once operating personnel understand the system, and have confidence by seeing stable and consistent control of the unit, they will allow safety margins to be reduced. The objectives should be to have the advanced control systems commissioned continuously. The design of a comprehensive computer control system is shown in Fig. 3. In this system all of the systems described could be operating including multivariable control, limit control, feed forward and some simple control loops.

WHAT DOES A PROCESS COMPUTER COST?

Costs of computer systems vary enormously depending on the scope of the project and factors such as the type and age of the instrumentation, whether it is computer compatible, whether there is space in the control room, whether a major instrumentation revamp is contemplated, etc. The cost for an installation can be as low as 300 to 400 k$, or perhaps more than ten times this figure. When considering a project with a large scope versus a smaller project creaming off the most valuable computer applications, the 80/20 rule often applies. That is, you can achieve 80% of the benefits with only 20% of the cost. Computer projects can have a very fast payout; much less than two years when applied to highly profitable processes like the FCCU.

HOW TO DEVELOP A SUCCESSFUL COMPUTER CONTROL PROJECT

Like any other, a computer project needs careful planning and preferably the involvement of personnel who have done it before. The scope and objectives of the project need to be thought out and operations personnel need to be involved in the project from the start. A proven computer hardware and software system should be purchased that can be operated in a production environment. The equipment should be procured via a tight contract on supply, installation and commissioning of the system with a retention clause.

The most important factor in the success of the project is the necessary early involvement of personnel skilled in four major disciplines; namely, process, computing, instrumentation and control applications.

CONCLUSIONS

The FCCU process achieves maximum profitability with a combination of advanced control using a process computer plus a monitoring and optimization system run offline to generate operating targets. With computer systems relatively cheap and the control technology available process computers make attractive low-risk investments.

37

Control and Optimization of Olefin-Cracking Heaters*

M. L. Sourander, M. Kolari, J. C. Cugini, J. B. Poje, and D. C. White

Neste Oy's ethylene unit and the overall design of its computer control are described in Chapter 39 (M. Nasi et al.).[1] The unit, which has a nameplate capacity of 200,000 tpa ethylene, contains nine heaters, eight of which are modified Lummus SRT 1 heaters. The ninth is a Lummus SRT 2 heater. A major function of the computer control system is to control the operation of the ethylene cracking heaters. This chapter addresses the details of these controls.

DESIGN CONSIDERATIONS

Feedstocks

The feedstocks to the ethylene unit, which come from the nearby Neste refinery, include heavy feed from the hydrocracker, naphtha, n-pentane, n-hexane, LPG (propane/butane) and an off-gas stream. Ethane and propane are recycled, and ethane from the benzene plant is also processed. So far naphtha has been the unit's main feedstock, but it is increasingly being replaced by more profitable feeds. The recycle ethane from C_2-fractionator bottoms and the ethane stream from the benzene unit are fed preferentially to the ethane heaters with the excess going to gas heaters along with LPG, recycle propane, and gas stream.

Heater Operation

All heater feedstocks are preheated before being fed into the heater's four feed passes (coils). In the heater's convection section the feed is preheated, after which dilution steam is added to each feed pass and the mixture is fed to pyrolysis coils located in the heater's radiant section. At the coil outlet, the temperature is 800 to 850°C depending on the type of feedstock and heater cycle age. To stop the secondary reactions before equilibrium is reached, the heater effluent is cooled rapidly to 350 to 500°C in two transfer line heat exchangers (TLE's).

Basic Control System

Each of the four coils has a feed flow controller and a dilution steam flow controller of its own (Fig. 1). Burners are divided among four zones, one for each coil with one fuel gas pressure ratio controller each.

The outlet temperature controller of coil C (target coil) adjusts all four fuel gas pressures simultaneously by changing the setpoints of the pressure controllers. The heat transferred to the individual coils can be balanced by adjusting the settings of the fuel gas pressure ratio controllers.

All computer controls are setpoint supervisory controls, which means that the computer can only alter the setpoints of analog controllers. When the computer is off-line, these controllers work as normal analog controllers with operator setpoint entry.

The heaters have four chromatographs (GC's) with 18 streams for effluent analysis and one chromatograph with two streams for gas feed analysis. The GC's have their own dedicated microcomputers for basic functions. The GC system has a minicomputer that provides independent man/machine communications, supervises the functions of the GC's, and sends analysis data to the main process computers via a serial link.

* Nokia Electronics of Helsinki, Finland, provided the computer system including the basic control and operator interface software. Special acknowledgement is due to Matti Nasi, the Neste project manager, to Aimo Sorsa, the Nokia project manager, and Carl Wetzel, the Setpoint project manager, for their diligent efforts in support of the project. Other major contributors hae been Matti Alikoski, Pentti Paurola, and Jari Vettenranta of Neste; Pentii Tynnila and Esa Nieminen of Nokia; Chuck Graham and Tim Billings of Process Computers; and Pierre Latour, Lynn Richard, and Mark Darby of Setpoint.

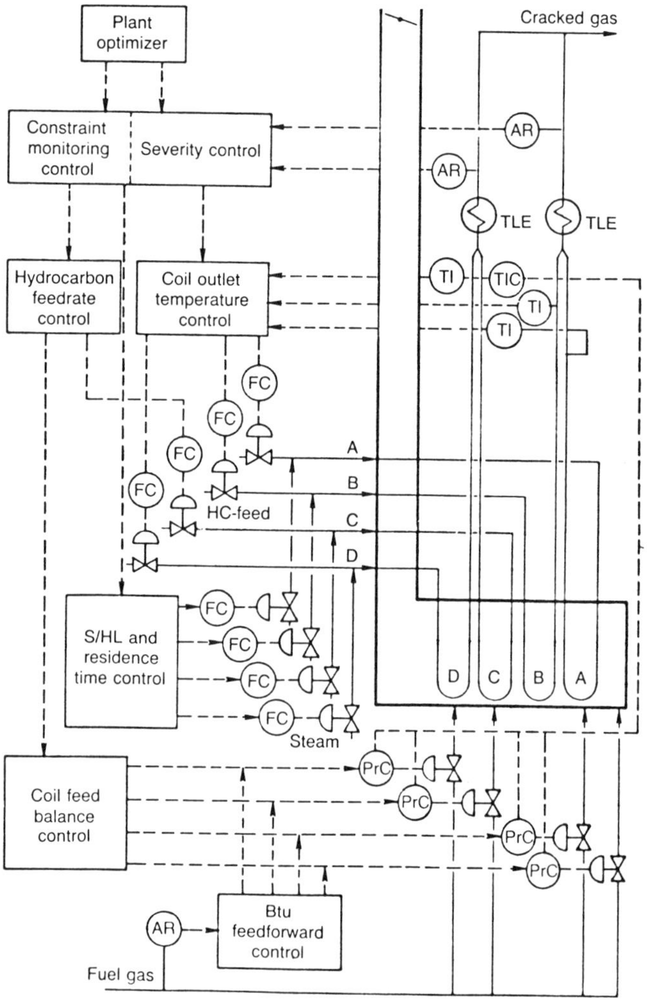

Figure 1. Furnace control logic.

Production Strategy

Neste Oy's ethylene unit can use seven different feedstocks (liquids and gases) and three different recycles as feed to nine heaters at varying rates to meet the production demand for its seven products. The feedstocks are selected according to their availability and relative profitability. No simple formulation of production strategy except for satisfying the market need, can be given for this highly interacting production plant. Therefore, maximization of gross margin, subject to market demand, feed availability and plant constraints, was selected as the overall production objective. It was decided to implement the production and control strategies supporting this objective by a hierarchical closed-loop optimization and control structure of computer control. This implies that the highest control level module in the hierarchy is a closed-loop plant optimizer, which at scheduled intervals converts the economic data together with plant variables to new controller setpoints—automatically without operator assistance—to reach the final production and control objective. The closed-loop optimizer targets are implemented via the pyrolysis heater controls.

PYROLYSIS CONTROL AND OBJECTIVES

Effect of Process Variables on the Pyrolysis Reaction

In ethylene cracking heaters, as in any reacting system, it is useful to distinguish between two concepts—severity and selectivity. Severity can generally be thought of as the overall conversion of the feedstock. Selectivity refers to the fraction of the feedstock converted that becomes valuable products rather than less valuable byproducts. For ethane and propane cracking, severity can be directly measured on-line as conversion of the feedstock. However, for butane and heavier feedstocks, direct on-line measurement of severity is impossible. It has been found that the methane/propylene ratio is a reliable indication of severity for a variety of heavy feedstocks over a wide range of normal operating conditions. For ethane and propane feedstocks, the selectivity can again be measured directly as the fraction converted to ethylene. For heavier feedstocks, selectivity can be related to the methane/ethylene ratio.

Since the design and geometry of an existing olefins cracking heater are fixed, the heater effluent yields of different components are dependent on the following pyrolysis parameters:

- Hydrocarbon feedstock composition
- Cracking severity or conversion, which is a function of coil outlet temperature (COT) and the equivalent average residence time (RT). A given severity can be attained by an infinite number of COT-RT combinations.
- Cracking selectivity, which is a function of hydrocarbon partial pressure and the equivalent average residence time.

The manipulated variables that affect the severity in the heater besides the COT are coil feed rate and steam rate; other variables affecting the severity through the residence time are heater outlet pressure, coil inlet temperature and reaction mix molecular weight. The cracking severity is most strongly dependent on coil outlet temperature. At a specified severity and feedstock composition, the effluent composition is also dependent on the cracking selectivity. High selectivity is favored by short residence times and low hydrocarbon partial pressures. For a heater whose severity is fixed, selectivity can be practically affected only by manipulating the steam-to-hydrocarbon ratio, which affects the hydrocarbon partial pressure and residence time, and by hydrocarbon feed adjustments affecting the pressure drop, residence time and heater outlet pressure.

Since changes in the COT, the feed rate and steam rate affect the severity and selectivity simultaneously, there is a strong interaction between the variables, which makes the yield and severity modeling and control difficult.

Objectives of Pyrolysis Heater Control

The overall objective of the heater control is to stabilize the heater effluent composition from a given feed to yield the maximum gross plant margin without violating the heater's physical constraints. In developing the control strategy, the interactions must be considered and decoupled to reach the overall objectives. Stabilizing the yields means stabilizing the severity and selectivity at a constant feed composition. Figure 2 shows a typical plot of yields versus severity.

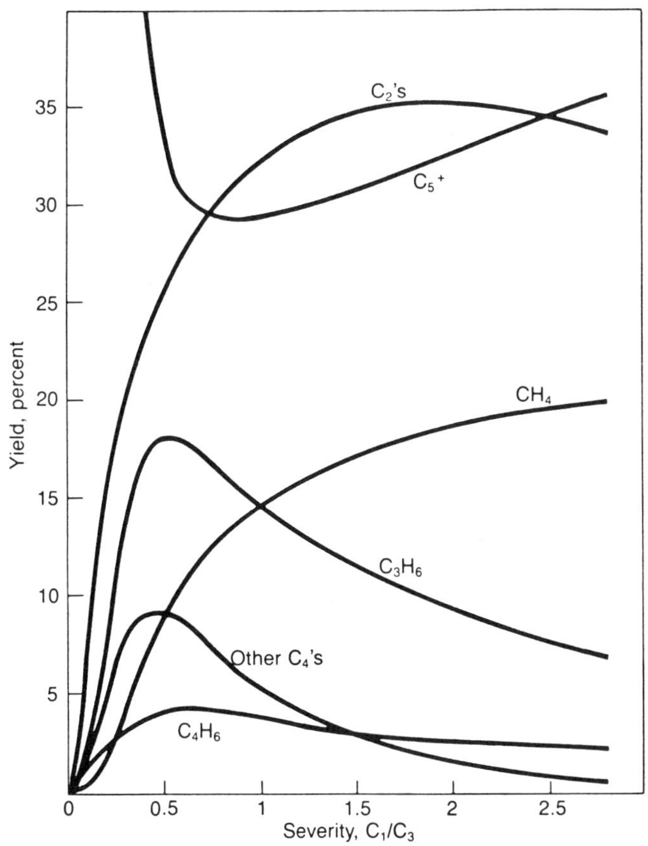

Figure 2. Yield versus severity.

Because all heaters in Neste's ethylene unit have four coils, this should imply that each coil severity and selectivity should be controlled to individual targets. There are, however, two sample lines per heater taken after each TLE, which means that analysis feedback data from heater effluent are available only for two coil pairs. Therefore, each heater is considered as two "subheaters" on the process side. The conditions in each subheater coil should be equalized because the true severity and selectivity in each coil cannot be measured.

HEATER COMPUTER CONTROL SYSTEM DESCRIPTION

Heater control consists of regulatory and constraint control loops as shown in Fig. 1. The functions of the regulatory control loops are to equalize COT's, mass feed rates and steam rates during ordinary process disturbances. The most significant external disturbances are those induced by fluctuations in feed or fuel composition. Because of the interaction between heater firebox zones and coil outlet temperatures, all COT's, feed and steam rates cannot be simultaneously controlled to long-term targets during transients. Therefore, control priorities have to be established according to process performance criteria to minimize the effect of disturbances on the final control goal.

The operation of the heaters is constrained by numerous physical conditions, which cannot be violated without risk of damaging the heater. This necessitates close attention from the operator when using conventional control systems. However, pushing heater performance near its production and efficiency limits brings an increase in profit. Therefore, a control system implementing the feed rate and severity targets determined by the plant optimizer must include a constraint control logic, which prevents the heater from violating constraints.

A comprehensive error-checking system is provided within the computer system which minimizes the time any control module is off control. The system is divided into modules so that the failure of critical variables will only partially degrade the control but not cause the control functions to be totally taken off-line. Although this type of design and error checking expands the size of the programs, it is necessary to maintain high control availability. Each control module is designed to track the process when off control, which allows the operator to switch to computer control with little effort.

The operator can make large changes to the heater operating conditions without upsetting the furnace control because control moves have been selected that minimize the effect on interacting controls yet provide reasonable response.

Severity Control

The quality of severity control is dependent primarily upon three basic functions:

- Coil regulatory controls, especially outlet temperature control
- Feedforward severity control model characteristics
- Feed and effluent analysis and analysis error checking. These functions are discussed in detail below.

Regulatory Controls. The sensitivity of severity to changes in the outlet temperature is so pronounced that without accurate and responsive outlet temperature control it is not possible to achieve good severity control regardless of the quality and quantity of analytical equipment. Heater feed rates, dilution steam ratio, coil outlet pressure, feed composition and other variables generally exhibit limited short-term variations. Therefore, cracking severity can be controlled by manipulating only COT's. Balancing the outlet temperature of individual coils (called coil balancing) is extremely important for severity control in larger multicoil furnaces.

If a deviation between the coil C (target) outlet temperature setpoint and measurement is sensed, all analog pressure ratio controllers for different firebox zones simultaneously adjust the burner pressures. This gives reasonable firing control of all coils in addition to accurate control of coil C outlet temperature during varying heater loading and firing situations. COT's for coils A, B, and D are computer-controlled to targets by manipulating the corresponding feed rates, which causes the coil feed rates to differ from the desired feed rates. These deviations are of lesser importance for severity control and are brought back to zero by adjusting simultaneously all pressure ratio controller setpoints. This multivariable control problem is difficult for the operators to handle manually because of the substantial interaction between the coils. The computer calculates the appropriate control action required to bring all coils to their feedrate target and simultaneously keep the pressure ratio controllers within their operational range. Total hydrocarbon thruput of a heater is maintained by adjusting the coil feed rate. Fuel gas heat of combustion variations are compensated by feedforward adjustments of the pressure ratio control.

Feedforward Model Control. The severity control algorithm is a combination of independent feedforward control and feedback control for each subheater. The feedforward control adjusts the coil temperature for changes in severity target, feed rate, feed composition and other variables around the furnace. Severity is maintained at a predetermined target for each "subheater" by manipulating COT targets independently. Different residence times in individual subheaters call for different coil outlet temperatures to maintain the same severity.

Gas feeds contain ethane, propane, propylene, and mixed C_4's, with the feed composition ranging from nearly all ethane to nearly all butane and subject to rapid changes. The severity control for the gas heaters automatically switches from ethane to propane conversion or to the butane severity indicator dependent upon the dominant material in the feed.

The severity control is designed so that a minimum of analytical feedback information is required. This calls for accurate modeling of the pyrolysis reaction severity as a function of feed rate, dilution steam, feedstock composition and other variables. The model is constantly recalibrated (updated) by the measured severity, i.e., methane-to-propylene for liquid feedstocks or ethane/propane conversion for E/P feedstocks. Because of the interaction with other variables, the severity model uses primary measured data directly rather than derived variables, such as selectivity and residence time. The updating cycle is independent of and asynchronous to control cycles with appropriate compensation for analyzer and transport lags. The feedforward severity control model is capable of maintaining severity control through periods when the effluent analyzer is not working properly.

Analytical Considerations. Analysis of heater effluent is a difficult analytical problem. Despite the quality of equipment, sample system design and maintenance, occasional errors in the effluent analysis do occur. Therefore, the error detecting capability of the computer system is extremely important for good severity control. If a bad analysis is not detected and the control model is updated, the severity control is in error regardless of the characteristics of the control model or the quality of outlet temperature control. Occasionally ignoring a good analysis through rigorous error checking incurs little penalty because of the feedforward control models.

Selectivity Control

With relatively small deviations in the feed to heater coils, maintaining a constant steam-to-hydrocarbon (S/HC) weight ratio affects both the hydrocarbon partial pressure and the residence time in offsetting ways to stabilize selectivity. However, true selectivity control, defined as a constant methane/ethylene ratio in the heater effluent for liquid feedstocks, cannot be maintained throughout the heater cycle by manipulating only the steam rate. There exists a relatively small practical and economical window for adjustments of the steam rate.

The coil outlet pressure increases during the run because of coking, and can be kept at a desired level only by reducing the heater feed rate, which again means a loss of production and is not normally acceptable. Therefore, it is anticipated that some selectivity is gradually lost during the heater run and that the selectivity is approximately maintained at a constant value during several severity control model updates and plant optimizer cycles (in fact for more than a day) by keeping the steam-to-hydrocarbon ratio constant, when operating at the same heater feed target. The disturbances from selectivity to severity are thus decoupled.

Constraint Monitoring Control

Constraint monitoring control serves two functions, first to provide an interactive control algorithm for feed rate, steam-to-feed ratio and severity control, and second to allow the thruput and severities of the furnaces to be maximized against constraints when desirable. Changes in feed rate, steam rate and outlet temperature affect the severity control. An interactive control algorithm is used to simultaneously implement changes in outlet temperature to compensate for the feed rate and steam to hydrocarbon ratio effects on severity.

About 10 constraints are monitored to prevent violation by the heater regardless of the feed rate, steam-to-feed ratio, or severity targets supplied by the operator or optimizer. The constraints describe the limits of pyrolysis reaction conditions, such as maximum severity, maximum temperature of tube metal, minimum residence time, maximum coil outlet temperature, maximum TLE outlet temperature and maximum feed rate. They also describe the operating limits of the heater, such as minimum excess air, maximum stack gas temperature, minimum enthalpy ratio, maximum valve positions and maximum burner pressure, where sonic velocity is approached. Each constraint uses projection techniques based on constraining variable models to predict the approach to a constraint. When a constraint is about to be violated, the feed rate, steam-to-hydrocarbon ratio and COT targets are relaxed in a priority order until the constraint is satisfied. Parameters are provided that allow the user to tighten or relax constraints. Also, the control displays are color-coded to allow easy detection of constraints or variables that are limiting operation of the furnace.

Heater Startup/Shutdown Control

A major cause of shortened furnace run length in ethylene plants is poor startup procedures. The startup of a furnace is very sensitive since the furnace operates at low volume thruput, high residence time and high temperature. During these periods the cracking severity is much more sensitive to temperature changes than in normal operation and temperature control is more difficult. Both the outlet temperature and the feed rate are controlled from the computer system during startup and shutdown.

The outlet temperature is ramped at a selected rate to the final startup target. After introduction of hydrocarbon feed, the feed rate is simultaneously ramped to its target, either with severity control on or off. The feed rate is also ramped at an operator selected rate to the specified feed rate target. The startup can be placed in hold at any time and restarted on demand by the operator.

Furnace shutdown is the reverse of furnace startup with the same features.

Heater Area Material and Energy Balance

A component material balance is calculated for each heater based on feed analysis and pyrolysis variables. The gas streams are analyzed by an on-line gas chromatograph.

The heater effluent composition is determined by combining data from the heater effluent analyzer and yield models for the effluent components not analyzed.

Using the heater material balance, fuel composition and rates and steam production data, a furnace energy balance is calculated on-line. This information has proved valuable in comparing relative furnace efficiency and evaluating overall furnace performance.

OPTIMIZATION

Implementation of maximization of gross margin in a complex plant, such as an ethylene unit, requires the use of a plant optimizer. The optimization for the unit downstream of the heaters is decoupled from the problem of total plant optimization by introducing incremental component recoveries, which are functions of separation area loading and performances.[1] The optimizer model estimates the extent to which the increased ex heater component rates will be recovered in the cold section and how the recycle streams are affected. It also takes into account the corresponding changes in utility costs in the cold section and the heater area. The plant optimizer balances the benefits of changes in component rates in the heater effluent due to changes in heater severity or feed rate against changes in energy needs, coking rate, recovery in cold section, product needs and plant constraints.

Optimization Structure

The manipulated (decision) variables for the plant optimization are the heater feed rates and severities target for each of the different plant fresh feeds, i.e., naphtha, n-hexane, n-pentane, heavy feed, LPG, or gas mix. As a further simplification, there exists a definite profitability ranking for each of the feedstocks and those that are most profitable should be, and are, run to the limits of their availability or to heater processing limits. Hence, one feed type is designated as a swing feed. It is generally the least profitable feed that is available in excess beyond plant demand. In the optimization algorithm, the rate for the swing feed is determined with the other feed rates set at their current rates. The cracked gas compressor suction pressure is not set by the optimizer because it always operates near minimum pressure.

Constraints considered for the optimization include current projected heater rate and severity maximum and minimum limits for each feed type as well as ethane recycle limits. Separation area constraints are included though it is normally expected that separation decisions will be based on the economics of incremental recovery rather than constraints. Move limits are included on each decision variable.

The objective function used for the optimization is a steady-state one based on the current operation of the plant and does not include any explicit function of time. Two objective functions are available, and either can be selected by the operator. The first leads to maximization of ethylene production against plant constraints. The second objective is an economic function that includes feedstock costs, product values, heater fuel usage and major steam usage.

Plant Model

To predict the effect of changes in the decision variables on the objective function and the approach to constraints, a model of the plant is used. Two key considerations affect the use of this model. The first is the nonlinearity of plant response. A unit change in heater severity will give different component yield changes at different severities. For example, a change in ethane heater severity will give a change in the ethane recycle and the mixed gas heater feed rate, which depends on both the ethane heater and the gas heater severity. The second consideration is the normal inaccuracy and steady-state bias inherent in any on-line model of plant operation. Controls based on models that attempt to predict absolute or positional outputs can generate occasional output moves in the wrong direction during transients in plant operation. For these reasons, the entire

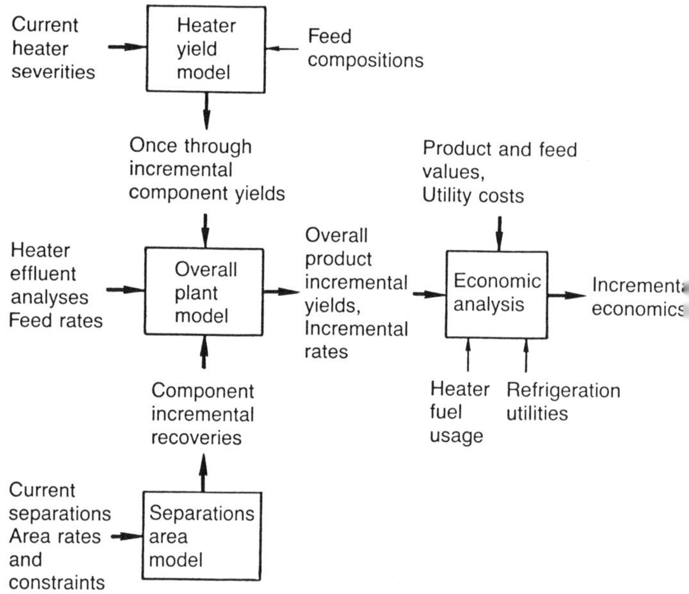

Figure 3. Plant model.

structure and use of the model is based on the prediction of the effect of incremental moves. Figure 3 illustrates the plant model functions.

Yield models are included for each feed type as a function of severity, feed composition and other appropriate unit process variables. Once-through heater effluent yields are predicted for the current measured severities. These yields are then used to calculate a complete component balance for the plant, including recycle streams. A new set of yields is calculated for each feed type at an incremental severity. In addition, expected marginal ex plant recoveries for key components are calculated for the cold section of the plant. These recoveries are based on the current operation of the separation columns and reflect individual column constraints if encountered. The heater effluent yields and incremental recoveries are used to directly calculate expected changes in ex plant product recoveries. Recycle flows are also calculated directly, without iteration.

The product and feed flows are used to calculate the economic objective function. Utility usage is predicted on the basis of expected change in key plant flows.

Optimization Algorithm

Among many possible algorithms, linear programming has been found to have advantages for ethylene plant optimization. First, linear programming handles constraints easily, and it can be expected that the optimum solution for many of the decision variables will lie at the intersection of constraints. Second, linear programming is computationally efficient. Although the optimization problem is nonlinear the plant response can be linearized through the use of the incrementals calculated by the plant model, in the vicinity of the current operating point. A sketch of the procedure is shown in Fig. 4. Recursive techniques are mandatory for this highly nonlinear problem.

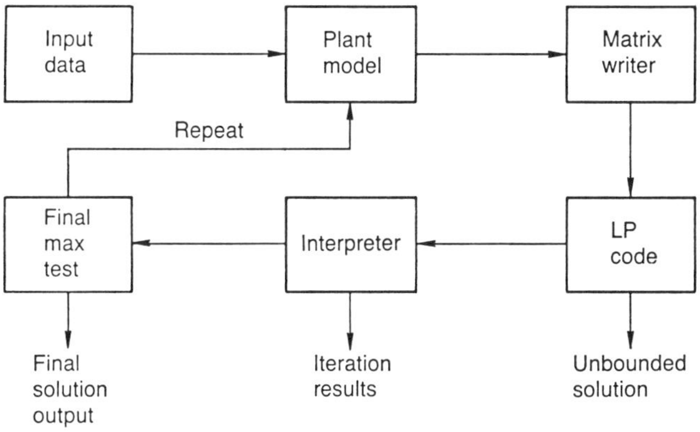

Figure 4. Recursive linear programming.

The optimizer calculates the most profitable set of moves for the decision variables as delta moves from the current targets. These new targets are implemented via the feed-allocation module and heater-constraint controls. Security checks are provided to prevent steady-state optimization moves from being implemented at a dynamically inopportune time.

Generally, the optimizer is executed every four hours. If desired, the operator can initiate an extra optimization at any time after an interval that provides time for the plant to stabilize.

Further discussion of optimization procedures can be found in reference 2.

RESULTS

A More Valuable Product Package

Because the production of Neste's ethylene unit is presently limited by market conditions, the thruput maximization feature in the computer control has not been used, except for a short trial period. The objective generally used has been the maximization of an economic function at fixed ethylene demand. This has led the optimizer to increase the average severity of operation. The net result from heater and separation area control shows that the computer control gives 2 percent more ethylene from the same amount of feed, at the expense of less valuable byproducts (Table 1). This has made it possible to produce the desired amount of ethylene at a lower naphtha feed rate than before. The increased heater fuel gas consumption due to higher severity is more than balanced by the significant reduction in the energy demand of the separation area. As a net effect, a slight decrease in the total energy consumption is seen.

Table 1
Summary of Results

	Before	After
1. Ex plant yields, per cent of feed		
—$CH_4 + H_2$	19	21
—C_2H_4	32	34
—C_3's	19	17
—C_4's	8	7
—C_5^+	21	20
2. Plant total energy consumption	Base	< Base
3. Heater control standard deviation (Based on targets)		
—Coil outlet temperature, °C	3 . . . 4	< 1
—Coil outlet temperature band (hottest to coldest) average, °C	5 . . . 8	1
—Coil mass flow rate, per cent		
—Liquid feed	0.8	0.5
—Gas feed	>5	2
—Coil S/HC, per cent		
—Liquid feed	1	0.5
—Gas feed	>4	1.5
—Severity, per cent	approx, 4	< 1.5
—Severity, using severity feedforward model control only, 24-h run, percent	N.A.	2
—Effluent methane-to-ethylene ratio, 72-h run of midcycle, percent of average	1.5	< 1
4. Heater operation changes		
—Maximum allowed coil outlet temperature, °C	Base	+ 3 . . . 5
—Maximum allowed feed rate, percent	Base	+ 4
—Severity target, percent	Base	+ 2 . . . 10
—S/HC target, percent	Base	Base
—Maximum heater duty, percent	Base	+ 10
—Heater thermal efficiency, percent	Base	+ 3
—Heater run length, days	25	33 +

Accurate Coil Outlet Temperatures

Exact and responsive control of outlet temperature in all coils is necessary for good control of cracking severity. Feed rates disturb the severity less and are easier to compensate. The COT control's standard deviation from the target is less than 1°C in all heaters and coils and in most cases about 0.5°C per coil. Before computer control, 3 to 4°C was considered good. The average temperature band between all coils in a heater (hottest to coldest coil) is now less than 1°C in all heaters as opposed to a band of 5 to 8°C, which earlier was regarded as good performance. Besides reducing the fuel and stabilizing the performance of downstream equipment, the COT's are now very responsive to target changes because the interaction to the firebox is much less pronounced than before computer control. To date, precise temperature control now allows a 3 to 5°C higher maximum coil outlet temperature than before, because high temperature excursions are much more rare. As more experience is gained, it is expected that these maximums will be increased further.

Balancing of Coil Mass Flow Rates

Accurate control of mass flow rates requires density compensation, because the flow elements give signals of pressure differential only. The lack of compensation prior to computer control made it difficult to set the coil mass thruput accurately, particularly for gas heaters where feed densities vary substantially because of widely changing compositions. In the past this has occasionally led to heater overloading and mechanical problems.

Although the coil feed rates have been made manipulated variables in computer control to hold the COT's, the interactive balancing control of coil feed has decreased the deviation in the control of coil mass flow rates. For gas heaters, the computer has made it possible to control the heater thruput accurately, which prior to computer control varied according to the heater's feed composition. The balancing of coil mass flow rates equalizes the coil feed rates and keeps the heater's total feed rate at target. In liquid feedstocks the coil balancing is performing with a 0.5 percent control deviation, because of a good response between coil feed rates and firebox fuel pressure ratio. Gas feeds are much less responsive to changes in pressure ratios, and consequently the control quality is not nearly so good (about 2 percent) as for liquid feedstocks, but there is a significant improvement from the previous situation. Occasionally, even the maximum bias between coils does not equalize all coil feeds to a desired feed rate.

Control of S/HC Ratio and Selectivity

The control of steam-to-hydrocarbon mass ratio requires an accurate compensation of both feed flow and steam flow measurements. Computer control has improved the control by reducing the standard deviation to less than half of the previous standard deviation.

Coking causes the pressure at coil outlet to increase gradually, which also increases the hydrocarbon partial pressure. This means a gradual loss of selectivity, which can only be compensated for slightly during the heater cycle by increasing the S/HC ratio. The loss of selectivity can be seen as an increased methane-to-ethylene ratio in liquid feed effluents.

Increasing the S/HC ratio causes extra cost for dilution steam and heater fuel, which is only partially recovered as useful energy. This means that large deviations from the optimum S/HC ratio are uneconomical. Therefore, the loss of selectivity is allowed and only stabilized to the extent required for decoupling by stabilizing the S/HC ratio. Experience on liquid feedstocks suggests that the selectivity is practically constant for a day or two in mid-cycle, but is sensitive to rapid increases in coil outlet pressure at end of the heater cycle. This suggests that the optimum run length for a heater is, when unused heater capacity is available, a function of selectivity loss rather than the maximum temperature tolerance of the equipment.

Sampling and Analyzing Heater Effluent

Successful control of pyrolysis requires analyses from the reaction product mixture in the heater effluent. Continuous sampling of the gaseous effluent stream throughout the heater cycle traditionally has been difficult, especially for liquid feeds because of plugging, liquid carry-over and contamination. Therefore Neste decided to develop a new sampling system that would give representative as well as repeatable gas samples to be analyzed by gas chromatographs. The key issues in this new sampling system are accurate separation of reaction products to avoid liquid problems, careful selection of equipment geometry, prevention of plugging and coke buildup through proper design, sample conditioning, and sample pressure and flow control in the long sample transport lines up to the point of injection to the chromatograph.

The results are encouraging: The sampling systems have operated satisfactorily since installation without cleaning for longer than one heater cycle, and in many cases over two cycles without noticeable change in sample quality. The components methane through propane are analyzed by the GC's.

Stable Severity even without Analysis

The control deviation of severity is reduced to less than half of the previous one as a result of the computer control. It is now less than 1.5 percent of target. A typical plot of severity control is shown in Fig. 5. Earlier it was attempted to keep the COT's at the desired value regardless of changes in feed composition or, for instance, heater outlet pressure. This obviously caused overcracking, especially near the end of cycle. Because different coil outlet pressures often occur between two subheaters, different temperatures are needed for reaching the same severity. This is also true between individual heaters. The COT thermocouples are occasionally defective, and without severity control this would lead to

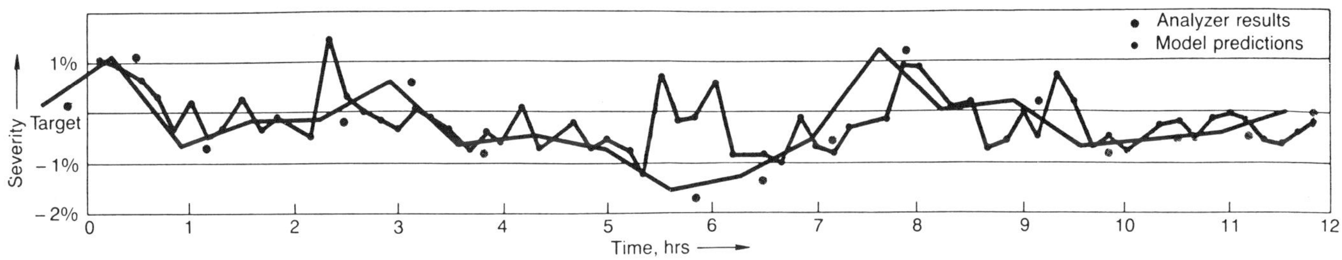

Figure 5. Severity control performance.

severe over- or undercracking. In fact, the severity control has led to identifying defective thermocouples.

Now the severity control is switched on immediately after the introduction of feed during startup control, and is used throughout the heater cycle until shutdown.

In spite of the severity now being 2 to 10 percent higher, the coking rate is slower and the overcracking situations previously occurring, particularly during startup, are now avoided. The run lengths of the heaters have thus increased by more than 30 percent, bringing considerable savings in decoking costs.

Although the severity control of the heaters uses the effluent analysis to update the severity model, actual control moves are made by the severity feedforward control model, as can be seen from the sample of severity control history in Fig. 5. The model predicts accurately the severity, and makes proper control moves even without analysis feedback. Once the severity model is updated with a few representative feedback analyses, it can operate at least two days with less than 4 percent severity drift during normal steady operating conditions and steady feed composition. The control deviation during severity model control (without effluent analysis update) is 2 percent of target during a 24-hour run. This feature is particularly useful, since the GCs need regular calibrations and preventive maintenance, so the severity control is not lost during these periods.

Control Against Constraints

The heater constraint control is especially useful for the ethylene unit which is not market- or feedstock-limited. The maximization of thruput against heater constraints while meeting the desired heater run length then often becomes the optimum policy. Neste has not yet benefited from this feature because of the current market-limited conditions. On the other hand, heater protection while reaching a constraining efficiency is a constraint control function constantly in use. Because the heaters are natural-draft operated, the control of excess air becomes difficult in changing load situations. The constraint control quite frequently protects the heaters from incomplete combustion by relaxing the heater load, until proper manual adjustments on the burners and stack dampers are made on the basis of operator messages. Thermal efficiency of the heaters has increased about 3 percent as a result of more accurate control of excess air. The gas heaters are often constrained by maximum allowed COT when operating at high severities, despite the fact that the maximum COT has been increased by 3 to 5°C since computer control was introduced.

Towards the end of heater cycle, many variables, such as the temperature of the tube metal or of the TLE effluent, become constraining or approach the maximum, or the burner pressure increases to the point of sonic velocity. When these constraints are met, the heater's feed rate is relaxed accordingly, until the heater is shut down for decoking.

As a link and check between plant optimizer and heater manipulation, the constraint control is of vital importance because the optimum severities given by the optimizer occasionally are near the prevailing maximum performance of the heater. In these situations the constraint control temporarily relaxes the severity or the feed rate.

Optimizing Economy in Closed Loop

The plant optimizer has been in use whenever the computer has been in use. Its yield correlations have proven to be in good agreement with earlier experimental data. Predictions of the plant model have been thoroughly tested against actual plant operation and found to be in very close agreement. The control moves made by the optimizer have been easy to understand and accept. Generally a trend towards higher severities has been dominant, despite the increased heater duty demand. The feed rate of the swing feedstock varies according to the demand for the various products. The resulting component yield changes (Table 1) are partly due to operating at optimum severity. The optimizer has not required any updating efforts or maintenance personnel with extra skills; in fact it is constantly operated by the operators only.

Operating the Plant

The operating personnel accepted the computer system from the beginning of commissioning. The control system now solves problems that the operators were constantly confronting previously. Details of control availability, system training and operator acceptance have been discussed in the earlier article by M. Nasi et al.[1]

During the introduction of the computer control, a deeper understanding of the pyrolysis heaters and their control developed among the personnel, partly because of repeated training and partly as a result of their own observations on heater performance. Especially the balancing of coil feed was readily understood and accepted because its control had been an unsolved problem. This development has led to new, more efficient concepts in heater scheduling and feedstock management, including the planning of heater loading and cycle lengths.

Converting the quantitative benefits of computer control to economic terms, a project payback time of less than two years is calculated.

References

1. Nasi, M., Sourander, M., Tuomala, M., White, D. C., and Darby, M., "Experiences With Computer Control of an Ethylene Plant," *Hydrocarbon Processing,* Vol. 62, No. 6, 1983.
2. Latour, P. R., "On-Line Computer Optimization 1: What it is and Where to Do It," *Hydrocarbon Processing,* Vol. 58, No. 6, June 1979, p. 73, "On-Line Computer Optimization 2: Benefits and Implementation," *Hydrocarbon Processing,* Vol. 58, No. 7, July 1979, p. 219.

38
Feedforward Control for Claus Units

G. Taggart

For operations with less than maximum efficiency, decreasing efficiency with turndown, and no continuous index of actual performance are tolerable, feedforward control for Claus units is relatively simple and economical. However, as complexity increases, the need for tail gas analysis increases. In fact, at some point feedback control is preferable. There is ample experience with tail gas analyzers, both photometric and chromatographic, to conclude that the measurements can be made reliably and without excessive maintenance.

Feedforward control is the scheme by which "the principal factors affecting the process are measured . . . and used in computing the correct output . . ."[1] It is an extremely attractive technique, particularly for Claus sulfur recovery units where results are in excellent agreement with theoretical predictions. Feedforward control of Claus units is not a new idea. It was first done successfully in the mid-1960s, when reduced operator attention and increased yield were the justification. Literature[2] describes a recent effort and articulately states the case for feedforward control.

The fundamentals of the Claus reaction must be used as the base for a feed forward control. In its simplest terms, the chemistry of Claus sulfur recovery units involves partial oxidation of hydrogen sulfide to sulfur dioxide and the catalytically promoted reaction of hydrogen sulfide with sulfur dioxide to produce sulfur. The partial oxidation reaction may be written as:

$$3H_2S + 3/2O_2 \rightarrow 2H_2S + SO_2 + H_2O \quad (1)$$

and the catalytic reaction as:

$$2H_2S + SO_2 \rightleftharpoons 2H_22H_2O + 3S \quad (2)$$

The simple relationship of acid gas and air implicit in the partial oxidation reaction is reflected in the conventional air to acid gas flow control of Claus units (Fig. 1) and emphasizes the potential for feedforward using analytical measurements (Fig. 2).

In the application of feedforward techniques to Claus sulfur recovery units, it is necessary to consider some variables not ordinarily of concern in process control and it is necessary to define acceptable (non-optimum) limits of unit performance.

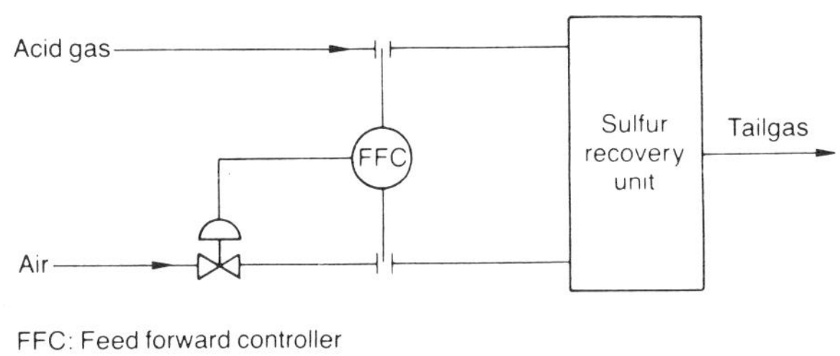

Figure 1. Conventional Claus unit acid gas to air flowrate controller.

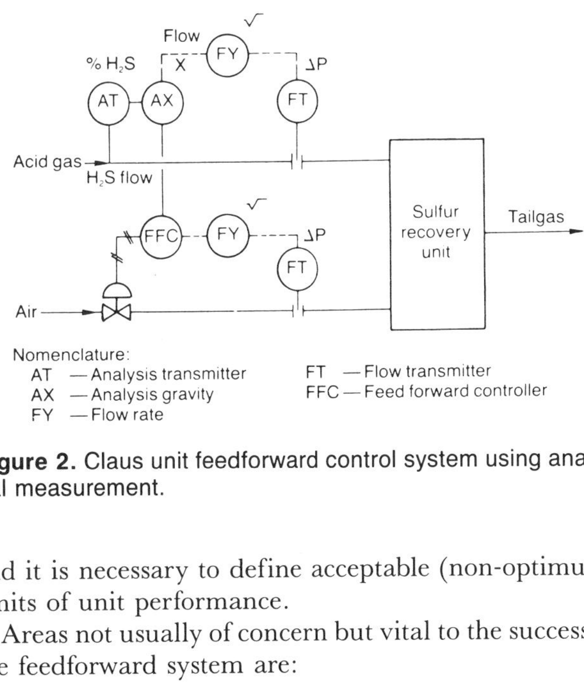

Figure 2. Claus unit feedforward control system using analytical measurement.

Areas not usually of concern but vital to the success of the feedforward system are:

- Measurement and signal processing accuracy
- Effects of minor feed components
- Effects of other influences on measurements
- Process variability
- System complexity

In order to assess the impact of the first four of these, Fig. 3 will be of value. The abscissa of Fig. 3, r/r_s, is the value of air-acid gas ratio normalized to stoichiometric air-acid gas ratio or percent deviation of r from the de-

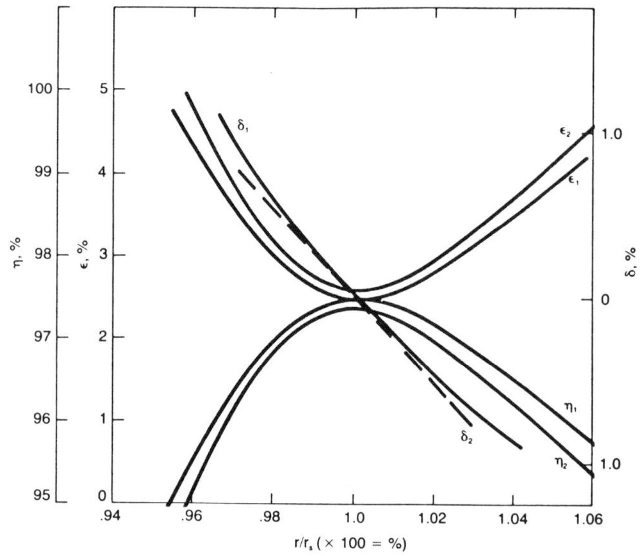

Figure 3. Effect of normalized air to acid gas flowrate on recovery efficiency, tail gas unit feed, and tail gas per cent difference.

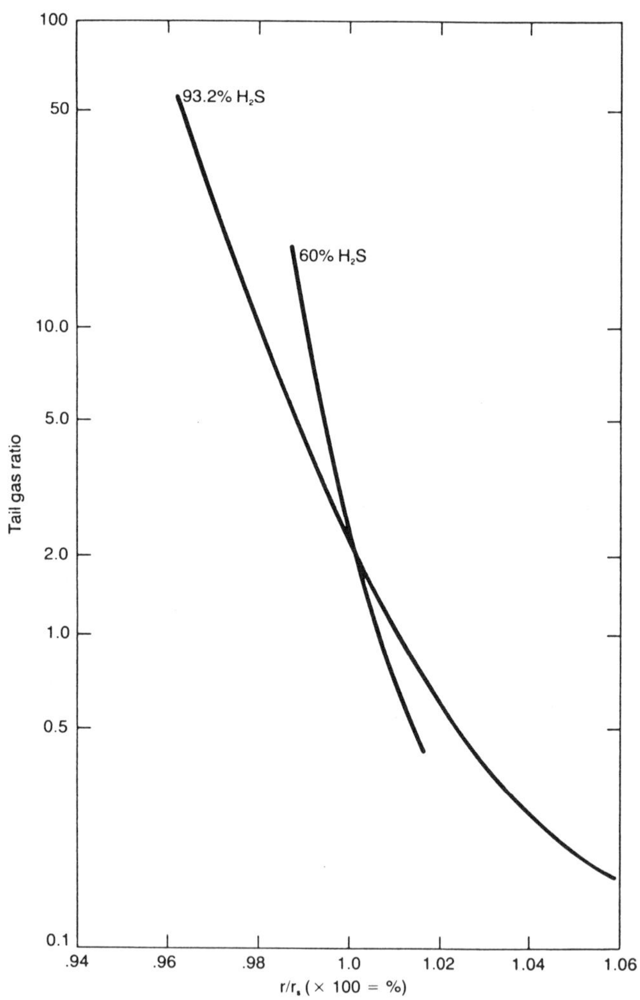

Figure 4. Relationship between tail gas ratio and normalized air-acid gas flow ratio.

sired value r_s. The ordinates are η, recovery efficiency of H_2S to S, and ϵ, percent sulfur in tail gas (load on the tail gas treating unit). Figure 3 also plots δ, tail gas mol percent difference[3] and Fig. 4 plots the tail gas ratio against r/r_s. These data are included to provide a more familiar context in which to judge performance. One set of curves (subscript 1) is for a 93.2 percent H_2S acid gas; the second set (subscript 2) is for a 60 percent H_2S gas. The data are from computer runs on idealized gases. The basis for judging impact is to consider expected or potential deviations from $r/r_s = 1$ and consider their effect on unit performance.

MEASUREMENT AND SIGNAL PROCESSING ACCURACY

A typical analytical measurement accuracy for a wide range instrument (0–100 percent) is ± 2 percent made up of some range of repeatability error plus errors resulting from fluctuations in the sample condition: pressure, temperature, water content, etc.[4] A system consisting of orifice plate, flow transmitter and square root extractor can be expected to have an accuracy of about ± 0.75 percent.[5] The multiplier further compounds the errors.

The potential range of deviation for an analysis/flow product then is roughly ± 2.75 percent and the ultimate error of a controlled signal based on that as a set point could easily be ± 3 percent. Statistically, it is likely to be smaller but probably not smaller than about ± 2.1 percent. For both 93.2 percent and 60 percent feeds a ± 2 percent variation of r/r_s corresponds to about 0.5 percent of unit efficiency, probably an acceptable penalty. For larger variations, the efficiency penalty is marginally acceptable at best.

EFFECTS OF MINOR FEED COMPONENTS

Claus unit feeds, particularly in the refinery or chemical industries, tend to be quite complex and may contain significant quantities of hydrocarbons and other flammables. "Significant" in this sense relates to the oxygen consuming potential and can be stated in terms of oxygen demand and H_2S equivalent. Table 1 shows values for some typical compounds and reflects the reaction of hydrocarbons to CO and water.

These numbers are not of themselves important but must be considered in light of potential variability Variability of either minor components or H_2S concentrations must be considered since the H_2S to combustible ratio is important. Certainly judgments must be made about the need to measure these components. A reasonable rule may be that any component present in sufficient concentration to account for 2 percent of the total oxygen demand should be measured.

Table 1
Oxygen Demand of Typical Minor Feed Components

Compound	Oxygen demand Mols O_2/Mol	H_2S equivalent Mols H_2S/Mol
H_2S	0.5	1.0
CH_4	1.5	3.0
C_3H_8	3.5	7.0
NH_3	0.75	1.5

Clearly any variability adds to other sources of error to produce operating inefficiency.

EFFECTS OF OTHER INFLUENCES

There are a surprising number of ways in which apparent flowrates can be different from actual flowrates; most can be identified and compensated or corrected. The most common and most readily identified source of error is the static pressure; if measurements are made at amine regenerator pressure and blower discharge pressure, errors can be significant (in terms of r/r_s). Pressure is easy, however, since measurements can be made on the thermal reactor side of control valves and, though variable, the ratio has a quite small deviation. Temperature, on the other hand, is generally not considered critical, although a 10°F change produces an error of about 1 percent; diurnal and seasonal changes in temperature must be considered for possible automatic correction if feedforward is to be successful.

In the case of a sour water stripper gas, temperature affects not only the basic flow equation but it also has a significant effect on water content, and thus on specific gravity. Again, a detailed consideration of these effects may lead to correction of measurements for these variables.

PROCESS VARIABILITY

It was suggested above that the usual oxidation products of hydrocarbon compounds are CO and water. In some process configurations, these products are CO_2 and water. The amount of hydrocarbon reacting to CO_2 is frequently variable depending on the nature of the process configuration. Conditions of this type can be accounted for only by sets of measurements which ordinarily are not made.

SYSTEM COMPLEXITY

With today's process instruments, the performance of any of the above suggested functions is fairly straightforward. Some units, however, will require that many of these functions be performed. It is certainly not uncommon to find feeds which contain several "significant" components. Clearly, such units can be handled only by systems which are complex and costly; which require a good deal of engineering support; and which isolate the plant operator almost completely from the process. There is some point of diminishing return, and judgments must be made by the engineer or owner to define that point.

Feedforward control is sensitive to errors in input data. There are many sources of error which should be considered in designing the feedforward control system. Table 2 is a brief semi-quantitative resumé of some sources of error. It is apparent that a total error of several percent is possible.

Table 2

Source of error	Approximate error, %	Notes
Analyzer accuracy	±2%	1, 2
Flow measurement and computation system accuracy	±.75%	2
1% change in feed HC concentration	⩾3%	3, 5
Errors in apparent flow due to		
10°F temperature change, constant composition	±1%	3
specific gravity change, acid gas from 120° ⇄ 110°F	±.3%	3, 4
specific gravity change, sour water gas from 180° ⇄ 190°F	±.7%	3,4
50% change in relative humidity at 85°F	±.4%	3

NOTES:
1. Accuracy is higher for minor components.
2. Accuracy stated at full scale.
3. Additive to errors in measurement accuracy.
4. Not independent of temperature change error.
5. May be several percent depending on hydrocarbon distribution.

References

1. Shinskey, F. G., *Process Control Systems,* Second Edition, McGraw-Hill Book Co., 1979.
2. Dougherty, J. F., and Morgan, D. J., "Claus Plants Equipped with New Control Scheme," *Oil and Gas Journal,* Vol. 78, May 26, 1980, pp. 68–70.
3. Taggart, G. W., "Optimize Claus Control," *Hydrocarbon Processing,* April 1980, Vol. 59, No. 4, pp. 133–137.
4. Bucher, R. E., Private communication.
5. Shinskey, F. G., "Accuracy and Rangeability of Head Flow Meters," *Application News,* Foxboro Publication No. 230.

39
Ethylene Plant Computer Control*

M. Nasi, M. Sourander, M. Tuomala, D. C. White, and M. L. Darby

Neste Oy's plant at Porvoo, Finland includes a large oil refinery, petrochemical works, polymer plants, plus polystyrene and industrial chemical units.

The ethylene unit, a part of the petrochemical works, was commissioned in late 1971. The unit's feedstocks come through pipelines from the oil refinery. The main product—ethylene—is fed to the polymer plants. Other byproducts go to other petrochemical units or back to the refinery.

Fig. 1 shows the simplified flow diagram and the most important process values of the ethylene unit.

The unit contains nine cracking heaters, and its nameplate capacity is 200,000 tpa ethylene. The feedstocks—naphtha, n-pentane, C_3/C_4 hydrocarbon cuts and certain recycle streams—are cracked in the heaters at a temperature of 800 to 850°C into ethylene and byproducts, which are separated in the downstream process units. Pyrolysis fuel oil and pyrolysis gasoline are separated from the furnace effluent in the hot section. The pressure of the remaining charge gas is raised in the compressor section from 0.5 to 36 kg/cm^2.

After drying, the charge gas is refrigerated to approximately −100°C and partially liquified in exchanger systems using propylene and ethylene as coolants. In the cold section of the unit, hydrogen, methane (fuel gas), ethylene, ethane, a C_3 cut, and a C_4 cut are separated. Before the separation of ethylene from ethane, acetylene is hydrogenated in the acetylene converters.

COMPUTER CONTROL PROJECT

The project's feasibility was studied by Neste engineers in collaboration with the engineering company which designed the unit. The aim was to find out how and where the computer would benefit the unit through improved control and optimization. During the study it was found through process experiments, extensive laboratory analyses, and computer simulations, that computer control should yield the greatest profits in the following areas:

- Stabilization and optimization of the cracking heaters.
- Increases in product recovery through better control of the main columns in the cold section.
- Decreases in internal recycle streams and energy cost.
- Decreases of ethylene losses in the acetylene converters.
- Increases in availability to operating personnel of more complete information on the production situation and the functioning of the process.

The project planning was similar to that of Neste's previous project on crude unit computer control discussed in an earlier paper.[1]

The feasibility study was begun in June 1979. On the basis of its results, Neste decided in March 1980 to carry out the computer control project.

The additions and alterations to the instrumentation were carried out by Neste's own engineering department, Neste Engineering. The cost of this section of the project was roughly half the total cost. Neste's total manpower investment in the complete study and project

* This project was truly a team effort and would not have been successful without the hard work of many individuals from Neste, Nokia, who supplied the computer system, Process Computers, and Setpoint. Special acknowledgement is due to Aimo Sorsa, the Nokia project manager, Carl Wetzel, the Setpoint project manager, and John Cugini, the Process Computers' project manager, for their diligent efforts in the support of the project. Other major contributors include Matti Alikoski, Matti Kolari, Pentti Paurola, and Jari Vettenranta of Neste: Pentti Tynnila and Esa Nieminen of Nokia; Chuck Graham, Tim Billings, and Jim Poje of Process Computers; and Pierre Latour and Lynn Richard of Setpoint.

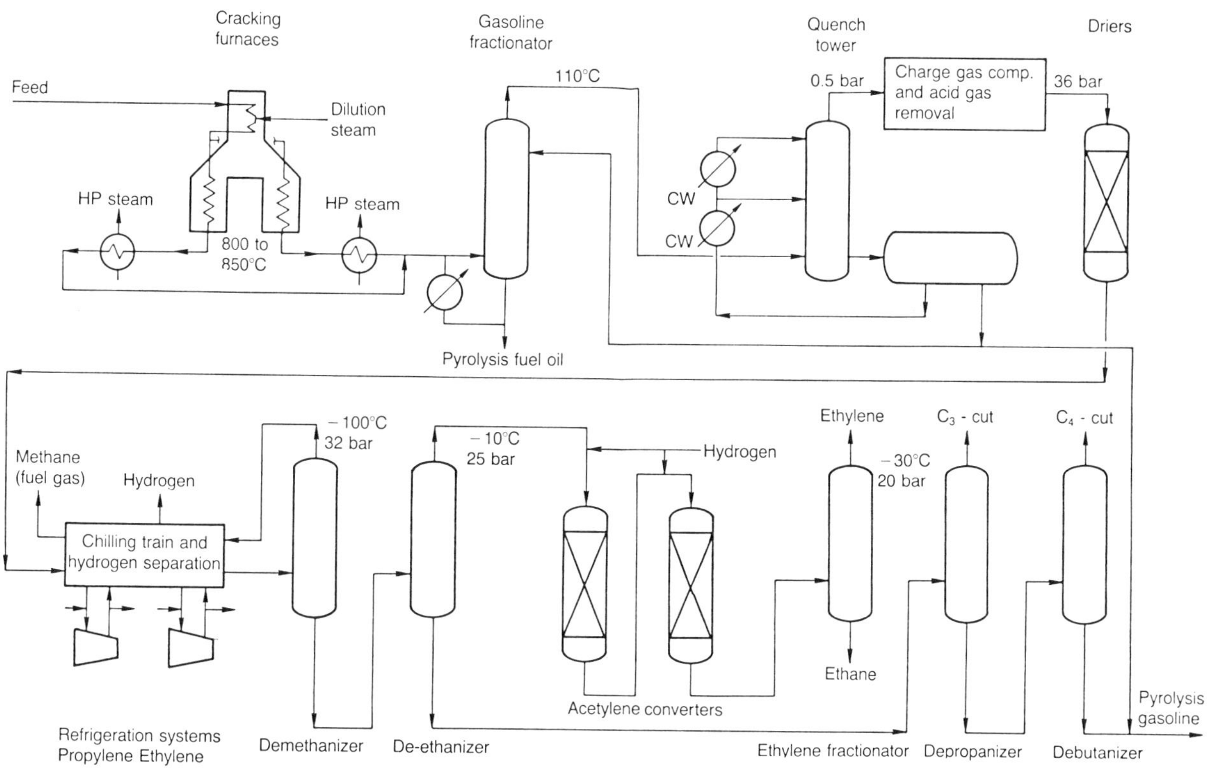

Figure 1. Ethylene unit flow diagram.

was about 30 man-years including manpower requirements for instrumentation modifications and installation.

The purchase agreement for the computer system was signed in October 1980. The system was installed and connected to the plant instrumentation in May 1982. Final tuning and testing were completed by mid November 1982, so the project was completed successfully within the budget and time schedule.

OVERVIEW OF CONTROLS

The unit was originally equipped with conventional electrical analog instrumentation including about 350 control loops. During the feasibility study several alternative computer/process interfacing instrumentation modifications were studied, among them a partial or total replacement of the existing instrumentation by using distributed digital instrumentation systems. The alternative finally chosen was replacement of 160 analog controllers with similar controllers of computer-compatible, stepping motor driven type. The main reasons for this selection rather than a digital instrumentation system were:

- Smaller investment cost and amount of installation work.
- Reduced requirement for changes in control room panel boards, which minimized the need for operator training.
- Digital instrumentation showed no additional benefits beyond those given by analog instrumentation and computer controls.

Gas Chromatograph System

The feasibility study made it evident that the old gas chromatographs (GC's) could not be used for closed-loop control. A new chromatograph system was therefore installed consisting of four GC's for analyzing 18 heater effluent streams, one GC for analyzing two gas feed streams and four GC's for eight cold section streams. The GC's are microcomputer controlled and linked via the gas chromatograph computer with the process computer system. Extensive validity checking is performed on the analyzer and analyses.

Operator Interfaces

The computer control system consists of subsystem A, dedicated to heater control and optimization, and subsystem B, which controls and optimizes the cold section. Subsystems A and B, which are mid range fourth generation computers, interchange data through a high-speed data link.

The control room has two independent operator stations for heater side control and two for cold side control. Each operator station has a color CRT and a dedicated pushbutton panel. The operator can also take hard copies of all displays—an excellent tool during commissioning and upsets.

For process control, operators use process displays, control displays and trend displays. Process displays give a clear picture of all main instantaneous variables of a process, e.g., a heater. Control displays are representations of computer control strategy diagrams. Several control displays are needed to incorporate all the controls of one process unit; for example, each heater has, for normal operation, seven different displays. Operator commands are given through sequences of pushbutton calls. A cursor is used to identify the controller and operation. An extensive history of values for every calculated and measured variable is available for operators in trend displays.

Computer Control Strategy

The production strategy of the ethylene unit determines the computer control strategy. This means that there must exist a hierarchically organized computer control system supporting the production objective—maximization of gross margin—supported by the analog control system. The hierarchical system is shown in Fig. 2. For any control to function it is mandatory that the controls below it function. Thus the regulatory control is the most important of the computer controls. Their tasks are to keep the controlled variables at their targets by manipulating the analog controllers' set points. Constraint controls use a projection technique to keep the desired process variable against the physical limits of the process or at its optimum target by making proper moves to the corresponding regulatory controller target. An optimizer (local or plantwide) calculates the optimum target values, which in turn are implemented by the constraint control.

The computer regulatory control system functions by stabilizing the performance of different processes.

The optimizers balance the benefits in any unit against losses in others according to economic criteria.

The plant optimizer is an online closed-loop module using the projected cold section recoveries in the plant model to estimate how much of the increases in component rates in heater effluents will be recovered in the cold section. It also takes into account the corresponding

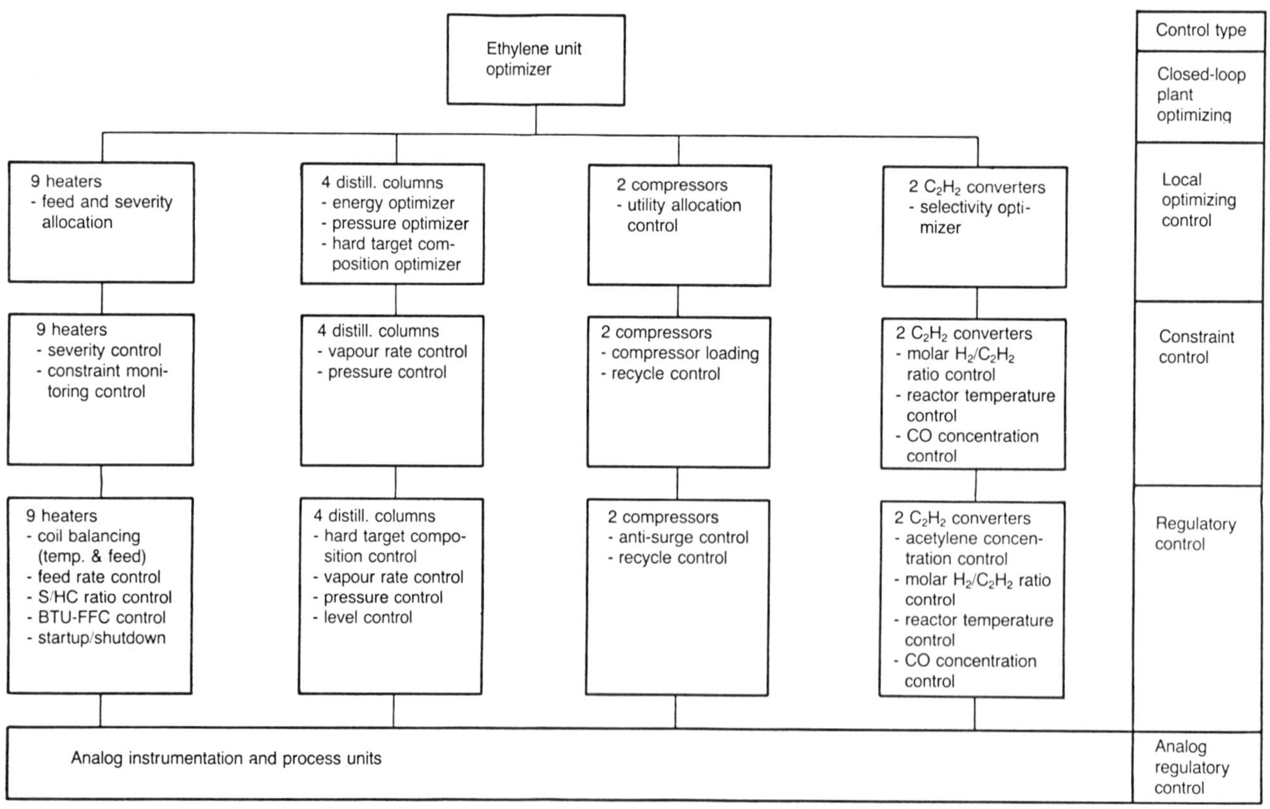

Figure 2. Control system hierarchy.

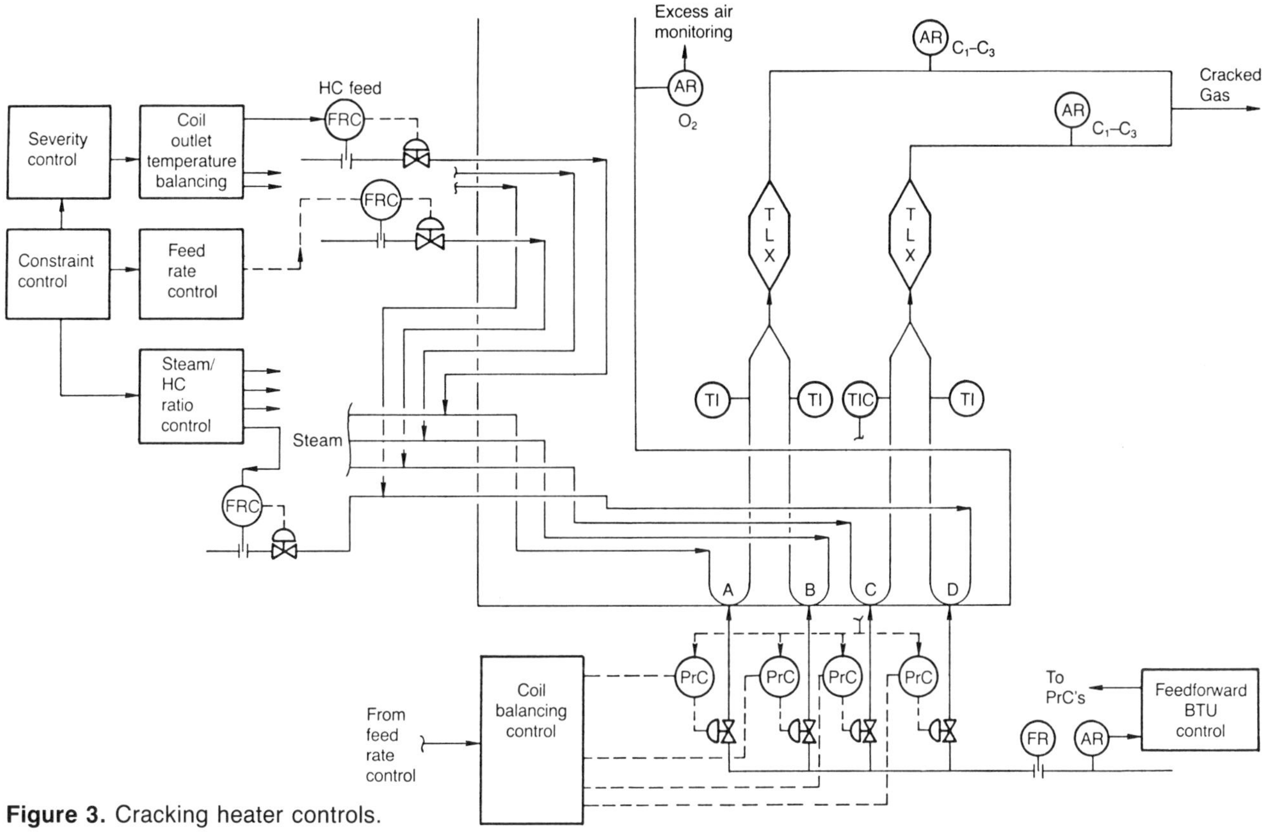

Figure 3. Cracking heater controls.

utility cost changes in the cold section, as well as incremental heater energy consumption.

HEATER CONTROLS

A typical Neste cracking heater is shown in Fig. 3. The hydrocarbon feed rate to each coil is regulated by a flow controller. Steam is mixed with the feed through four steam flow controllers. After cracking, the heater effluent is cooled rapidly in two transfer line exchangers (TLX). An analog temperature controller adjusts the overall fuel flow, through pressure ratio controllers, to hold the outlet temperature of the target coil constant. Four pressure ratio controllers on the fuel gas are provided to allow balancing of the firing in the different coils of the heaters. During the heater cycle, coke deposits on the coils and TLX's eventually constrain the operation of the heater and force a shutdown for decoking.

Control Objectives

The operation and control of the cracking heaters is central to the overall economics of plant operation since feedstock costs are the largest component of production costs.

The general objectives for heater control are to meet the desired severity and feedrate targets without violation of any operating constraints. Severity fluctuations are reduced by reducing the variations in steam to hydrocarbon ratio (S/HC), balancing the coil outlet temperatures (COT), equalizing the coil HC feeds, and compensating for fuel composition changes. Furnace constraints are monitored and control targets selected that do not violate the constraints.

Control Strategy Description

An overview of the furnace computer control modules is given in Fig. 3. The temperature of the control coil outlet is controlled through the analog TIC by manipulating heater fuel gas addition. The other coil outlet temperatures are controlled at their targets through adjustment of the hydrocarbon feedrate to the coil. This control structure provides fast and responsive temperature control during changes in heater operating conditions.

Steam addition to the coils is controlled to hold a target ratio of steam to hydrocarbon flow with an override based on maximum coil residence time limits. This control reduces selectivity and stabilizes heater effluent yields.

To minimize disturbances to the other controls, an interactive multivariable feed balancing control is used. Because of the substantial interaction between the coils during changes in firing, all pressure ratio stations are simultaneously adjusted to bring all coil feed rates to their targets. In addition, this module also adjusts the ratios to keep the overall average of the ratios a controllable range. Feedforward compensation is added to the pressure ratio stations based on fluctuations in fuel gas composition. The overall feed rate to the furnace is controlled through adjustment of the target feed rate to the target coil.

The coil outlet temperature control, the steam to hydrocarbon ratio control, and the feed rate/coil balancing control module, are basic regulatory control modules that can be operated independently with operator entered targets. However, the manipulated variables for these modules also affect, in a highly nonlinear and interacting way, heater yields, coking rates, and heater constraints such as tube metal temperatures, TLX outlet temperatures, draft, fuel flow, etc. Two higher level control modules are used to compensate for these interactions by adjusting the targets of the lower regulatory loops. The first is severity control. For ethane/propane (EP) feedstocks, the severity factor is directly calculated from feed and effluent analyzers as conversion of the feedstock. For heavier feedstocks the effluent methane/propylene ratio is used as a severity factor. This ratio is directly related to yields over a wide range of operating conditions. Severity control is implemented through a model-based severity estimate, using measured severity feedback for model updating. With this configuration, analyzer problems do not result in loss of severity control, but only a less frequent updating of model constants. The target for severity control comes from the plant optimizer module and the output from the module is the target for the coil outlet temperature controls.

The second higher level control module is the heater constraint control. The targets for this module are severity, feedrate and S/HC ratio. They come either from the optimizer or from operator entry. Projections of the effect of changes in manipulated variables on the constraining variables are made and the set of targets for the manipulated variables are calculated which keep the heater from violating any constraints. If possible, all desired setpoints for feedrate, severity and steam to hydrocarbon ratio are implemented. If constraints intervene, these desired setpoints are implemented on a priority basis, with severity targets generally having the highest priority.

In addition to the control functions, a number of performance calculations were performed around the furnace with excess air monitoring and transfer line exchanger outlet temperature monitoring of particular importance.

DISTILLATION COLUMNS CONTROLS

The distillation columns under computer control are a condensate stripper, demethanizer, deethanizer, and ethylene fractionator, as shown in Fig. 4.

Control Objectives

The computer control objectives for each distillation column include: operate close to the most important product purity targets, ensure that column constraints are not violated and optimize the use of energy. Modular control functions and single variable optimizers are used to meet these objectives. An overview of the major computer loops are shown in Fig. 4.

Control Strategy Description

The hard target is the most important composition target for a column. The computer control maintains the hard target under most conditions. The soft target is the composition target for the other end of the column. The composition of this stream is maintained as long as no physical constraint limits are violated. When a constraint such as the flooding limit is approached, the soft target purity must be relaxed if stable operation is to be maintained.

Hard target controls adjust the column material balance to control composition through manipulation of either reboiler flow controller or product flow controller targets. Online models are used on the demethanizer and deethanizer to predict hard target compositions from temperature and pressure measurements with model updates from gas chromatograph analyses. Use of the predictors minimizes the impact of measurement lag and allows for tighter control of hard targets.

The hard target controls for the ethylene fractionator are configured differently from the other columns. Methane plus ethane concentration in the product ethylene is the control variable. The vent flow is manipulated to control the amount of methane; the ethylene product flow is adjusted to control the amount of ethane to a value consistent with the current methane level and overall target.

Feedforward controllers are used on all the columns to help preserve the material balance with changes in the feedrate. On the demethanizer and ethylene fractionator a one-way decoupler is used to minimize the impact of energy balance changes on the hard target controller. This controller adjusts the downstream loop

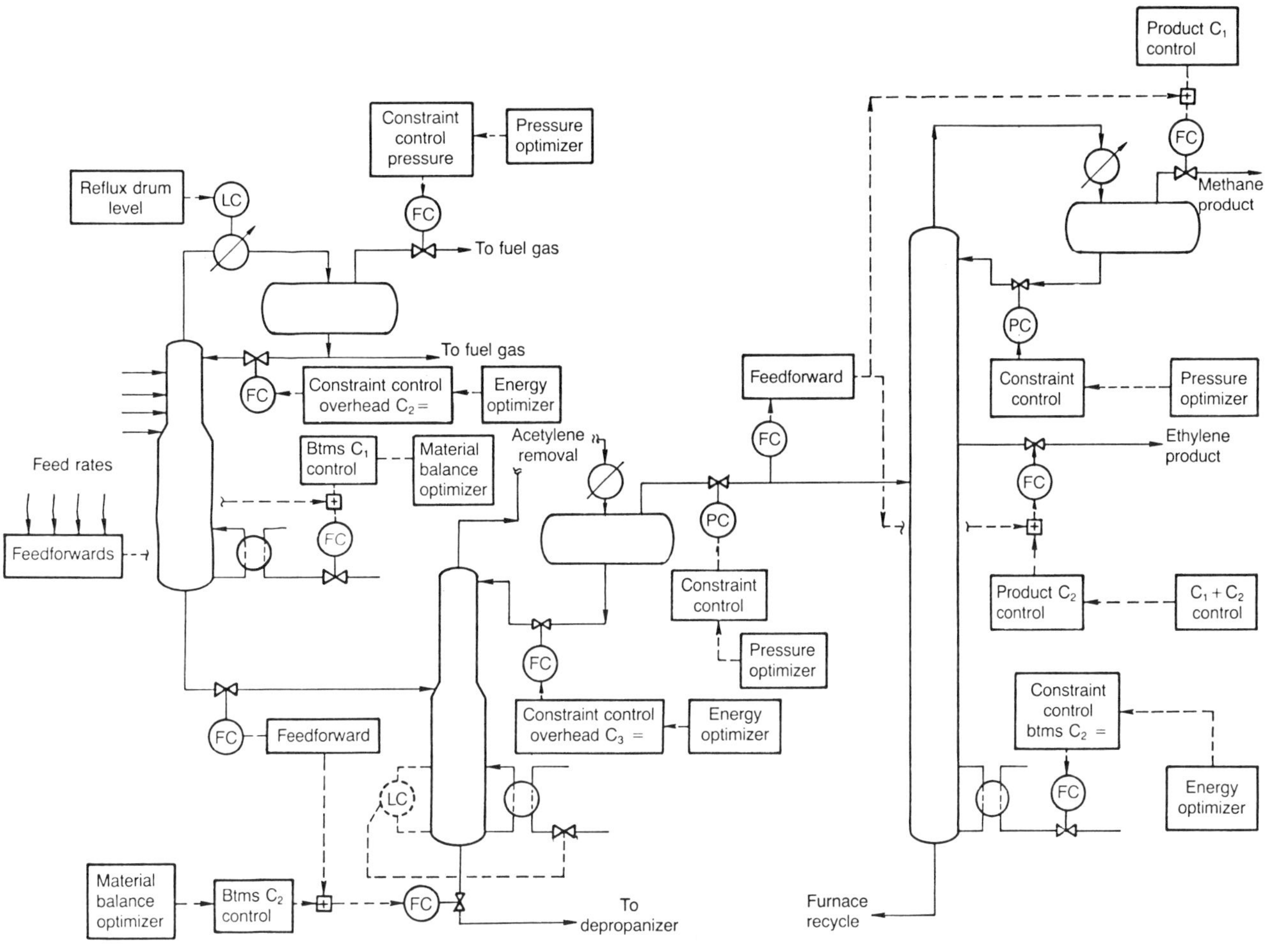

Figure 4. Distillation columns controls.

of the hard target controller whenever a change is made to the energy balance.

The hard target optimizer calculates a distillate-to-feed ratio such that profit on products is maximized. The optimizer composition outputs are checked against a purity target to ensure that specifications are not exceeded. The material balance optimizer for the methane in ethylene product also adjusts the methane concentration target for the demethanizer bottom based on differential product values and plant constraints.

Soft target controls manipulate the energy balance by adjusting either the reflux flow controller or reboiler flow controller setpoints. The control variable for the soft target controllers is an internal vapor rate, calculated at either the top or bottom of a column. A constraint control method is employed to select the vapor rate target. A variety of potential constraints on the columns are measured, compared with preset limit values and projected to an equivalent limiting vapor rate. The most limiting constraint becomes the target for the vapor rate controller. Potential operating constraints include reboiler and condenser valve positions, pressure drops, condenser temperature approaches and soft target compositions.

The energy optimizer calculates an optimum internal vapor rate such that product values are balanced against energy costs[2,3]. It includes a model that correlates the separation factor to column internal vapor rate. The optimizer includes an energy balance, component and total mass balance, and accounts for changes in relative volatility and tray efficiency. The compositions of the column streams are treated as pseudo-binary mixtures consisting of a heavy key and a light key. Product differential values around each column are calculated based on the current pricing of ethylene and byproducts. Values for intermediate streams are calculated from the final product values and calculated downstream column processing costs. The optimum vapor rate is only implemented if it will not violate any limits.

Pressure control on the distillation columns is integrated with the vapor constraint controls and energy optimizer. It considers the current limiting vapor constraint and selects the direction that pressure should be moved to reduce variable costs. The optimum pressure is not always at a minimum. If a constraint such as the approach temperature on a condenser becomes limiting, pressure is increased to relieve the constraint if economics so dictate.

The material balance and energy optimizer work together with the pressure controls to optimize the separation for any possible operating condition. By using individual optimizers, the frequency of solution for each optimizer is adjusted to fit the disturbances which affect it. The execution period of the optimizers generally fall between one and three hours.

REFRIGERATION ALLOCATION CONTROLS

Two steam turbine driven multistaged centrifugal compressors supply refrigeration to the ethylene plant. The propylene system supplies four temperature levels of refrigerant, and the ethylene system supplies three temperature levels. Recycle streams from the compressor discharge to each stage are adjusted through an analog control system to prevent surge. The computer system process display of the ethylene compressor is shown in Fig. 5.

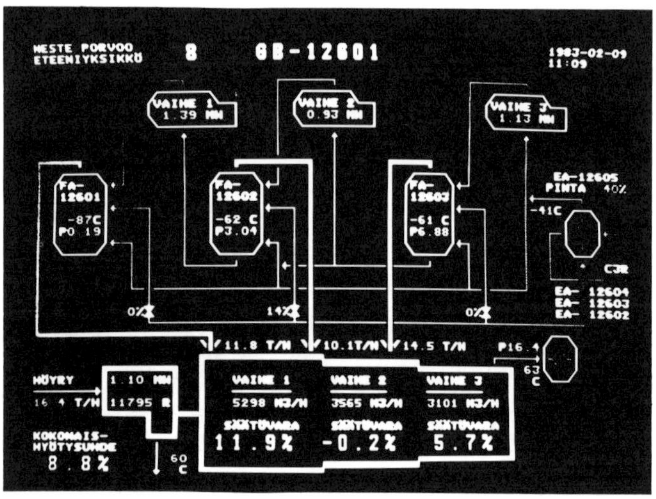

Figure 5. Ethylene compressor process display.

The objectives for the computer control of the refrigeration systems are to minimize steam usage by reducing excess recycle and to determine refrigeration costs for the column energy optimizers while preventing compressor overloads.

Minimum flow targets for each stage are calculated based on compressor performance curves. Volumetric surge flows are calculated as functions of both rpm and stage pressure ratio. The larger of the two predictions plus a safety margin is selected as the surge flow.

Refrigeration costs are calculated for each stage and then modified based on compressor constraints. Mass and energy balance, a steam turbine balance, steam cost and condenser cost calculations provide the inputs for the individual refrigerant level cost calculations.

Refrigeration costs are decreased toward a preset minimum over time as the recycle valves open. This encourages the column optimizers to use more refrigeration since it is less expensive.

To keep the refrigeration users from demanding more refrigeration than the refrigeration compressor can handle, the computer adjusts the cost upward as maximum compressor capacity is approached.

ACETYLENE CONVERTER CONTROLS

The acetylene converters are adiabatic fixed bed catalytic hydrogenation reactors which remove enough acetylene from the deethanizer overhead so that the ethylene product will meet the maximum allowable acetylene specification. The reactions that occur simultaneously in the reactors are the hydrogenation of acetylene to ethylene, the hydrogenation of ethylene to ethane, (which is undesirable) and a third reaction, (also undesirable), which produces a C_4 and heavier polymeric material called "green oil."

A two-stage reactor system, with preheat and intermediate heat exchange, is used for the reaction. As shown in Fig. 5, independent control of hydrogen addition and inlet temperature is permitted for each bed. Also, the level of CO addition to the hydrogen makeup can be controlled through adjustment of the methanator bypass hydrogen. Carbon monoxide is a moderator of reactor conditions, which increases selectivity by encouraging acetylene hydrogenation over ethylene hydrogenation.

Control Objectives

The primary objective in operating the converters is to ensure that the acetylene in the ethylene product does not exceed the specification limit. The secondary objective is to minimize the hydrogenation of ethylene while meeting the first objective. The primary disturbances to the reactors are fluctuations in feedrate and acetylene concentration. In addition, the purity of the hydrogen and the CO content of the unmethanated hydrogen varies. An added problem is the dynamics of the reactor response are often faster than the frequency of acetylene analysis.

Control Strategy Description

In designing controls for the converter, it is important to realize that there is a partial conflict between the primary and secondary objectives of operation. Increasing the hydrogen make-up to the reactors decreases the outlet acetylene but also increases the ethylene losses. Increasing temperature results in decreases in outlet acetylene and increases in green oil production. Adding CO to the reactor system increases the selectivity of the reactions but reduces the activity of the catalyst which can require compensating increases in hydrogen addition and temperature.

To compensate for changes in feed rate, analog flow ratio controllers are used to ratio hydrogen addition to the first and second reactors to the reactor feed rate. The inlet hydrogen/acetylene ratio control module reduces the effect of fluctuations in inlet acetylene and hydrogen composition. Final reactor effluent acetylene concentration is controlled by manipulation of reactor inlet temperature. Overrides are included to increase hydrogen addition in the event of large excursions in outlet acetylene concentrations. In addition, feedforward compensation to second reactor inlet temperatures is included based on inlet acetylene concentration changes.

With the above controls, the disturbances to the acetylene converter have been reduced. However the proper targets for hydrogen and CO addition and inlet temperature for the first reactor have yet to be determined. There is a very wide range of targets for these variables which will yield on specification product. The optimizer in Fig. 6 calculates the proper targets for the regulatory control loop based on an objective function of minimizing ethylene loss while ensuring that the outlet acetylene composition target is met. Kinetic models of the reactor system are used to calculate the effect of changes in manipulated variables on acetylene and ethylene change and standard optimization procedures are used to calculate optimum setpoints. Deactivation of the bed during the run is modeled as a decrease in effective catalyst volume.

UNIT OPTIMIZER

With the column energy optimizer, the refrigerant allocation, the ethylene impurity optimizer, and the pressure optimizers discussed previously, the separation area has been optimized for any given set of heater feed and severity conditions. The setpoints for these variables, feedrate and severity, should be determined in a way that meets overall unit objectives.

The production objective for the unit is to maximize the gross margin that can be taken as the difference between the product revenue and the variable operating costs. Two objective functions in the optimization algorithm are provided to implement this objective. This objective function is based on changes in product values, feedstock costs, and utility costs. The first is an economic index, representing gross margin, which is to be maximized at fixed ethylene demand. The second objective function permits maximization of ethylene production against unit constraints.

A swing feedstock is selected as the feedstock currently least profitable to run. Its rate is determined by

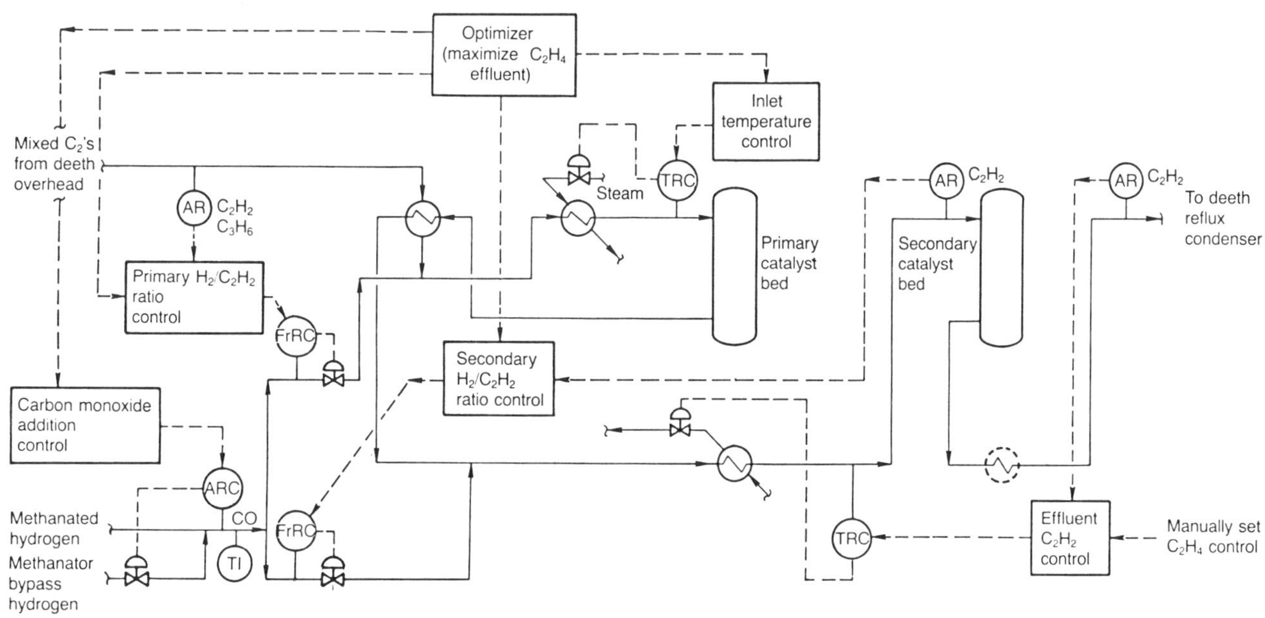

Figure 6. Acetylene converter controls.

the optimizer. The processing rates of the other feedstocks are fixed at the current rate that should be their maximum availability. Hence, the independent or manipulated variables for the optimizer will be the severity of operation for each feedstock type and the feedrate of the swing feed type.

In general, this problem can be considered a nonlinear constrained optimization problem. It is nonlinear because a unit change in one of the manipulated variables will give changes of different magnitude to the process variables depending on the current value of the severities and feedrate. For example, a unit increase in naphtha cracking severity will give different component yield changes for different severities. A change in ethane heater severity will give a steady state change in the ethane recycle and the mixed gas heater feedrate that depends on both the ethane heater and the gas heater severity.

In addition, there are numerous constraints on possible changes in the manipulated variables including absolute limits on heater feedrates and severities and limits imposed by other heater conditions such as maximum allowable temperatures. There will seldom be hard physical limits in the separation area. The tower controls have been designed to sacrifice recovery to avoid violation of constraints. Hence, the final separations area optimum operating point will reflect the incremental recovery of ethylene more than constraints. However, provisions are included for the special case where absolute limits in the separations area are encountered. Maximum change limits are included for all manipulated variables in the optimizer.

To handle the nonlinearities and constraints, the optimization problem is solved by a recursive or sectional linear programming procedure. Recursive linear programming is the technique of solving the nonlinear problems through linearization around the current operating point, finding the linear programming solution of the resultant problem, substitution of the new solution in the original problem and relinearization until convergence to the overall optimum answer is obtained.

The plant model is based on a complete material balance for the plant. Heater effluent yields are projected for each feedstock at the current severity of operation and an incremental severity. The differential yields from these two cases are used to calculate the linearized coefficients which reflect the expected change in constraint variables and the objective function for incremental changes in the independent variables. The effect of recycle is calculated directly, without iteration. Incremental recoveries expected for the cold section are used to calculate the net explant product changes due to changes in heater effluent component rates. The economic index used for optimization includes costs for incremental fuel and utilities as well as product and feedstock prices. Allocation of heater feedrate and severity changes is made to equalize percent loading for the controlled heaters.

For more discussion of optimization procedures see reference 4.

RESULTS

Although the computer control system contains a feature to permit maximization of throughput against plant constraints, Neste has not been able to use this feature because of the current market limits. However, it has been observed that the computer control at optimum severity gives 2 percent more ethylene from the same feedstocks, at the expense of less valuable byproducts, as seen in Table 1. This has made it possible to produce the desired amount of ethylene at a lower naphtha feed rate.

Table 1
Results

		Before	After
1	Change in explant yields, weight percentage of feed		
	H_2+CH_4	19	21
	C_2H_4	32	34
	C_3's	19	17
	C_4's	8	7
	C_5+	21	20
2	Heater run length, days	25	33+
3	Energy consumption, MW	Base	≤ Base
4	Standard deviation of control, fraction of target		
	Cracking furnace severity	Approx. 0.04	< 0.015
	Column hard target composition	> 0.4	< 0.10
5	Ethylene losses in acetylene converters, percent of feed	0.3 to 0.4	0
6	Ethylene losses in demethanizer, percent of distillate	> 2	< 0.3
7	Ethylene losses in ethylene fractionator, percent of bottom product	> 2	< 0.1

Heater severity control is used after the introduction of hydrocarbon feed during startup and throughout the heater cycle until shutdown. The cracking is stabilized, so that the severity deviation is now less than 1.5 percent of target. In spite of operating at a higher average severity than before implementation of computer control, the coking rate is slower and the overcracking situations previously occurring are now avoided. The run lengths of the heaters have increased by more than 30 percent, bringing considerable savings in decoking costs.

Because of the energy optimization in each column, maximum recovery against the physical limits of a distillation column has not been observed. A significant reduction of ethylene losses is attributable to computer control, due primarily to shifting of the material balance with only a small requirement for additional energy.

Ethylene losses have been reduced to less than 1/6 of their previous value in the demethanizer and to less than 1/10 of their previous value in the ethylene fractionator. The average recovery in the deethanizer has not increased, but its energy requirements have decreased by 25 percent.

The small standard deviation of the ethylene product impurities, which the computer control has brought about, has allowed an increase in the impurity target to three times the earlier average. For similar reasons even the target of methane impurity in demethanizer bottoms has been increased to twice the earlier average.

Traditionally acetylene has been over hydrogenated to ethane instead of ethylene because of manual operation "on the safe side." After introduction of the acetylene converter optimizer and control, ethylene losses have decreased by about 0.3 to 0.4 percent of converter feed and, when newly regenerated catalyst is used, even turned to net production of ethylene. The ethylene product has not been off spec due to acetylene concentration since completion of the commissioning of the controls.

The increased energy efficiency of the compressors and other machinery has decreased the demand for energy by more than 12 percent as a result of efficient utility allocation control and energy optimizers. This is due primarily to the reduction of excess recycling in the compressors.

Due to the higher severity in cracking, the fuel gas consumption of the heaters has increased. This is, however, more than balanced by the energy savings at the compressors, so that the unit's total energy need is virtually unchanged.

RESULTS

A New Way of Operating

From the beginning of the commissioning, the operating personnel accepted the system. Naturally this required a substantial amount of training before and during the commissioning. Before the commissioning all operating personnel were trained for two weeks in computer control and operator station use. The commissioning itself was no doubt the best training, and during this five-month period the computer control and operator action gradually became more and more routine.

When the computer control was familiar to the operating personnel, they made many valuable observations on "troublemaking" loops. Careful analyses of these problem areas led to more efficient control structures, which were quickly implemented. After commissioning the positive and even enthusiastic attitude towards computer control has been retained.

When computer control was implemented, operators could concentrate on true control objectives, such as severity control for heaters and composition control for columns.

Some other features beneficial to operator acceptance have been noted:

- There were minimum changes in the analog panel boards, and all computer control moves are actually seen as changes on the analog controller setpoint dials. This has made it easier to understand the computer/analog control interaction.
- Production strategy problems have been solved by introducing a control system with the same objectives that the operating personnel earlier tried to reach manually.
- All operator access to the computer is simple and easy to understand, although the control strategy and programs behind it may be complex. In fact, the control and process displays were designed in cooperation with the operating personnel so as to support the herarchical structure of the computer control.
- The computer performs the routine logging instead of the operators.

Efforts During Commissioning

During commissioning considerable time was spent verifying the computer calculations in various operating conditions. On average, five Neste engineers participated in the commissioning. The presence of these engineers and discussions of the computer controls' functioning facilitated acceptance by the operator. Process simulation, laboratory analysis, and hand calculations were used. When inconsistencies were found, the reasons were examined carefully. Most often this led to a poor measurement, which was then corrected by the instrument maintenance group. The few cases where the calculations or the concept behind them were not complete were rapidly corrected by close cooperation. This work gave a sound basis for closing the computer control loops and helped the personnel to place reliance in the system.

All computer control loops were thoroughly tested at different targets and conditions. Even column controls with notably long dead times and time constants performed satisfactorily.

The successful and continuous analyzing of heater effluent streams, which provides the update information to the severity control, has traditionally been difficult.

Therefore, Neste invested considerable effort in developing a sampling system which would give representative as well as repeatable gas samples of the heater effluent stream to be analyzed in the GC's. The results are encouraging: the sampling systems have always operated satisfactorily without cleaning for longer than one heater run cycle, and in many cases over two run cycles without noticeable change in sample quality.

Although many composition control loops successfully use composition predictor models, long analyzer downtime periods are not allowed. The work activities of the analyzer maintenance group has shifted more to preventive maintenance and less to fault detection. There has been no increase in the staffing of the instrument maintenance group assigned to the unit.

The plant optimizer and heater calculations use yield correlations between severity and individual components. These correlations have proved to be in good agreement with experimental data on total analyses of heater outlet streams. In fact, these calculated yields have also been used for future planning purposes.

Control Availability

About a month after commissioning, a major hardware fault was found in subsystem A's disc unit. Because of the built-in backup system the downtime of the heater controls (including disc testing) was limited to 48 hours. With this exception, there have been no system failures.

During the running cycle of a heater all controllers have been closed almost 100 percent of the time. The only exceptions are due to instrument failures, a monthly average being about five hours when not all controls are on. As mentioned previously, severity control uses a feedforward model for making control moves and effluent analysis is used only to update the model. Therefore, closed-loop operation is not directly related to the analyzer online time; in fact the severity control has been operated satisfactorily for more than two days without analyzer feedback. During this operation the actual severity was followed offline and showed only slightly greater variation compared to normal severity control. A database problem restricted the use of one heater's control when brought into use after a standby period. This was corrected without delay by making an additional condition check in the programs.

All cold section controllers have been closed for more than 99 percent of the time except for three loops: The methane controller in the ethylene fractionator has been on only about 70 percent of time because of the limited operating range of the controller. The nonlinear control of the ethylene fractionator bottoms level has not been on after commissioning because of process changes in the external LPG feed system. The control of the first stage in the ethylene refrigeration compressor has been online about 95 percent of time because of control valve reparation.

Experiences on Computer Control

The observations on control performance during different unit loading and using different feedstock types are discussed below.

The demethanizer and ethylene fractionator controls are not affected by differences in feeds. In the current product price and utility cost situation, the optimal performance has resulted in a reduction of ethylene loss from columns by using additional energy. The only exceptions to this are situations where the refrigerant compressors are overloaded, for instance when a large quantity of ethylene product is being liquefied and run into the cold storage. The deethanized optimum performance has on all feeds near the minimum energy consumption because of the relatively small benefit from increased recovery. Column optimum pressure varies according to column feed and refrigerant temperature. For low feeds and refrigerant temperatures, minimum pressure is preferred.

Acetylene converters reach optimum performance at high CO concentrations in hydrogen. The aging of the catalyst is compensated for by increasing the temperature in the first bed.

The heater coil balancing control is more responsive to liquid feedstocks than to gas feedstocks. In some gas heater cases, not even the maximum possible bias between individual pressure ratio controllers has been sufficient to equalize coil feeds. This has often been caused by inconsistent temperature measurements. The severity control module has identified problems caused by inaccuracies in coil outlet temperature measurements. In fact, a difference of 15°C has been seen between two identically-operating heaters, when operating at the same severity. After replacing the thermocouple elements, the difference was reduced to 2°C. The system has also helped locate other defective instruments and to avoid overcracking.

The unit optimizer has performed control moves which are easy to accept and understand. A trend towards higher severities has been dominant. Only when ethylene production is being restricted has the optimizer decreased severities to make more profit from the byproducts. High gas cracking severities are always preferred.

Because of the more exact control of coil outlet temperature, average heater outlet temperatures a few degrees higher than before are tolerated. Partly due to this and to the precise severity control the heaters are now operated at higher average severities than before. Although the severity was only occasionally measured before computer control, this change can also be seen by comparing the explant yields before and after implementing the computer control. Plant utility consumption has decreased at the same time.

The computer control system has been a success, and its expected economic justifications have even been surpassed. By converting the yield and utility consumption benefits due to computer control into monetary units, the payback time of the system is found to be less than two years.

The computer system contains features which are still not used to their fullest extent. To achieve this, the operating personnel must further develop their skills in controlling the process by computer. The management has also decided to start a study of further computer control in other units at the Neste petrochemical works.

References

1. Rinne, R. et al "Experience with Distillation Unit Computer Control," *Hydrocarbon Processing,* V61, N4, March, 1982, p. 141.
2. Matthews, S. A., "Practical Application of Distillation Column Control Systems," presented at Houston Energy Conference (1980).
3. Martin, G. D., Latour, P. R., Richard, L. A., "Closed-Loop Optimization of Distillation Energy," *Chemical Engineering Progress,* V77, N9, Sept. 1981, p. 33.
4. Latour, P. R., "On Line Computer Optimization 1: What is is and Where to Do it," *Hydrocarbon Processing,* V58, N6, June 1979, p. 73. "On Line Computer Optimization 2: Benefits and Implementation," *Hydrocarbon Processing,* V58, N7, July, 1979, p. 219.

40
Hydrotreater Computer Control*

G. D. Martin, L. D. Van Horn, and K. M. Cassaday

A computer-based control system was installed in Gulf Canada's lube oil plant at Clarkson, Ontario to produce 10 Mbpd of high quality lubricating oils using Gulf's new lube hydrotreating process. The following sections present a process description, a description of the control objectives, the control functions designed to meet those objectives, and a summary of the performance documented to date.

In June 1980, Gulf conducted a feasibility study for computer control of their lube oil process, and the results of that study indicated that control of the heart of the process, the hydrotreater reactor and fractionation section, would yield significant economic benefit. In October 1980, Gulf reviewed a detailed design based on the control functions defined in the earlier study. After the desired functions were refined to Gulf's specification, control programming began at the Houston computer staging site. The applications software was demonstrated in Houston in May 1981, and integrated into the same computer on-site at Clarkson in July 1981. The system linked to an existing microprocessor-based hardware data acquisition and control system.

Control commissioning and initial tuning were finished by mid-December 1981. Final tuning and testing were done in mid-January 1982, when the project was completed successfully, within budget and ahead of schedule.

PROCESS DESCRIPTION

The hydrotreater unit is shown schematically in Fig. 1. The key areas selected for computer control application were the first stage hydrotreater and fractionation. Some simple control functions were implemented on the hydrogen plant; these will not be discussed here.

The first stage hydrotreater is designed to process 10 Mbpd of mixed gas oils to yield a full range of products from fuel gas to the heavy waxy lube products.

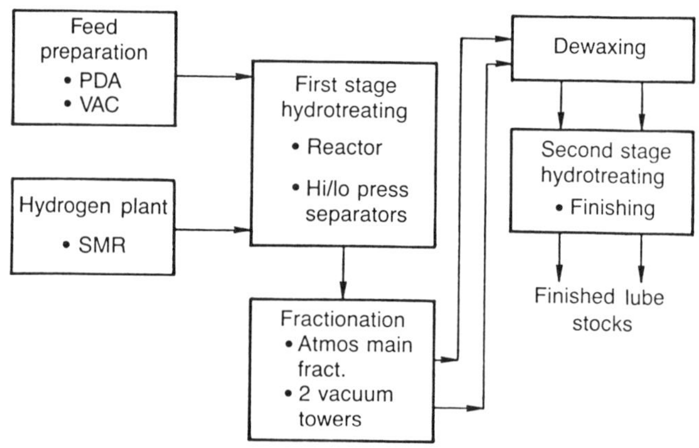

Figure 1. HTU process overview.

The unit runs in two blocked out operations; light block and heavy block. In each block, only two feeds are used. Light block operation charges a blend of light vacuum gas oil and heavy vacuum gas oil. Heavy block operation charges a mix of heavy vacuum gas oil and deasphalted oil.

The feed to the reactor is heated through a reactor feed/effluent exchanger, then through the reactor charge heater. The oil and hydrogen mixture passes through the reactor, which contains several catalyst beds. The reactor effluent goes to a separation section, the high-pressure vapor from separation is recycled, and the low-pressure vapor goes to fuel gas. Bottoms liquid becomes feed for the fractionation section. Recycle hydrogen is compressed to reactor pressure and joined by makeup hydrogen from the hydrogen plant.

Recycle gas is split into four streams. The first goes directly into the reactor feed, and the other three are quench flows introduced between catalyst beds in the reactor.

* Originally presented at the AIChE Spring National Meeting, March 1983, Houston, Texas.

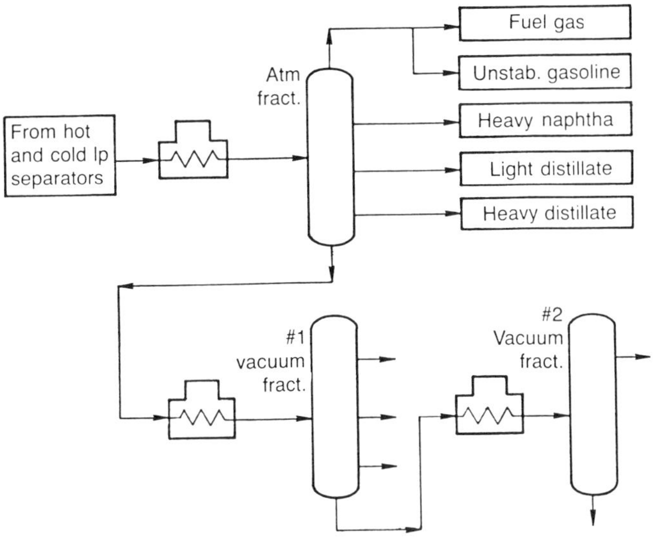

Figure 2. Fractionation section.

The lube fractionation section is shown in Fig. 2. It separates the liquid effluent from the first stage hydrotreater into a full range of products:

- Fuel gas
- Unstabilized gasoline
- Heavy naphtha
- Light distillate
- Heavy distillate
- Several waxy lube stocks

The lighter products (fuel gas through distillate) are made in the atmospheric column, subject to boiling range specifications. The waxy lube stocks are made in the vacuum columns, subject to viscosity specifications. Steam is used in the vacuum columns to lower the oil partial pressure and keep heater outlet temperature below lube oil stability limits.

CONTROL OBJECTIVES AND FUNCTIONS

There are four main objectives for this computer system: stabilize and maintain reactor severity for consistent product yield and quality, stabilize the distillation to control product quality, control viscosity, minimize fuel consumption, and reduce steam required in the vacuum columns. A set of coordinated control functions meets these objectives on line.

The reactor controls incorporate a new multivariable control methodology called IDCOM*. Transport delay through the reactor beds introduces significant delays which must be accounted for by the reactor controls. IDCOM is well suited for this type of problem. Reference 1 gives a detailed account of the IDCOM application.

The fractionation technology is based upon mass and energy balance calculations for tower internal liquid and vapor rates. These are repeated automatically every minute from on-line data. The resulting information is used to derive performance indicators for separation (internal reflux ratios), boiling ranges (true boiling point [TBP] cutpoints) and constraints (vapor velocities, liquid loads). Similar applications are discussed in References 2 and 3.

Reactor Severity

The yield and quality of the bulk waxy lube product from the first stage hydrotreater are very sensitive to the reactor severity. A considerable price is paid for over-reaction in terms of lost lube yield, and additional hydrogen consumption. There are also penalties for under-reaction, mostly in formulation costs for the finished lubes. Therefore, it is important economically to operate the reactor at exactly the desired severity.

Severity is a function of feed rate, reactor pressure, hydrogen purity, reactor weight average bed temperature (WABT), and catalyst activity. Feed rate and pressure are controlled so they are essentially constant and change only when a new setpoint is entered. Hydrogen purity varies little and slowly. Catalyst activity decreases over the run, changing so slowly that it is constant in the short term also. Hence, the key to controlling reactor severity is to control WABT.

The reactor computer controls are shown in Fig. 3. WABT is controlled by a three-level hierarchy: the first level microprocessor controls set each of the bed inlet temperatures; level two includes several IDCOM computer controls that maintain bed outlet temperatures, and level three, the WABT control, adjusts the bed one inlet temperature and the outlet temperature targets. In addition to controlling the reactor WABT, this control can enforce a profile on the reactor bed outlet temperatures. This profile is important for catalyst life. The IDCOM methodology can account for changes in feed rate and hydrogen partial pressure by feedforward compensation. The WABT control also has a severity feedforward from feed rate.

* The name IDCOM appears several times in control descriptions. It is a new multivariable control algorithm. The name is an abbreviation for IDentification and COMmand. IDCOM belongs to a new class of control techniques called "predictive control" and is often discussed in the literature. It is sometimes called MAC, for Model Algorithm Control. IDCOM has been used many times in Europe, but this Gulf hydrotreater application is the first process application in North America. Shell Oil Company has reported use of a similar algorithm called DMC, Dynamic Matrix Control, on an industrial furnace and on a refinery catalytic cracking unit. References 4 and 5 provide reviews of both IDCOM and DMC.

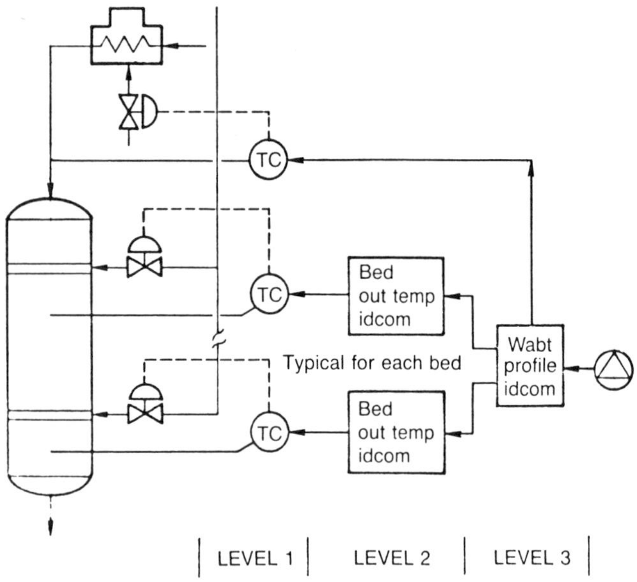

Figure 3. Reactor computer controls.

Product Quality

The fractionation section separates the separation section liquid product stream into distillates and waxy lube stocks. Advanced controls are useful and in fact necessary for accurate control because of the sensitivity of these separations. For example, the slope of the equilibrium flash vaporization (EFV) curve (percent yield per °F) of the bulk waxy lube stock is 3 to 5 times that for a typical long residue from crude oil.

The major fractionation control objectives are to stabilize and regulate the distillation processes and control the product quality parameters (viscosity for the lube stocks).

The fractionation computer controls are shown in Figs. 4, 5 and 6. To stabilize distillation, cutpoint controls are included on the atmospheric column and the first vacuum column side draws, and flash zone controls are included on the second vacuum column. On the second vacuum column, TFLASH is the cutpoint between the bottom sidedraw and bottoms, estimated from flash zone conditions. Product quality and viscosities are controlled by adjusting the distillation control targets.

In either block of operation, only three of the viscosities are controlled. The fourth one is left floating.

A control system is designed to optimize separation in the second stage vacuum column. The major objectives are to ensure adequate but not excessive liquid reflux across the separation trays in the deentrainment section and to limit the vapor velocity through the deentrainment section by controlling the stripping steam. This scheme gives good separation between side and bottom products, and limits entrainment.

Fuel Consumption

Other controls are designed to save heat by optimizing the three fired heaters. Feedforward controls and level controls are included to assist the operators in response to process changes.

Overflash is the internal liquid reflux flow between the bottom sidedraw and the flash zone (entry zone) of the atmospheric fractionator. It is the liquid rundown into the flash zone from above. Since excess overflash is just like excess reflux, it is minimized to conserve furnace fuel, but is limited on the low side to preserve HD/Bottom separation. The overflash control block shown in Fig. 4 slowly adjusts the feed heater outlet temperature to control overflash. In this case, overflash is calculated by heat and material balance.

The first vacuum column has similar controls, but since the excess heat in that column is indicated by a measured heater recycle (recycle flow back to the feed to

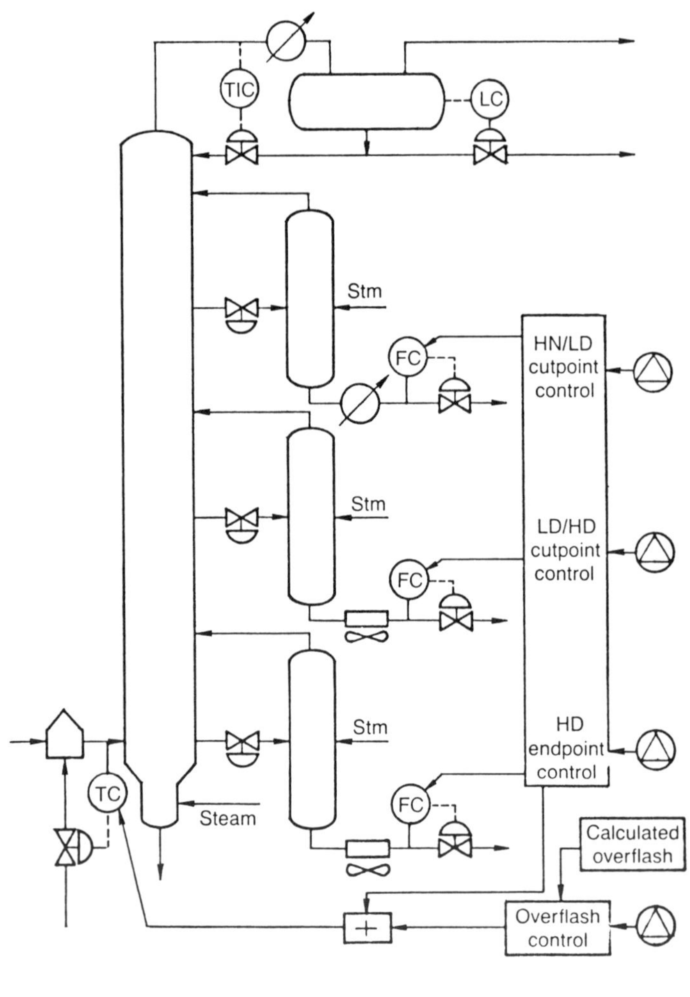

Figure 4. Atmospheric column computer controls.

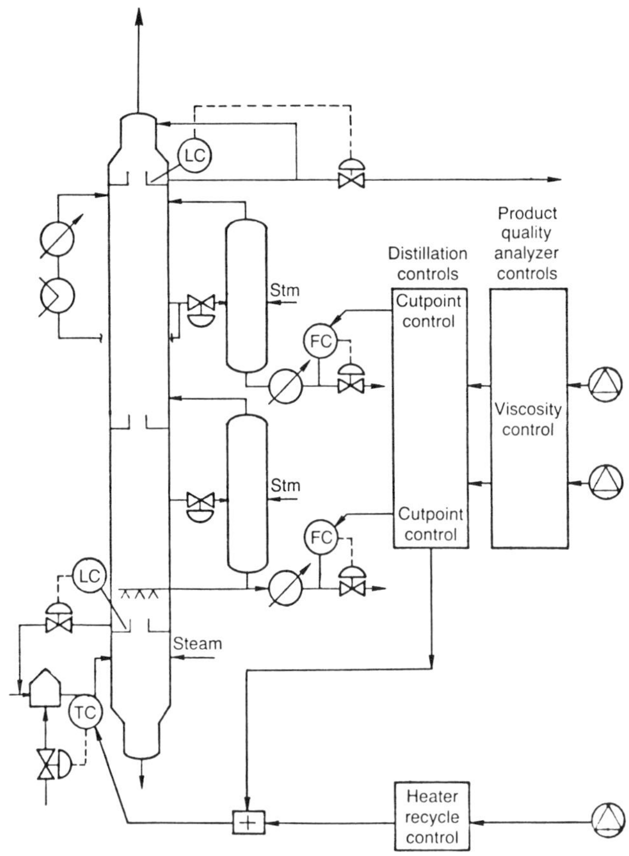

Figure 5. First vacuum column computer controls.

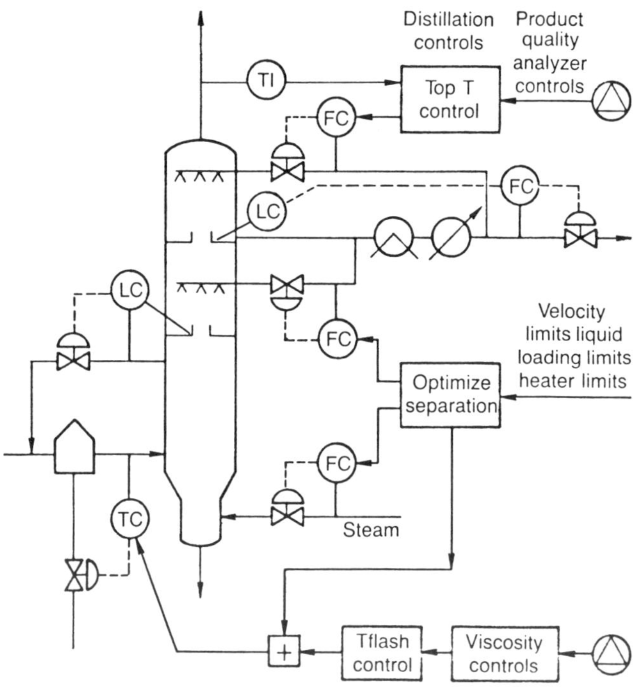

Figure 6. Second vacuum column computer controls.

the heater), a heater recycle control is used to adjust the heat input to the tower (Fig. 5).

Steam Consumption

The hydrotreated lube oil products are very sensitive to temperature, and high temperatures cause off-color product. Steam is injected into the flash zone of both vacuum columns to reduce the oil partial pressure and flash zone temperature required for proper separation. The minimum amount of steam required is set by the maximum allowable flash zone temperature, and by the effectiveness of the separation controls. Smoother operation also allows a significant reduction in steam.

Results

The computer significantly reduced the variations in reactor WABT and severity, thereby allowing improved yields and tighter distillation and viscosity control. Variations in product viscosity index have been reduced. Fuel and steam consumption are lowered. Each of these results is detailed later. Variations cited are standard deviations based on one-minute samples unless otherwise noted.

WABT Variations

The IDCOM computer controls produced a 75 percent reduction in variations of the key process variable, reactor WABT. Prior to control commissioning, typical WABT variations were ±3 to 4°F. Reactor temperature control reduced the average daily variation in WABT to less than ±1°F, with some daily variations as low as ±0.2°F.

Other than major process upsets, the most severe test for the IDCOM reactor controls is a block switch. Feed is changed immediately from one block to the other, and a significant swing in feed rate usually occurs. Furthermore, the required WABT for the two blocks differs by about 6 to 7°F. The result is that a block switch requires good regulatory control to compensate for the feed swing and strong servo control to track a large change in WABT target.

Before computer control was installed a block switch would cause a swing in the reactor WABT of up to 15°F that would typically line out after three hours. Several block switches have been made since the IDCOM controls were commissioned, and now the WABT swing from block switch is usually less than 8°F and lasts only 1.5 hours.

Product Quality

Product viscosities are measured on-line in SUV, and lab analyses for viscosity index are performed daily. Viscosity variations were reduced an average of 70 percent and viscosity index variations were cut by an average of 64 percent. The most impressive improvement was for one of the most important products, where variations have been reduced from ±9.8 SUV to ±0.9 SUV. This product could not be maintained within specification before computer control.

The improved yield and quality control has led to a significant savings by expanding unit flexibility. Since viscosity is an average property virtually any yield of any given product can be obtained. By working with specification ranges of the key product plus adjacent products, a significant increase in the yield of the key product can be obtained. In the case where the refinery crude thruput is driven by lube oil demands, increasing the yield of the key product (at the expense of the adjacent products) by one barrel can decrease the crude required by approximately 30 barrels.

Fuel Consumption

Fuel consumption is measured by a calculated fuel factor, which is basically the amount of fuel consumed per barrel of feed. The computer controls have reduced fuel consumption by reducing reflux in the three fractionation towers. Improved separation controls and the two heat input controls, overflash for the atmospheric column and heater recycle for the first vacuum column, have contributed to at least a 25 percent reduction in hydrotreater unit fuel factor.

Steam Consumption

The computer separation controls for the two vacuum columns have provided smoother operation and sharper product cuts. As a result, the steam requirement has dropped. Total steam to the second vacuum column has been reduced by 50 percent.

The computer control performance is summarized in Table 1.

Table 1
Lube Unit Control Performance

	Jan–May, 1981 viscosity variation	Range of acceptable viscosity	Mar, 1982 viscosity variation	Range of acceptable viscosity
Lt Blk #1	±4.0	3	±2.8	2
Vi	±5.6	4	±1.8	4
Lt Blk #2	±1.9	3	±1.2	1
Hv Blk #1	±5.4	2	±1.4	3
Vi	±6.2	4	±2.4	3
Hv Blk #2	±9.8	3	±0.9	3
Reactor WABT	Daily variations now better than ±1°F, as low as ±0.2°F. Before typical ±3 to 4°F.			
Fuel factor	Reduced by 25% or better Lower overflash in atmospheric column Lower heater recycle in first vacuum column reduced by 67% Lower total steam to second vacuum column reduced by 50%			

LONG-TERM PERFORMANCE

In the nine months since the computer controls were commissioned, there have been indications of improved long-term performance that can only be estimated at this time. First, catalyst deactivation appears to have been reduced, and second, improved thruput flexibility has already been used to significant advantage once.

Catalyst Deactivation

The IDCOM reactor temperature controls have led to a sizeable reduction in reactor temperature swings and upset excursions. The reactor bed outlet temperature profile has helped to spread the work evenly among the beds. By maintaining control within reactor temperature constraints, excessive catalyst deactivation appears to have been reduced. So far, this has been the best of the three catalyst runs to date, with no measurable catalyst deactivation through the first nine months of 1982.

Thruput Flexibility

Traditionally, maximizing thruput has been a computer control benefit area. In Clarkson's case, there was a period of time when it was necessary to minimize thruput to reduce inventories. No benefits have been claimed for this, but Gulf feels they were able to run at lower levels with computer control and avoid having to shut down to reduce inventories.

COMPUTER CONTROL ON-LINE TIME

A running record is kept in the computer for each computer control loop, a measure of its percentage online time. Since this measure can be updated only when the computer is operational, it only reflects the percentage of computer time that the controls are on-line. This measure has on average exceeded 85 percent since commissioning. The critical reactor temperature controls have an on-line percentage exceeding 90 percent. The computer has been available more than 95 percent of process stream time since commissioning.

OPERATOR ACCEPTANCE

Prior to commissioning, Gulf engineers were trained in the use of the computer controls and calculation functions. During staging, preliminary training was done in Houston, and for the IDCOM reactor controls, this included hands-on experience with a reactor dynamic simulator. Operations staff were trained during commissioning by Gulf engineers.

Since the controls had been designed to achieve engineering objectives in a manner very similar to the operators way, operator acceptance was quick. After two weeks of the last staging period in January 1982, all operators were comfortable with the computer controls. This was enhanced for the critical reactor temperature controls because IDCOM regulates in a way very similar to the approach an experienced operator would take. Soon after commissioning and verification of the controls, the operators not only accepted the computer controls, but would insist that they remain on.

References

1. Martin, G. D. and Van Horn, L. D., "IDCOM Hierarchical Multivariable Control Application," presented at the 1982 Canadian Conference on Industrial Computer Systems, McMaster University, Hamilton, Ontario, May 3-5, 1982.
2. Rinne, R., Sunell, H., Latour, P. R., and K. K. Paynter, "Experience with Distillation Unit Computer Control," *Hydrocarbon Processing,* Vol. 61, No. 3, March 1982, pp. 141-148.
3. Van Horn, L. D., "Crude Unit Computer Control—How Good Is It?," *Hydrocarbon Processing,* Vol. 59, No. 4, April 1980, pp. 145-148.
4. Martin, G. D., "Long-Range Predictive Control," *AIChE Journal,* Vol. 27, No. 5, Sept. 1981, pp. 748-753.
5. Marchetti, J. L., Mellichamp, D. A., and Seborg, D. E., "Predictive Control Based on Discrete Convolution Models," *I. &E. C. Process Des. Dev.,* Vol. 22, No. 3, 1983, pp. 488-495.

41
Advanced Control of Acetylene Hydrogenation Reactors*

M. Näsi, M. Alikoski, and D. C. White

Ethylene produced from commercial cracking units must often meet a very restrictive specification on the maximum allowed percentage of acetylene impurity. Neste Oy's ethylene plant at Porvoo, Finland, meets this requirement through the use of fixed bed hydrogenation reactors. Careful control of these reactors to minimize the undesired hydrogenation of ethylene to ethane while meeting the acetylene impurity limits was one of the primary objectives of the computer control project installed on the plant. Details of the hydrogenation reactors advanced controls are presented here.

A summary of this project and an overview of the computer control system is given in Chapter 39. The ethylene unit has a nameplate capacity of 200,000 tpa ethylene. Feedstocks, which come from the nearby Neste refinery, include a heavy feed from the hydrocracker, naphtha, n-pentane n-hexane, LPG (propane/butane) and a refinery offgas stream. Ethane and propane are recycled, and ethane from the nearby benzene unit is also processed.

PROCESS DESCRIPTION

The primary objective in operating the acetylene converter is to assure acetylene removal from the deethanizer overhead so that the ethylene product will meet its stringent acetylene specification. The secondary objective is to minimize ethylene losses in secondary reactions. A schematic flow diagram of the process is shown in Fig. 1. The process sequence consists of two fixed bed adiabatic reactors with interstage cooling. The hydrocarbon reaction takes place in the vapor phase. The catalyst used is palladium on an aluminum oxide pelletized support.

The normal acetylene content in the converter feed is 0.3 to 0.4 vol% and the ethylene product acetylene specification is 5 ppmv. The feed to the converter, a C_2 fraction, comes from the overhead of the deethanizer and flows through three exchangers, where the feed to the primary bed is heated against secondary and primary bed outlet streams, and with steam, if necessary. The temperature of the combined hydrogen and hydrocarbon inlet reaction mixture is controlled. A controlled quantity of hydrogen containing small quantities of carbon monoxide is fed to the reactor feed before the steam exchanger.

The primary bed effluent is cooled against the primary bed feed and then heated or cooled in two exchangers. Either heating or cooling may be required depending on the relative values of the first reactor outlet temperature and the required second reactor inlet temperature. Hydrogen is fed to this stream before the exchangers.

The secondary bed effluent is fed to the deethanizer overhead drum after heat exchange in the feed exchanger and the deethanizer overhead condenser. The feed to the C_2 fractionator comes from the uncondensed vapor from this overhead drum. Green oil, which is a high molecular weight material produced by undesirable side reactions, is drained from drums before and after the secondary bed.

Cracking furnace startups and shutdowns, changes in feed type to the cracking heaters and changes in the deethanizer overhead reflux ratio introduce significant disturbances to the rate and acetylene content of the feed to the converters.

During the course of a run the catalyst in a bed deactivates and the inlet temperature or hydrogen addition rate must be increased to compensate. When the operat-

* Nokia Electronics of Helsinki, Finland, provided the computer system including the basic control and operator interface software. Special acknowledgment is due to Aimo Sorsa, the Nokia project manager, and Carl Wetzel, the Setpoint project manager for their diligent efforts in support of the project. Other major contributors were Mauri Sourander, Matti Kolari, Pentti Paurola and Jari Vettenranta of Neste; Pentti Tynnila and Esa Nieminen of Nokia, and Pierre Latour, Lynn Richard and Mark Darby of Setpoint.

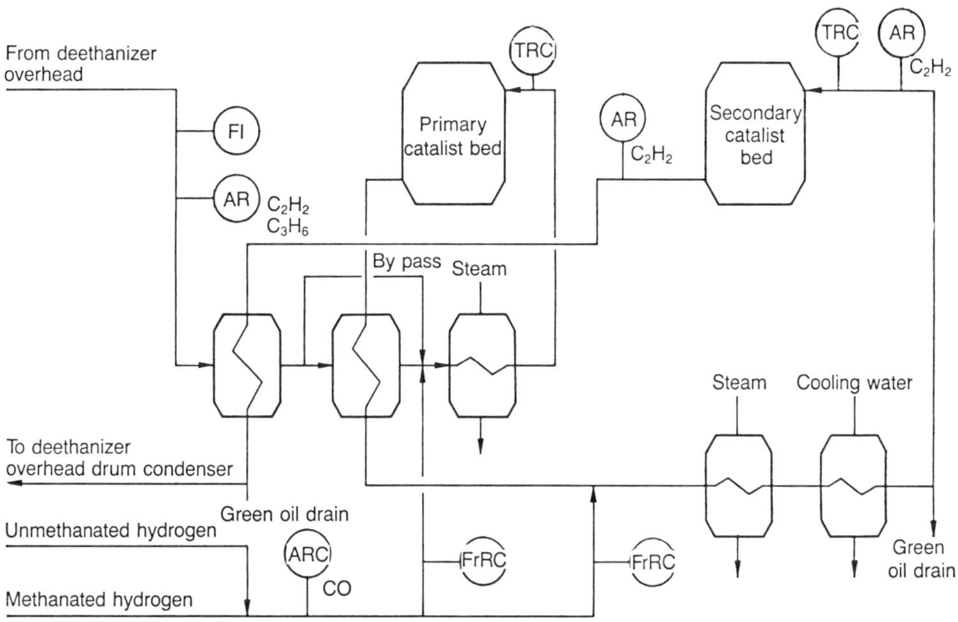

Figure 1. Acetylene converter flow diagram.

ing limit is reached, the deactivated catalyst is regenerated. There are three similar catalyst vessels. Only two of these are in service simultaneously, while the third is either on standby or being regenerated.

Basic Instrumentation and On-Line Analyzers

The basic control equipment and on-line analytical equipment includes:

Primary bed feed:

- Hydrogen addition is ratioed to feed flow
- The temperature controller can activate either of two valves; one allows feed to bypass the second exchanger against primary bed effluent, while the other controls the steam flow to the third feed exchanger.

Secondary bed feed:

- Hydrogen addition is ratioed to feed flow
- The temperature controller can activate either of two valves; one controls steam flow to a heating exchanger, the other cooling water flow to a cooling exchanger.

Hydrogen feed:

- The carbon monoxide content of hydrogen is controlled by a CO analyzer controller. This controller manipulates a valve that adjusts the amount of unmethanated hydrogen mixed with the hydrogen fed to the reactors.

For control of the converters there are two gas chromatographs:

- One chromatograph analyzes converter feed for acetylene and propylene content and the primary bed effluent for acetylene content. New analyses are obtained every 16 minutes.
- The other is used only for measuring converter effluent acetylene with an analyzer frequency of 8 minutes.

EFFECT OF PROCESS VARIABLES ON THE REACTION

The major reactions occurring in the reactor can be summarized as:

1. acetylene + hydrogen → ethylene
2. ethylene + hydrogen → ethane
3. acetylene + ethylene + hydrogen → C_4 + polymeric material (green oil)

Of these reactions, the first is desirable while the second two are undesirable. The last reaction is poorly understood and its exact stochiometry is unknown. When reference is made to the activity of the catalyst, it is the rate of reaction (1) that is being measured. The selectivity of the catalyst is measured by the relative rate of reaction (1) compared to reaction (2).

The three major manipulated variables affecting the rates of reaction are:

- Reactor inlet temperature
- Reactor inlet hydrogen composition
- Reactor inlet carbon monoxide composition.

Assuming adequate inlet hydrogen is available, then the effect of inlet reactor temperature on the reactor effluent acetylene is shown in Fig. 2. Note that the effect is nonlinear, i.e., at different inlet acetylene levels the change in acetylene across the reactor per unit change in inlet temperature varies.

The effect of hydrogen is shown in Fig. 3. Again the effect is highly nonlinear with large changes in acetylene per unit change in hydrogen observed at low hydrogen levels. However, the effect of hydrogen on the selectivity of reactions is also quite strong as shown in Fig. 4. Ethylene losses increase as hydrogen addition increases.

Carbon monoxide acts as a moderator of the system favoring reaction (1) over reaction (2). Mechanistically, this is explained by the selective adsorption of the carbon monoxide on catalyst sites where the hydrogenation of ethylene would take place. This effect can be seen in Fig. 5. Of course, the carbon monoxide also affects the observed activity of catalyst in a reversible fashion.

Green oil production is primarily a function of temperature with increasing green oil production at higher reactor temperatures.

REGULATORY AND CONSTRAINT CONTROLS

The regulatory and constraint controls are shown in Fig. 6. Their objective is to hold the last reactor outlet acetylene at its target value in spite of disturbances to the plant. This objective is to be met without violation of any unit constraints. To compensate for fluctuations in inlet acetylene or hydrogen composition, control of the inlet hydrogen to acetylene composition ratio for both reactors is performed. The hydrogen composition is calculated from hydrogen addition rates, makeup hydrogen purity, and current plant operating conditions. This calculated value has been found to closely track laboratory measurements on the stream.

There is a target for the hydrogen to acetylene ratio that can be entered by the operator. The hydrogen addi-

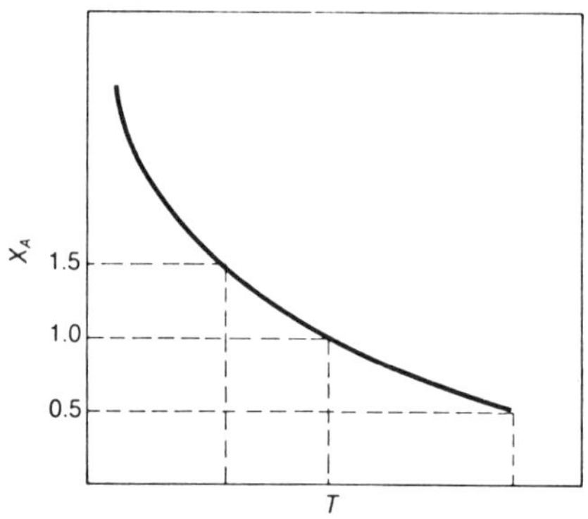

Figure 2. Outlet acetylene composition vs. temperature.

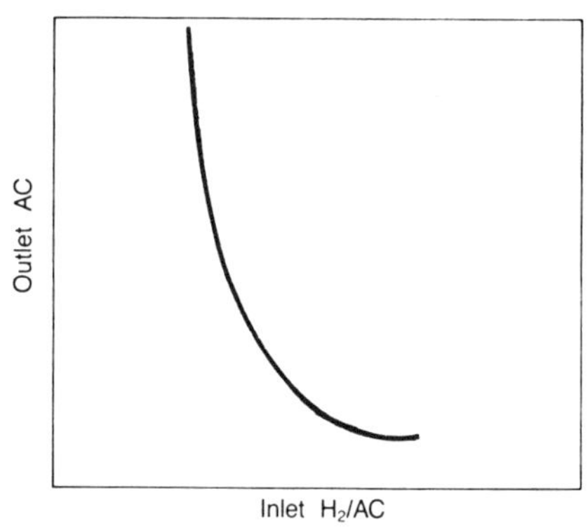

Figure 3. Outlet acetylene composition vs. inlet hydrogen/acetylene ratio.

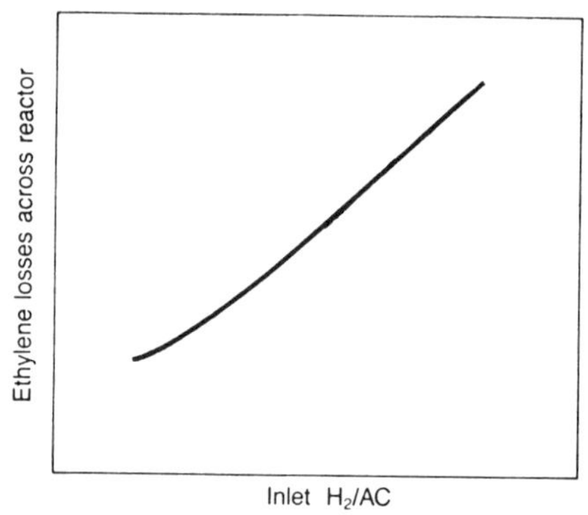

Figure 4. Ethylene loss vs. inlet hydrogen/acetylene ratio.

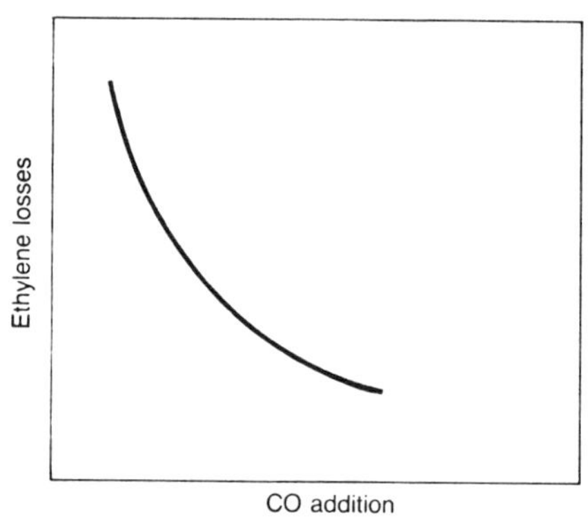

Figure 5. Ethylene loss vs. CO addition.

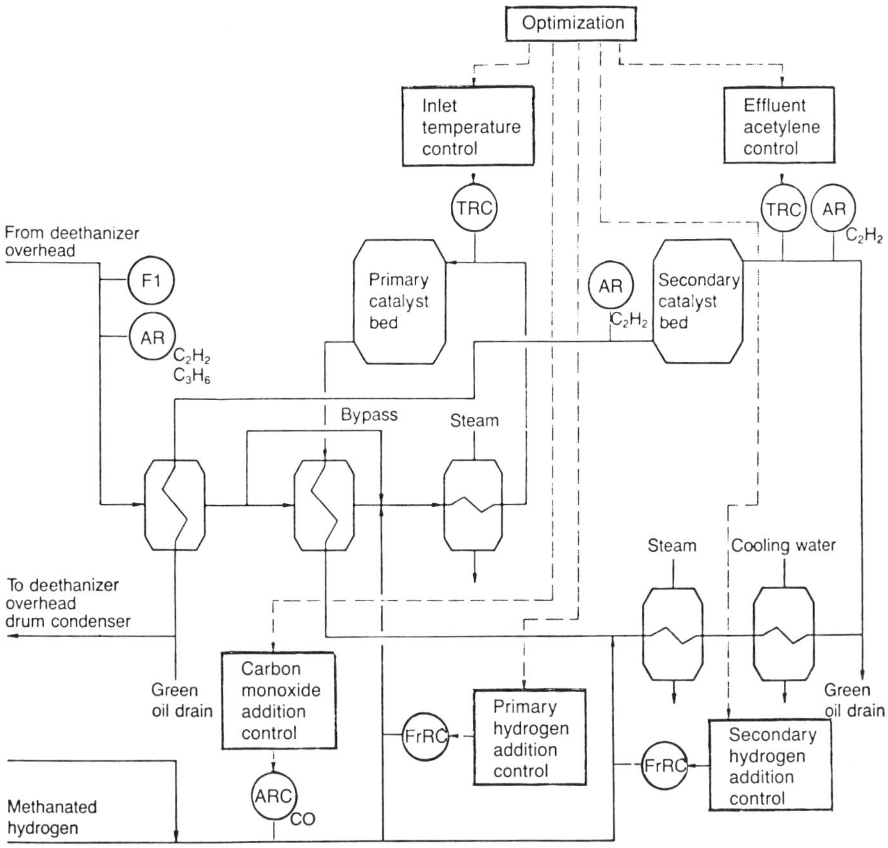

Figure 6. Advanced control strategy.

tion rate calculated from this ratio is checked against the projected maximum or minimum rates based on a number of constraints including maximum bed temperature rise, maximum outlet acetylene and others. If the target ratio does not violate any of these constraints, then it is implemented. If one of the constraints overrides, then the hydrogen addition is set by the limiting constraint. Feedforward is included for major measured disturbances to the reactor system. Feedforward action is nonsymmetrical. Larger compensation is made for disturbance directions that would cause increases in outlet acetylene than for those directions that would cause a decrease.

The inlet temperature of the second reactor bed is controlled to hold the outlet acetylene composition at its target value. A nonlinear gain is calculated for the control loop to compensate for the nonlinearity in process response referred to earlier. Constraint overrides on temperature controls are used as well.

OPTIMIZATION

For the reactor system there are five independent manipulated variables that affect the reaction—two inlet temperature controllers, two hydrogen addition controllers, and a controller on the carbon monoxide addition to the hydrogen. There is one specification—second reactor outlet effluent acetylene composition—that must be met. The available degrees of freedom immediately suggest the opportunity for optimization, i.e., selecting the targets for the set of manipulated variables such that they minimize or maximize some objective function in addition to meeting the specification. For the acetylene hydrogenation reaction, maximizing the net ethylene gain (or minimizing the loss) across the reactor was chosen as the objective function.

Model Formulation

The first requirement for optimization is development of a suitable model which reflects the effects of manipulated variables on the objective function. Extensive studies have been reported in the literature for the individual acetylene or ethylene hydrogenation kinetics over noble metal catalysts. Generally, the results have been found to follow a Langmuir-Hinshelwood formulation, i.e.,

$$R_1 = \frac{K_1 PH^{Y_1} PA^{Z_1}}{(1 + BHPH^{1/2} + BAPA + BCPC)^M} \exp\left(\frac{-DE_1}{RT}\right)$$

$$R_2 = \frac{K_2 PH^{Y_2} PF^{Z_2}}{(1 + DHPH^{1/2} + DEPE + DCPC)^N} \exp\left(\frac{-DE_2}{RT}\right)$$

where R_1 = rate of reaction (1), acetylene hydrogenation
R_2 = rate of reaction (2), ethylene hydrogenation
DE_1 = Arrhenius constant, reaction (1)
DE_2 = Arrhenius constant, reaction (2)
R = gas constant
T = temperature
PI = partial pressure of component I
BI = absorption coefficient for reaction (1), component I
DI = absorption coefficient for reaction (2), component I
Y_1, Y_2 = order of reactions with respect to hydrogen
Z_1, Z_2 = order of reactions with respect to unsaturated components
M = constant relating to absorption term, reaction (1)
N = constant relating to absorption term, reaction (2)
Component I = H–hydrogen
A–acetylene
E–ethylene
C–carbon monoxide

Each reactor is modeled as a plug flow, adiabatic system with no significant dispersion or mass transfer effects. The kinetic expressions for the combined system of acetylene hydrogenation in the presence of ethylene at commercial concentration levels are simplified to permit an approximate analytical solution to be developed.

To test the validity of the reactor model, a series of tests were run on the unit. The predictions of the model matched the measured responses of the plant.

The catalyst in the reactors deactivates with time. To model this deactivation and compensate for other process effects, parameters are calculated from on-line reactor performance data. These parameters are then used in the reactor model as the effective catalyst reactor volume. The deactivation of the catalyst implies that the effective reactor volume should decrease with time and this effect is indeed observed.

The optimization procedure is to:

1. Calculate the effect of incremental moves in independent (manipulated) variables such as CO concentration, temperature, H_2/AC, on the ethylene change across the reactor at constant last reactor effluent acetylene.
2. Calculate optimum setpoints using partial derivative information from (1). The optimization problem is a nonlinear one and a conjugate gradient search technique is used to solve the problem.
3. Implement the optimization solution setpoints through the regulatory and constraint control structure discussed previously.

RESULTS AND EXPERIENCES

Before the computer control was implemented, the converters were operated "on the safe side" by adding too much hydrogen and using unnecessarily high temperatures, which caused over-hydrogeneration of ethylene. Operating without on-line analyzers also contributed to the problem, because laboratory analyses were performed only six times a day.

After the introduction of the acetylene converter controls and optimizer, the hydrogen/acetylene ratio and ethylene losses have decreased. To minimize ethylene losses, the catalyst runs have become shorter. However, the net economic benefit gained by computer control and optimization has been significant.

A summary of the results is shown in Table 1.

Table 1
Results

	Before	After
Ethylene losses, percent of feed	0.3 to 0.4	0
Hydrogen/acetylene ratio at feed inlet	3.1	1.7
Off spec laboratory analyses of ethylene product (acetylene), %	0.8	0.4
Run length, months	10	4
Green oil	Base	Slightly increased

In addition a remarkable advantage is the more complete information on the process conditions available to the operating personnel. This gives them a more thorough understanding of the functioning of the acetylene converters and is very useful in day-to-day monitoring and operation.

At present there is a control problem caused by slow frequency of analyses from the gas chromatographs. The dynamics of the converters' responses are often faster than the frequency of analysis and one of the regulatory targets or constraints may override the optimum target. Further improvements in control would be gained by more frequent or continuous analyzers.

42

Economic Benefits of Advanced Alkylation Unit Control

G. L. Funk and J. A. Feldman

Alkylation units are getting a more critical review because of their growing importance in making larger volumes of "no-lead" gasolines. This chapter shows an economic benefit of $1.0 million per year by using advanced control concepts on a typical plant of older design (1960–70). The basis for this comparison is a hydrofluoric acid (HF) alkylation unit providing 390 metric tons per day (3,300 bpd) of alkylate. The use of advanced control upgrades the product and reduces acid consumption for units of both older and modern designs.

THE NEED FOR BETTER CONTROL

The HF alkylation process makes high octane gasoline blendstock from liquid isobutane and olefins (propylene, butylenes, or amylenes) in the presence of the catalyst, liquid HF. The principal chemical reaction is the conversion of butylenes and isobutane into isooctanes:

$$\underset{\text{Isobutane}}{iC_4H_{10}} + \underset{\text{Butylenes}}{C_4H_8} \longrightarrow \underset{\text{Isooctanes}}{C_8H_{18}}$$

Unfortunately, this reaction is not perfect, and some paraffins that boil above and below the desired octanes are formed. Impurities in the feed, acid, and normal operating practices all can contribute to additional side reactions. Normally, between 30% and 40% of the fresh feed from a catalytic cracking unit is butylenes with a similar amount of isobutane. The remaining portion of the feed (Table 1) promotes or accelerates undesirable side reactions. The harmful nature of the side reactions are: increased isobutane consumption, increased acid consumption, increased acid soluble oil (polymer) formation, equipment fouling, and/or corrosion problems.

Alkylation plants are designed to operate smoothly, to be stable and to require very little human operator intervention. The process operations are very straightforward. The apparent ease and smooth operation of the plant does not necessarily mean the plant is operating optimally. The designers of alkylation plants usually assume that only limited instrumentation, very few calculations, and limited information are available to the hu-

Table 1

Range in Olefin Feed Composition

HF alky feed composition	Weight percent
Propane and propylene	0.0– 0.8
Isobutane	30.0–42.0
Normal butane	14.0–22.0
Butylenes	26.0–45.0
Isopentane	1.2–14.0
Normal pentane	0.1–1.5
Amylenes	0.3– 4.5
Water, [ppm]	30.0–80.0
Mercaptan sulfur, [ppm]	5.0–40.0

Incremental Value of Components

Component	Dollars per metric ton*
$i\text{-}C_4$	96
$n\text{-}C_4$	90
$i\text{-}C_5$	160
Alkylate	185
Ton-octane value = $4.50	

* Incremental value above propane value.

Benefits from Computer Control

Item	Thousands of dollars/year
Product quality improvement	484
Increased product recovery	204
Acid system improvement	53
Fractionation energy savings	403
Total benefits	1,144

man operators of the plant. Thus, the designs include only fail-safe features and those control loops that are easy for operators to manage and understand. The control loops generally are not designed to be efficient for use with automatic computer control. Typically, a reevaluation and redesign of the control system are necessary to gain maximum benefit from the newer advanced technology.

KEY FACTORS TO BE CONTROLLED

The successful operation of an HF alkylation plant depends on these principal factors:

- Presenting a proper composition mixture to the reactors. This includes appropriate volumes of isobutane, olefins and the HF acid. Each of these component streams should be as free of impurities as possible.
- Maintaining the proper reaction environment, which includes adequate contacting, controlled temperature, freedom from surges, etc.
- Making a proper separation of the reactor effluent into its various components.This includes separation of a high-purity isobutane recycle stream (note that impurities decrease capacity, promote undesirable side reactions and tend to draw the unit away from the most efficient operating conditions). The separation into products may include many if not all of these steps—deisobutanizing, depropanizing, debutanizing, deisopentanizing, butane-isopentane splitting and alkylate splitting. In addition, an isobutane-normal butane splitter may be used for feed preparation.

These operating factors indicate the essential and large number of separate stages in the process which must be coordinated or brought to a high level of performance for an efficient HF alkylation plant.

One basic operating principle requires that the reactor feed be maintained with a high ratio of isobutane-to-olefins (IC4/OLE) to enhance alkylate quality, to increase overall alkylate yield, and to suppress the undesirable side reactions, such as those that produce heavy alkylate and acid soluble oil or polymers.

The octane number of the alkylate has a non-linear relationship to the IC4/OLE ratio, as shown in Fig. 1. The octane number increases with increasing ratio, but at a continuously decreasing rate. When the ratio is increased from 4 to 6, the octane number increase is about 1.25 while the same increment of IC4/OLE increase from 10 to 12 produces only a 0.5 octane number increase.

Acid soluble oil (ASO) formation is related to IC4/OLE, as shown in Fig. 2. This relation also is non-linear showing a declining trend in value as the IC4/OLE ratio increases.

Alkylate yield (weight percent) at a fixed process charge rate is related to the IC4/OLE ratio, as shown in Fig. 3. As the external IC4/OLE is increased, the amount of fresh feed, i.e., butanes-butylenes mixture, which the plant can accept must be reduced because of physical constraints. This presents an optimization problem in selecting the best IC4/OLE ratio, since lower charge rates can be used to increase the IC4/ OLE ratio, thus yielding a higher octane product.

Reactor temperature variation has a near linear effect on alkylate octane number, as shown in Fig. 4. This effect is a depreciation of about 1.0 octane number for each 12°C rise in reaction temperature. The alkylation reaction is exothermic so that heat removal is a critical part of the process. Alkylation reactor temperature control by conventional means shows a standard deviation of temperature of 2.5° to 4.0°C. Some individual reactors showed even greater variations.

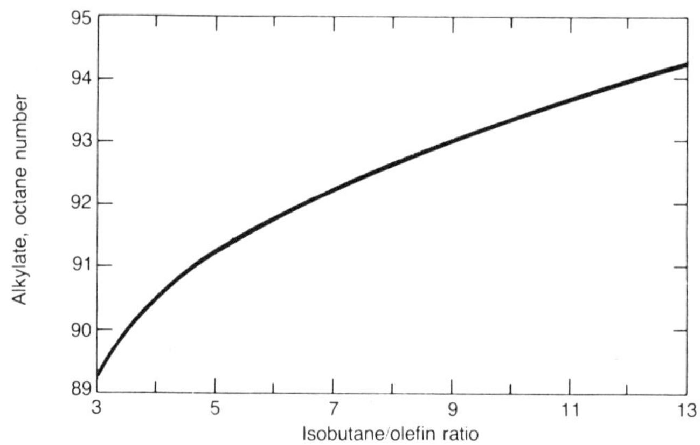

Figure 1. Product quality is non-linear with reactants ratio.

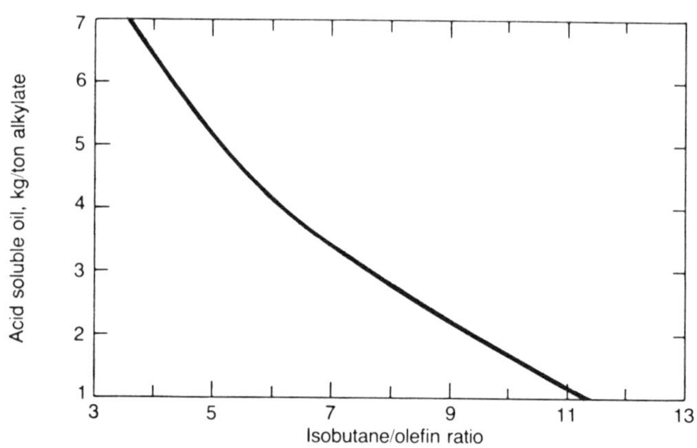

Figure 2. Formation of acid soluble oil also is non-linear.

A quality factor (F) is frequently used to indicate relative quality of alkylates measured by Research Octane Number, Aviation Blending Octane Number or Supercharge Blending Index Number. The F factor is defined[1] as:

F = (IC4/OLE)(IC4)/(SV)

where IC4/OLE = isobutane-olefin volume ratio
IC4 = volume fraction isobutane in reactor effluent
SV = olefin space velocity, volume/reactor volume/hour.

Note that as either IC4/OLE ratio or the fraction isobutane in reactor effluent (or both) increases, the F factor increases. An increase in olefin space velocity produces a contrary effect. The F factor relates to quality (octane), with higher values for F indicating higher octane quality. A typical variation shows an F factor standard deviation of 0.85 units with values ranging between 2.0 and 8.0. This represents a deviation of octane number of about ±0.25. Typically a change in the F factor of 2 indicates a clear octane change of about 0.55 in this F factor range.

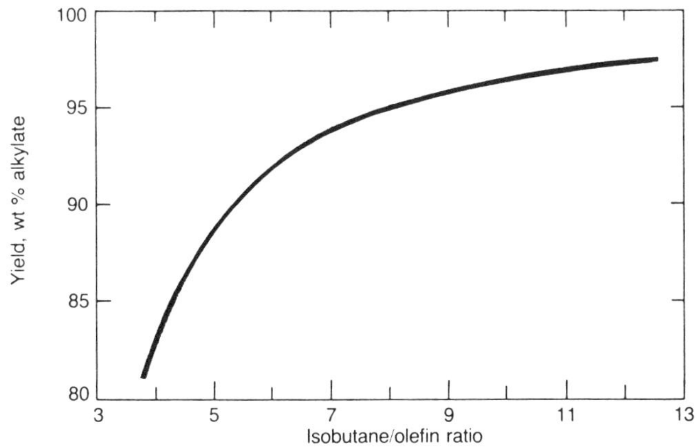

Figure 3. Feed constraints can limit yield advantages.

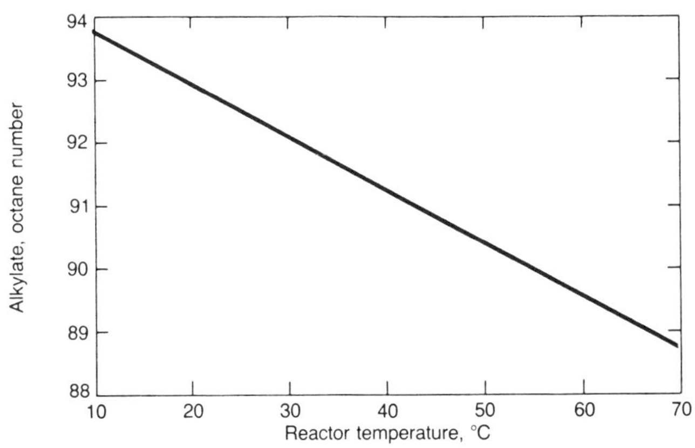

Figure 4. Lower reaction temperatures give better alkylate.

Table 1 shows that the butylenes contained in the B-B feed typically vary over a wide range—from 26% to 45% of the stream. Further, the isobutane content of the recycle stream ranges between 80% and 95% because of inadequate control of the downstream fractionators. These two factors—variation in B-B composition and purity of the isobutane recycle—make it very difficult for an operator to effectively hold the uniform IC4/OLE ratio that is required for top plant performance and high and stable octane quality.

Sometimes the source of process heat is a furnace. Furnace control by conventional means presents considerable variability in coil outlet temperatures. It is common for the variability between coils and for each separate coil to have a standard deviation of 5–7°C.

The task of balancing a multiple coil furnace manually is not an easy one. Maintaining efficient operation while holding this furnace unit near constraint levels manually is a formidable task. Panel board operators require a "comfort zone" some distance from desired process limits in which to operate. This removes the process far from optimum conditions. With proper computer control, the standard deviation of temperature is reduced typically to less than 1.5°C.

The preceding discussion has outlined the problems associated with manual or conventional "panel board" operation. Typical process variability has been shown for principal operating parameters. Increased production, improved quality, decreased costs—all are management's goals. The operator is a critical key to reaching them. A "better operator" is required. This leads directly to greater application for advanced supervision computer control. The benefits are substantial and achievable with the new technology available.

BENEFITS OF ADVANCED CONTROL

Application of advanced supervisory computer control in an HF alkylation plant has provided these benefits:

- Increased yield of alkylate per ton of feed stock
- Increased octane number quality of the alkylate
- Increased unit capacity
- Increased efficiency in the use of energy
- Reduced acid consumption
- Reduced acid soluble oil/polymer production
- Increased purity of separated products (through control of separation units).

The control system gives the capability to increase the IC4/OLE ratio. Consequently, the alkylate as a fraction of total products yielded is increased by about 1.7 percentage points with a corresponding decrease in non-alkylate products. The effect on the unit material balance is shown in Fig. 5. Most of the 1.7% gain is due to increased alkylate formation in the reactors. The remainder is attributable to computer controlled enhancement of recovery in the fractionation train. The alkylate yield from the reactors is increased by about four percent of the available B-B feed. Simultaneously the octane number of total alkylate produced is upgraded by 3/4 to 1.0 number. Acid consumption is significantly reduced by 10 to 15%.

The basic advanced supervisory computer control functions derived and recommended for application were:

For the alkylation reactors—

- isobutane-to-olefin composition ratio control
- acid (HF)-to-hydrocarbon ratio control
- reactor load optimization control

For plant furnaces—

- coil outlet temperature balancing control
- enthalpy control of heat input
- efficiency control
- constraint control

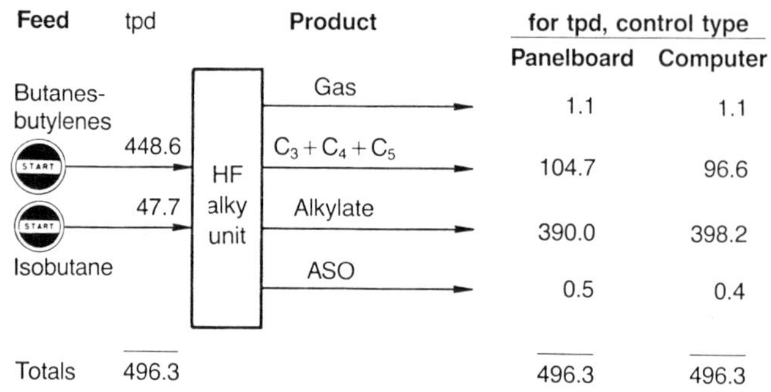

Figure 5. More alkylate is obtained with computer control.

For fractionators (including HF stripper)—

- product specification and composition purity control
- efficiency control
- constraint control

Several references[2,3,4] are available which contain good descriptions of application of advanced supervisory computer control to fractionators.

HF ALKYLATION

The feed is a mixture of butanes and butylenes (B-B) supplied from a fluid catalytic cracking (FCC) unit and some outside isobutane. The feed composition varies considerably depending on upstream operation of the

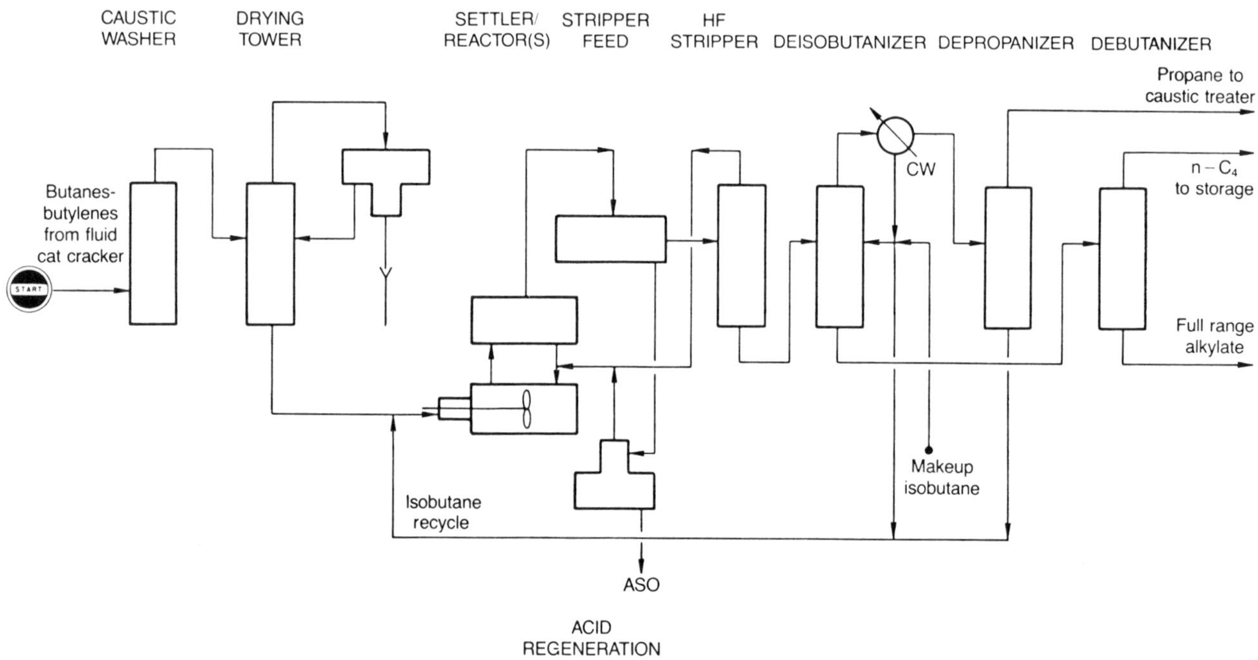

Figure 6. A typical HF alkylation flow diagram.

FCC unit and its associated gas plant. Table 1 shows typical ranges in composition of feed processed.

The B-B mixture received from the FCC unit is caustic washed, water washed and dried by distillation. The water content of dry feed is usually maintained below 50 ppmw. The dry B-B is held in the reactor feed surge tank (see Fig. 6).

The dry B-B charge is mixed with the isobutane recycle stream and charged to several parallel reactors. The fresh HF, regenerated acid and HF stripper overhead acid streams are mixed and pumped directly to the reactors. The reactors are cooled to maintain as low a reaction temperature as possible. The reactor effluent circulates from each reactor area to the stripper charge tank. The combined hydrocarbon mixture from all the reactors goes to the HF stripper charge tank where acid is separated from the hydrocarbons. The acid is returned to the reactors after being combined with a small stream of purified acid supplied by the acid regenerator. The mainly alkylate-isobutane mixture (with some acid) goes to the HF stripper from the charge tank.

The reboiled HF stripper tower removes any remaining free HF. The HF acid with some isobutane is sent overhead from the HF stripper and is returned to the reactors. The stripper bottoms product is fed directly to the deisobutanizing tower where the isobutane-propane portion is taken overhead and passed through a partial condenser. The vapor fraction is finally fed to the depropanizing tower. The depropanizer overhead eventually goes to the plant fuel gas or LPG recovery system. The depropanizer bottom product, mostly isobutane, joins the isobutane recycle stream to the reactors.

The deisobutanizer bottom product is the feed to the debutanizer. This tower sends overhead the normal butane leaving as the bottom product a debutanized full range alkylate stream. The normal butane stream may be later used for blending motor fuels and aviation gasoline and alkylate may be fractionated (split) into light and heavy alkylates.

References

1. Funk, G. L. and Smith, D. E., "Estimating economic incentives for a computer control system: An applications approach," *IEEE Transactions on Industry Applications,* Vol. IA-15, No. 4 (July/Aug. 1979), pp. 394–398.
2. Smith, D. E. et al, "Distill with composition control," *Hydrocarbon Processing,* Vol. 57, No. 2 (Feb. 1978), pp. 99–107.
3. Griffin, D. E., "Fractionator controls can save energy," *Oil & Gas Journal,* Vol. 79, No. 10 (March 9, 1981), pp. 140–154.
4. Funk, G. L. and Smith, D. E., "Optimizing fractionation economics through advanced control," *Proceedings I.S.A. Conference,* Houston, Tx., Vol. 35, Part 2 (Oct. 20–23, 1980), pp. 599–609.
5. Gray, James H. and Handwerk, Glenn E., *Petroleum Refining: Technology and Economics,* Marcel Dekker Inc., New York, NY (1975), pp. 142–161.

43

Control of NH_3 Process With Varying Natural Gas LHV

F. Yazhari

Prior to the discovery of natural gas in Oregon and its use in Reichhold's ammonia plant, the gas supply had a fairly constant Btu value of 1045. The control of critical process variables was accomplished by assuming a constant gas composition. Due to the difference in composition of the natural gas discovered in Oregon (930 to 950 Btu/scf), and the gas from the utility, and since the two gases are blended per area demand, the heating value of the incoming gas can now vary from 930 to 1045 Btu within a very short time. The control schemes already in use could not compensate for the wide swings in gas composition.

The drastic and frequent changes in composition of incoming natural gas could cause a complete plant shutdown if the process were to operate at maximum capacity. Therefore, we reduced maximum ammonia production by 2.5%. We also determined that under these frequent composition changes our gas consumption was increased by about 3.0%.

To eliminate these problems and improve efficiency, we decided to acquire an instrument that would provide our process control computer with the composition and heating value of the incoming gas in the least amount of time possible.

Due to their long analysis time, the combustion calorimeter and the gas chromatograph would not meet our objectives. Therefore, we purchased a Calsensor natural gas heating value analyzer—which uses infrared absorption techniques and provides the heating value and composition of the gas on a continuous basis—and interfaced it to our process control computer. The algorithms for the control of the following variables were modified:

Natural Gas Used As Feedstock. Rather than controlling the volumetric flow rate of the gas used as feedstock, the rate of therm usage is now under control. The volume of the gas is varied to compensate for changes in gas composition.

Steam to Carbon Ratio. Since the composition of the gas is continuously analyzed, the computer now maintains the desired moles of steam per mole of carbon fed to the primary reformer.

Hydrogen to Nitrogen Ratio. To maintain a hydrogen to nitrogen ratio of three to one in the synthesis loop, an on-line gas chromatograph is utilized. The chromatograph provides the computer with the composition of the process gas. The computer calculates the H_2/N_2 ratio and, by using a dead-time compensating algorithm, sets the ratio of therms of natural gas used as feedstock to total moles of nitrogen introduced to the front end of the plant. The significant portion of the required nitrogen is introduced through air to the secondary reformer. But, as Btu's of the incoming gas decrease its nitrogen content increases, since the Oregon gas contains 5.0–6.0% nitrogen.

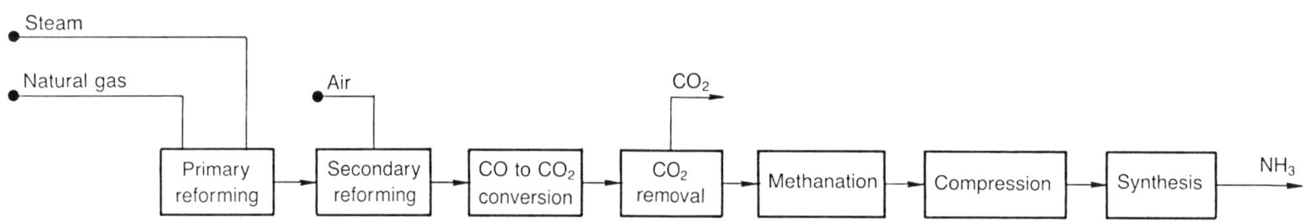

Figure 1. Block diagram of a typical ammonia plant.

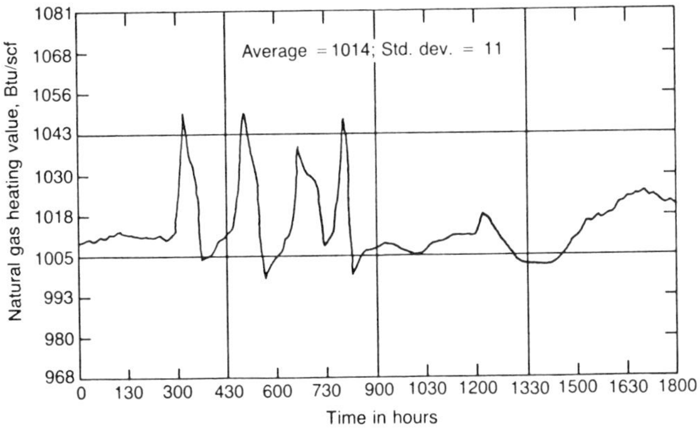

Figure 2. Natural gas heating value, Btu/scf variations.

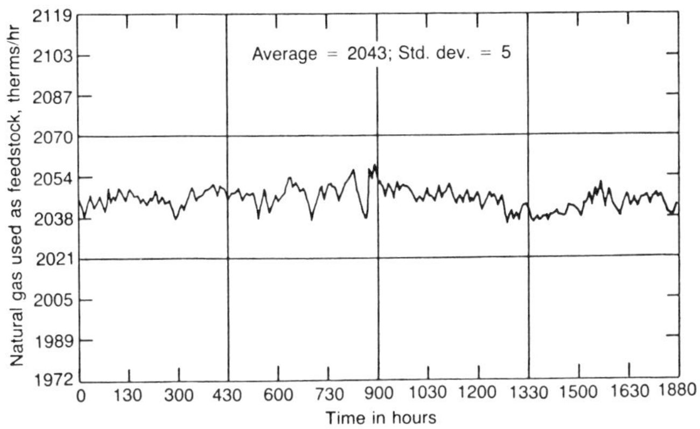

Figure 3. Natural gas usage as feedstock variations in therms per hour.

The computer continuously calculates the total nitrogen introduced to the process and adjusts the air flow to maintain the desired therms per nitrogen ratio required to maintain a H_2/N_2 ratio of three in the synthesis loop.

Primary Reformer Temperature. The temperature of the process gas exiting the primary reformer is now controlled by adjusting the total therms of fuel fed to this unit. Because of a long dead-time in the control loop, a dead-time compensating algorithm is used to set the required therms of fuel usage. The volumetric flow rate of natural gas is then adjusted to compensate for any deviation in the therm usage caused by a change in the natural gas composition or variation in the flow rate of the purge and flash gases burned as fuel.

Figure 1 is a block diagram of processes in the ammonia plant. Figure 2 is the plot of the natural gas heating value over an 18-hour period. Figures 3 through 6 are the plots of the critical variables under computer control during the same period.

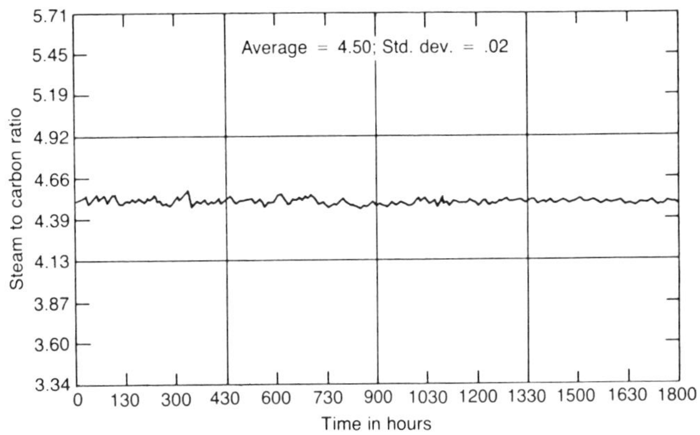

Figure 4. Steam to carbon ratio variations with natural gas changes.

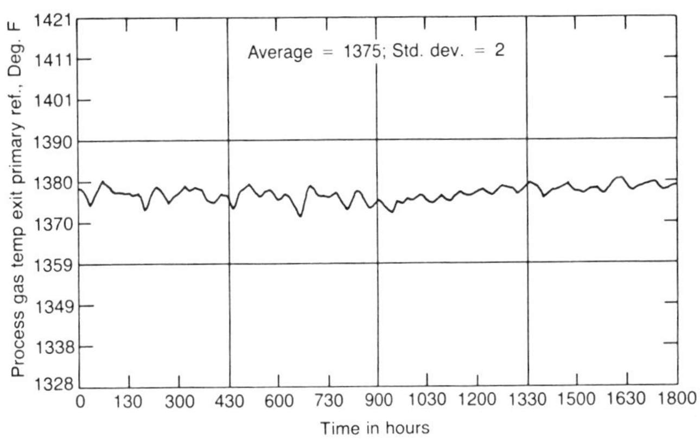

Figure 5. Process gas temperature exit primary reformer °F variations.

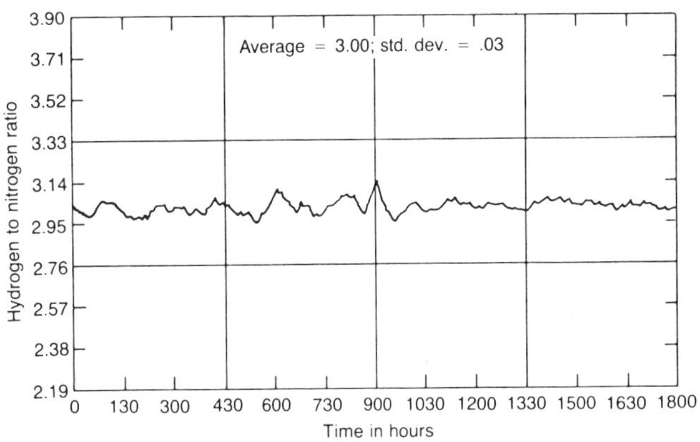

Figure 6. Hydrogen to nitrogen ratio variations.

With the use of the on-line gas analyzer and modification of the control algorithms, we are now able to again operate at the highest production rate of 250 tons per day. In addition, we have been able to eliminate the 3.0% increase in gas consumption which resulted from frequent and drastic changes in the gas composition.

44

Adaptive Gain Improves Reactor Control*

M. J. Whatley and D. C. Pott

An interesting application of the modern control theory technique called adaptive control is presently in use on a process located at a plant in the Standard Oil Co. (Ind.) system. The results of this application are: first, the transformation of an uncontrollable process to a controllable one; and, second, a significant economic savings to the corporation. It also shows that a detailed analysis of the mechanical, chemical, and control systems can provide both the basis for revising an existing control system and some of the reasons why that system is inadequate.

The contrast between classical and modern control theories is seen in four major areas. In classical systems, the controller manipulates the error to calculate its output. The error is subjected to a limited number of analog computing functions such as multiplication by a constant and integration. Each controller has one specific strategy—to hold the measured variable at the setpoint. And each controller controls only one variable, independent of all others.

In a modern system, the controller can manipulate many different variables in addition to the error to compute its output. Its computing power is unlimited in a practical sense. It can do all classical computations plus many others, including a table look-up. The controller can independently change its strategy as a function of time or the condition of the process. It can also control many variables at one time to accomplish a complex objective.

The implementation of a modern control theory project requires a good understanding of the dynamic, as well as the steady-state character of the process. As you can see we are limited only by our understanding of the process, our imagination, and the economics of the situation.

THE PROCESS

Now let's take a look at the process: a polymerization unit built in 1968. The unit contains three sections: feed preparation, reaction, and finishing. The reaction section, which is the part focused upon, is shown in Fig. 1. It contains three similar reactors. Each reactor is divided, for control purposes, into two sections, referred to as A-zone on the left and B-zone on the right. The design capacity of the unit, when built, was 25 million lb per year. In 1972, it was debottlenecked to between 60 and 90 million lb per year, depending upon product. There were no changes to the reactor internals.

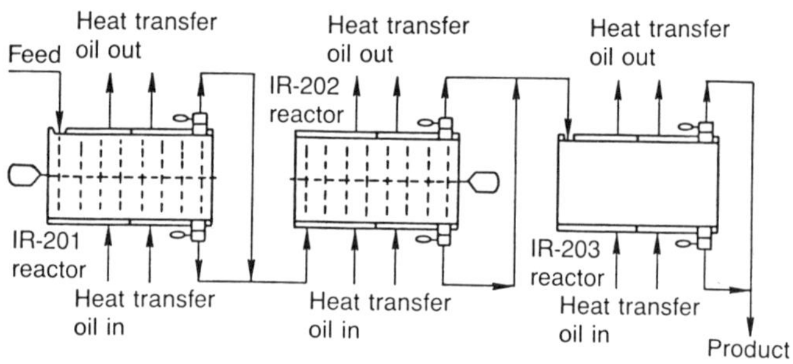

Figure 1. Polymerization reaction section.

The polymerization reaction is highly exothermic. Tight reaction temperature control $\pm 5°F$ is necessary, with $\pm 1°F$ desirable, to produce a product whose physical properties are within specifications. The reaction temperature is controlled by transferring a portion of the heat of reaction from the reactants to a circulating heat transfer oil. Control is made difficult by the high viscosity and limited agitation of the reactants, which contribute to poor heat transfer. Temperature control has been a persistent problem in the first two reactors. Since most of the reaction occurs in the first two reactors, the third reactor has few problems and was not part of our study.

* Originally presented at the API 49th Midyear Refining Meeting, New Orleans, Louisiana, 1984.

THE PROBLEM

The nature of the problem is three-fold. First, each of the many grades of polymer made in the unit require a different set of values for the controller tuning constants. Some of these products are made for only 12 to 24 hours. This precludes retuning for each change. Second, even for a particular product the controller cannot be tuned for all conditions. If it is tuned for normal operation, the system response to upsets is very poor, increasing the probability of a runaway reaction. Conversely, if it is tuned to handle upsets, normal operation suffers. This is analogous to operating the accelerator and brake of a car as if in an emergency situation. It would be somewhat difficult to control the speed of the vehicle. And third, setpoint changes must be made extremely slowly to preserve some system stability The slow setpoint changes increase the amount of off-spec material produced during the transition between different products.

In view of these problems, the controller tuner has only one course of action—compromise tuning. This results in normal operations which are poor, but better than in a system tuned for upsets. Upset response is also poor, but better than in a system tuned for normal operation. And the probability of runaway reactions remains with us.

Remembering that temperature control to ±5°F (±3% of scale) is necessary and to ±1°F (±2/3% of scale) is desirable, let's look at some graphic evidence of the problems confronted.

Figures 2 through 4 are redrawn graphs from process chart recorders. Figure 2 shows a poor but not unusual

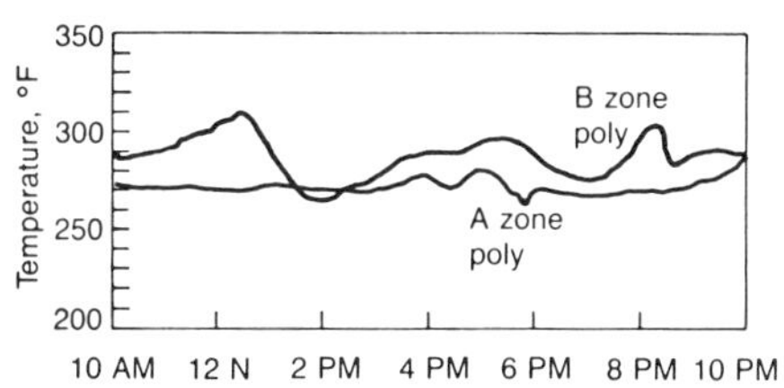

Figure 2. Poor but not unusual control in the first reactor.

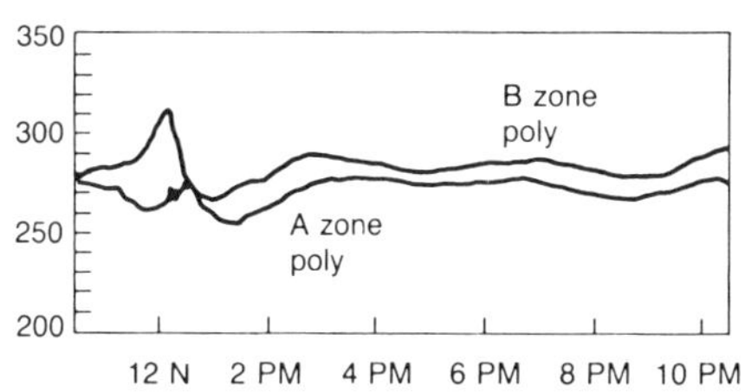

Figure 3. Response to an upset.

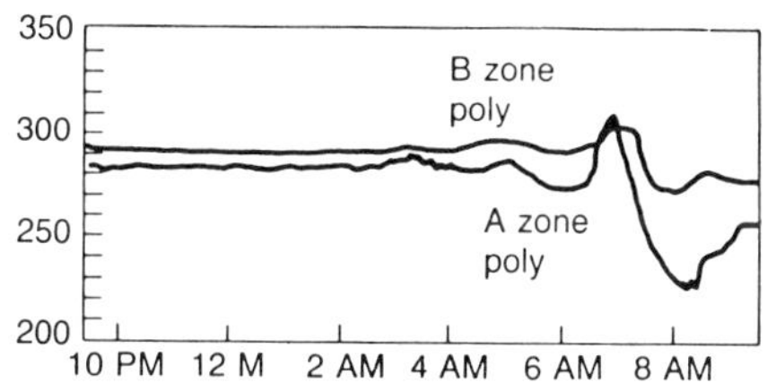

Figure 4. Response to setpoint changes.

operation in the first reactor, zones A and B. In the A-zone the temperature deviation is an acceptable ±5°F, but in the B-zone it is greater than ±20°F.

On the left-hand side of Fig. 3 is evidence of the system response to an upset. In this case it was necessary for the operator to intervene at noon and transfer the B-zone to manual control to prevent a runaway It was fortunate that an operator was watching the chart since no alarm had sounded.

Figure 4 depicts at attempt (at 6:00 a.m.) to raise the temperature in the A-zone and lower it in the B-zone. At 6:30 a.m. it was again necessary for an operator to intervene to prevent a runaway.

The result of temperature control problems in the process in 1981 was the production of 3.0 million lb of off-spec material with an economic penalty of 5 to 10 cents per lb.

SYSTEM ANALYSIS

In 1981 a detailed system analysis program was begun. The objectives were to determine the real causes of the temperature control problems and, if possible, to design a control scheme to eliminate them. An early phase of the program was to determine the interrelationships of the important variables in the reactor control process. Since these processes are identical in all reactor zones, the analysis can be generalized and we need look at only one zone of one reactor.

Figure 5 is a simplified process and instrumentation diagram of one reactor zone. The polymer feed enters the reactor on the left, is polymerized, then product is withdrawn on the right. Within the reactor the reactants are mildly agitated—about to the degree of passing a knife through cake batter. A portion of the heat of reaction is removed by the heat-transfer oil. This controls the reaction temperature and thus the reaction rate.

The classical cascade control scheme shown here regulates the reaction temperature by varying the temperature of the circulating heat-transfer oil. The reaction temperature controller sets the setpoint of the oil temperature controller. That controller modulates either the hot or cool oil valves to adjust the circulating oil temper-

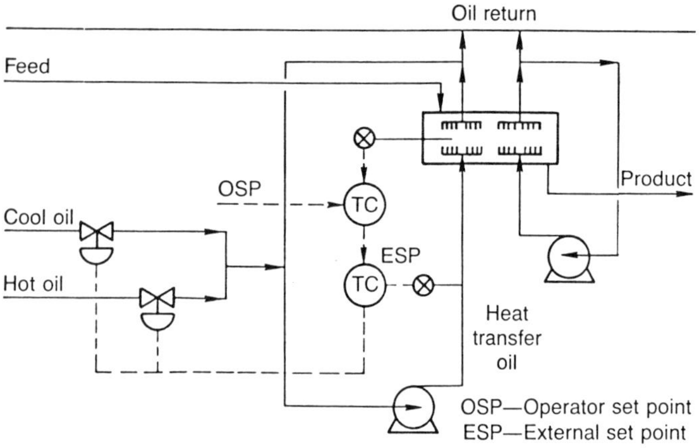

Figure 5. Simplified reactor zone P&I diagram.

ature to match its setpoint. We use two controllers here to minimize the effects of upsets in the oil system.

By inspecting the diagram we can see four of the five major areas that our analysis must cover. They are: the heat exchange mechanism between the oil and the polymer in the reactor, the relationship between the circulating oil temperature and the inlet oil flow, the relationship between the temperatures of the circulating and inlet oils, and the relationship between the fractional opening of the oil control valves and the flow through the valves. The fifth area is the need to control both upset and normal operations.

Now let's take a look at the results of the detailed analysis of these relationships.

Figure 6 shows the result of the analysis of the reactor heat-exchanger mechanism. We model the reactor as a simple heat exchanger where T_c is the circulating oil temperature and T_p is the polymer temperature. Not surprisingly we find a straightline relationship between changes in the oil circulation and polymer temperatures. In fact, observations in the field show this to be nearly one-to-one.

Analysis of the oil-circulation loop modeled as a mixing pipe shows two relationships between the flow of inlet oil and circulation temperature. The first is that the change in inlet oil flow is directly proportional to the change wanted in oil circulation temperature.

The second is that the change in inlet oil flow is inversely proportional to the difference in temperature between the inlet and circulating oils.

Control valves are available with several different flow characteristics, that is, the relationship between flow through the valve and fractional valve opening. The analysis must specify the particular one needed. Since the pressures in the circulation loop and inlet oil headers are relatively constant and there is a straightline relationship between the control and measured variables, a linear valve characteristic is required.

The final point we need to address in our analysis is the necessity to control both normal and upset conditions. To accomplish this, the polymer temperature controller must adapt itself to the condition of the process. With a constant-gain controller, the gain for normal operation is too low to react to runaways or mechanical upsets with the necessary speed. The deviation adaptive tuning scheme shown in Fig. 7, increases the controller gain from a base to a fixed maximum, as the actual polymer temperature moves above the setpoint. This higher gain quickens the controller response to temperature excursions preventing runaways.

In summary, the analysis of the system reveals the following:

1. The temperature of the circulating oil must be closely controlled since there is a one-to-one relationship between it and the polymer temperature.
2. The flow of the inlet oil must be controlled by a linear valve characteristic and there must be driv-

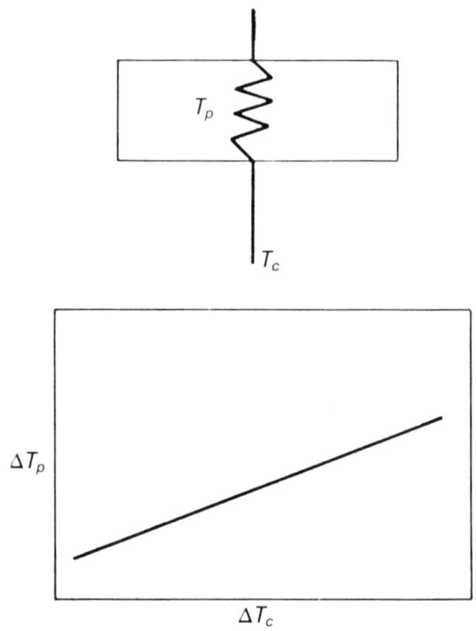

Figure 6. Reactor heat exchange model.

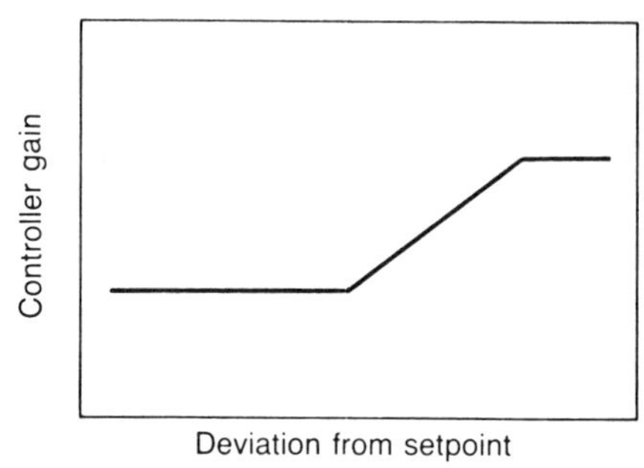

Figure 7. Polymer temperature control adaptive gain.

ing force compensation in the oil-temperature controller.

3. Adaptive polymer temperature control is necessary to adequately control both normal and upset conditions.

THE INADEQUATE EXISTING SYSTEM

The existing system was analyzed to find where it deviated from the ideal system. There were three main areas.

The oil control valves have a highly nonlinear equal percentage characteristic, not the desired linear characteristic. This is a major reason for the different tuning necessary for different products. As the product properties vary, the heat of reaction varies. This drives the system to a new operating point on the valve curve, changing the overall process gain.

The existing controllers are constant-gain devices. The gain set by the tuner does not change on its own. They cannot compensate for oil temperature driving force differences and must be tuned for some average operation.

Additionally, their constant gain does not provide the runaway and upset handling of adaptive tuning of the polymer temperature control.

SYSTEM MODIFICATION

Now that we have learned why the existing system is inadequate and how an ideal system should be configured we ask, "How do we modify the existing system to incorporate what we have learned?"

We first check the hardware requirements. The existing measurement devices are thermocouples and are quite adequate for the job. Conventional control instruments do not allow their gain to be a calculated variable, therefore, they are unsuitable. A new microprocessor-based controller is ideally suited for this application since most allow gain calculations.

The micro-based controller selected for the experimental system was fairly typical of the type. It was purchased in anticipation of projects like this one because it is an excellent device for experimental plant applications. It provides an informative operator interface as a standard feature. It is small, light and portable, but rugged since it is a plant not a lab instrument.

It has both classical and modern control capability, built-in memory, can do complex calculations, make decisions and change control strategy based upon process conditions, and has extensive alarming capability, to provide early warning of impending disaster. In fact, it is a special-purpose digital computer.

Using the micro-based controller to replace the existing controllers makes converting the existing system to an ideal system a simple matter.

The technique of adaptive oil temperature control shown in Fig. 8 is used to provide the necessary driving-force compensation in the oil circulation loop temperature controller. The gain of the controller is varied between a fixed minimum and a fixed maximum, based upon the positive temperature difference between the inlet and circulating oils.

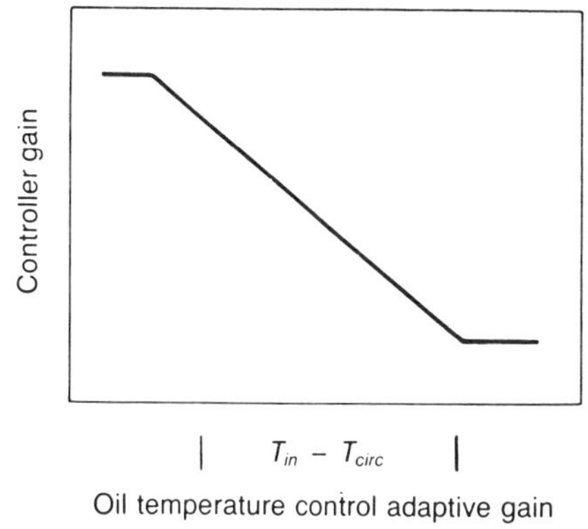

Figure 8. Driving-force compensation.

We can retain the range and high resolution at low flow rates of the existing control valves by linearizing their characteristic in the controller, instead of replacing them with linear valves. We simply multiply the controller output by the inverse of the valve characteristic and send that signal to the valves.

The deviation adaptive gain needed for polymer temperature control is a standard feature in the microbased controller. We tell it the values of the minimum and maximum gains, where to start the transition, the slope of the increase, and it does the rest.

RESULTS

To answer the question, "How well does it work?," we installed the micro-based control system at the unit in August 1982. The following figures show the answer.

Figure 9 shows a comparison between the existing and experimental systems under normal conditions in the first reactor. No, the pens which made the traces for the experimental system were not tied into place. This is just how well the system works. Temperature control is within $\pm$ ½°F.

Figure 10 shows a similar comparison between the two systems. This time it is in the second reactor—traditionally the most difficult to control. The pen recording the A-zone temperature of the existing system was not

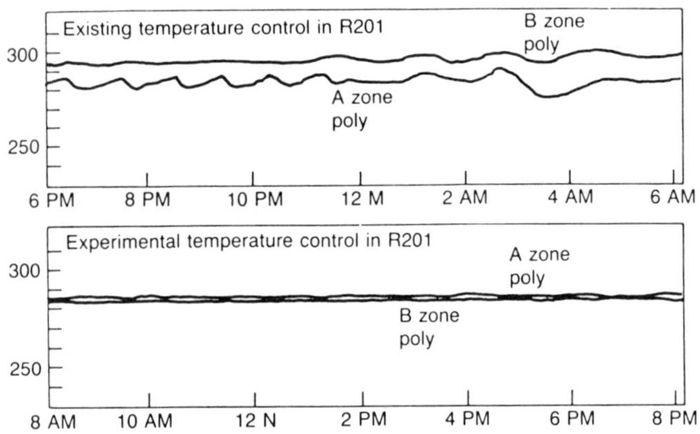

Figure 9. Existing and experimental control comparison.

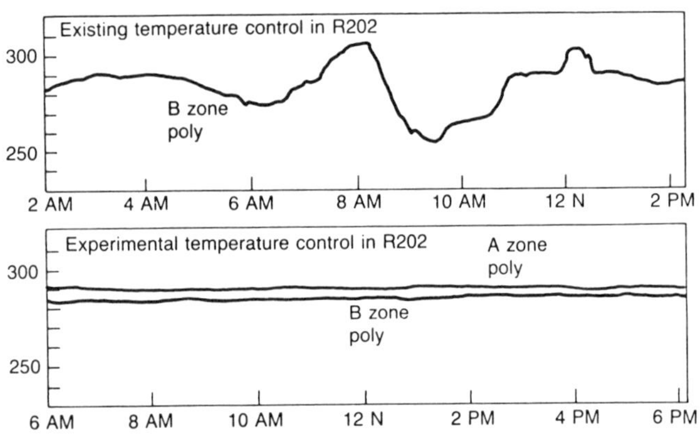

Figure 10. Similar comparison for the second reactor.

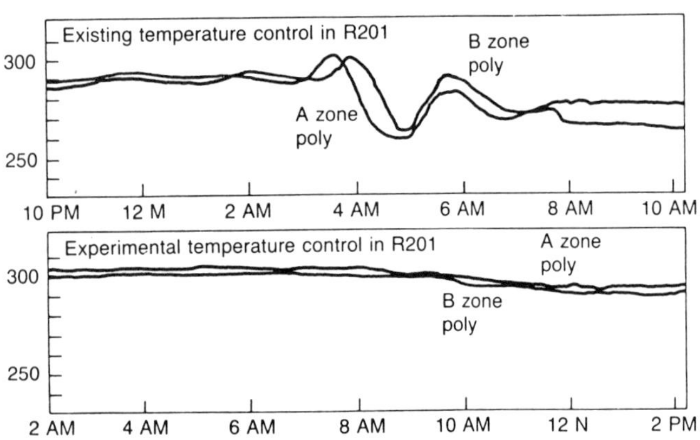

Figure 11. Setpoint change comparison.

working. Again temperature control is within $\pm 1/2$°F and no artificial pen restrictors were in use.

Figure 11 shows a comparison of the systems undergoing a setpoint change. Notice the smooth response of the experimental system between 8:00 a.m. and noon. However, this change was made much too slowly.

Figure 12 shows an example of the response of the experimental system to a setpoint change of ± 25°F made

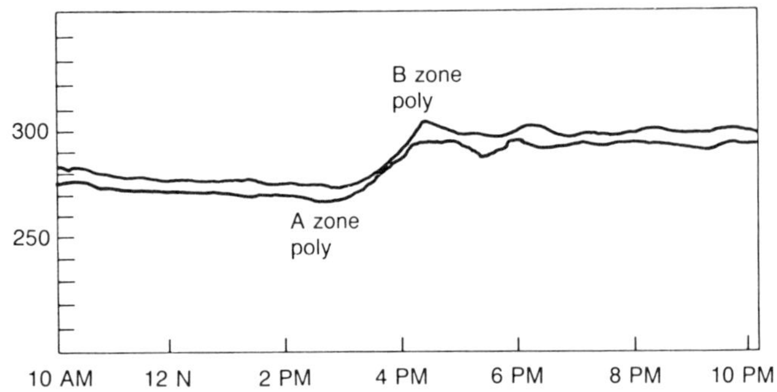

Figure 12. Response to a step setpoint change.

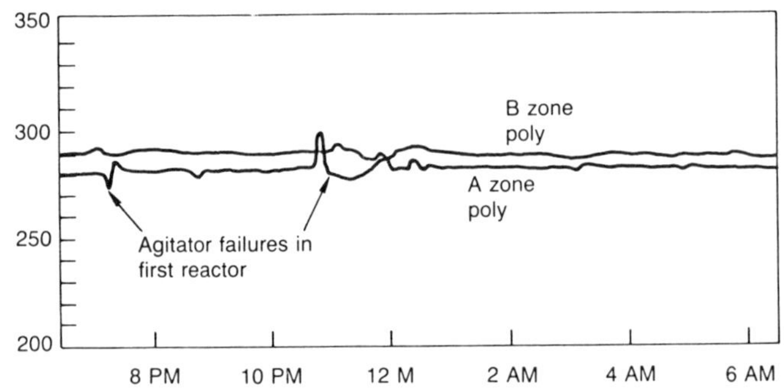

Figure 13. Response to a mechanical upset.

in a single step at 3:00 p.m. Acceptable operation was achieved by 5:00 p.m., only two hours later. With the existing system this change would either have caused a runaway or have had to be made over a five-hour period.

Figure 13 shows an example of the response of the experimental system to a mechanical upset. The agitator failed at 7:30 p.m. and again at 11:00 p.m. With the existing system a runaway would have occurred. With the experimental system there was no off-spec material produced; a graphic example of the value of deviation adaptive gain tuning.

Obviously the experimental system controls temperature, in fact, there has been no temperature-control-caused off-spec material produced while it was in operation. But, is there an economic benefit?

A recently approved appropriation request for the duplication of the experimental system at the plant presented the following figures. For an installed cost of less than $100,000 the corporation expects an annual saving of one-quarter of a million dollars, a return on investment in excess of 150 percent.

In addition to this economic advantage, there is the potential to produce polymer products which were previously impossible. Both our research and manufacturing departments are working on this and several new contracts have been obtained.

45

Exchanger Revamp Saves Energy

P. Cambier, G. Choux, and M. Schäfer

The Saarland refinery had a nominal capacity of 3.6 Mtpy in 1981. Its operation was based on two crude distillation units, a gas oil desulfurizer and a sulfur plant. The refinery conducted a revamping feasibility study to modernize its number one distillation unit. Details of the resulting heat exchanger revamp and control and optimization system are presented here.

REVAMPING DESCRIPTION

The originial process flow diagram of the distillation column exchanger train is shown in Fig 1. The heat exchanger train was of conventional "in-line" layout; only one pumparound reflux was provided at the gas oil draw level. The overhead condensing system consisted of an aircooler and a water trim cooler at ground level below the reflux drum. This design resulted in large use of vaporizing reflux and a low energy recovery efficiency, and a high operating pressure.

The revamping feasibility study was carried out on the basis of 36 and 40°API crudes. The installed modification resulted in a new process layout as shown in Fig. 2. The primary modifications were:

- Addition of a top pumparound reflux and removal of the water trim cooler, resulting in the elimination of a two-phase vertical flow to the reflux drum.
- Addition of a steam generator on the bottom pumparound (steam was previously purchased at high cost).
- Modification of the heat exchanger train by adding new exchangers and by modifying some others.

A particular feature of the revamped heat exchanger train was the installation of split flow configuration exchanger trains (three parallel branches) upstream and downstream of the desalter to maximize heat recovery with minimum surface area.

The revamping project was based on the following expected gains:

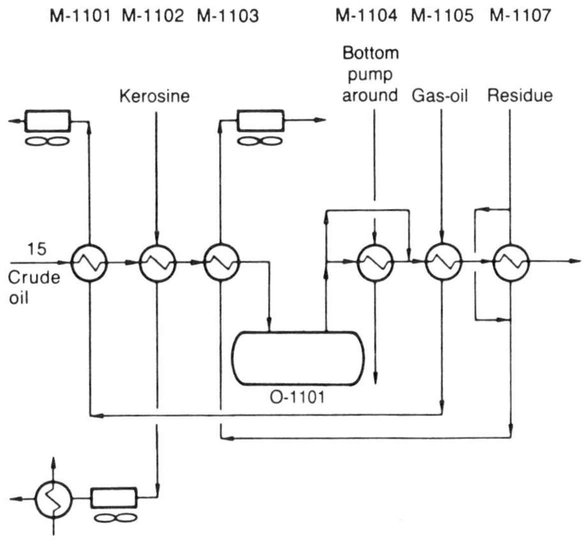

Figure 1. Original preheat train.

- An increased crude preheat temperature of 40°C which leads to fuel savings of 1.04 tph at the heater.
- Fuel savings of 0.17 tph due to the lower operating pressure obtained by the removal of the water trim cooler and elimination of a two phase vertical flow to the reflux drum.
- Production of seven tph of medium pressure steam on the bottom pumparound.
- Added improvements were expected in the gas oil recovery.

Because of the increased complexity of the double parallel arrangement of the heat exchangers, special care was given to control of the preheat train.

OBJECTIVES OF THE CONTROL SYSTEM

The control system was designed to monitor and control the heat exchanger train operation with the following targets:

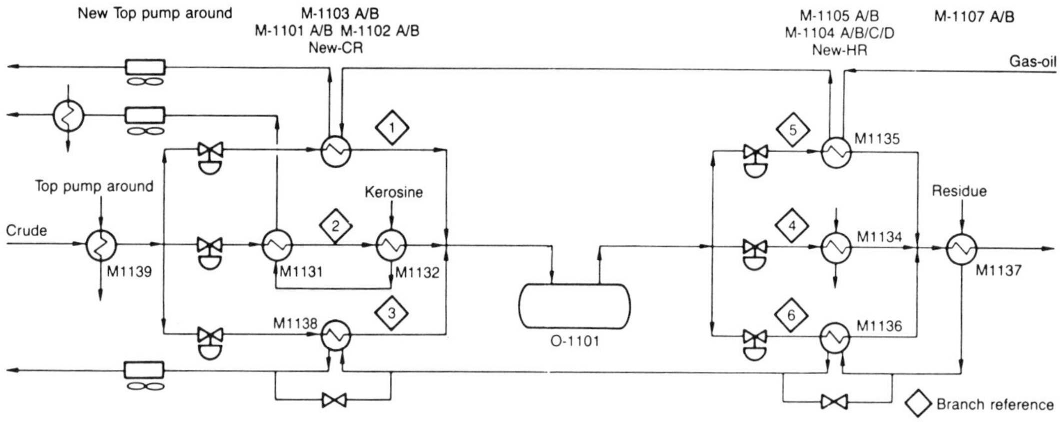

Figure 2. Revamped preheat train.

- Maximize the preheat temperature.
- Control the desalter temperature between upper and lower limits set by process considerations.
- Maintain sufficient pressure within the desalter and downstream exchangers (suppression of premature vaporization).

The above objectives had the following constraints:

- Ensure that the temperature limits of products to storage are respected.
- Ensure a minimum velocity in each exchanger to avoid excessive fouling, especially on the crude oil side of the preheat train.
- Maintain the heat-removal duty of the lowest pumparound as set by the operator.
- Maximize heat recovery on the top pumparound, as compared to heat removal at the overhead condenser.
- Keep operating pressures within the design pressures of the heat exchangers.

In addition to these constraints, the operation and optimization system had to handle the following conditions:

- Operational changes including lighter or heavier crudes than design, minimum thruput operation, crude oil changes.
- Progressive exchanger fouling.
- Manual operation of some controllers.
- Control mode during unit start-up and shut-down.

PARAMETRIC STUDY

To check the unit behavior over the whole operating envelope defined above, 36 operating cases were selected for the parametric study. These operating cases are the following:

Type of Crude. The refinery generally processes crudes whose API gravity ranges between 36 and 40°API. These two crude qualities were chosen for the revamping study and for checking the unit performance. In addition to these crudes, two other crudes, one lighter (44.2°API) and one heavier (31.2°API) were considered to widen the unit flexibility.

Unit Thruput. It was expected that unit thruput would range between 35 and 100% of the nominal value. Thus, three cases were selected corresponding to 35, 70, and 100% of the nominal thruput for each design case.

Preheat Exchangers Fouling. To check the unit in all operating cases, especially to verify the desalter temperature and the products battery limit temperatures, three levels of fouling (low, medium, heavy) were selected.

Two approaches were considered: The first one might be termed conventional or purely automation, and would have led to a mathematical and global study of this problem which, in this particular case, is one of a constrained optimum research.

The second one, called "process approach," uses existing process knowledge and allows one to greatly simplify the problem.

This last approach, therefore, was adopted. The process control strategy was elaborated according to the process approach through an in-depth analysis of the process. The different analysis steps are the following:

Perturbations Analysis requires a detailed analysis of the process in terms of input-output analysis to define all the possible perturbations having a significant effect on the process and to classify them into two main types: those which can be eliminated by appropriate means, and those which cannot be eliminated and eventually require feedforward compensations.

Constraints Analysis leads to a list of all the process and mechanical constraints.

Sensitivity Analysis leads to determining the process sensitivity to operating variables and disturbances, to determine feedforward necessity and to choose operating variables leading to the most effective control.

In the particular case of the Saarland refinery, the analysis of disturbances, constraints and sensibility were conducted through a parametric study of static process responses to a number of deviations from basic cases, and resulted in a further simplification of the problem, as will be described later.

The perturbation analysis was performed according to:

- Thruput changes.
- Crude quality changes.
- Exchangers fouling evolution.
- Topping operating conditions (pumparound duty changes, cut yields changes).
- Operating desalter temperature limits.
- Battery limit temperatures of products.
- Feed temperature.
- Partial manual mode.

The unit behavior according to these different perturbations was checked through the simulation, with a steady-state program of the 36 defined operating cases. With this analysis, it is possible to measure their effects on the process, to decide on feedforward compensation and to select control schemes.

The different constraints, already described, were integrated as limits or alarms in the elaborated control strategy. Figure 3 summarizes the different process constraints considered.

The sensitivity analysis was performed to determine the heat exchanger train behavior for the different operating variables, in this particular case, the crude flowrates in the different branches.

It has to be noted that before this sensitivity study, the system had, from a theoretical point of view three degrees of freedom:

- The crude flow in two of the three branches of the first exchanger group, the third one being fixed by material balance.
- The crude flow in either branch five or six. The crude flow in branch four is fixed for BPA duty removal. The last branch crude flow establishes the material balance.

Results of this sensitivity analysis, performed with a steady-state simulation program, are summarized in Table 1. These results show that the crude preheat is very insensitive to the crude allocation in the branches before

Table 1
Crude Preheat Temperature Variation Versus Crude Allocation in First Exchanger Group

Crude flow variation as compared with optimal allocation*	Branch 1	Branch 2	Branch 3
Wt *percent variation			
−20	− 0.1°C	+0.3	−0.9
−10	0	+0.2	−0.4
+10	−0.1	−0.3	+0.2
+20	−0.4	−0.8	+0.4

* The crude was balanced equally in the two other branches to compensate for the variation in the considered branch.

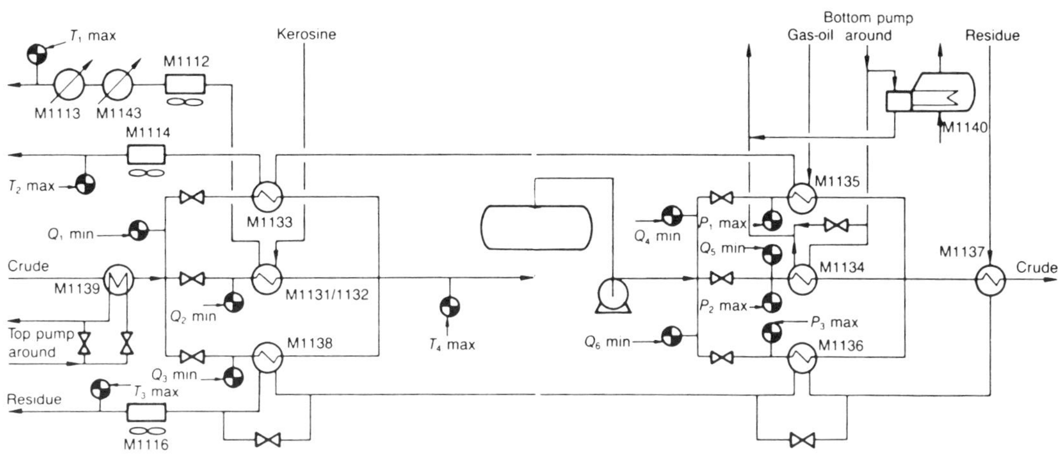

Figure 3. Preheat train process constraints.

the desalter. This conclusion allows the problem to be simplified by imposing an equal allocation of crude in the three branches before the desalter when there is no limiting condition on product or desalter temperatures. This example shows clearly the usefulness of this type of analysis in that the problem has been simplified from three to one degree of freedom.

CONTROL AND OPTIMIZATION STRATEGY

The overall control scheme is presented in Fig. 4.

Since a good process knowledge of the problem had been established, it was possible to define an operation and optimization strategy for the heat exchanger train and associated distillation columns. As it seemed unrealistic to describe all the possible unit operating parameters with a file containing optimized values for all the crudes, it was decided to develop an adaptive model. Such a model might have been either a random-search optimum method of heat recovery or a simplified adaptive model of the process.

This last approach was chosen to avoid problems associated with control noise that appear when a random search is applied. In addition, we thought that the optimum would be reached faster with a direct optimum search using an adaptive process model.

Control Strategy Upstream of the Desalter. According to the conclusions of the parametric study, the normal crude allocation between branches 1, 2, and 3 is set to one third of the crude flowrate. This is obtained by a ratio controller giving priority to the desalter pressure control. Operating constraints, desalter temperature limits and minimum flow in heat exchangers may modify the normal crude allocation according to the following strategy:

The high desalter temperature is satisfied by bypassing exchanger M1138. In this situation the preheat temperature is reduced and the non-recovered heat is transferred to the residue air cooler. If the maximum residue battery limit temperature is reached, the system reacts by modifying the crude allocation to saturate the other aircoolers on kerosine and gas oil in the same manner, the crude being balanced on the residue branch.

In case of low desalter temperature no automatic compensation was foreseen because this situation only appears in a few particular cases. Thus, only a manual bypass was installed on the crude-residue exchanger M1136.

Control and Optimization Strategy Downstream of the Desalter. Heat exchanger M1134 has a fixed duty imposed by the column operation. The optimal crude

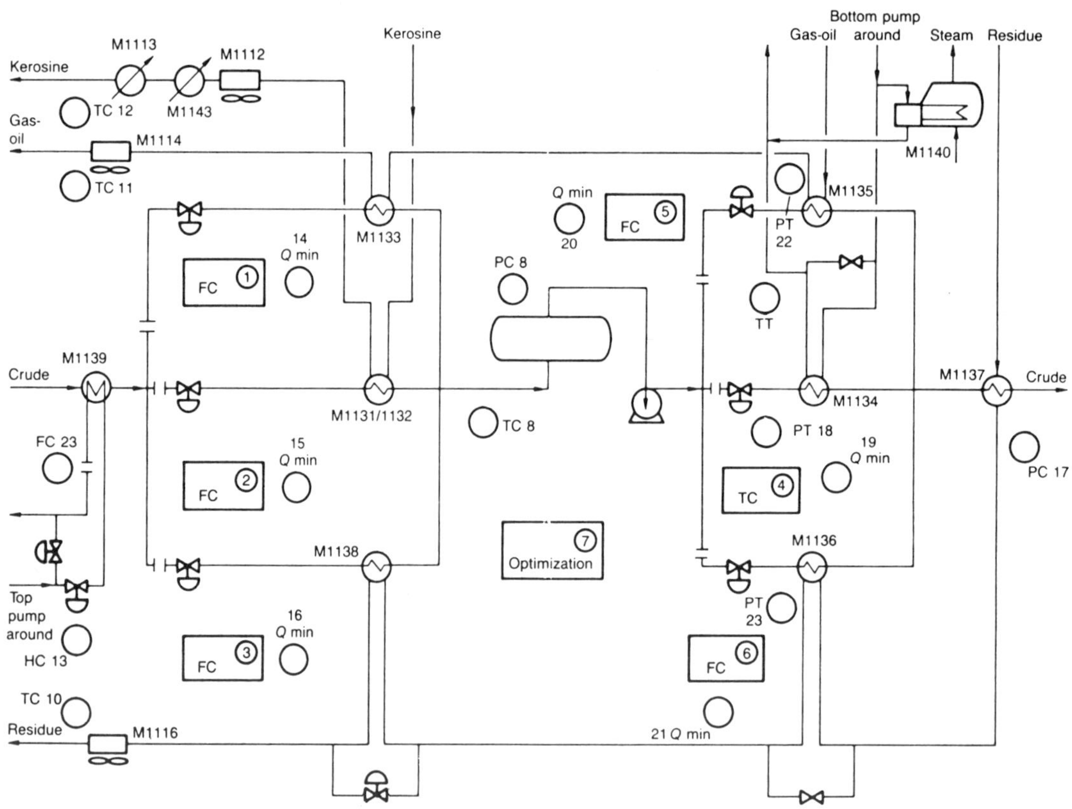

Figure 4. Exchanger train control system.

preheat is obtained by maximizing the use of this exchanger, i.e., by imposing a maximum pumparound flow and by adjusting the crude flow to reach the imposed thermal duty. The bottom pumparound duty is adjusted by controller TC4 with priority given to the pressure control on the crude side. As for the first exchanger group, minimum flow constraints ensure that minimum flows in each branch are respected to avoid any excessive exchanger fouling.

The crude split between branches five and six is calculated every 10 minutes to keep the crude preheat close to the optimum value. This calculation is based on the knowledge of a coefficient related to the global heat transfer coefficient of each of the last four exchangers. Then, supposing that this global heat transfer coefficient does not vary too much on a relative area of flow temperature variation, the calculation algonthm gives the flow rates for each branch corresponding to the maximum preheat. These results are applied to the heat exchanger train, and actual global heat transfer coefficients are calculated. These new values are used for a new iteration of the optimizing program, with adaptation of the model to the real heat exchanger train performance.

The optimization program is detailed later.

CONTROL SYSTEM DESCRIPTION

The objectives of the system were defined according to the following criteria:

- Use of microprocessor-based system technology.
- Well proven system for easy maintenance.
- Operator-system interface based on a controller-like operation with graphics display.
- Hierarchical control.
- High computation capability at low level in the hierarchy.

According to that definition, a system where all the calculations are performed on the basic control cards was selected.

The control functions were split into two parts. One is related to the control and constraints protection level, and is made on cards with high security. The other level is related to the optimization level that resets the control level, and is performed on separate cards. The information required by the optimization level is prepared continuously (three times per second) by the control level. The optimization subsystem uses the capability of the cards to simulate the preheat train to find the optimum from the present status by a stepping algorithm.

When a maximum is obtained, the value of the crude allocation is transferred to the control level through a ratio between the different branches. The system limits the modification to a maximum value to limit the disturbances. The system is fitted with an automatic on/off control that checks the validity of the parameters and allows the optimizer to work only when the unit is stable.

Standard calculation blocks only were used for the computation. Three original features were used:

- A functions generator to reduce the number of blocks and provide calculation limits when variables are outside realistic values.
- Use of special subroutines to simulate each heater exchanger.
- Use of a pseudo controller with a voting system that determines whether to select a new step of simulation or to stop the algorithms when convergence is achieved.

Operating Experience

All the objectives of the revamping have been reached: steam generation, pressure decrease in the column and increased crude preheat temperature.

The implemented controls are fully satisfactory, and the overall investment has been paid out in less than one year.

The distillation unit and heat exchanger train are operated entirely from the visual display unit, although the unit could be operated from the existing front panel. The optimization program is used online more than 80% of the time, the remaining 20% representing mainly "condensate—crude mix" treatment during which it is difficult to use the optimization program online.

CALCULATION DETAILS

Heat Transfer Coefficients. The heat transfer coefficients are calculated with the "NTU" method:

$$r = mc/Mc \tag{1}$$

$$e_{overall} = (ts - te)/(Te - te) \tag{2}$$

$$e_{shell} = (B - 1)/(B - r) \tag{3}$$

$$\text{with } B = \sqrt{(1 - e_{overall}r)/(1 - e_{overall})}$$

$$ntu = \frac{1}{\sqrt{1 + r^2}} \ln \frac{2 - e(1 + r - \sqrt{1 + r^2})}{2 - e(1 + r - \sqrt{1 + r^2})} \tag{4}$$

$$\text{Thus } U = (mc)(ntu)/A \tag{5}$$

where A = exchanger area for one shell
m, M = weight flowrates of cold and hot streams, respectively
c, C = C_p of cold and hot streams, respectively
te = inlet temperature for cold stream
ts = outlet temperature for cold stream
Te = inlet temperature for hot stream
Ts = outlet temperature for hot stream

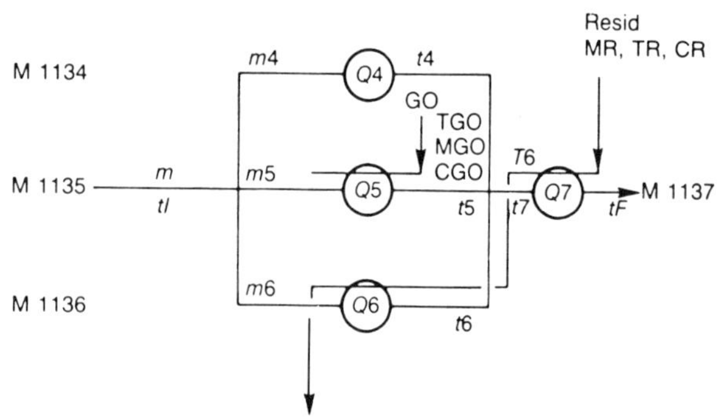

Δm = mass flow variation for optimum research
Δm_o = initial mass flow variation for optimum research.

Figure 5. Symbols description.

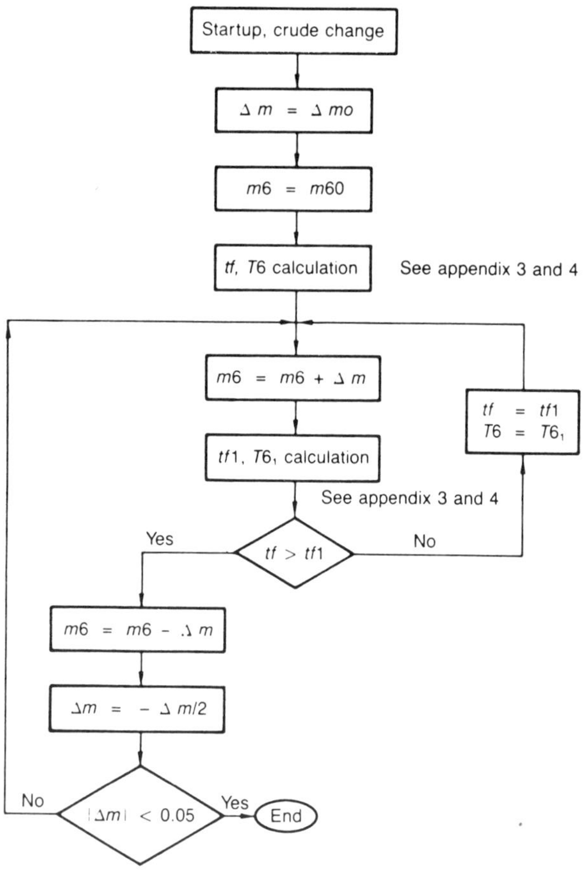

Figure 6. Optimum crude allocation calculation.

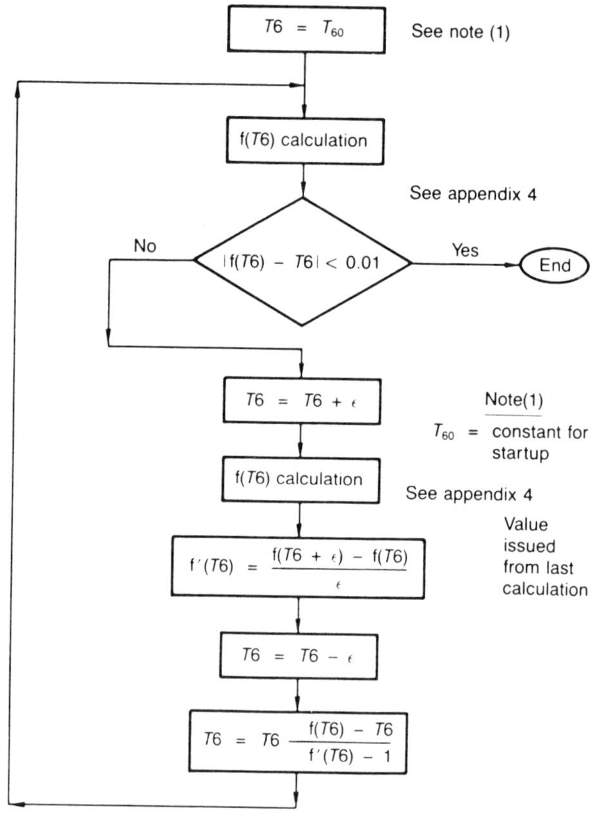

Figure 7. Calculation of T6.

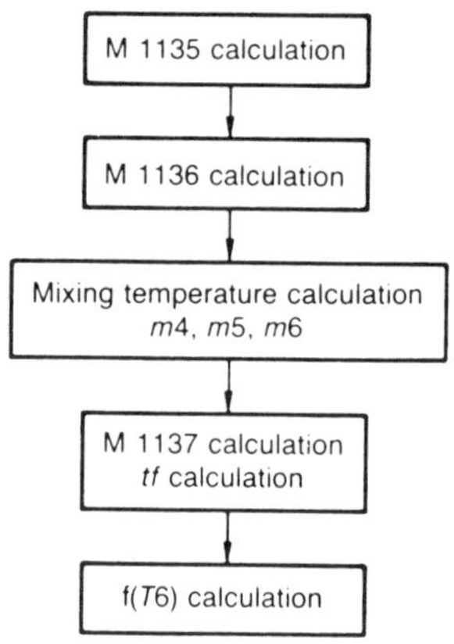

tf calculation
$tf = e7(TR - t7) + t7$
with $t7 = m4t4 + m5t5 + m6t6$ supposing equal C_p values.

f(*T6*) calculation
$T6 = TR - E7\,(TR - t7)$
$E7 = e7\,(r7)$

Figure 8. f(T6) calculation.

Optimal Crude Allocation. The principle of the calculation is presented in Figs. 5–8. The symbols used in the calculations are defined in Fig. 1. The calculations for each exchanger use the calculation described previously (NTU method).

46
Feedforward Air Control For Fuel Btu Changes

E. Vicknair

Fuel flow to a boiler is normally measured on a volume basis, and assuming a constant Btu value of the fuel, the volumetric measurement represents a Btu flow rate proportional to the amount of combustion air required. When the Btu value of the fuel varies, the volume flow no longer represents an accurate measurement of heat input to the boiler, nor does the signal remain proportional to the combustion air required. Excess air control corrects the fuel-to-air ratio, but not until the fuel is burned. Large deviations in fuel Btu values cause the excess oxygen control to be inadequate as the only correction for the fuel/air ratio.

Such a problem existed on a 600-Mpph boiler located at Union Carbide's Taft, La., plant. The normal fuel supplied to the boiler consisted primarily of methane and was measured using a conventional orifice run. As an energy conservation measure, a hydrogen waste stream from a production unit was introduced to the fuel gas supply. The differential pressure across the orifice in the fuel stream no longer represented a Btu flow since the composition and Btu value per cubic foot of gas varied. Figure 1 illustrates the change in heat input as percent hydrogen is increased, while maintaining a constant orifice differential.

Existing measurements of orifice differential, temperature, pressure and density of the fuel stream were used to determine not only the Btu flow to the boiler, but also the amount of air required to combust the measured fuel. Figure 2 illustrates the change in combustion air flow as the hydrogen content varies. Changes in fuel pressure and temperature are also used to compensate for the Btu and air flows (Fig. 6, 7).

As illustrated in Fig. 4, deviations in the fuel stream hydrogen content produced a nonconstant Btu/air ratio requiring the introduction of a feedforward signal to adjust the air flow. The feedforward signal is derived from the calculated theoretical air flow and then ratioed to the air flow measurement (Fig. 5).

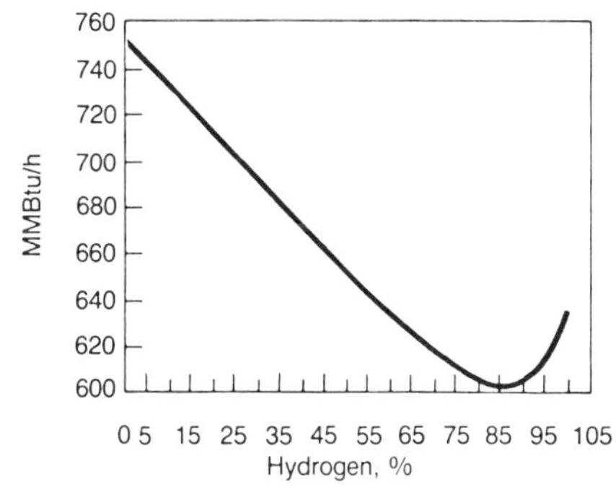

Figure 1. Heat input vs. percent hydrogen.

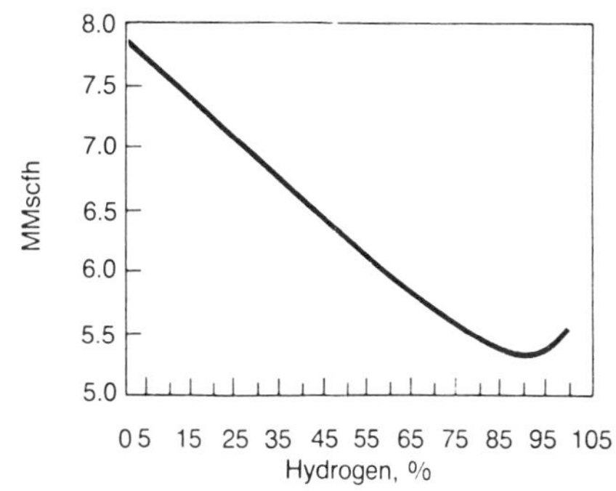

Figure 2. Theoretical air flow vs. percent hydrogen.

CALCULATION DERIVATION

The gas flow to the boiler is assumed to be a H_2CH_4 mixture. The measured variables on the gas stream are pressure, temperature, density, and the differential pressure across an orifice run.

Let Q = heat input, Btu/h
W = mass flow, lb/h
ρ = density, lb/ft^3
Δh = differential pressure, %/100

ΔH = heat of combustion, Btu/lb
P = pressure, psia
T = temperature, °R
R = gas constant, 10.73
V = air flow, scfh

$$Q(\text{Btu/h}) = W(\text{lb/hr})\Delta H(\text{Btu/lb})$$

The formula for mass flow with density correction factor is:

$$W = (W_1/\sqrt{\rho_1})\sqrt{\Delta h \rho}$$

where W_1 = flow at calculated conditions, lb/h
ρ_1 = density at calculated conditions, lb/ft^3

Since W_1 and ρ_1 are known constants and Δh and ρ are measured, the mass flow can be calculated.

The formula for determining the heat of combustion for a gas mixture is:

$$\Delta H = \Sigma X_i \Delta H_i$$

where X = mass fraction of H_2
$1 - X$ = mass fraction of CH_4
ΔH_{CH_4} = 21,520 Btu/lb
ΔH_{H_2} = 51,623 Btu/lb

Therefore, $\Delta H = X(51{,}623) + (1 - X)21{,}520$
$\Delta H = 30{,}103X + 21{,}520$

The value of X cannot be measured directly and is related to the measured values as follows:

$$MW = \Sigma \tilde{X}_i MW_i$$

where $\tilde{X}$ = mole fraction of H_2
$1 - \tilde{X}$ = mole fraction of CH_4
$MW_{H_2} = 2$
$MW_{CH_4} = 16$

Therefore $MW = \tilde{X}(2) + (1 - \tilde{X})16$
$MW = 16 - 14\tilde{X}$

Converting from mole fraction to mass fraction:

$$= \frac{\text{moles}_{H_2}}{\text{total moles}}$$

$$= \frac{\tilde{X}/MW_{H_2}}{\tilde{X}/MW_{H_2} + (1 - \tilde{X})/MW_{CH_4}}$$

$$= \frac{X/2}{X/2 + (1 - X)/16} = \frac{8X}{7X + 1}$$

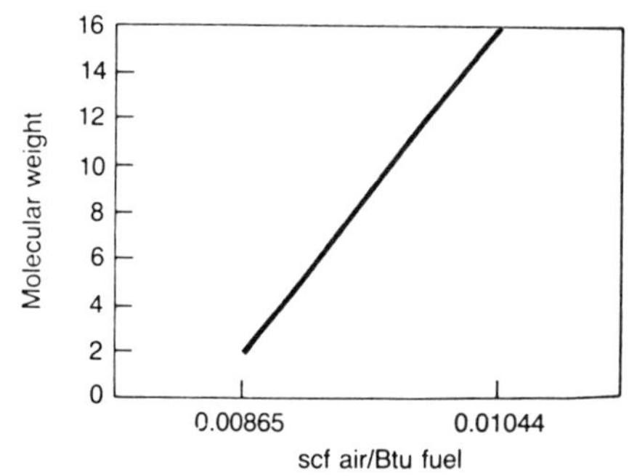

Figure 3. Molecular weight vs. scf air/Btu fuel.

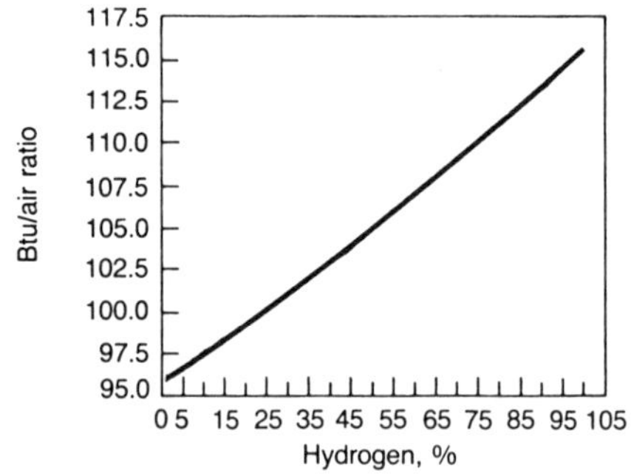

Figure 4. Btu/air ratio vs. percent hydrogen.

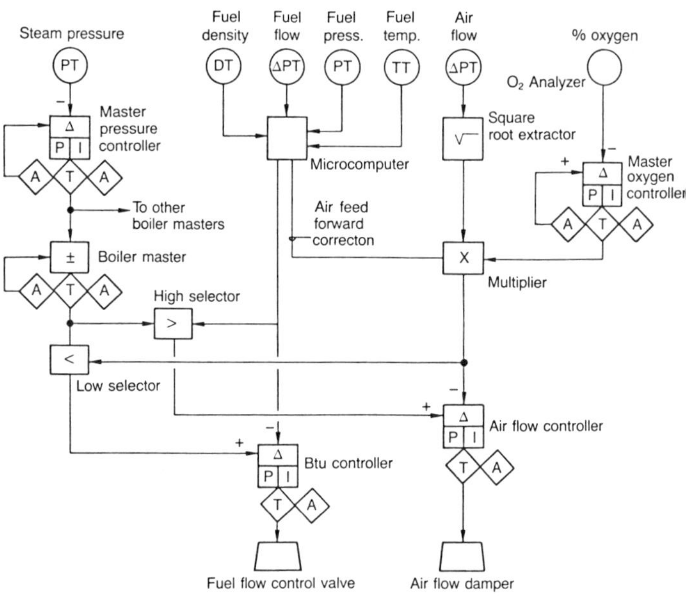

Figure 5. Boiler control scheme.

Substituting $8X(7X + 1)$ for $\tilde{X}$ in the previous equation for molecular weight:

$$MW = 16 - 14\tilde{X}$$
$$= (16 - 14)(8X/(7X + 1))$$
$$MW = 16/(7X + 1)$$
$$MW = \rho RT/P$$

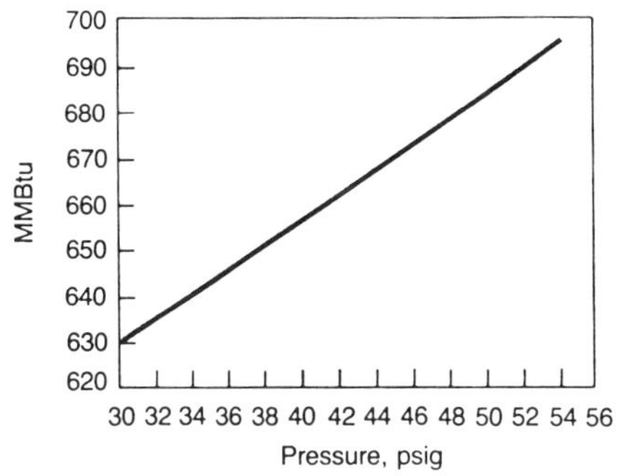

Figure 6. Fuel pressure vs. heat input.

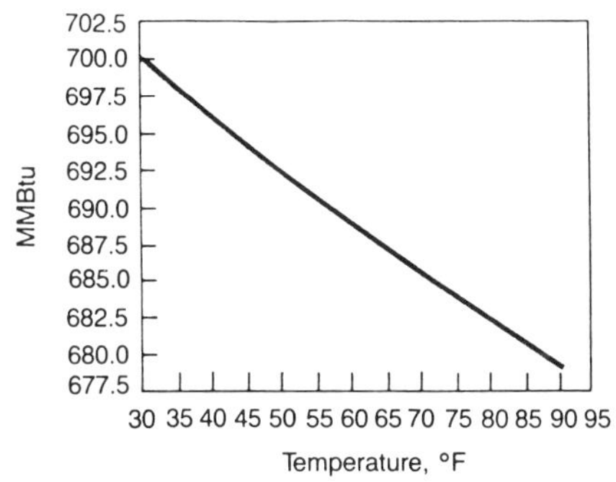

Figure 7. Fuel temperature vs. heat input.

From the gas law.

Substituting 16/(7X + 1) for molecular weight in the previous equation:

$$16/(7X + 1) = \rho RT/P$$

Solving for X,

$$X = 0.213021P/T_\rho - 0.14286$$

Substituting for X in the equation for heat of combustion,

$$\Delta H = 30{,}103X + 21{,}520$$
$$= 30{,}103(0.213021P/T_\rho - 0.14286) + 21{,}520$$
$$\Delta H = 6{,}412.6P/T_\rho + 17{,}219.49$$

Solving for heat input,

$$Q = W \cdot \Delta H$$

$$Q = \frac{W_1}{\sqrt{\rho_1}} \sqrt{\Delta h \rho} \left(\frac{6{,}412.6P}{T\rho} + 17{,}219.49 \right)$$

This equation is used to calculate the total Btu flow to the boiler. The theoretical air flow required for completed combustion is calculated as follows:

$$V = \frac{sft^3{}_{Air}}{Btu_{Fuel}} \left(\frac{Btu_{Fuel}}{h} \right) = \frac{sft^3{}_{Air}}{Btu_{Fuel}} \cdot Q$$

The amount of air required to combust a Btu of fuel is directly related to the molecular weight of the gas, and

$$\frac{sft^3{}_{Air}}{Btu_{Fuel}} = \frac{sft^3{}_{Air}/sft^3{}_{Fuel}}{Btu_{Fuel}/sft^3{}_{Fuel}}$$

Using reference tables

2.38/275 = 0.00865 for H_2 with MW = 2
9.53/913 = 0.01044 for CH_4 with MW = 16

Showing that the relationship between molecular weight and $sft^3{}_{Air}/Btu_{Fuel}$ is a linear one (Fig. 3),

$$sft^3{}_{Air}/Btu_{Fuel} = 0.0001279(MW) + 0.0083943$$
$$= 0.0001279(\rho RT/P) + 0.0083943$$
$$= 0.0001279(10.73\rho T/P) + 0.0083943$$
$$sft^3{}_{Air}/Btu_{Fuel} = 0.0013724\rho T/P + 0.0083943$$

Therefore,

$$V = Q(0.0013724\rho T/P + 0.0083943)$$

SUMMARY

The Btu calculation provides improved boiler control and becomes a necessity with variations in fuel composition, pressure and temperature. The air flow correction introduced as a feedforward signal allows the excess oxygen control to fine-tune the combustion process with greater savings in the cost of energy.

47
Digital Systems for Better Boiler Control

C. L. Hughart

Steam generation in petrochemical plants and refineries is in a state of change. Expensive fuels have resulted in greater use of waste heat recovery boilers and other energy conservation measures. As a result, many conventional boilers have been mothballed. Improved flue gas analyzers and digital controls are replacing less efficient and less reliable control hardware. As the production of steam becomes decentralized, control systems needed to meet expanded plant objectives must be installed. Production, engineering, and maintenance personnel are finding increased need to learn more about this specialized control area. This chapter discusses conventional control systems common in industrial boilers plus improvements made possible with currently available hardware.

COMBUSTION CONTROL

Modern boilers generally use a full metering, parallel positioning, cross-limiting combustion control system (Fig. 1). The system can be understood easier if broken into three sections:

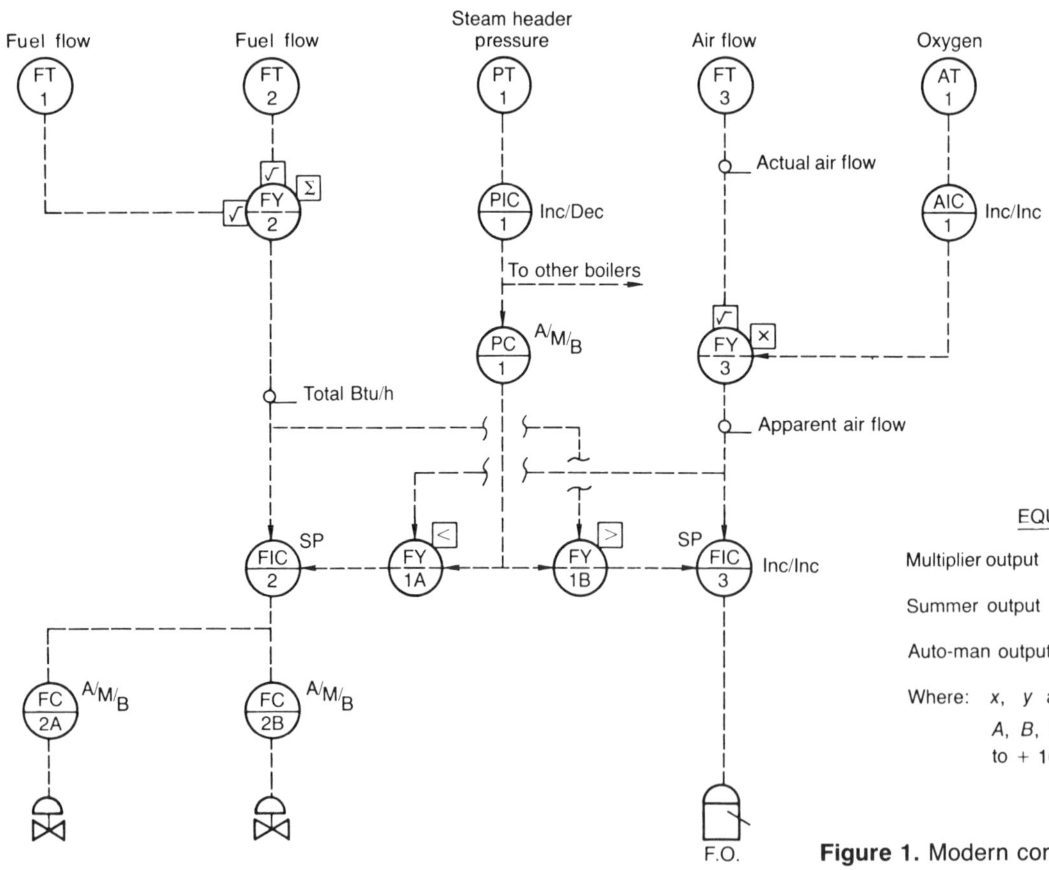

EQUATIONS

Multiplier output $= \frac{A}{100}(x)(y) + B$

Summer output $= A(x) + B(y) + C$

Auto-man output $= (x) + B$

Where: x, y are variables 0-100
A, B, C are constants -10 to + 10.

Figure 1. Modern combustion control system.

- Fuel
- Combustion air
- Steam pressure or demand

This particular boiler uses two fuels whose linearized flow signals are added on a Btu/h basis in summer FY-2, before passing to the measurement input of fuel flow controller FIC-2 and high selector FY-1B. The fuel control signal passes through parallel automatic-manual bias stations to the fuel control valves.

The linearized combustion air flow signal goes to multiplier FY-3 where it is modified by the control signal from a flue gas analyzer as needed to maintain the desired flue gas composition. The "apparent" air flow signal from FY-3 is connected to the measurement input of the air flow controller FIC-3 and low selector FY-1A in the fuel control circuit. The air flow controller positions a damper or varies the speed of the forced draft fan.

Steam header pressure is controlled by PIC-1, which sends a signal to the fuel/air controls on each of the boilers supplying steam to the header. This signal passes through the boiler master controller PC-1 on each boiler, which allows individual boilers to be biased, placed on manual with a fixed load or taken out of service. The boiler demand signal from PC-1 goes to high and low selectors FY-1B and FY-lA mentioned earlier. These selectors provide a "cross-limiting" feature that ensures sufficient combustion air for the fuel being admitted to the boiler.

FEEDWATER

Three-element feedwater control is standard on most boilers (Fig. 2). Drum level control alone is not satisfactory. Changing steam demand (resulting in changing steam header pressure and boiler firing) produces a temporary inverse response in the level measurement

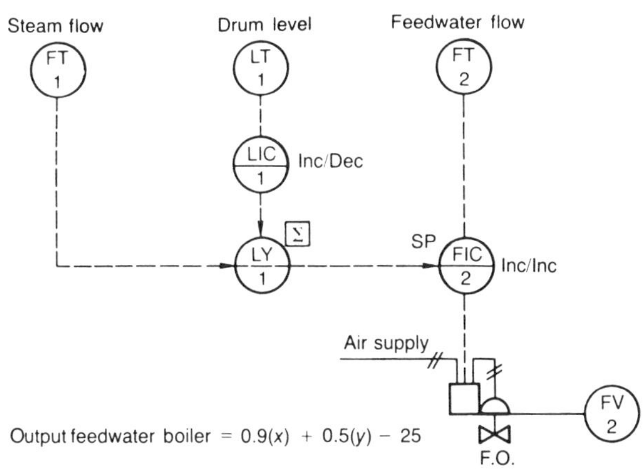

Figure 2. Three-element boiler feedwater control.

signal, prompting the feedwater valve to move in the wrong direction.

The boiler steam flow signal passes through summer LY-1 to the feedwater controller set point. When calibrated correctly, the system provides a pound of feedwater for each pound of steam and blowdown leaving the drum. The level controller is tuned to correct for long-term drift in level but to have little effect on feedwater flow during short-term excursions in level.

COMBUSTION TESTS

Combustion and feedwater controls are calibrated by running combustion tests on the boiler, recording data across a range of boiler loads and using these data to calculate the gain factors for the Btu summing relays, the relay calculating the boiler feedwater control set point, etc. The final span of the air flow transmitter, the total Btu measurement and the characterization of dampers and control valves are also developed from these data.

Combustion tests should be carried out any time the control system malfunctions and the problem is not obvious. Copies of data taken during previous combustion tests are helpful in diagnosing problems. Tuning of the controllers occurs after the control system is calibrated.

CONTROL HARDWARE

Figures 1 and 2 may be implemented using pneumatic, electronic, or digital control hardware. Digital controls have proven so successful however, that they are recommended for all industrial boilers. Advantages include:

- Better operator interface.
- Better control.
- Greater reliability.
- Control functions that require no calibration and do not drift.
- Reduced maintenance.
- Less need for the control engineer to compromise the design to avoid sacrificing reliability.
- Total installed cost less.
- Operators and operating technical staff have better understanding of the controls.
- Easy to revise control configurations to meet changing requirements.

Figure 3 demonstrates some of the advantages of digital boiler controls. This system is being installed to control two 350,000 lb/h, 600 psig auxiliary fired boilers which, augmented by other waste heat recovery equipment, will supply the steam for a large petrochemical plant. These boilers use hot exhaust from gas fired tur-

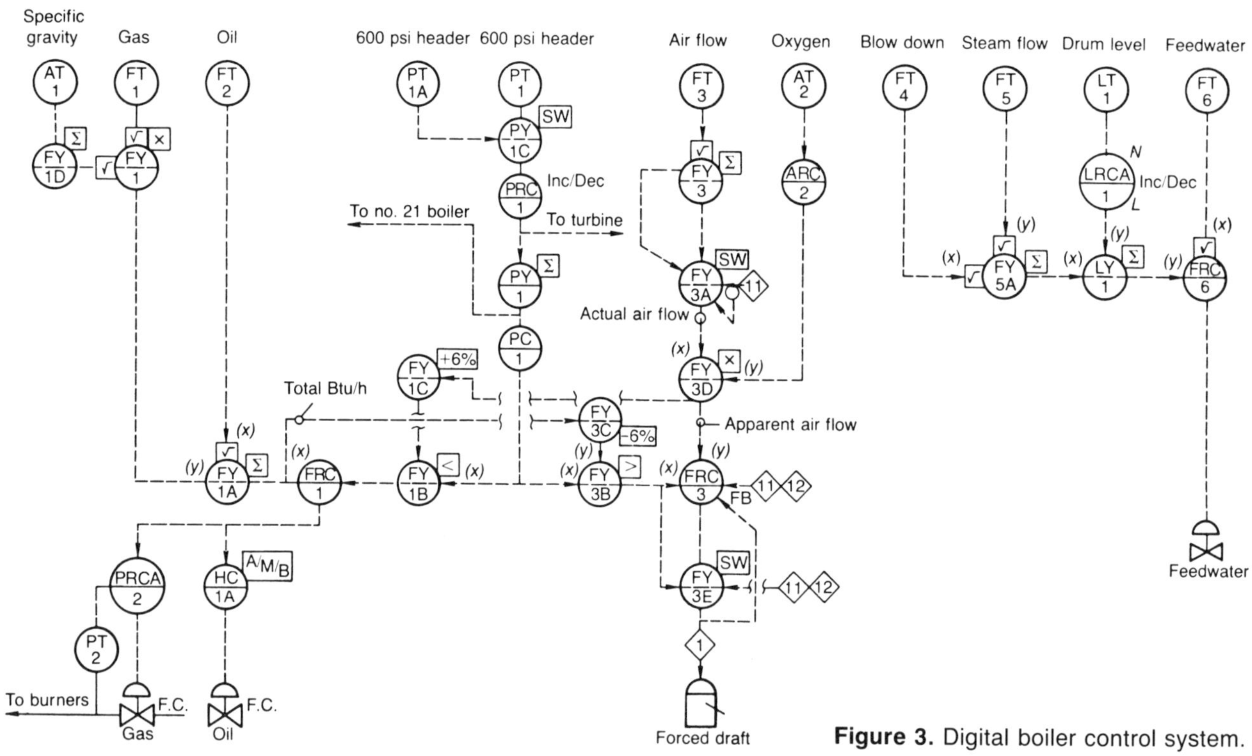

Figure 3. Digital boiler control system.

bines to recover heat and to supply the oxygen needed for auxiliary firing. All of the functions shown, except transmitters and control valves, are configured in the digital control system. During normal turbine operation, the forced draft fan will be off and isolated from the boiler by isolation dampers. When the turbine trips, the control system is designed to sense the shutdown, start the forced draft fan, open the isolation dampers, adjust the forced draft damper to its proper position for the quantity of fuel being fired at the time, and then switch the air flow controller to cascade control. The oxygen controller will also come into control and adjust the combustion air to maintain a predetermined flue gas oxygen content. This switch from turbine exhaust to forced draft fan will occur without assistance from the operator.

Other features shown in Fig. 3 include:

- A circuit to correct the gas flow signal for changes in heating value using a gravitometer and two computing functions (FY-1 and FY-lA).
- Bias on the cross-limiting relays of the fuel/air controls to reduce the interaction and instability in the control system caused by noisy fuel and air flow signals.
- Provisions to switch the span of the air flow transmitted signal automatically when the turbine trips and the forced draft fan starts.
- Blowdown flow is added to the steam flow in the feedwater controls.
- The addition of a burner pressure controller on the fuel gas to reduce upsets when burners are added or removed from service.

The drum level control shown in Fig. 4 increases reliability and accuracy compared to traditional installations. When differential pressure transmitters are used to infer drum level, the transmitter signal is a function of the water density, steam density, and the water height above the bottom connection. Transmitter span and zero adjustments are set to give a correct level indication at normal operating temperatures. At other temperatures however, the transmitter signal will be in error. The greatest error between the transmitted signal and the sight glass occurs during startup when the boiler is cold.

Figure 5 shows the indicated level for an 800-psig boiler from a cold startup to normal operation. The amount of error is dependent upon the operating pressure of the boiler and is greater for higher pressures.

Traditionally, a datum column has been used to correct the transmitter signal for density changes. This device develops a reference head whose temperature (and density) vary with boiler pressure providing compensation to the measurement signal. Datum columns provide an accurate level indication from a cold startup to normal operating pressures at the normal operating level, however they must be positioned precisely in relation to the boiler drum, and require extra piping and insulation. They are difficult to understand and troubleshoot and will read incorrectly if leaks occur in the reference piping.

The level control system in Fig. 4 uses a drum water temperature measurement and two computing functions, a multiplier and a scaler, to compensate the level transmitter for changes in steam and water density. The compensated level indication is shown in Fig. 6. An error of 2.5% of span still exists in the compensated signal due to nonlinearities in the transmitted signal. Substituting a function generator, available in most digital systems, will reduce the error to a negligible amount. Simple calculations such as these are ideal for digital control systems.

Union Carbide's Chemicals and Plastics Division started up their first digital boiler control system in late 1980. Four locations now have a total of nine boilers operating with digital controls, three more plants have systems in the construction phase and studies are underway at other locations. All systems are expected to pay for themselves with fuel savings. Additional benefits include improved safety, improved morale, reduced upsets and reduced stack emissions.

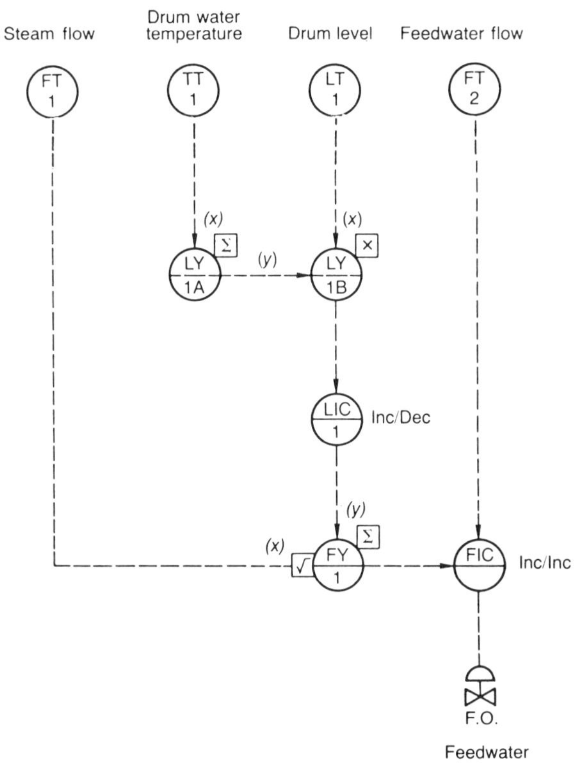

Figure 4. Improved drum level control.

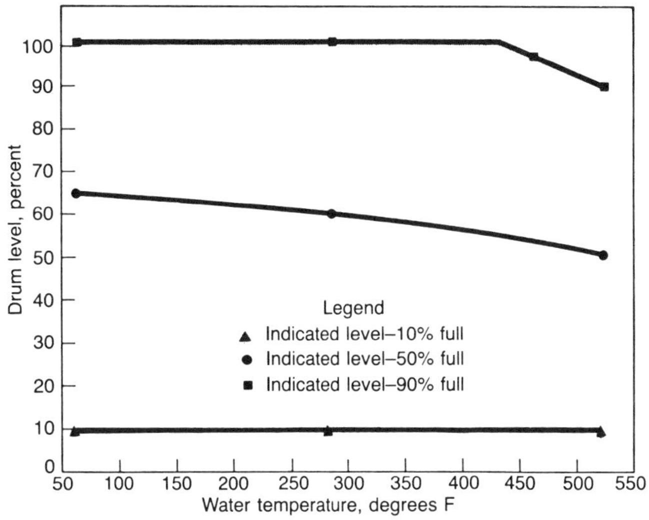

Figure 5. Uncompensated drum level vs. drum water temperature.

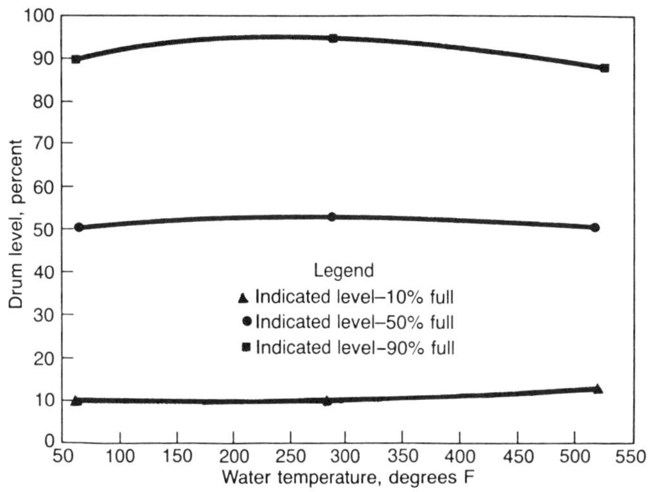

Figure 6. Temperature compensated level vs. water temperature.

In another example, a digital control system is installed to replace direct connected pneumatic instrumentation on three boilers firing coal as the primary fuel augmented by gas and byproduct fuels. In-situ, zirconium oxide analyzers measure the oxygen content in the flue. These analyzers respond in seconds to changes in the fuel/air ratio and have proven to be accurate and reliable.

The digital system has been on control on one coal fired boiler for six months. Flue gas oxygen is controlled at 2%, more than 95% of the time. A portable carbon monoxide analyzer and analyses of the carbon content of the ash were used to determine the oxygen controller setpoint. There are no visible emissions from the stack.

48

Cold Cycle Optimization

M. Lucet, J. P. Toulet, J. L. Gaultier, and J. F. Michel

Most complex cold cycles that remove heat at different temperature levels with one or more parallel compressors consume more energy than required. Presented is a control solution employing a computer to generate the cold cycle setpoints. The result has been about a 20% reduction in energy consumption without sacrificing performance.

Control of complex cold cycles is difficult because:

- Process cold requirements vary.
- Discharge pressure fluctuates due to variations in condensing temperature.
- Thermal loads distribution may be very different from the design values of the unit. Consequently, the compressor stages have working conditions very different from design conditions and some stages work at "surge limit."

Due to these conditions, operators select "low-risk" setpoints; compressor recycling is above average and energy consumption is unduly high.

The application presented was carried out on the ethylene and propylene cold cycles of our 400,000 T/year steam-cracker. Improvement in our process has been very important.

- Energy consumption of the compressors dropped by 20%.
- Running incidents are not so critical due to the automatic correction of settings during transient states. Formerly, each running hitch was monitored by two operators who supervised the cold cycles. Now, computer control remains in action even when there are incidents. Operators no longer worry about cold cycles. The automatic system reacts upon sudden, important variations of thermal loads better than operators.

The computer is a means of acquiring a very accurate knowledge of the production processes. Optimum settings for a unit can be calculated with it. By coupling the computer to the unit, the best settings can be calculated and applied at all times.

This methodology definition of a decision algorithm to determine the settings and automatic use of the algorithm at all times, gives rise to improvements in the process that are extremely profitable.

We shall discuss the example of a complex cold cycle that acts as a utilitarian in the process. It supplies at all times the cold required by the process; all fluctuations within the unit affect it directly. Thus, operators who cannot ensure constant attention to the settings allow large safety margins that result in wasted energy.

COMPLEX COLD CYCLE

Figure 1 shows a schematic of a complex cold cycle. This cold cycle works at three temperatures—the temperatures at which the coolant fluid vaporizes at the suction pressures of the three compression stages. Compression is obtained by two compressors in parallel. The discharge gases of the last stage are condensed in an exchanger. The liquid obtained is fed through the different stages from drum F.0. F.0 feeds the E.3 exchangers and the third stage suction drum F.3 with liquid. F.3 feeds the E.2 exchangers and the second stage suction drum F.2 with liquid etc. The recycling lines start from the third stage discharge and supply the F.1, F.2, and F.3 suction drums.

Generally, a complex cold cycle is characterized by:

- Cooling liquid supply at several temperature levels.
- Compression obtained by one or more compressors.

Off line mathematical model of a cold cycle with pure coolant fluid

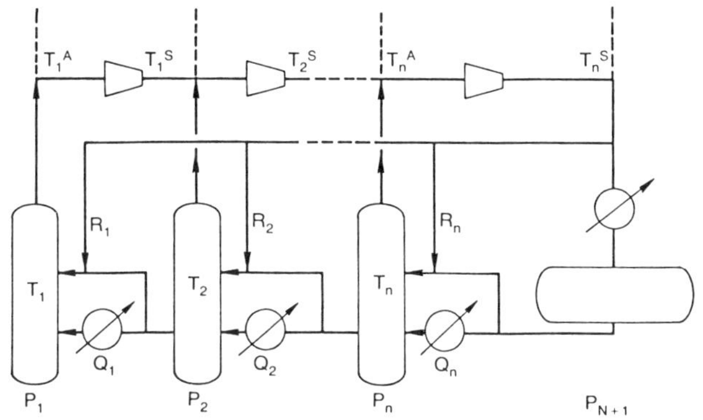

Inputs $P_1\ P_2 ------ P_n\ P_{n+1}$ drum pressures
$Q_1\ Q_2 ------ Q_n$ thermal loads on the different stages
$S_1\ S_2 ---- S_K$ rotating speeds of the compressors
T_n^S discharge gas temperature

1st step thermodynamic calculations for every drum

$T_i = f(P_i)$ vapor pressure
$h_i = f(T_i)$ mass liquid enthalpy
$H_i = f(T_i, P_i)$ mass vapor enthalpy
$H_r = f(T_n, P_n + 1)$ discharge gas mass enthalpy

2nd step calculations of energy compression of every stage i. They are recurrency calculations from the first stage.

For every compression stage i we achieve the following operations:

T_i^A gets an initial value

(1) $V_i = f(T_i^A, P_i, P_{i+1}, S_1 ---- S_k)$ } volumetric suction flow - characteristic curves of the compressors

$M_i = V_i/\rho_i(T_n^A, P_i)$ mass flow through the stage i

L_{i+1} and R_i are computed solving the following two equations

$(M_i - M_{i-1}) + L_{i+1} + R_i$ } material balance around the drum i

$(M_i - M_{i-1})H_i + Q_i + L_i h_i = L_{i+1} h_{i+1} + R_i H_R$ } enthalpic balance around the drum i

$T_i^A = (M_{i-1} T_i^S + (M_{i-1}) T_i) / M_i$ } we obtain temperature T_i^A by mixing 2 flows

The process repeats from line (1) until convergence on the value of T_i^A

$T_i^S = f(T_i^A, P_i, P_{i+1})$ output temperature of the stage i

$W_i = f(V_i, P_i, P_{i+1})$ compression work of the stage i - characteristic curves of the compressor

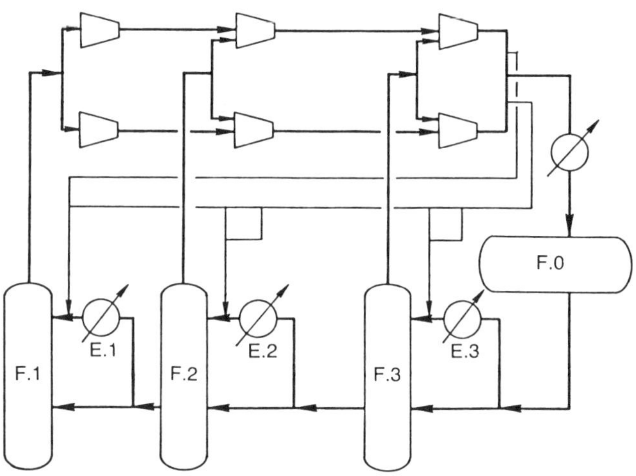

Figure 1. Complex cold cycle.

- Liquid distribution to the different exchangers. This can be either a "cascade" type, as in Fig. 1 or a direct supply; the F.0 drum supplying each of the exchangers, E.1, E.2, and E.3, directly.
- Recycling lines at the compression stages. These lines can either go from the last stage discharge to the suctions of each stage or merely link the thruputs and suctions of each stage.

Traditional Controls

The operators have a certain number of controllers whose setpoints they apply in the control room.

Standard settings include compressor speeds and suction pressures. *Compressor speeds* are maintained at the setpoint level by an appropriate setting. *Suction pressures* are used to control the boiling points of the cooling fluid. Boiling points of cooling fluid in the exchanger must be controlled to ensure that the exchangers using this fluid work satisfactorily. As a rule, it is not the temperature but the pressure at which it boils that is adjusted. Pressure controllers set the pressures of the suction drums. They act on the recycling valves and open them to increase suction pressure.

Some cold cycles have more sophisticated regulators (Fig. 2). The pressure controllers act in split range on the compressor suction valves, then on the recycling valves. To increase pressure in the suction drums, the pressure controller first closes the suction valves of the stage. Then, when these valves are almost shut, leaving a small passage, it opens the recycling valves.

With these settings, fluctuations in the process mean excessive energy consumption by the compressors. Let us examine some of these variations.

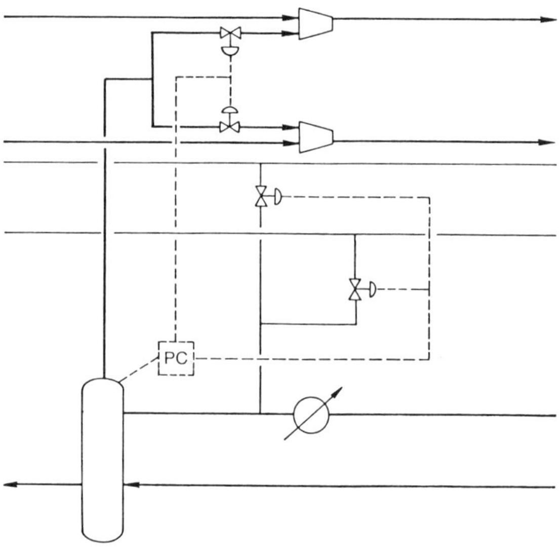

Figure 2. Suction pressure control.

Variation in Thermal Load at One of the Stages. Let us imagine a decrease in the heat load at one stage (E.1, Fig. 1, for example). Pressure in drum F.1 tends, in this case, to decrease. The pressure controller reacts by closing the suction valves and then by opening the recycling valves (Fig. 2). An energy-wasting recycling is thus introduced. To get rid of it, all the pressure and compressor speed setpoints must be changed. This is a difficult task which the operators will avoid. In addition, they will naturally tend to select "low-risk" settings which they do not have to change very often. Such settings result in a high recycling rate.

Variation in the Compressors Discharge Pressure. When the compressors discharge, the fluid is condensed in an exchanger. This cooling fluid, which is often water or air, is not at a constant temperature. The variation in the compressors discharge pressure is related to the variation in cooling fluid temperature. Since suction pressures and compressor speeds remain constant because of the settings, the compression ratio varies and flow in the last stages of the compressors varies considerably. Flow can become insufficient and reach the compressors surge limit or it can become very strong and result in a high recycling rate for that stage. To avoid this, the operators would have to modify the settings constantly. In this case, too, the principle of "always using low risk setpoints" is applied.

USING THE COMPUTER

First, we looked at the propylene cold cycle of our latest steam cracker and made a model of it.

Propylene Cold Cycle Model

This mathematical model can be summarized as follows:

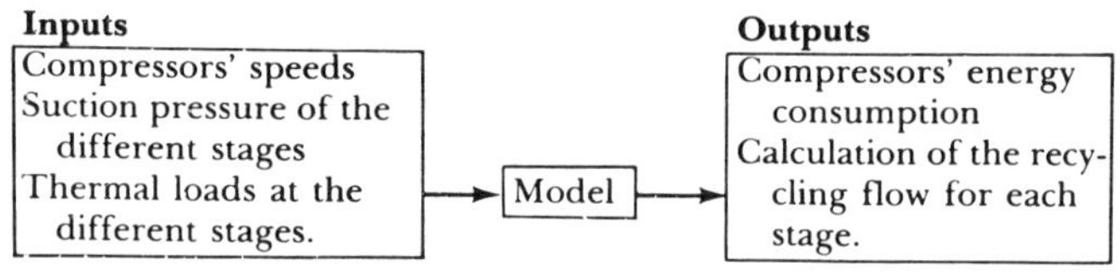

The program uses a model of the compression: characteristic curves, surge limit and energy consumed, and propylene thermodynamic models: vapor pressure and enthalpy.

A more detailed explanation of the mathematics is shown in Fig. 3.

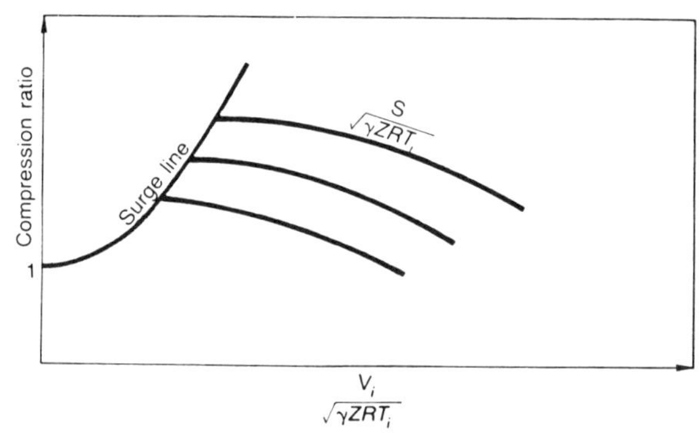

Figure 3. Compressor surge limit.

With the help of this simulator model, we became aware of the financial importance of accurate settings on this cold cycle; it then helped us to develop the algorithm defining the cold cycle settings.

Using the Algorithm

The propylene cold cycle model showed that an accurate and permanent adjustment of the setpoints available to the operators resulted in considerable energy saving. Of course, the on-line computer will determine the cold cycle settings, compressor speeds and suction drum pressures.

The aim of the algorithm is to minimize the compressors' energy consumption. However, this reduction must not hamper smooth running of the process and the cold cycles. Thus, a certain number of restrictions are considered by the algorithm. They are:

- The smooth running of the suction drums pressure setting. The algorithm should not fix setpoints that make the pressure settings impossible.

- The compressors surge limit (Fig. 4). Flow through each of the compression stages must be higher than the surge limit flow.
- Smooth running of the exchangers on the cold cycle. The exchangers only work properly if the cold side is at a sufficiently low temperature, i.e., if the vaporizing pressure is low enough.
- Limits on the pressure setpoints. To avoid air entering the cycle, the first stage suction pressure must be higher than atmospheric. The difference between inlet and outlet pressures of each exchanger must be big enough to overcome load losses.
- Critical speeds of the rotating equipment. Maximum and minimum compressor speeds are limited by the critical speeds of the rotating equipment.

Variables used in the on-line calculation of settings are shown in Table 1.

Antisurge Security

The computer may break down without warning and no longer ensure antisurge protection. Therefore, for increased security, we have added a purely analog antisurge protective device. It is a standard type of protection[1,2] which opens the recycling valves when flow through the compressor becomes too low. In view of the position of the flow transmitter orifice plates and the shape of our compressors' surge curves, we have had to design new antisurge protection plans.[3,4]

RESULTS

The decision algorithm was applied in September 1977 to the propylene cold cycle of our largest steam cracker. This cold cycle is characterized by three compression stages and two parallel compressors. From the very start, a considerable reduction was noted in energy consumption of the two compressors. Subsequently, we improved the algorithm and it was adopted in its final form in mid January 1978. Reduction in energy consumption was about 20% of the total.

At the end of January 1978, we used the algorithm which we had developed to control the ethylene cold cycle on the same steam cracker (four compression stages and two compressors in parallel). In view of experience acquired on the propylene cold cycle, we did not go through the mathematical model phase with the ethylene cold cycle. The algorithm is applied directly to the unit. After the first month in service, most of the coefficients had reached their definite value and reduction in energy consumption was about 20% of the total. Figure 5 shows how the ethylene cold cycle energy consumption varied over a period of time.

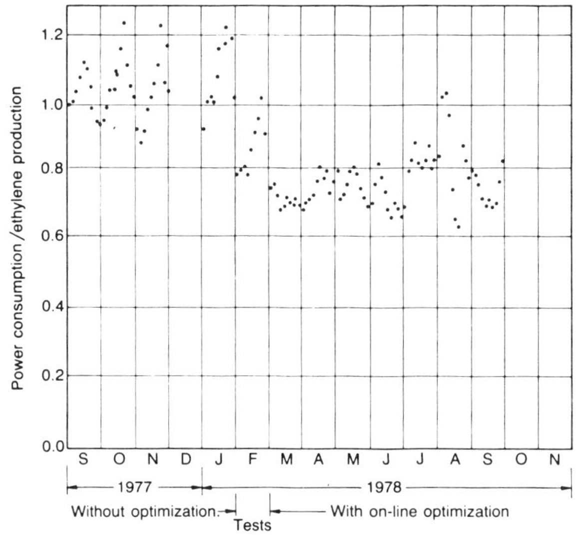

Figure 4. Power consumption results.

Table 1
On-line Calculation of the Settings

Unknowns	
Variation of speeds	ΔS_j
Variation of drum suction pressures	ΔP_i
Function to minimize	
energy consumption of the compressors	
$W(S_j, P_i)$ minimum	
$j = 1,k \quad i = 1,n$	
Constraint to be respected	
— correct settings of the suction drum pressures. The pressure controllers define valve opening 0_i between -100% and + 100 percent the range [- 100; 0] is for the suction valves (-100 fully open valves, 0 valves shut to a minimum). The range [0 - 100] is for recycling valves (0 shut valve, 100 fully open valve). Mathematically the constraint is written as $0_i > 0$ minimal	
— Surge limits $D_{ij}(S_j, P_i, P_{i+1}) > D\ \text{limit}_{ij}(S_j, P_i, P_{i+1})$	
— Correct running of exchangers (sufficiently low pressure) $P_i < P_i$ maximum	
— Limits on the pressures $P_i > P_i$ minimum	
— Limits on the speeds S_j minimum $< S_j < S_j$ maximum	

A calculator was fitted to the unit for handling the distillations and cold cycles. Data logging software had previously been written in house. The cold circuit running algorithm is in FORTRAN and has three steps, each using less than 4 K words.

Long-Term Running

This system has been working for more than two years without interruption; it ensures at all times minimum energy consumption compatible with the smooth running of the unit.

This automatic determination has a second advantage when handling the cold cycles during highly unstable periods on the process, such as production hitches, sharp variations, etc.

Here are examples of two typical incidents which we have had:

- A failure in the naphtha pumps—the unit remained unsupplied for several minutes, then, when it started up, there was a sudden acceleration in working conditions. The cold cycles had to run in very erratic conditions throughout this period.
- Our steam cracker produces ethylene in gas or liquid form; the latter is intended for storage purposes. We have frequently had last-minute cancellations from big ethylene gas consumers. Immediately, production of ethylene gas drops while production of liquid ethylene increases considerably. The ethylene cold cycle which supplies coolant for condensing the ethylene has an increased demand for cold at its first stage. The propylene cycle which supplies coolant to the ethylene cycle also has to come into action.

Every time there is an incident, we observe the same scenario. The operators keep the computer connected to the cold cycles and have complete faith in it. Pressures and speeds change very rapidly to adapt perfectly to new developments in the situation. Formerly, in similar circumstances, there was an operator permanently keeping watch over each of the cold cycles. His handling of the situation was less efficient than the present automatic system.

To sum up with respect to production incidents, the automatic system has resulted in:

- Better handling than with manual settings.
- Two operators, relieved of the task of watching over the cold cycles, can now devote themselves entirely to settling the production incident.

Nomenclature

D_{ij}	suction volumetric flow of the stage i, compressor j
h_i	mass liquid enthalpy in the drum i
H_i	mass vapor enthalpy in the drum i
H_R	mass vapor enthalpy of the discharge gases
L_i	mass liquid flow leaving drum i
M_i	mass flow through the stages i of the compressors
O_i	valve opening defined by the pressure controller of stage i
P_i	pressure in the suction drum i
Q_i	thermal load on the stage i
R_i	recycling flow on the suction drum i
S_j	speed of compressor j
T_i	temperature of drum i
T_i^A	temperature of inlet fluid for the stages i of the compressors
T_i^S	temperature of outlet fluid for the stages i of the compressors
V_i	suction volumetric flow of the stages i of the compressors
W_i	compression energy of the stages i of the compressors
ρ_i	fluid density of inlet flow for the stage i
γ	specific heat ratio of the gas transported
Z	supercompressibility factor
R	perfect gas constant (joules/°K)

Subscript

i	index of the stage of compression
j	index of the compressor
k	number of parallel compressors
n	number of compression stages

References

1. Magliozzi, T. L., "Control System Prevents Surging in Centrifugal-flow Compressors," *Chemical Engineering,* May 8, 1967, pp. 139-142.
2. Gendrot, M., "Variante de régulation antipompage pour les compresseurs centrifuges à vitesse variable," *La Technique Moderne,* Mars/Avril, 1971 pp. 105-109.
3. Gupta, B. P. and Jeffrey, M. F., "Optimize Centrifugal Compressor Performance," *Hydrocarbon Processing,* June, 1979, pp. 88-92.
4. Staroselsky, N. and Ladin, L., "Improved Surge Control for Centrifugal Compressors," *Chemical Engineering,* May 21, 1979, pp. 175-184.

49
Process Control of Turboexpander Plants*

C. G. Guffey and W. A. Heenan

The turboexpander process (Fig. 1) uses the high pressure of the natural gas stream to drive a rotating expander wheel to extract mechanical work from the flowing stream. The removal of energy and the reduced pressure result in a large refrigeration effect that condenses ethane, propane and heavier components in the natural gas as liquids for recovery and sale. The recovered work is transmitted directly to a shaft which usually drives a gas recompressor. This compressor increases the pressure of the dry residue gas after liquid recovery. Operation of the demethanizer or de-ethanizer at the low pressure of the main gas stream eliminates the requirement for a separate demethanizer overhead gas booster compressor. There are many variations of this simplified process including supplemental inlet refrigeration, parallel trains of heat exchangers, side reboilers on the demethanizer and multiple expanders that must be controlled and affect the dynamic response to variable changes. This paper excludes these complications in the analysis of the system.

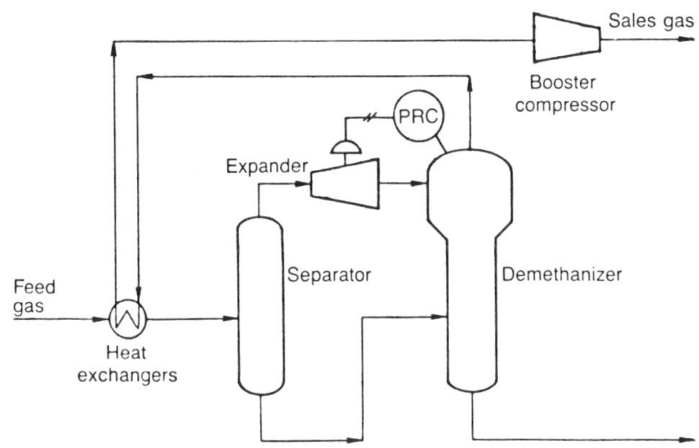

Figure 1. Expander process with "free" pressure drop.

PLANT DISTURBANCES

Process disturbances to turboexpander plants occur principally from the feed gas and usually consist of flow rate and/or pressure changes. The feed gas frequently originates from a producing operation where normal gas field operating problems with wells, dehydrators and compressors result in large and frequent changes. If the gas processing plant has feed gas compressors, considerable variations in feed rate can result from feed compressor operating problems. Variations in feed gas composition and temperature are usually small and take place in a seasonal pattern over long periods. The process plant operating further downstream on a large pipeline delivery system ("straddle" plant) can have very large flow rate changes, but other factors are usually constant or change very slowly.

PROCESS DESCRIPTIONS

The simplest turboexpander operation uses a "free" pressure drop and has no feed or residue gas compression. This situation occurs frequently at pipeline transmission line terminals or high pressure gas fields where the gas pressure must be reduced to enter a lower pressure pipeline system, resulting in a "free" pressure drop. The most common scheme for control in this situation, shown in Fig. 1, allows the pressure of the demethanizer to be controlled by the inlet guide vanes on the turboexpander. These vanes change the incidence angle of the gas entering the expander blades and produce changes in the expander speed and power. These guide vanes have an effect similar to a control valve in some respects.

With this control scheme, pressure sensing is fairly rapid for both upstream and downstream changes as the

* Special thanks are due to the following people who provided information and background assistance in preparing this chapter: Joe Chapman, Key Engineering, Jeff Dailey, Exxon USA, Bill Dynelt, Champlin Petroleum, and Lewis Lee, Fish Engineering.

Originally presented at the 63rd Annual GPA Convention, March 19-21, 1984, New Orleans.

pressure sensor is centrally located in the process equipment. There is no major equipment volume between the expander and pressure sensor so that the control response is also rapid.

Most process plants do not have a "free" pressure drop and must provide a residue gas compressor to boost the gas pressure back to the plant inlet or pipeline pressure. A process control scheme that is reported to work quite well for this situation is shown in Fig. 2. As in Fig. 1, the expander controls the demethanizer pressure, but additional pressure control on the outlet residue gas is available using the residue gas compressor speed to control suction pressure. This control scheme provides for excellent protection against downstream disturbances.

If there is concern about process disturbances originating from the feed, it is possible to locate the pressure sensors closer to the upstream part of the plant as shown in Fig. 3. The pressure control is very rapid as the sensor on the separator is just upstream from the expander. However, the use of the residue compressors to control the demethanizer pressure introduces a considerable lag time due to the exchanger and recompressor volume between the sensor and the compressors. This system can also result in pressure relief problems on the demethanizer if a residue compressor should fail.

A compromise, using the best features of the two previous control schemes, is presented in Fig. 4 and is reported to operate very successfully. A signal selector is used to permit either the upstream separator or downstream demethanizer pressure to be controlled by the expander. The controller with the greatest output or error signal is selected to control pressure with the expander. The possibility of demethanizer overpressure due to residue compressor failure has been eliminated, and fast control of the demethanizer pressure is possible with the expander.

In the case of a "straddle" plant operating on a large transmission pipeline, it is important that the gas pressure be completely restored to the inlet pressure so as not to reduce pipeline delivery capacity. A process control scheme used by Dome Petroleum at its Empress plant is shown in Fig. 5 as reported by Garvey and Griffin.[1] Large flow rate changes can result in either an over- or underload of the expander/recompressor and booster compressor. In order to maintain a fairly constant pressure drop across the plant, a plant bypass and the expander guide vanes are used to balance the flow through the plant. In the case of low flow rates, the bypass valve operates to recycle gas through the plant to maintain minimum flowrates. The residue compressors are loaded to the maximum level. The plant bypass is designed with a large dead band so that it does not open and close frequently. The expander controls to a fixed

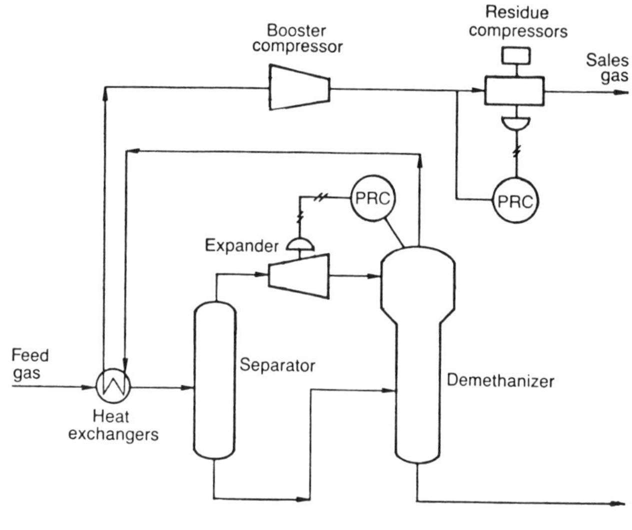

Figure 2. Protection from downstream disturbances.

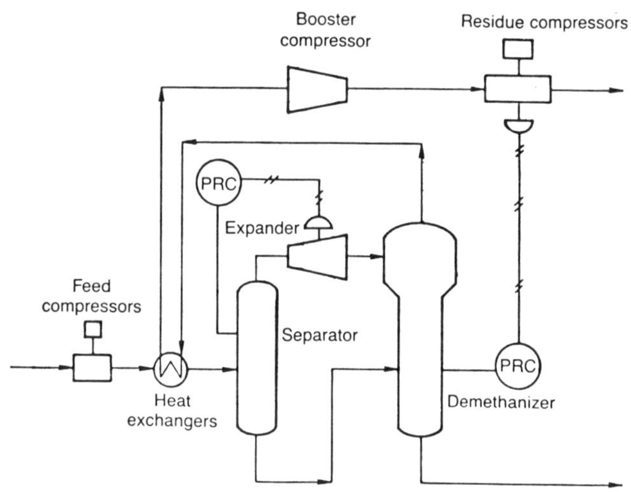

Figure 3. Protection from upstream disturbances.

setpoint so that it makes most of the smaller control actions. This schematic does not provide for demethanizer pressure control. The plant is controlled by a process control computer and uses a feedforward control scheme and chromatograph to control the demethanizer so that pressure control is not as important on this column as when bottoms temperature control is used.

Another process control scheme (Fig. 6) that does not control demethanizer pressure has been proposed by Shinskey,[2] but has not been tested in operation to our knowledge. The control scheme action is similar to that in Fig. 5 although it uses direct flow control instead of a bypass. Most plants would require a bypass to permit pipeline flow to continue around the plant. The residue compressors are shown controlling the discharge pressure instead of suction pressure. The pressure of the de-

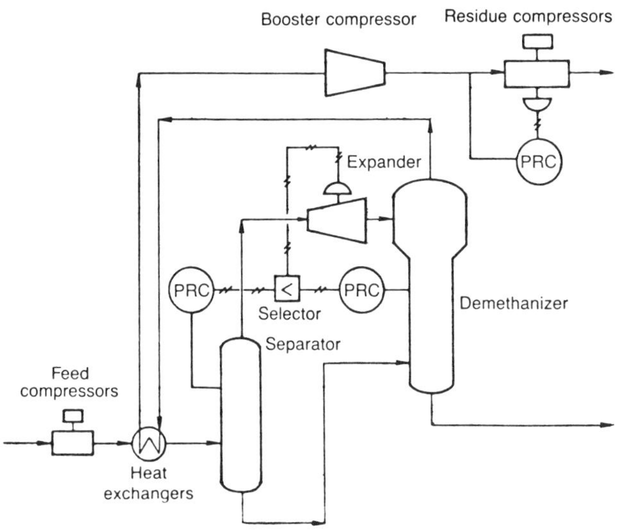

Figure 4. Protection from both upstream and downstream disturbances.

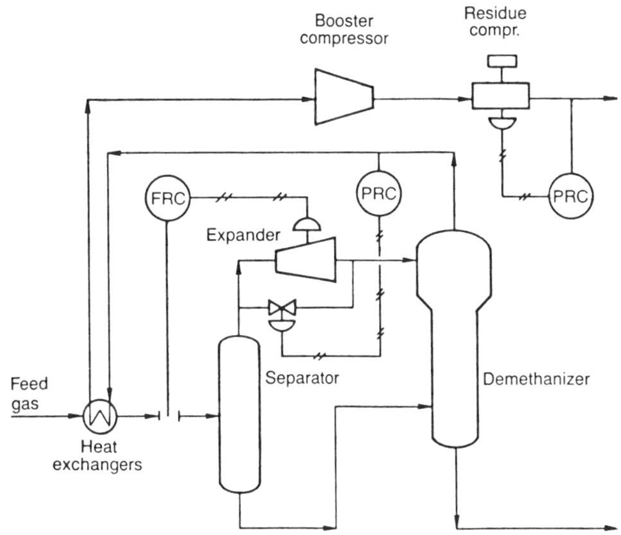

Figure 6. Expander bypass floating pressure control[2].

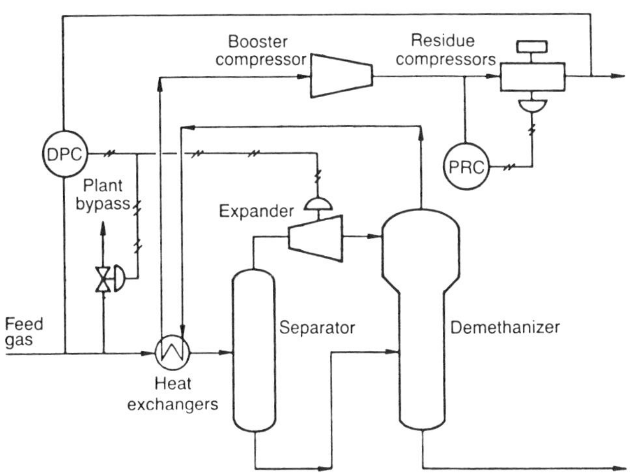

Figure 5. "Straddle" plant pressure drop control[1].

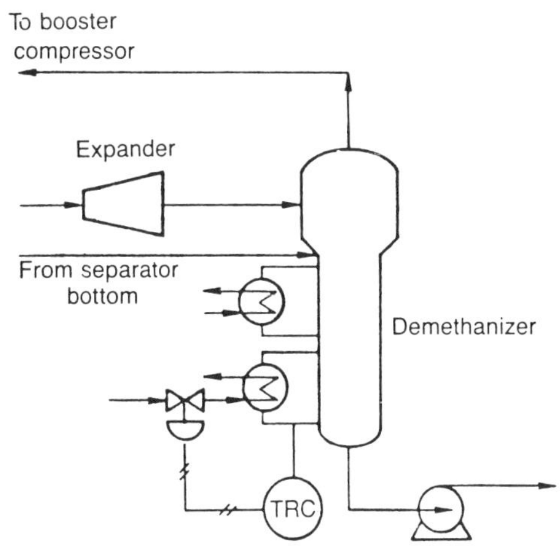

Figure 7. Conventional feedback temperature control.

methanizer is controlled using a bypass valve around the expander with a "floating control" method. This control method uses very slow instrument tuning so that there is essentially no short-term control and only a very long-term "drift" action to maintain the pressure of the tower in the general area of a setpoint. This system would probably also require chromatographic demethanizer bottoms analysis and control. Shinskey maintains that this system results in minimum demethanizer pressure and hence minimum energy consumption.

DEMETHANIZER CONTROL

As mentioned previously, the control of the demethanizer (or de-ethanizer) is interrelated with the control of the remainder of the process. Most plants have used simple feedback control systems to control this important tower, but with improved chromatographs and computers, more advanced feedforward control is now being used.

Feedback Temperature Control

The simplest feedback control scheme used to control the bottoms composition of the demethanizer infers composition from some temperature near the bottom of the column. This temperature is controlled by manipulating the flow of feed or residue gas to the reboiler or heat exchanger at the bottom of the column as shown in Fig. 7. Since changes in both pressure and composition affect the correlation between inferred temperature and bottoms composition, the bottoms composition needs to be manually checked frequently and the temperature setpoint adjusted accordingly.

Major disadvantages of this simple feedback scheme are its inaccuracy and sluggishness. Before any control action is taken, the disturbance must propagate all the way to the temperature transmitter. For example, consider a sudden change in feed rate to the demethanizer. No feedback control action will be taken until the change is noticed by the temperature transmitter (10 to 45 minutes delay depending on plant size) near the bottom of the column. Once control action is taken, it is slow because the mechanism of control involves heat transfer. Therefore, large, frequent feed rate disturbances will cause large oscillations in the bottoms, C_1/C_2 composition.

Cascade Control With Analyzer

One method of improving the accuracy of the temperature feedback control scheme is to use an analyzer such as a chromatograph (Fig. 8). The controller either directly regulates the flow of residue gas to the column reboiler or adjusts the temperature controller set point, which in turn regulates the flow of residue gas.

Chromatographic analyzer controllers, by their nature, are discrete devices. That is, they take discrete samples every few minutes. Therefore, it is advantageous to cascade a discrete analyzer controller to a continuous temperature controller, which can control while the analyzer controller is idle. Also, since the temperature transmitter can be located higher up in the column, it senses minor changes before the analyzer at the bottom. Therefore, usually the analyzer controller is cascaded to the temperature controller.

Many plants have difficulty tuning discrete chromatographic controllers. In many cases the problem originates with the control strategy and not with the reliability of the analyzer. A controller's output should not continue to be driven when it can have no immediate effect. A sampling or discrete controller is more compatible with a sampled or discrete process such as a chromatograph than with a continuous controller. The control strategy has been outlined by Shinskey.[2] When the controller first sees new information, it is enabled to operate for a very short ΔT, the control interval, using only integral action. Then the error signal is removed, preventing further integration until the next interval. This controller action is like a sample and hold circuit. By means of a sampling controller, critical damping will be accomplished on a sampled dead-time process using integral action only. For a given error, E, the controller output, M, changes according to:

$$\Delta M = E\Delta T/I$$

where I is the integral time. Then the process will respond within the sample interval by the amount ΔC, and

$$\Delta C = K_p\Delta T$$

where K_p is the process gain. The error, E, will be canceled by the change, ΔC, if

$$I = K_p\Delta T$$

A sampling integral controller is very effective on processes dominated by dead time. It requires only that the sample interval must exceed the process dead time.

While is it possible to perform analyzer discrete control using analog pneumatic or electronic equipment, the authors have had greater success using digital equipment to perform this control due to simplicity, less maintenance and ease of tuning. However, even when well-tuned, this cascade control scheme will be accurate but slow and will not handle large upsets well. For small infrequent upsets in feed rate, this scheme works well.

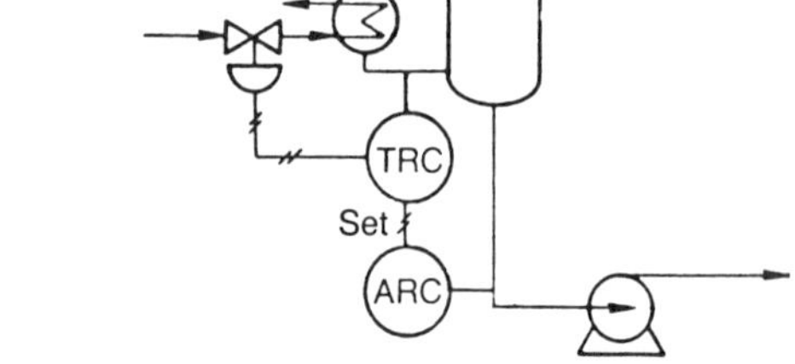

Figure 8. Cascaded analyzer feedback control.

Feedforward Control With Cascaded Analyzer Feedback Trim

For large, frequent upsets in feed rate to the demethanizer, feedforward control has been found to be extremely helpful (Fig. 9). In this diagram the feedforward controller is shown receiving its input from the vapor feed rate only. It is assumed that the liquid flow rate is proportional to the vapor feed rate. If this is not the

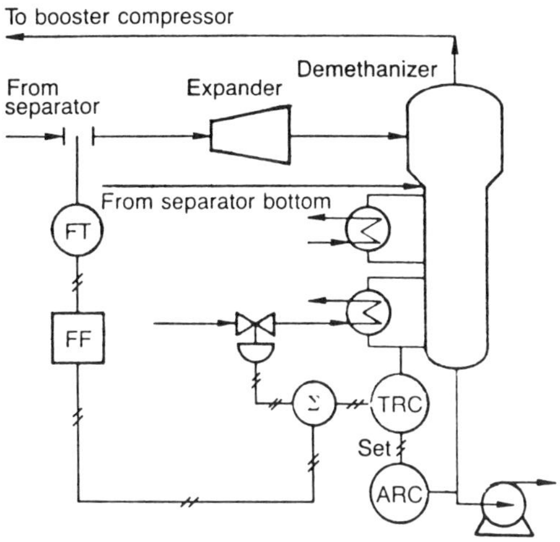

Figure 9. Cascaded analyzer feedforward control.

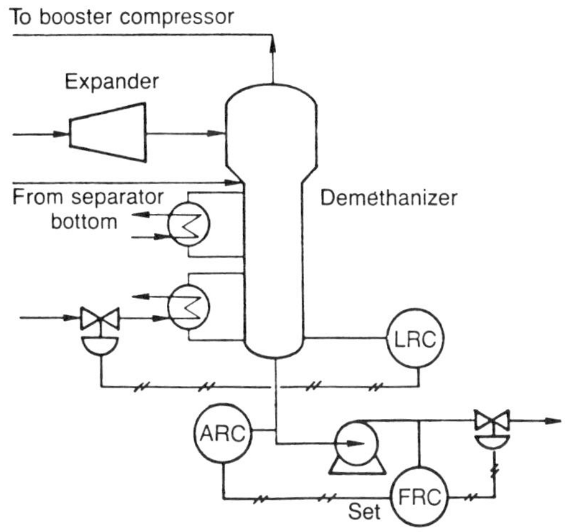

Figure 10. Floating pressure control.[2]

case, both vapor and liquid feed rates could be input to the feedforward controller.

Theoretically, feedforward control could provide near-perfect control for feed upsets except for the inaccuracies in obtaining exact transfer functions. Experience has shown that a simple first order model with dead time can make very significant improvements in demethanizer control. Large upsets in feed rate are handled by the feedforward controller while minor offsets are controlled by the feedback analyzer cascade scheme. Both electronic and pneumatic analog simple lead/lag controllers are available. However, digital equipment has proven superior and easier to implement.

More elaborate feedforward control schemes using several inputs to compute enthalpies have been used successfully. Digital equipment is essential for this control method. In most cases feedforward control of demethanizer feed should provide extremely good control.

Floating Pressure Control

If the objective of the demethanizer is to maximize recovery of bottom product with a specification of its methane-ethane ratio and only about thirty percent of the feed gas is condensed, bottom product flow will usually be smaller than the boilup rate. In this case, the control scheme in Fig. 10 is recommended by some. The scheme uses an analyzer controller cascaded to a flow controller to control bottoms composition by regulating bottoms flow rate, while bottoms level is controlled by regulating the residue gas flow to the reboiler-exchanger. The authors are not aware of anyone using this control scheme. It should minimize the effects of the heat balance on product quality. Perhaps difficulty with discrete analyzer controllers in general has hampered an evaluation of this scheme.

CONCLUSIONS AND RECOMMENDATIONS

While each plant will have its own dynamic characteristics and therefore its own optimum control scheme, it appears that the majority of plants would benefit from (1) the pressure control scheme that uses a signal selector to permit either the upstream separator or downstream demethanizer pressure to be controlled by the expander and (2) the feedforward control scheme with cascaded analyzer feedback trim to regulate the fractionator bottom composition. Both of these schemes are best implemented using small digital controllers or computers.

Ideally, it would be extremely beneficial to be able to dynamically simulate the entire process. Each plant could be analyzed to determine the control scheme that would optimize responses to various disturbances. No such digital simulation package is generally available. Differential equations involving material and energy balances along with transport equations could be written for each major piece of equipment in the plant; i.e., separator, demethanizer, expander, etc. A generalized program that could be readily adapted to a particular plant situation and used to determine the optimum control scheme would be a benefit to the natural gas industry.

References

1. Garvey, C. A., and Griffin, D. E., "Computer Control Increases NGL Recovery," *Oil and Gas Journal,* p. 156, Aug. 24, 1981.
2. Shinskey, F. G., *Distillation Control,* McGraw Hill Book Company, New York, 1977.

Part 3
ADVANCED AND CONVENTIONAL CONTROL AIDS

50
AI and MAP in the Processing Industries*

L. A. Kane

Artificial intelligence (AI) has the potential to do more for process control in the next 10 years than microprocessors have in the past 10 years. Manufacturing automation protocol (MAP) could do for digital systems what 3 to 15 psi did for pneumatic systems and 4 to 20 ma did for analog electronic systems. This chapter reviews these technologies, gives real examples of their use in the processing industries, and shows how they are related.

In the broadest sense, AI is the science of developing computer systems to emulate human intelligence (reasoning, decision making, language, learning, etc.). Although this includes several technologies, the branch of AI currently used in the processing industries is "expert systems."

The intent of an expert system is to simulate the decision-making process an expert performs to solve a problem. The decision-making process is converted to heuristics (rules of thumb) and deep knowledge (process model) that are put into the expert system's data base (knowledge base). The rules of thumb are usually translated into statements of the type "IF this happens THEN take this action" (examples to follow).

The major advantage is assisting those with only limited knowledge of a subject by giving them access to the decision-making process used by someone very knowledgeable in the subject.

The disadvantage is that the expert system is only as knowledgeable as the chosen expert/s and the skill of the knowledge engineer in interpreting and inputting that knowledge into the knowledge base. Gaps in this process will result in knowledge gaps in the expert system, and the possibility for incorrect decisions if unaccounted for situations arise. Expert systems, however, do have the ability to learn and adapt to provide the correct action. But perhaps the greatest problem in implementing expert systems will be user acceptance. How will people perform when a computer consistently comes up with better solutions to problems in their own areas of expertise?

A good application for expert systems is diagnostics. An expert system for automobile repair would request the symptoms of the problem and suggest one or more parts that may be at fault. The decision-tree structure of an automobile repair manual is analogous to the structure of an expert system knowledge base.

The development of expert system knowledge bases is called "knowledge engineering." A knowledge engineer is one who, after considerable consultation with experts, inputs the experts' knowledge into the knowledge base. These people are in great demand.

DATA VS. KNOWLEDGE PROCESSING

Computer hardware and software requirements differ greatly for data processing and knowledge processing. The recent availability of reasonably priced computer systems tailored to AI processing is the reason AI is beginning to be used now. The principles of AI have been known for many years.

To illustrate the different requirements for data processing versus knowledge processing, consider the way we make decisions. Several factors are evaluated in coming to even the simplest conclusions. The result of all this input is only one output. The process of using several inputs to provide a few outputs is convergent logic. Data processing involves divergent logic. Anyone familiar with computers knows their capacity for generating tremendous output from only a few inputs.

Humans are poor calculators, but have the ability to process inexact parameters and symbols rapidly. Data processing computers are excellent calculators, but can't relate to inexact parameters. Humans can process several pieces of data simultaneously (parallel processing), whereas computers usually process data one piece at a time (serial processing). Hence, for the speed required

* I wish to thank Dr. K. Kumar Gidwani of LISP Machine, Inc., Los Angeles, Calif., for supplying information and reviewing this chapter.

to do on-line applications, AI computers require a parallel structure and large amounts of memory (some off-line applications will be given later). Since the memory must be searched quickly and parameters must be interpreted symbolically, a different type of programming language is used.

The most popular language for AI processing is LISP (LISt Processing). Most languages place data in fixed memory locations, whereas LISP moves data around in memory as needed for processing. Few of us in the processing industries will need to know LISP, because AI suppliers provide user interfaces that allow application programming to be done in a natural language in the form of If-Then statements.

Another term you'll probably want to mention at your next cocktail party is "fuzzy logic." It is mentioned here to illustrate how computers can handle inexactness, i.e., If the reactor is *too* hot, then *slowly* close the valve. Fuzzy logic[1,2] provides a way to describe how well something fits into a group in terms of "grades of membership" from 0 to 1. It is defined as "a kind of logic using graded or qualified statements rather than ones that are strictly true or false. The results of fuzzy reasoning are not as defined as those derived by strict logic, but they cover a larger field of discourse."[1] For example, if Andy is a 48-year-old member of a technical society, we may want to know to what degree the label "young member" applies to Andy. A typical answer might be 0.3 indicating a grade of membership of 0.3 for Andy in the group labeled "young members."

REQUIREMENTS FOR ON-LINE APPLICATIONS

Figure 1 shows the structure of a system for on-line process control applications.[3] The knowledge base includes three types of knowledge: deep knowledge (process model), heuristic knowledge (rules of thumb) and meta knowledge (how to use the data). The deep knowledge provides information on the structure of the process and how various components and variables interact; just as our process models do now. The heuristic knowledge provides the information necessary for making decisions about what to do during process upsets.

The inference engine decides how best to search and use the knowledge base (meta knowledge), and is really what separates an on-line system from an off-line one. Since an on-line system must operate fast enough to perform real-time functions, it must focus only on pertinent areas of the plant during an upset, rather than searching the entire knowledge base. The inference engine also executes forward chaining (taking known data and coming to a conclusion) or backward chaining (assuming this

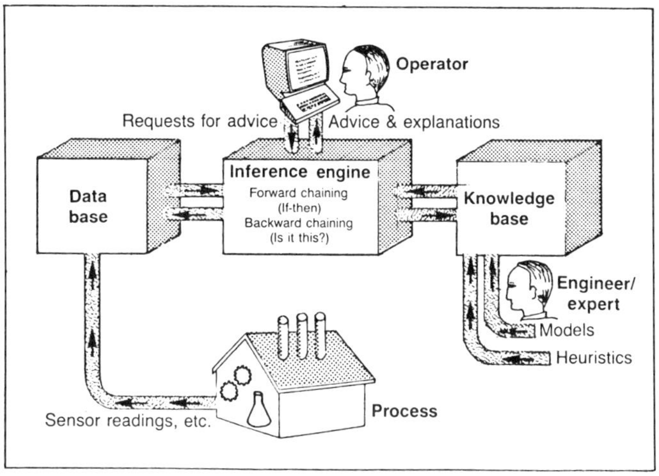

Figure 1. Structure of an expert system.[3]

conclusion, what data do we have to support it?) depending on the situation.

Unlike conventional systems, expert systems are modular so the knowledge base can be changed as new information becomes available without any programming changes.

Although this technology is in its infancy, this system was installed at Texaco Chemicals in Port Arthur, Texas, in May 1985 and at Exxon Chemicals in Bayway, N.J., in September 1985.

On-line controller tuning, a form of adaptive control, is another presently used application for expert systems. Several manufacturers offer self-tuning controllers. Basically two approaches are used. The first approach emulates the popular Ziegler Nichols tuning method. In this case, pattern recognition is built into the controller's knowledge base. The user selects the desired response in terms of damping and overshoot. The knowledge base then monitors process response to disturbances and adjusts the PID tuning values until the desired response is achieved.

A second method uses a process modeling approach. Loop response is compared to a loop model response and differences are used to update the model. Both methods have received good acceptance.

ON-LINE PROCESS INDUSTRY EXAMPLES

Most on-line applications currently in use are based on the "How would our best operator and process engineer handle the situation?" approach. The systems installed at Exxon and Texaco use two processors in parallel: a LISP processor to handle the expert system and a 68010 processor for real-time data acquisition (Fig. 2).[4] The 68010 uses an internal MULTIBUS to communicate with the distributed process control system.

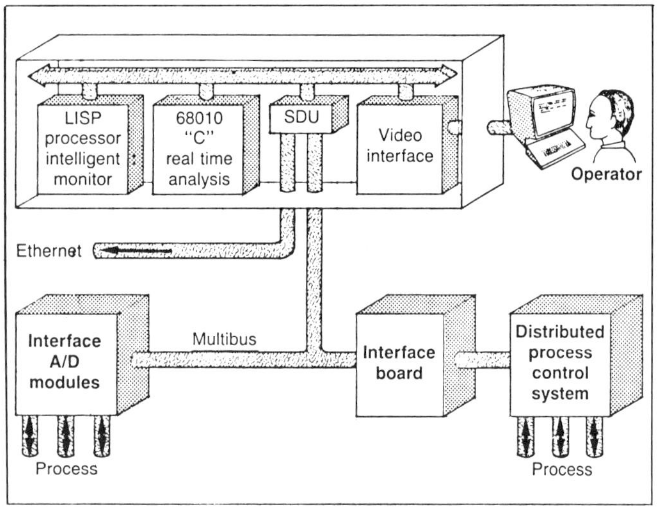

Figure 2. Expert system for process control.[4]

The first step in system application is inputting the deep knowledge (process model). This is done schematically by selecting plant components (Fig. 3).[5]

Heuristics are inputted via a natural-language interface that allows the user to input rules as If-Then statements (Fig. 4).[5] Details on system configuration procedures are given in References 5 and 6.

Unfortunately, but understandably, Exxon and Texaco are very secretive about the specifics of their applications (which is why I wrote this chapter instead of them).

The Texaco installation[4] involved four types of experts: a chief operator, a foreman, a process engineer and an application engineer. The problem they are trying to solve is how to achieve the same level of productivity during the evening and night shifts as they do during the day. During the day these experts work together to solve problems, but at other times the operators alone have trouble achieving the same performance.

The expert system considers the effects of advanced control strategies on the process. For example, when the column feed rate is changed, reboiler heat duty is changed automatically to maintain a specified feed/reboiler heat duty ratio. The control strategy cannot handle a large upset in the reboiler heating medium. If this occurs, the expert system advises the operators on how to compensate for an abnormal change in the reboiler heat.

An example of rule capture in an If-Then format is:

IF
REFLUX TO CHARGE RATIO R1 (CALCULATED VARIABLE)
IS-GREATER-THAN
1.2
AND
TOLUENE ANALYZER A70-TOL

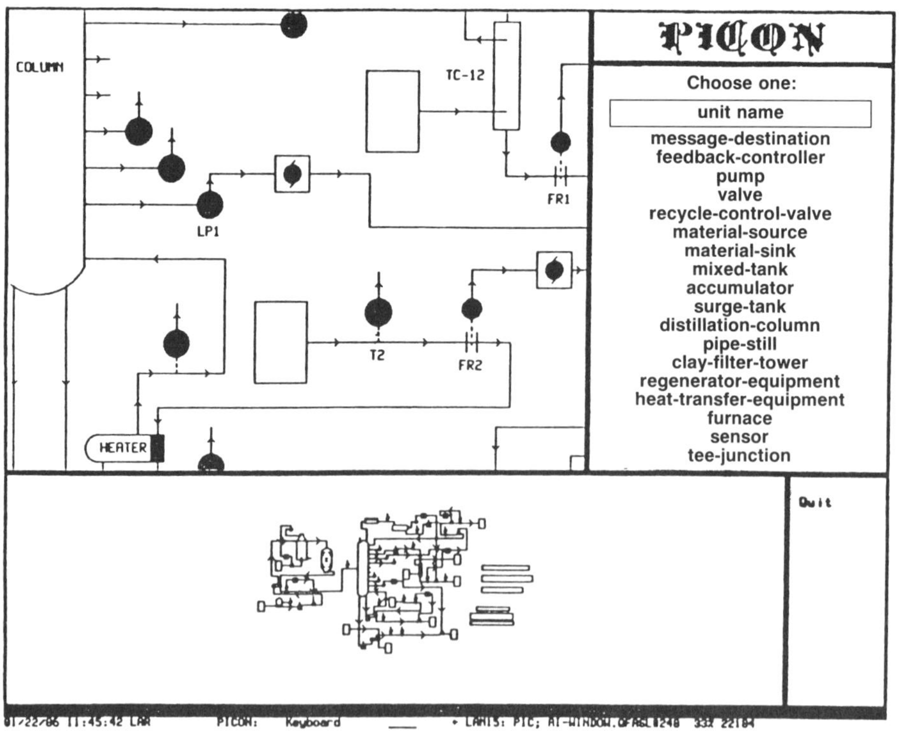

Figure 3. Schematic capture of process knowledge.[5]

IS LESS THAN 200 PPM
AND
TR25
IS GREATER THAN 250 DEG-F
THEN
SEND
"Lower heat-to-feed ratio (Q/F) by 0.5
MBtu/bbl"
TO
OPERATOR

The message for this rule is presented to the operator as:

DATA 1) Reflux to charge ratio = ***
2) Both ends are on spec.
3) Toluene analyzer A70-TOL = ***
4) TR-25 = ***
SUSPECTED PROBLEM: Excess Heat to benzene tower
OBJECTIVE: Save utilities
ACTION: Lower heat-to-feed ratio (Q/F) by 0.5 MBtu/bbl

A distillation column example for the separation of benzene and toluene in a petrochemical complex (Fig. 5) is given in Reference 6.

The process is difficult for an operator to control. Significant process upsets occur because:

- Feed composition varies considerably when supply barges are changed.
- Feed flow varies as much as 50% due to upstream processes.
- The clay filtration towers (one shown) are switched every few days to reload filter media.
- Excess hydrogen from other processes is dumped into the fuel gas supply causing hot water temperature to the reboiler to vary.
- Vapor cooling varies with the weather (particularly significant during rain).

Process control is made more difficult because:

- Dynamic response times vary from minutes to hours.
- Control actions can have opposite effects depending on column condition.
- The cooling fans are only adjustable in gross increments by turning one of four fans on or off.

Two types of experts were used to define the knowledge base: operations and engineering.

The general capabilities of the system are:

- Intelligent alarming, particularly on complex combinations of conditions

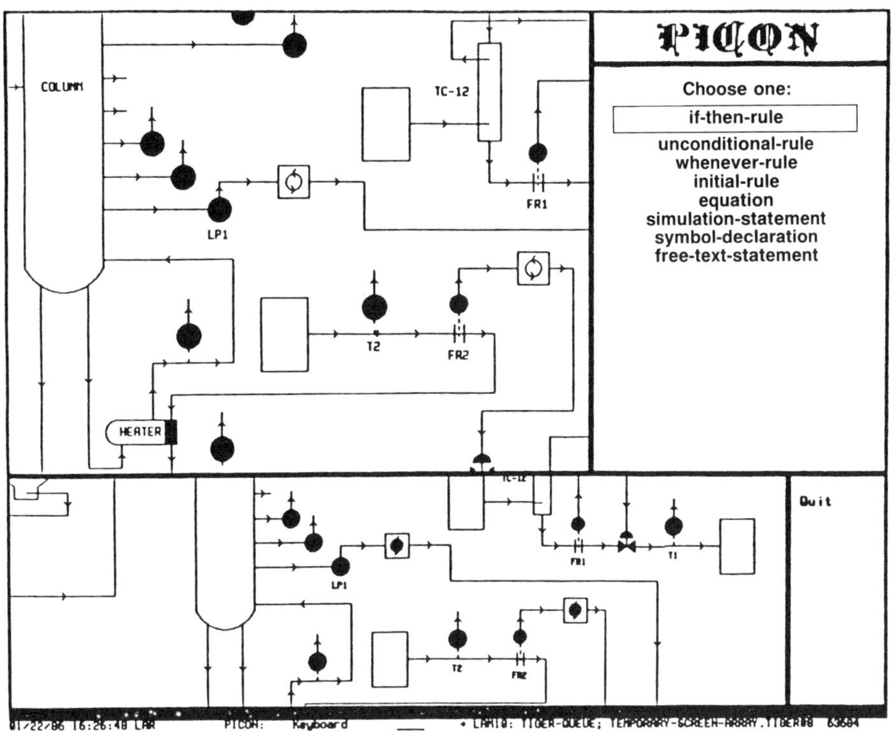

Figure 4. Schematic capture of heuristic knowledge.[5]

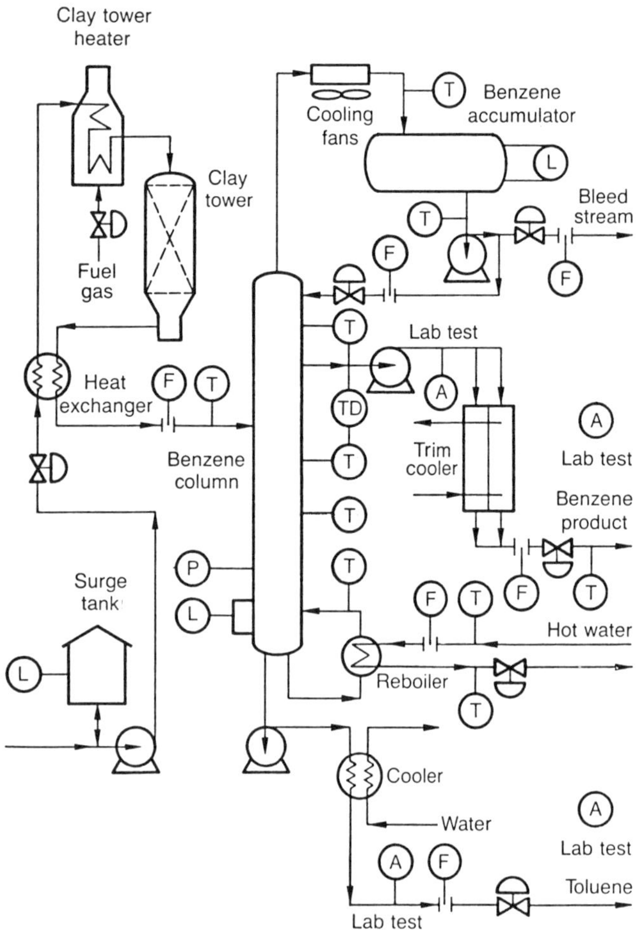

Figure 5. Distillation column example.[6]

- Detection of possibly significant events by inference applied to heuristic rules about dynamic process conditions
- Focus inference. In a typical case, a possibly significant event would trigger a focus on the process unit gathering information required for all inference around the process unit.
- Diagnosis. An explanation is given of the resulting inference path.

The following rule shows how a different control procedure is initiated depending on tower conditions.

IF benzene off specification
AND tower-flooded
AND reflux-low-limit
THEN reduce-reboil-procedure
IF benzene off specification
AND tower not flooded
THEN increase-reboil-procedure

In summary, current on-line applications include:

- Alarm management
- Operator aid on difficult to control processes (e.g., startup and shutdown)
- Adaptive control (e.g., controller tuning)
- Process and equipment troubleshooting.

OFF-LINE APPLICATIONS

Hardware and software demands are not as great for most off-line applications as they are for on-line ones. I have even found a $50 expert system program for personal computers that can be used to build rudimentary, but useful knowledge bases. Some examples from the literature illustrate present applications.

Control System Design

Distillation is a good example of complex control due to interaction between the controlled and manipulated variables.[7] Shinskey[8] is developing a knowledge-based system that will draw a recommended control system for two-product distillation columns. The designer enters process and column data in a question and answer format as follows:

Mol. Frac. Lt. Key Top? ___, Btms? ___
Mol. Frac. Ht. Key Top? ___, Btms? ___
Mol. Frac. Lt. Key Feed? ___, Feed % vapor? ___
Number of trays? ___, Reflux/Dist. Ratio? ___
Tray Eff.? ___, Rel. Volatility? ___
Plot (P)? ___, Calculate RGA (R)? ___

The system recommends pairings of controlled and manipulated variables and calculates the integrated error associated with each pairing. Pairings with the least integrated error have the least interaction. The system then asks about column hardware details and the economic objective of the column and draws the recommended control system.

Process Control Troubleshooting

Schlager[9] describes how calls at 3 a.m. to solve problems with computer controlled fermenters at Miles Laboratories prompted him to develop a microcomputer-based expert system to aid troubleshooting. To learn as much as he could about expert systems, he decided to write the program directly rather than use any of the expert system tools available. The program was written in PROLOG.

To use the system, the operator asks the program about possible faults. A dialog begins until all possible faults are reported. A possible dialog for a pH problem follows (bold type is operator input):

& which (x : possible-fault x)
X a problem ? **ans (phe 301 low)**
X a problem ? **no**
((bad-electrode phe 301))
(phe 301 in auto yes / no ? **yes**
(out-of-base phe 301) a problem ? **no**
((bad-base-pump Ferm #3))
No (more) answers
&.

The ampersand indicates the system is ready for input. In this example, only one of the two pH loops (phe 301) reports a low alarm. Possible fault Rule #2 (I/O card failure because two alarms are on the same signal conditioning card) fails because there is only one alarm. Rule 3 (Is the problem related to pH?) matches and the pH rules are called. Since only one pH loop is in alarm, it is possible that the pH electrode may be bad and this is reported to the operator. The next pH rule is then tried. It asks for confirmation that the loop is in automatic. The operator replies "yes" and the rule then asks if there is base in the reservoir. When the operator answers "no" the rule concludes that there is something wrong with the base pump. This can trigger other rules that diagnose pump failures.

The author concludes that "the system represents a concise way to store a troubleshooting manual, thus allowing a process operator to rapidly diagnose a failure."

MANUFACTURING AUTOMATION PROTOCOL

MAP is a group of developing standards to allow communication between digital equipment from different manufacturers. It is included here because of a potential problem that implementation of MAP, or any type of industrial network in the processing industries, is likely to cause: data incompatibility. Artificial intelligence is a probable solution to this problem.

Although MAP was instigated primarily by General Motors to provide networking in discrete manufacturing, efforts by ISA (Instrument Society of America), SME (Society of Manufacturing Engineers) IEEE (Institute of Electrical and Electronics Engineers) and others show that interest in it is growing in the processing industries as well.

The MAP/TOP User's Group is a committee of SME that meets regularly to discuss problems and applications (TOP, meaning Technical Office Protocol, is a group of standards being developed for office automation). To increase awareness of MAP in the processing industries. SME has formed a MAP/TOP Process Industry Working Group. ISA, working with IEEE, has developed ISA-S72-01-1985 entitled, "PROWAY-LAN Industrial Data Highway." This standard addresses compatible interconnections for a local area network using a token-bus format for process control applications. It addresses the International Standards Organization-Open Systems Interconnection (ISO-OSI) layers 1 and 2, which are an outgrowth of the MAP effort.

ISO/OSI Levels

Standards for seven levels of computer compatibility are under development. Table 1 shows the hierarchy of the levels and gives a brief description of their applications. As of this writing, only levels 1 and 2 have usable standards (ISA-S72-01-1985 and IEEE 802.2 through .5). Standards for the others are under development.

Table 1
OSI Layered Architecture[10]

Layer 7 **Application** (User application process and management functions).
Layer 6 **Presentation** (Data interpretation, format and code transformation).
Layer 5 **Session** (Administration and control of sessions between two entities).
Layer 4 **Transport** (Transparent data transfer, end-to-end control, multiplexing, mapping).
Layer 3 **Network** (Routing, switching, segmenting, blocking, error recovery, flow control).
Layer 2 **Link** (Establish, maintain and release data links, error and flow control).
Layer 1 **Physical** (Electrical, mechanical, functional control of data circuits).

While these standards can solve the problems of compatible physical connections between products and proper software protocols for exchange of data, the problems of "sorting data" will remain. Deciding what data need to be sent to which computers and how often requires intelligent decisions. Unless this problem is addressed, the frustrations we now feel about too much computer-generated data will get worse. Many of the examples shown for AI hint at its ability to function as an intelligent data sorter.

CONCLUSION

AI and MAP are developing technologies but they are not futuristic technologies. AI is being used in a few processing plants and some plants do use networks to exchange data between process control, engineering, maintenance, storeroom and corporate data processing computers.

With computers finally receiving wide acceptance in the processing industries, the major obstacle to implementation of the more advanced control methods that have been touted at universities for many years has been removed. It is certainly not unreasonable to expect these technologies to become widespread within the next decade, especially considering the rapid advances in computer hardware that have, and will continue, to occur.

References

1. Zadeh, L. A., "Making computers think like people," *IEEE Spectrum,* Aug. 1984.
2. Goff, K. W., "Process Control Applications of Artificial Intelligence," API 51st Midyear Refining Meeting, San Diego, Calif., May 1986.
3. PICON Product Summary, LISP Machine, Inc.
4. Gidwani, K. K. and Dalton, W. S., "Innovative Knowledge Engineering for Real Time Expert Systems," SOCOCO '86 Conference, Vienna, Austria, May 1986.
5. Moore, R. L., Hawkinson, L. B. and Knickerbocker, C. G., "Expert Systems Applications in Industry," ISA '84 International Conference and Exhibit, Houston, Texas, Oct. 1984.
6. Moore, R. L., Hawkinson, L. B., Knickerbocker, C. G. and Churchman, L. M., "A Real-Time Expert System for Process Control," IEEE 1984 proceedings of the first conference on artificial intelligence applications.
7. Myron, T. J., "Expert Systems—Directions and Expectations for Process Control," API 51st Midyear Refining Meeting, San Diego, Calif., May 1986.
8. Shinskey, F. G., "An Expert System in the Design of Distillation Controls," Proceedings of the Third Chemical Process Control Conference, Asilomar, Calif., Jan. 1986.
9. Schlager, S. T., "An Expert System for Process Control Troubleshooting," AIChE Spring National Meeting, New Orleans, La., April 1986.
10. Jones, R. L., "Standard Networking Protocols and Gateways for Industrial Applications," *Significant Contributions in 1984 to the Computer Technology Division, Instrument Society of America.*
11. Moore, R. L. and Kramer, M. A., "Expert Systems in Online Process Control," AIChE Conference, Boston, Mass., Nov. 1985.
12. Meier, F., "Investigation Into ISA's Involvement and Potential Involvement in MAP/TOP Technology," ISA's President's Winter Meeting, Monterey, Calif., Feb. 1986.
13. Moreley, R. E. and Taylor, W. A., "What is Artificial Intelligence," Gould, Inc.

51
Using Performance Indices for Better Control

V. A. Bhandari, R. Paradis and A. C. Saxena

New instrumentation permits recording and presentation of all process information in a convenient and effective manner. Interpretation of this information, however, is no easier. In fact, it may be more difficult due to an increase in the information handled. Since the incremental cost is small, many process measurements that were previously ignored, except under abnormal conditions, are now included in the process database.

To illustrate the point, consider the dashboard of a modern automobile equipped with a large number of gauges to provide instant information, digital in many cases, on more than ten different variables. It is difficult, however, to quickly estimate the performance of the car, say, in terms of gasoline usage. If an instant reading of miles/gallon of gasoline used was displayed, it would be much easier to "manage" the performance of a car relative to any gasoline consumption related objective. The effect of variables such as driving habits, car speed, and road conditions on the gasoline usage would be known quantitatively at all times.

This chapter explores the concept of using carefully defined performance indices to facilitate management of a process. Examples from an actual distillation process are given to communicate the usefulness of the concept and how it is implemented in a manufacturing process.

DEVELOPING PERFORMANCE INDICES

A performance index is defined in terms of measured process variables. The value of this calculated variable results from the fact that it communicates more about the performance of a process than any of the individual process variables used in its definition. The definition of a performance index must reflect what the process engineer/manager considers to be an important measure of the process performance.

It is possible to define a performance index in one of two ways: an absolute measure, e.g., production cost in $/ton or a relative measure that compares actual performance with theoretical, design or expected performance, e.g., energy usage as a percentage of design value. In general, a process unit requires multiple performance indices. In addition to indices for individual process operations, it is useful to define an overall unit performance index on the basis of cost, productivity or product quality considerations. The utility of a performance index as a management tool is highly dependent on the meaningfulness of its definition.

Distillation Process Example

We have used the concept of performance indices to improve the management of an extractive distillation process. The feed to the process consists of a mixture of C_4 hydrocarbons. In the extractive distillation tower, the feed is contacted with a butadiene selective solvent. The butadiene is separated from the solvent in the solvent stripping tower and lean solvent is recycled to the extraction tower. The butadiene is further purified in the finishing tower into a high purity 1,3-butadiene product. The schematic arrangement of various towers is shown in Fig. 1.

The economics of the unit are strongly dependent on the performance of the three distillation towers. The extraction tower's operation is especially important to maintenance of unit productivity and quality. The lowest unit operating cost does not necessarily coincide with lowest-cost operation of individual towers. This results from the fact that, with some constraints, there is flexibility to apportion the total separation job among the three towers.

For effective management of the unit, it was necessary to define performance indices for individual towers as well as an index for the overall unit performance. In defining an index for the extraction tower, use was made of the concept of a tower separation factor,[1] S, defined in terms of a pseudo-binary (1,3-butadiene and all other components) as:

$$S_E = [(1 - y)/y][x/(1 - x)] \quad (1)$$

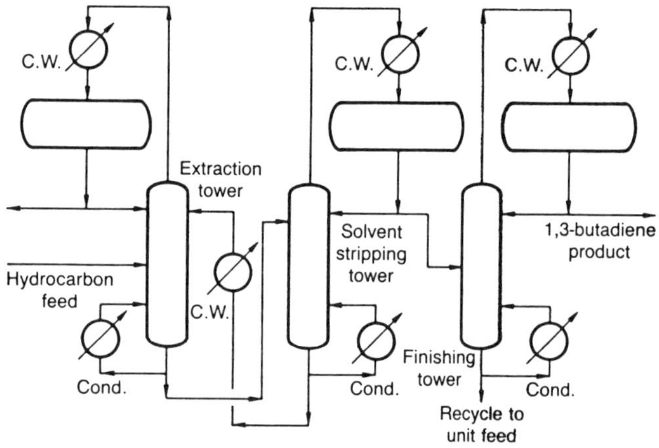

Figure 1. 1,3-butadiene extractive distillation process.

where x and y are respectively mole fractions of 1,3-butadiene on a solvent free basis in the bottom and top product streams. The separation factor defined in this manner is a measure of the actual work of separation performed by a tower. S_E has a value of one when no separation takes place and reaches its maximum value for the given design and operating conditions at total reflux. If Q_E is the net amount of energy usage in GJ per ton of feed supplied to the tower to achieve this separation, a performance index, I, analogous to the miles per gallon measure of a car's performance can be defined as:

$$I_E = S_E/Q_E \quad (2)$$

In a similar manner, for the finishing tower:

$$S_F = [y/(1 - y)][(1 - x)/x] \quad (3)$$

and,

$$I_F = S_F/Q_F \quad (4)$$

The separation performed in the solvent stripping tower is relatively easier (relative volatility = 25) and a performance index, I_S, based on tower design values was considered adequate.

$$I_S = Q_a/Q_d \quad (5)$$

where Q_a and Q_d are, respectively, actual and design energy consumptions for the tower, at equivalent feed rates.

The overall unit performance is measured by an index, I_T, based on unit productivity and energy usage in the three towers.

$$I_T = Q_TC/f \quad (6)$$

where f is the fraction of 1,3-butadiene in the feed that leaves the unit as butadiene product; Q_T is the total energy consumption of the extraction, solvent stripping and finishing towers; and C is the cost of energy in $/GJ.

Thus, I_T is simply the energy cost in a/ton of finished product.

The process performance indices I_E, I_S, I_F, and I_T are treated as continuous process variables in a manner similar to the variables from which they are calculated. Their instantaneous and various average values (hourly, daily, etc.) are reported in process variable report groups for individual towers and unit overview.

A typical process variable report for the extraction tower during a startup is shown in Table 1. The evolution of tower behavior to full steadystate is reflected in S_E and I_E values. The effect of some typical process variations on the performance of individual towers and overall unit economics are shown in Figs. 2 to 4.

Table 1
Extraction Tower Startup Summary

Time (h)	Feed rate, % max.*	Reflux rate, % max.*	Tower pressure, % max.*	Reboiler steam, % max.*	Bottom temp., (°C)	S_E	I_E
10	67	0	53	22	80	—	—
12	67	0	55	24	82	—	—
14	75	0	64	26	86	—	—
16	92	29	69	30	94	3	8
18	100	63	74	35	98	8	25
20		79	85	44	108	12	38
22		96	91	59	110	16	55
24		96	96	74	115	40	74
26		96	99	89	122	100	100
28		96	100	94	124	300	283
30		96	100	96	126	803	741
32		96	100	98	128	1,503	1,364
34		96	100	98	129	2,501	2,273
36		100	101	99	130	5,090	4,464
38		100	101	100	130	5,503	4,867
40		100	101	100	130	5,600	4,956
42		100	100	100	130	5,606	4,956

* Max. stands for the final steadystate value of the variable.

A long excursion in 1,3-butadiene content of the feed on the performance of extraction tower and overall unit economics is shown in Fig. 2. The separation difficulty in the extraction tower is more than doubled and the unit operating cost has increased by over 10%. In Fig. 3 a reduction in the flow rate of the solvent, a mass separating agent, is shown to reduce energy used for solvent stripping by about 25% and reduce overall unit operating cost by about 4%. As a result of upsets in an upstream unit, the content of an impurity in the feed to the extraction process has changed significantly. As shown in Fig. 4, this reduced recovery of 1,3-butadiene product in the finishing tower increased energy usage in this

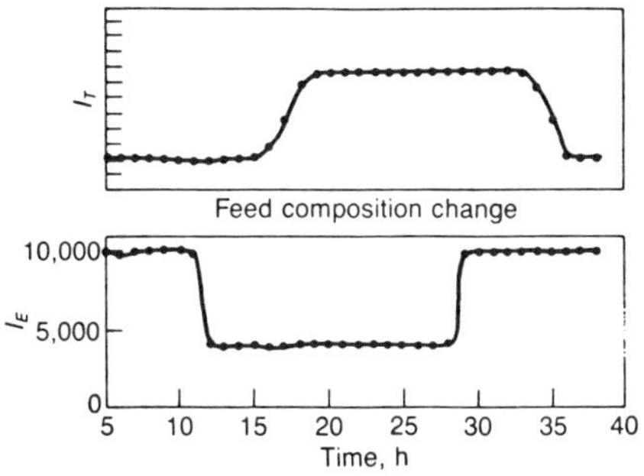

Figure 2. Extraction tower and overall unit performance indices.

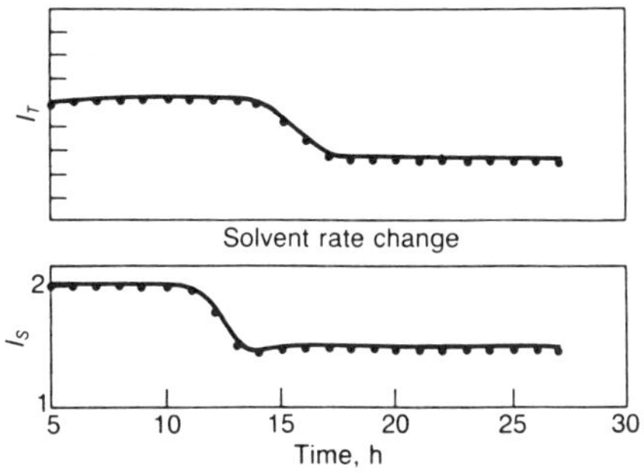

Figure 3. Solvent stripping tower and overall unit performance indices.

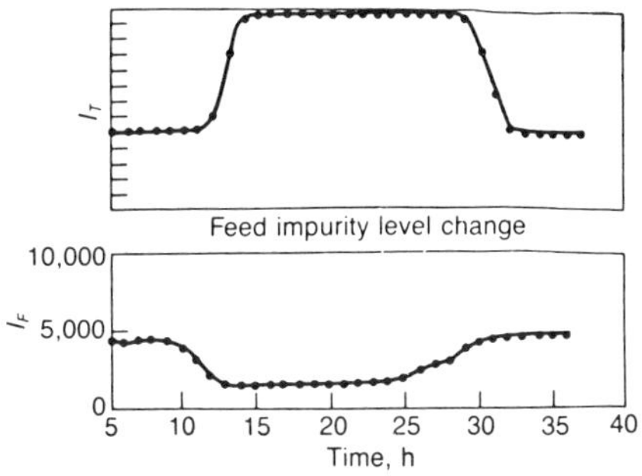

Figure 4. Finishing tower and overall unit performance indices.

tower and had a major impact on unit economics. In this instance, changes in the finishing tower performance index actually led to the discovery of abnormal operation of the upstream unit. Additional information on how these indices were used to optimize the process variables is available in a separate publication.[2]

Developing Performance Indices

The previous examples show that process indices are transformations of normal process information which provide a more useful picture of process performance for decision making purposes. This is considered to be a necessary requirement for the definition of a process index. For advisory control purposes, a performance index is equivalent to a process measurement; when its value exceeds the normal operating range, a process change is necessary. The corrective action and its extent are determined from a knowledge of process response. It is possible, and desirable in some cases, to close a control loop using a performance index just as with any other process measurement.

The use of the concept of performance indices is not limited to a distillation process. Any process operation in which some process measurements can be used to define a reliable criterion of performance can benefit from its use. For all other separation operations governed by an equilibrium process, similar methodology to that described here can be used to define a performance index based on the degree of separation achieved and the quantity of separating agent used. For rate processes such as chemical reactions and mass, heat or momentum transfer operations, both rate and extent of change may be important in addition to the usage of the change agent. Often, the quality of the product of an operation is variable enough to be included in the definition of a performance index. It is also possible to use transfer coefficients and dimensionless numbers in the definition of a performance index. There is much room for ingenuity; all that is necessary is that a useful index reflecting process characteristics be developed.

When a performance index is developed, it is important to give adequate thought to the end-user and how the information will be used. The operating personnel who use process variables such as reflux flow and reboiler steam flow as measures of process performance would feel most comfortable with simple indices such as reflux flow per unit flow of a key feed component or simple yield/efficiency indices. A process engineer would make better use of indices based on variables such as transfer coefficients, separation factors or fouling rates. The unit manager would be more interested in overall unit production cost and efficiency related indices. A performance index that is used as a controlled variable in closed-loop control, e.g., tower separation per unit of reboiler energy to manipulate feed rate, should be based on very reliable measurements and sound knowledge of process behavior.

Implementation of Performance Indices

The concept of performance indices is very conveniently implemented on a distributed control system. The performance index, a calculated variable, is treated as just another process variable for purposes of recording, historical analysis, graphic displays or report generation. The calculation of a performance index can often be carried out using configured mathematical operations or, in cases of high mathematical complexity, with the use of an on-line computer. The frequency of reporting the performance index, e.g., hourly or daily, is dictated by its end use and user.

A well-defined performance index should provide greater insight into the operation of a process than normal process data and should enable better process management for any given process objective. The use of performance indices thus represents better utilization of an existing distributed control system. Some selling of the concept to the operating personnel, as well as the unit manager, is essential. In our experience, once implemented, the acceptance of the concept and its value to process management are quite high.

USES OF PERFORMANCE INDICES

We have benefited from the use of performance indices in many different ways. Some of the more important areas of present or proposed utility are summarized here:

- *Process monitoring and optimization:* Process personnel are able to move the operation through a series of small changes towards more desirable/economical conditions.
- *Troubleshooting:* Process fouling, leaks and even abnormal operation of upstream units is reflected in the process performance indices. They are often an early warning of undetected process changes.
- *Learning tools:* Performance indices constantly provide a quantitative measure of the effect of any intentional or inadvertent process change. As a result, they are powerful learning tools for process personnel merely through day-to-day observations. They are useful for establishing cause-effect relationships and for steadystate or dynamic process model development.
- *Management tools:* They provide a constant focus on the performance and economics of individual operations as well as those of the overall process. They also reflect quantitatively the productivity-quality-operating cost tradeoffs. By quantifying the effect of feed purity on overall unit economics, performance indices can improve feed purchasing/pricing decisions.
- *Technology development:* Increased process knowledge and quantitative process descriptions often lead to changes in process control systems and strategy as well as in other process methodology. With the increasing use of artificial intelligence programs in the process industry, a database consisting of process variables and performance indices may prove suitable for technological development efforts.

CONCLUSIONS

Performance indices quantitatively reflect the effect of process changes. They enable fast and meaningful interpretation of the process behavior. They are powerful management and learning tools and are easily implemented using a distributed control system or an on-line computational device.

Nomenclature

C -cost of energy, \$/GJ
f -fractional 1,3-butadiene recovery
I -performance index
Q -energy usage, GJ/t
S -tower separation factor
x -mole fraction of 1,3-butadiene in the bottom product
y -mole fraction of 1,3-butadiene in the overhead product

Subscripts

a -actual
d -design
E -extraction tower
F -finishing tower
S -solvent stripping tower
T -overall unit

References

1. Shinskey, F. G., *Distillation Control for Productivity and Energy Conservation,* McGraw Hill Book Co., New York, N.Y., 1977.
2. Saxena, A. C. and Bhandari, V. A., "Reducing Energy Usage in Extractive Distillation," paper presented at the Seventh Annual Industrial Energy Technology Conference and Exhibition held at Houston, Texas, May 13–15, 1985.

52
Characteristics and Control of Process Disturbances

B. D. Stanton

The objective of automatic control systems is to hold processes at fixed operating levels that are preset by operators or computers. As in all things, there are forces (disturbances) at work trying to defeat this objective. This chapter discusses: 1) methods used to hold processes close to desired levels when faced with disturbances, and 2) the characteristics of control and disturbances. Only after these concepts are understood and implemented is optimal control possible.

The attenuation of disturbances is accomplished in two ways: first, by providing process filtering (i.e., tanks); second, by manipulating sources of energy to counteract the disturbances. Both of these ways cost money in the form of capital equipment and energy costs. In contrast to conventional control schemes, optimal control schemes normally push the present operating levels close to the brink of acceptable operation. Two simple examples are: (1) To reduce energy costs in operating large centrifugal compressors, the speed is reduced to a point where the machine is close to its surge point. Operating in surge, which is really blade stall, drastically reduces throughput and creates undesirable mechanical stresses. (2) A reactor normally has a nonlinear temperature vs conversion curve. It is desirable to operate at as high a temperature as possible without going into thermal runaway or decreased selectivity. To be safe, the temperature often is held below the optimum value, thereby reducing yield.

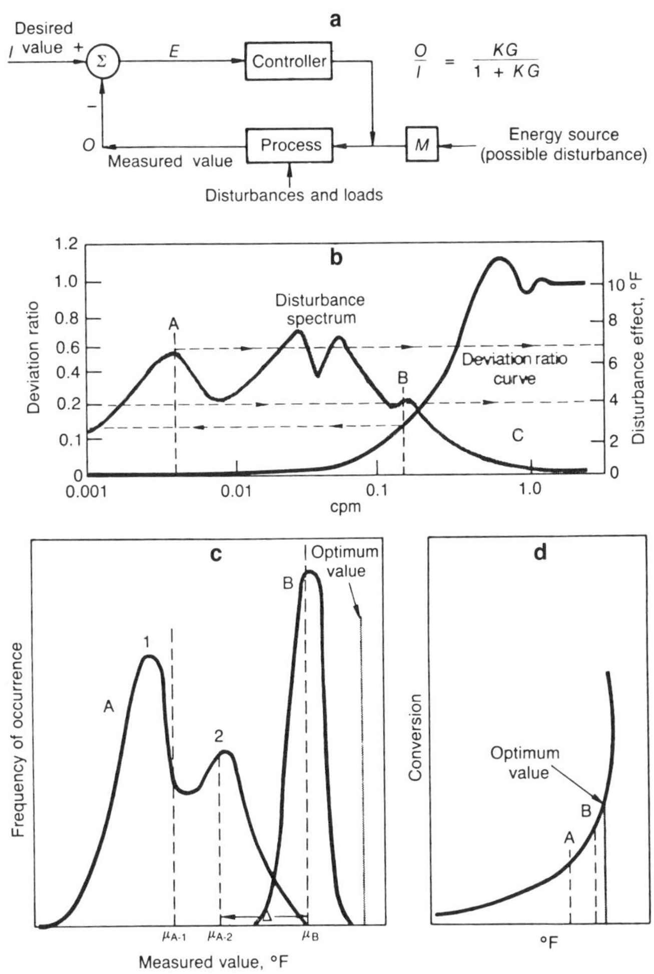

Figure 1. Single loop control and its characteristics.

CONCEPTS OF FRONTLINE CONTROL

First, however, several technical points must be discussed as they will be used throughout the chapter. The single loop controller is the front line device used to attenuate disturbances and to accept the inputs from advanced control systems. These control loops are an absolute necessity. They rapidly compensate for disturbances and hold the process in a fixed position so that the advanced control scheme, operating at a much slower rate, can make decisions about the best operating levels based on averaged or smooth data.

The basic control loop and its closed loop equation is shown in Fig. 1a. The objective of this loop is to hold a measured value, O, equal to the desired value, I. This reduces the error signal, E, to zero. The term G, in the equation represents all the dynamic terms in the loop. In time this term will go to unity. The term, K, is the gain around the loop. It is apparent that the larger the value of K, the closer O comes to I. The trick that the controller performs is to vary the gain, K, with time. The way it does this depends on its tuning adjustments, which are a function of the process dynamic and static terms. The latter normally do not change. The object is to hold K as high as possible at all times and when the process has lined out and G has approached unity, the controllers contribution to K is very large. The reason K must be lowered over certain periods is to keep the control loop stable preventing undesirable oscillations.

EVALUATING CONTROL LOOP EFFECTIVENESS

Assuming I is fixed or changed infrequently, the disturbances are the major problem that keep the measured value from the desired level. The effectiveness of the control loop can be analyzed using dynamic, static and statistical methods. Fig. 1b which is a deviation ratio curve[1], shows the dynamics for a single loop controller, and the spectrum of the disturbance. It works like this: at point "A" the disturbance would cause a 7°F variation in the measured value. This is without control. If the control system is in service and tuned as shown by the colored line, the disturbance at "A" would cause a zero°F change in the measured value. At point "B" the disturbance with no control would be 4.2°F, and with control, 0.5°F. In this frequency range disturbances would be attenuated by 0.12. Using the formula in Fig. 1a, the dynamic term, G, is about one and the gain term, K, is high, therefore, the error signal, E, is quite small. As will be shown later, keeping the error signal low is costly. The cost is the energy required to correct for the disturbance. It is very simple to run a plant test to evaluate the deviation ratio: run the loop on manual for eight hours and then on automatic for eight hours. Keep the records of the measured values and of the energy flow (i.e., steam, hot oil control valve position, pressure drop, etc.). With these records the effectiveness of the control, or its ability to attenuate actual disturbances, and the cost in energy to do this, can be evaluated.

A good way to present the statistical data is in the form of a histogram, as shown in Figs. 1c and 2. Curve "A" in Fig. 1c shows control via an operator who is holding the mean value safely below the optimum. It also shows that a second operator decided the desired value at point "1" was too low, so he moved the loop set point to point "2." This curve illustrates the danger in calculating the mean and standard deviations from data without looking at the histogram. Curve "B" shows the predicted histogram from a control study of an im-

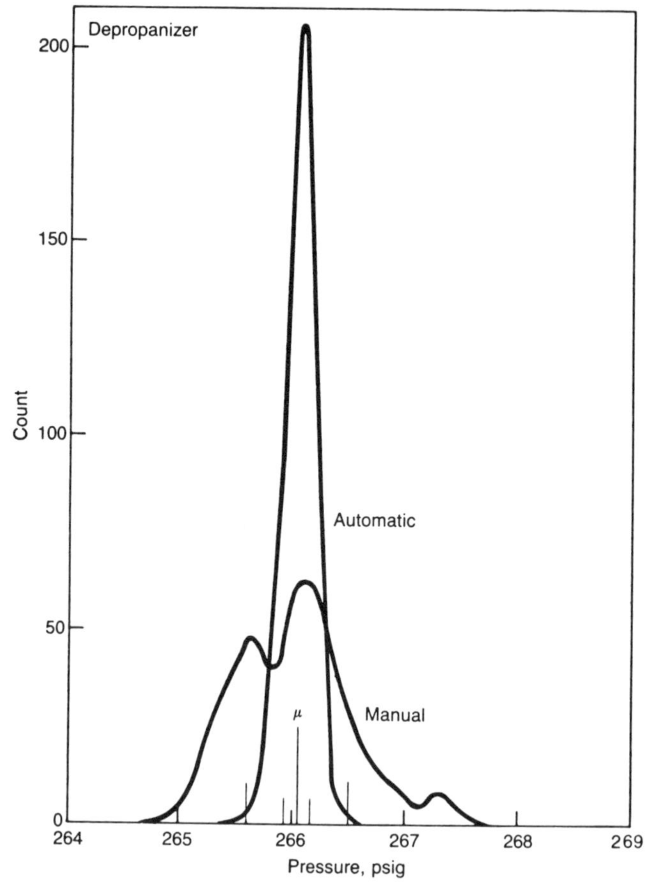

Figure 2. Amplitude probability distribution of top column pressure.

proved control loop. If the mean value was set at the peak of point "2" on curve "A," the delta improvement in operating near optimum could be evaluated and the cost of improved control could be analyzed. These data are from a reactor temperature control. They were taken over a seven-day period and represent a total of 7,000 data points. This is about a typical time span and number of data points necessary to evaluate a reactor temperature control scheme. The reactor temperature vs conversion curve is shown in Fig. 1d. Here, operating points "A" and "B" are shown on a static characteristic curve.

Another example of the statistical procedure is shown in Fig. 3.[2] This case illustrates another point. The product flow was from the bottom of one column to a fired heater and into another column. The controlled temperature was from a tray in the second column. In this case 3,000 data points were taken over a 50-hour period (a sample rate of one per minute). The objective here was tight control of the column temperature: never to exceed 730°F but to run as close as possible to this value. It can be seen from Fig. 3c that the controller did an excellent job. The operational chart record was a straight line. To accomplish this when faced with the disturbance of variable feed flow (shown in Fig. 3b), the heater temperature controller had to swing the fuel gas flow over the range shown in Fig. 3a. These data do not show the speed with which these gas flow changes occurred. This was analyzed using auto- and cross-correlation calculations, which can be found in reference 2. This heater was located at the end of a gas header which supplied three other gas fired heaters. At times the surges in gas flow reduced the header pressure affecting all the other heaters. This is a disturbance from the energy source, as shown in Fig. 1a. To reduce the product flow variations the bottom level control of the first column was improved by better measurement. The heaters' energy sources were decoupled by placing pressure controllers at each branch of the supply header as shown in Fig. 4. These pressure controllers were valve mounted and replaced the valve positioners. They held the fuel gas pressure to the burners constant when faced with fast (10- to 120-second period) disturbances and thereby prevented high speed disturbances from passing into the slowly responding process heater via the energy source. This would be attenuating the disturbances shown at point "C" of the deviation ratio curve, Fig. 1b. At this point the temperature controller could actually amplify the disturbances. The most significant improvement in this example was the attenuation of the process disturbance by better use of the first column's bottom surge volume. Decoupling via attenuating the effect of energy disturbances also was beneficial.

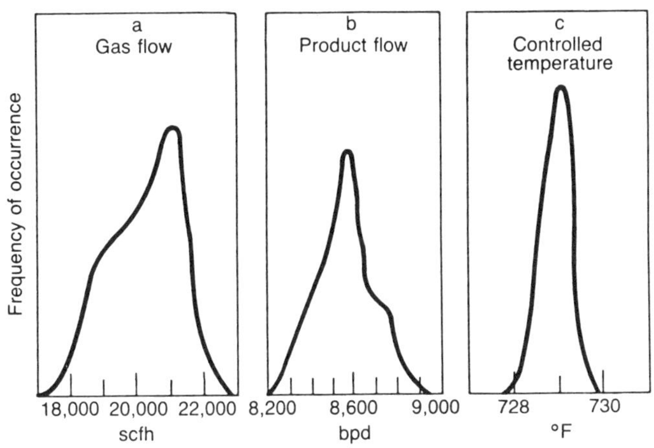

Figure 3. Histograms of measured variables.

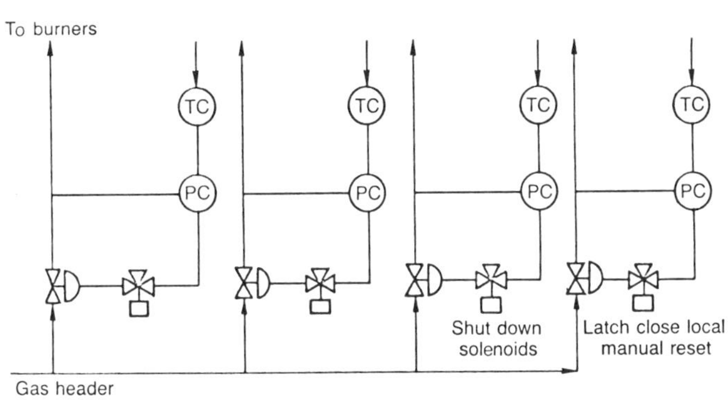

Figure 4. Decoupling heater energy sources.

The previous discussion can be summarized in some rules:

1. Prevent disturbance from entering critical systems by process filters or improved upstream control.
2. Study the static and dynamic characteristics of disturbance.
3. Be careful when using statistical results calculated by an online computer.
4. Evaluate improved control by considering the cost of the energy as well as the hardware.
5. Solve the problems in the basic control loops before attempting advanced control.

CHARACTERISTICS OF PROCESS DISTURBANCES

The objectives of the operations department are to run safely, on specification and at optimum. Normally, if the process is well-controlled, running at design is near optimum. There are two types of optima: (1) a technical optimum, which is to get the most through the plant without consideration of cost, and (2) an economic optimum, which makes the product just at specification at minimum cost. Sometimes these two optima are the same. The things that keep the operations department from achieving optimum results are disturbances. These vary from catastrophic (equipment failure) to daily weather changes. The remainder of this chapter discusses the characteristics of these disturbances.

The Fig. 5 chart shows a large upset entering a debutanizer column at point "A." The bottom temperature goes badly off control for four hours. There is no butane in the bottom of the column. It is assumed that bottom product is being blown over the top of the column, causing the top product to go off specification. At point "B" the temperature is back on control and at approximately the correct level. About 3 hours later at point "C," the bottom product is back on specification and the operation is normal. Even with normal operation the steam flow is moving with a frequency span of 0.15 to 0.02 cpm. At point "D" there is a large swing in the steam flow and not too large a change in temperature; while at point "E," the steam flow variations are as large as the temperature swings. Starting at point "C" there is a low frequency swing in the steam flow (over 3 hours long) with little swing in the average temperature. Superimposed on this low frequency swing are some high fre-

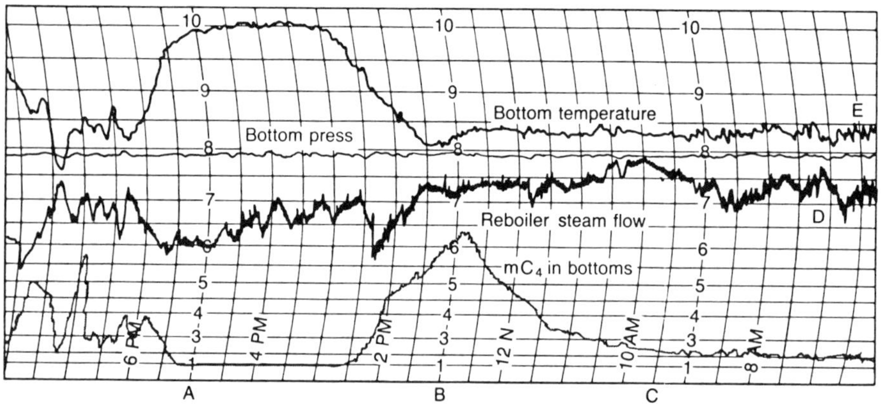

Figure 5. Effects of an upset entering a debutanizer.

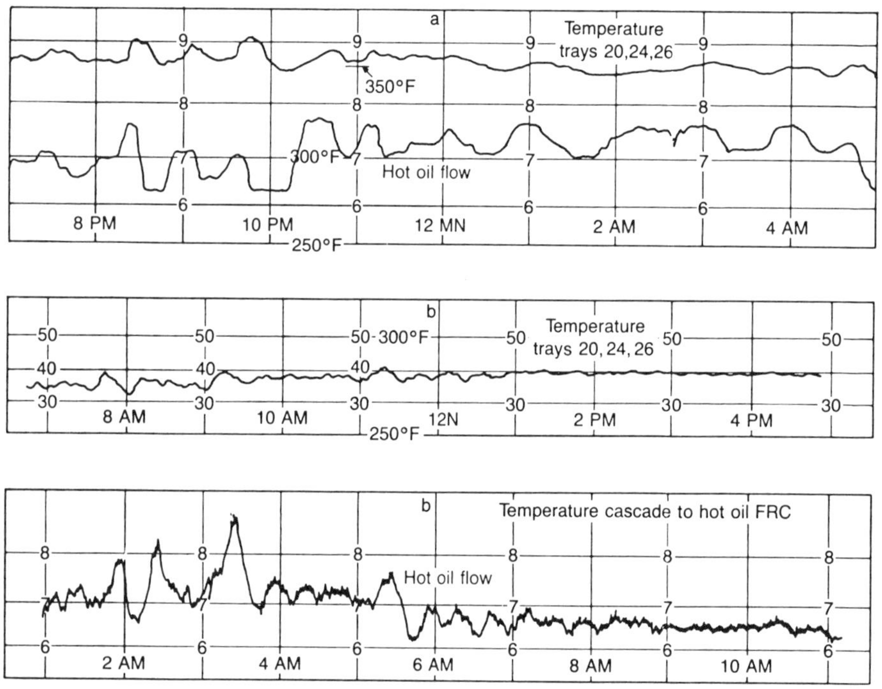

Figure 6. Reactions of column bottom variables.

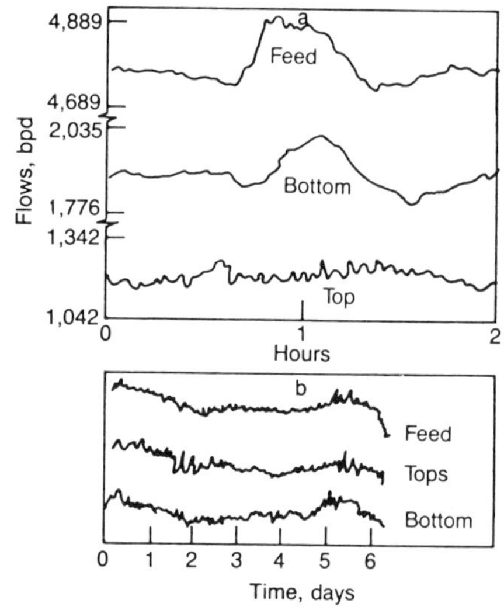

Figure 7. Material balance examples show how statistical data can be misleading.

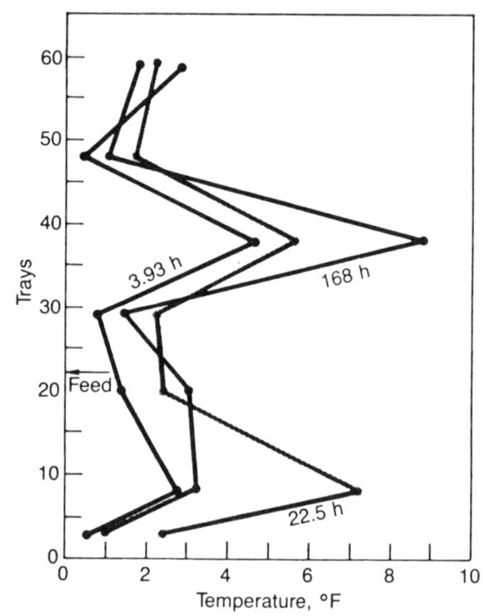

Figure 8. Standard deviation in temperature vs. trays.

quencies. From these data a deviation ratio curve could be developed and the pros and cons of controller tuning evaluated. Notice the thermal swings from points "C" to "E" show very little effect on the composition. This indicates temperature response is much faster than composition response, and that disturbances in the frequency range of 0.1 to 0.02 cpm can upset a temperature controller but not a composition controller.

Figures 6a and b show another case of a column bottom temperature controller and the hot oil flow. In Fig. 6a the temperature controller is directly adjusting the hot oil valve which is sticky and slow to respond. In Fig. 6b the temperature controller is cascaded to a hot oil flow controller. The flow controller speeds up the flow response, the deviation reduction curve is shifted to the higher frequencies and temperature error is reduced. Note in Fig. 6b there must have been a large thermal disturbance around 8 to 9 a.m. with a period of about 1 hour. The temperature controller almost totally corrected for this disturbance. Notice also the increase in frequency characteristics of the hot oil flow in the cascade case. It is a much faster responding control; therefore, deviations from the desired values are much smaller.

Figures 7a and b show two interesting cases of material balance. In Fig. 7a the feed changes of a 1 to 2-hour cycle reflect almost totally in the bottom product and not in the top. This means that if feed composition is fairly constant (we have seen that it does not vary this fast), there is valuable top product being lost in the bottom. Yet look at the same picture over a 7-day period and it is apparent that the feed changes are being reflected correctly in the top and bottom flow rates.[3]

Figure 8 shows another interesting point regarding false statistical indicators. If the 22.5-hour test was the only one run, it appears that tray 8 or tray 38 could be used as the control tray. Looking at short time span data and long time span data for high and low frequency components, tray 38 becomes the obvious tray to use in control of this tower.

Figure 9 is a good example of the effects of ambient temperature on equipment. There are two large 20,000 hp gas turbines A and B, operating compressors in parallel. It is important to keep the exhaust temperatures as high as possible because of downstream waste heat boilers and improved efficiency of the machines. There is a high limit where the turbine wheel will be damaged. It is apparent from these charts that the exhaust temperatures follow the ambient temperatures; therefore, it was necessary to temperature compensate this system. Note also that the two compressors need load balancing controls. (Outlet temperature correlates to load.)

Figure 10 shows records of the stack gases from a 250 million Btu/h process heater. Here it is apparent there are large amplitude, high-frequency variations in CO. The O_2 measurement was much smoother at about 1.2% O_2, and O_2 control by a damper was possible from this signal, whereas control of CO by a damper was impossible. On this large heater, O_2 control at 1.2% was possible because of the CO record. As long as there was no CO at 1.2% O_2, the operator knew the burners, louvers, etc. were all in correct adjustment. The savings were large.

Figures 11a, b, and c are records from a large boiler. The boiler is the largest, most expensive utility system. It is essential to a large number of plants and is, therefore, normally over-sized in capacity, or there are extra boilers. The boiler used in this example was one of three. The plant could run with two boilers at full load. Good boiler operation is essential to the safety and production of the processes. It is also a pure cost function and this cost is mostly in fuel. Therefore, efficient combustion control is desired. Figures 11a and b are two related figures of a load change to this boiler. Unfortunately, the two charts run opposite in time. Figure 11a shows the three element boiler feedwater control. Note the level controller has adjusted the water flow-steam flow ratio control so precisely that there is almost no bump in the drum level. Note also, the oxygen controller has adjusted the fuel air ratio so precisely that there is very little upset in the oxygen (approx. 6:30 to 7:30). Also, this control works so well that the oxygen is being held at a 1% level. To accomplish the fast basic control, the high-frequency noise is not greatly filtered. This control has been highly developed and works quite well. The deviation ratio for the steam header pressure is zero for all practical purposes. The disturbances were intro-

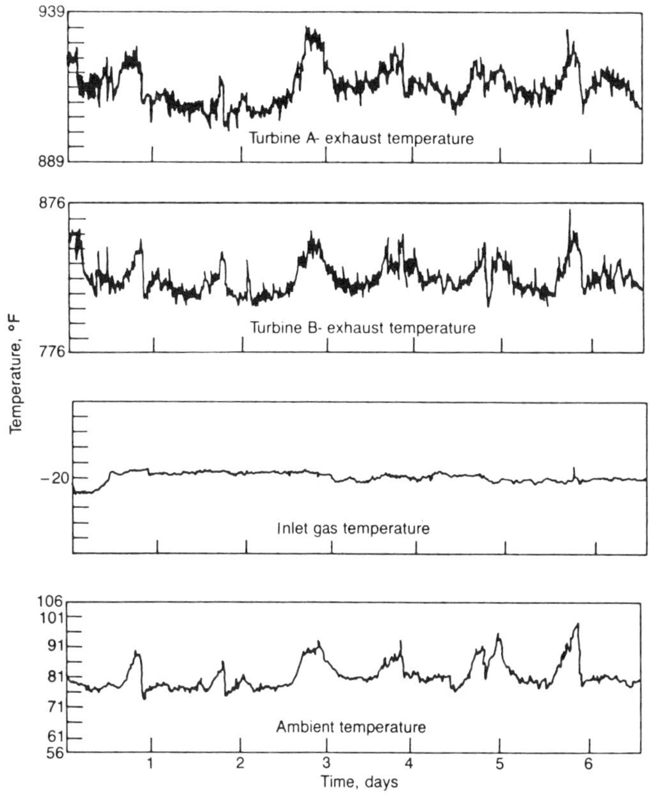

Figure 9. Effects of ambient temperature on two gas turbines.

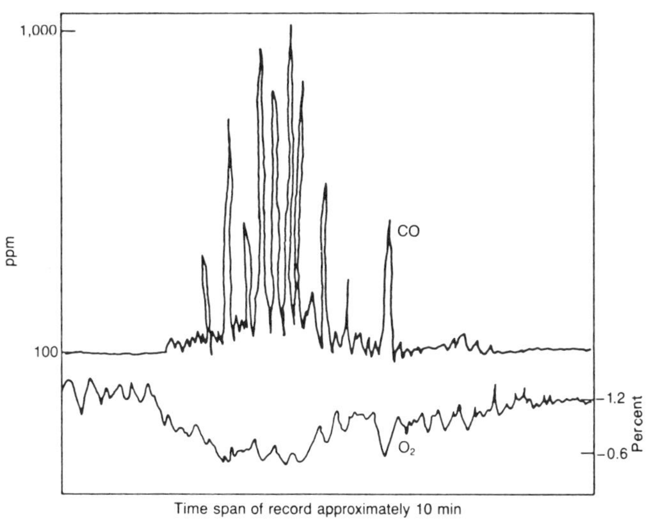

Figure 10. Stack gas records from a process heater.

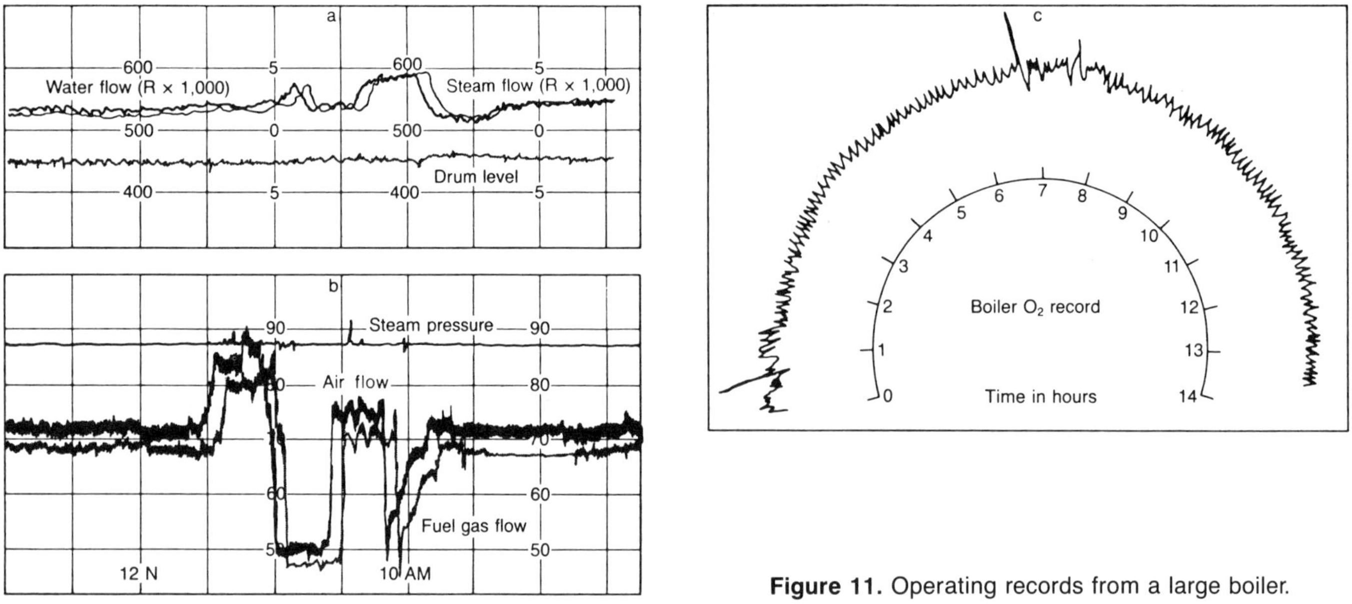

Figure 11. Operating records from a large boiler.

duced by changing the loading of one of the boilers running in parallel with this boiler.

Reactor operating data are normally held as confidential; therefore none are available. It is normal to minimize disturbances to a reaction system by control system design. An example of this is a direct chlorination reactor to make ethylene dichloride. The feeds were pure ethylene and chlorine on flow control. A slight excess of chlorine assures complete reaction of all the ethylene. This is the optimum. This is detected by a chlorine analyzer in the E.D.C. from the reactor. A target of 300 ppm chlorine in the E.D.C. is desired. Until the feeds were mass flow-ratio controlled with the chlorine analyzer trimming the ratio, the chlorine loss could not be reduced below 1,000 ppm, and the chlorine analyzer output was noisy. After the mass flow-ratio controls were installed to eliminate feed mass flow variations, the chlorine loss could be controlled at 150 ppm.

This chapter has attempted to bring attention to the types and characteristics of disturbances in petrochemical plants, and to the attenuation of these disturbances by automatic control system design, and/or process filtering. There is profit to be found in pursuing this subject in any plant. A basic rule to be followed is: "Maximize disturbance attenuation by process design rather than control system design." Also, it is better to find (and eliminate) the source of a disturbance.

References

1. Janssen, J. M. L., "Control system behavior expressed as a deviation ratio," ASME Transactions, Nov. 1954.
2. Stanton, B. D., "Application of system engineering by troubleshooting existing processes," 19th Annual ISA Conference, Oct. 1969.
3. Stanton, B. D. and Barkelew, C. H., "A normal operating distillation column," 1969 Joint Automatic Control Conference.

53

Energy and Product Savings Through Advanced Control

G. F. Hall

Generally, most of the economic incentives for a control system upgrade are derived from the use of advanced control techniques.[1] This is because advanced controls frequently improve product yield by 2 to 5% and often provide energy savings at least as large. Such improvements satisfy refinery managers who typically require a payback of one year or less in these situations. Following are examples where improved control provided energy and product savings, and how to estimate economic benefits.

Control system upgrade may involve moving to a microprocessor-based distributed control system; replacing panelboard instruments with those of new-generation, multifunction capabilities; or supplementing primary control with the power of process computers.

Prime examples for use in advanced control system justification are refinery units that are energy-intensive and/or involve distillation. Table 1 lists several such units, along with their calculated annual savings using advanced controls. (These figures were developed from feasibility survey data.)

Table 1
Case Studies of Calculated Annual Savings

Process unit	Charge rate (bpd)	Annual savings ($)
FCCU main fractionator	60,000	1,600,000
FCCU gas concentration	690,000 (lb/h)	3,000,000
Atm./vac. distillation	120,000/84,000	2,900,000
Vapor recovery	410,000 (lb/h)	1,900,000
Alkylation	29,000	3,200,000
Delayed coking	16,000	2,500,000
Hydrocracking	16,000	612,000

ECONOMIC BENEFIT ANALYSES

The first step in upgrading a system with advanced controls involves developing an economic benefit analysis. Consultants are available from private firms, software houses, engineering contractors and instrument manufacturers to assist in the performance of these tasks. Obviously, selection will depend on the application, the consultants' past performance in similar situations and their reputation in the industry.

With direction from the refinery engineering staff to ensure a mutual understanding of the scope of work to be performed, they offer various levels of services as follows:

Field Hardware Survey—a brief plant-site visit to determine hardware replacement, expansion and installation requirements. A hardware (and perhaps system software) quote is usually the resulting service.

Feasibility Survey—a short-term plant-site service to identify areas for improvement in process measurement and control. Recommendations, with associated economic incentives, are documented in a short report to establish a preliminary project scope. A budgetary quote for hardware, software and application engineering may accompany the report.

Engineering Study—generally associated with a control system project to finalize or further define control strategies, system data bases, benefits, equipment needs and application services. The results are detailed in a formal report, which is frequently included as a portion of the system documentation.

DETERMINING ECONOMIC BENEFITS

A fixed format for preparing economic analyses is precluded by the variety of process operations. Instead, the techniques are diversified, depending on both the data available and the consultant. They range from statistical methods using polynomial functions[2] to "gut feel." Several conventional techniques are given below.

Simplified Statistical Approach. Heater pass balancing[3] is being considered for a vacuum unit charge heater. Basis data for the dual-fuel, eight-pass heater are as follows:

Charge rate = 83,000 bpd
Average outlet temperature = 726°F
Average pass temperature deviation = −4°F, +5°F
Fuel oil consumption = 175.8 bpd
Fuel gas consumption = 3,980 Mcf/d, 1,123 Btu/cf
Fuel oil value = $34.50/bbl
Average % net deviation of fuel oil to charge = +0.73
Average % net deviation of fuel gas to charge = +3.50

Assumed net deviation in positive direction is an indicator of fuel loss.

Fuel oil loss:

$$\frac{175.8 \text{ bbl}}{\text{day}} \; \frac{0.73 \text{ bbl loss}}{100 \text{ bbl}} \; \frac{\$34.50}{\text{bbl}} \; \frac{350 \text{ days}}{\text{year}} \approx \$15{,}500/\text{y}$$

Fuel gas loss:

$$\frac{3{,}980 \text{ Mcf}}{\text{day}} \; \frac{1{,}123 \text{ Btu}}{\text{cf}} \; \frac{3.50 \text{ cf loss}}{100 \text{ cf}} \; \frac{1 \text{ bfoe}}{6.3 \times 10^6 \text{ Btu}} \; \frac{\$34.50}{\text{bfoe}} \; \frac{350 \text{ days}}{\text{year}} \approx \$299{,}800/\text{y}$$

Projected annual savings with this basis is: $15,500 + $299,800 = $315,300

This contribution is important, but a more significant benefit may be provided by tighter pass temperature control. Assume that the vacuum tower provides feed to a fluid catalytic cracking unit and that the bottoms are fed to a delayed coking unit. For this example, if approximately 30% of the crude charge is gas oil, then a relationship of 20°F/% gas oil charge (slope of TBP curve) is established for a boiling range of 650 to 1,250°F.

It is evident that the outlet temperature should be set as high as possible, i.e., to a high constraint, which is usually a temperature somewhat lower than where tube coking becomes significant. Unfortunately, though, operators tend to build in two or three times the necessary safety margin, and conversion to the more valuable cat cracker feed suffers unnecessarily.

Automatic heater pass outlet temperature control will take the swings out of pass temperatures and automatically compensate pass flows for charge-rate fluctuations. These features should be enough to gain the confidence of the operators, thus allowing a decrease in the safety margin by at least 10°F.

Assuming that the difference in value between cat cracker and coker feeds is $4.50 per barrel, the annual savings generated by increasing the heater outlet temperature by 10°F would be:

$$\frac{83{,}000 \text{ bbl}}{\text{day}} \; \frac{1\% \text{ gas oil}}{20°\text{F}} \; \frac{10°\text{F}}{1} \; \frac{\$4.50}{\text{bbl}} \; \frac{350 \text{ days}}{\text{year}} \approx \$653{,}600/\text{y}$$

This calculation does not include the cost of extra fuel; however, most of the cost is cancelled by the gain of heat transferred to crude prior to its being charged to the atmospheric distillation unit heater.

Historical Operations Approach. This example involves a graphical analysis used to define improved control benefits. Automatic crude tower side-cut end-point control using a decoupling technique[4] is to be studied.

Tower cuts are identified in Fig. 1. Typical daily operation of the crude tower reveals the average kerosine/light gas oil TBP overlap designated as O_1 in Fig. 2. The average overlap of the same products at similar operating conditions during test runs shows that more kerosine is slipping out the light gas oil stream than should be permitted (O_2). Endpoint control will, at minimum, maintain test run operation characteristics, while draw

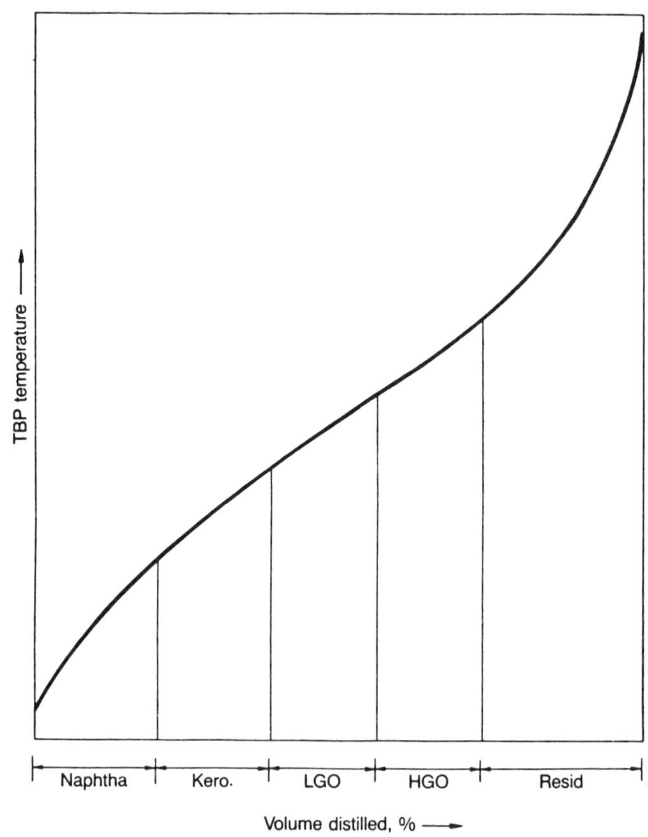

Figure 1. Typical distillation curve for atmospheric tower feed.

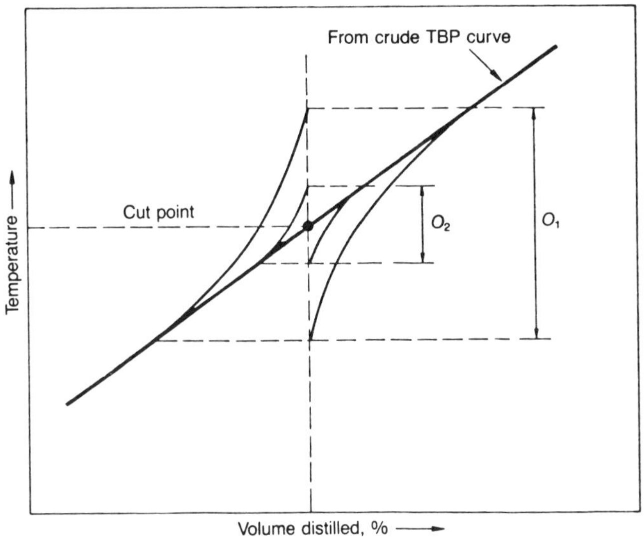

Figure 2. Comparison of atmospheric tower fraction TBP overlaps.

decoupling will minimize losses caused by tower disturbances.

Straight-line approximations of the overlap deflections will make calculation of lighter product loss easier. Once the loss has been established, simply multiply it by the difference in stream values and convert the product to annual loss in dollars.

If crude tower upsets are common, then losses because of product slippages out of lower tower draws should be added. Historical data may disclose "typical" losses to lower draws and overflash as a result of upsets caused by crude switches, flowrate changes or equipment failures. Perhaps hypothetical data could be used to present a case when disturbances are common.

Computer Program. Where applications are commonplace and independent, such as those directly involving process equipment, operating performance may be evaluated via a computer program. Table 2 shows the output of a program that can be used to readily calculate fired-heater or boiler combustion efficiency, along with estimated yearly savings with improved combustion controls.

HARDWARE CONSIDERATIONS

Since the rate of return on a control system upgrade depends heavily on the instrumentation selected, a careful evaluation of both current and future needs is necessary. Other important considerations are operator and maintenance personnel training, system maintenance and installation requirements, spare parts inventory and system data base management.

System selection prerequisites should include proof of the vendor's ability to provide training, technical support, continued system maintenance of both hardware and software, and complete system and application documentation. Field commissioning and startup services are also important.

It is necessary to put much thought into the selection of the type of system hardware, e.g., whether to perform advanced control with a process computer or with a microprocessor-based shared controller. Such a decision should result from inputs that include manpower availability, control security, operator acceptance, total cost and extent of upgrade.

Considering this particular question, very powerful microprocessor-based controllers are available today that can execute most advanced control schemes.[3] Several offer full redundancy for added control security, something rarely available in process computers. Savings are often made by avoiding a more expensive computer system.

On the other hand, a process computer is relatively easy to justify when its power is used for implementing advanced controls via conventional control panel instruments. Conversion of panel instruments to either computerset (supervisory) or DDC stations is usually accomplished by instrument retrofit. Pneumatic signals for recording or indicating process variables may be inexpensively converted to electric signals with the use of pneumatic multiplexers or high-density transmitters.

If data logging, active graphic displays and sophisticated calculations (like tray loading calculations) are not vital to system justification, then the most cost-effective implementation of advanced controls is by direct panelboard instrument replacement with new-generation, microprocessor-based panelboard controllers. Some offer more than 20 function block types in addition to PID; several also offer a self-tuning capability.

Another option is to use a self-contained unit controller, which is a shared controller offering multifunction capabilities. Several unit controllers, perhaps in conjunction with a personal computer, yield enough control capacity to manage a small process unit. Supplementing panelboard instruments with unit controller strength is a feasible approach in providing advanced control in selected unit area, e.g., fired heaters and fractionation towers.

SUMMARY

Economic incentives based on improved process control using advanced techniques are generally sufficient to warrant control system upgrade projects. Knowledgeable consultants are accessible for complementing a refinery's engineering staff in developing these incentives.

Table 2
Output of Computer Program Used to Determine Fired-heater Performance*

FIRED HEATER OR BOILER COMBUSTION EFFICIENCY PROGRAM

THIS PROGRAM USES 80 DEG. F AND 60 PERCENT RELATIVE HUMIDITY AIR (WHEN THE ACTUAL TEMPERATURE FOR THE COMBUSTION AIR IS NOT ENTERED) FOR COMBUSTION CALCULATIONS. CASE I REFERS TO EXISTING CONDITIONS. CASE II REFERS TO EXPECTED CONDITIONS WITH IMPROVED CONTROL EQUIPMENT AND APPLICATION EXPERTISE.

CALCULATIONS USED IN THIS PROGRAM ARE BASED ON THOSE DESCRIBED IN THE ASME POWER TEST CODE 4.1-A(1964) UNDER THE SECTIONS APPLICABLE TO THE ABBREVIATED EFFICIENCY TEST UTILIZING THE HEAT LOSS METHOD.

DO YOU PREFER:

1. TO INSERT VALUES YOURSELF

OR

2. TO HAVE THE PROGRAM AUTOMATICALLY INSERT VALUES FOR FLUE GAS AND FUEL ANALYSIS?

INDICATE PREFERNCE

2

ENTER APPROPRIATE FUEL TYPE 1 THROUGH 6:

1. NATURAL GAS
2. #2 FUEL OIL
3. #6 FUEL OIL
4. COAL
5. BAGASSE
6. WOOD CHIPS
7. MIXED FUELS (YOU WILL ENTER YOUR OWN VALUES)

1

ENTER EXIT STACK TEMPERATURE IN DEG. F FOR CASE I
790

ENTER CASE I % OXYGEN BY VOLUME
9.12

% CARBON DIOXIDE FOR CASE I: 6.700
% OF NITROGEN FOR CASE I IS: 84.170
% OF EXCESS AIR FOR CASE I: 69.549%

ENTER LOSS IN UNBURNED CARBON IN BOTTOM ASH IN % BY WEIGHT

0

ENTER UNIT FIRED DUTY IN MILLION BTU/HR
123

ENTER FUEL COST IN DOLLARS/MILLION BTU
5.45

POUNDS OF DRY AIR PER POUND AS FIRED FUEL = 26.285

POUNDS OF MOISTURE IN FLUE GAS PER POUND OF FIRED FUEL = 2.361

POUNDS OF DRY GAS PER POUND AS FIRED FUEL = 25.329

POUNDS OF WET GAS PER POUND AS FIRED FUEL = 27.690

ENTER ACTUAL COMBUSTION AIR TEMPERATURE
60

HEAT LOSS EFFICIENCY CALCULATIONS ARE AS FOLLOWS:

HEAT LOSS DUE TO DRY GAS = 18.645

ENTHALPY OF VAPOR AT STACK TEMPERATURE = 1425.300

HEAT LOSS DUE TO MOISTURE IN THE FUEL = 0.000

HEAT LOSS DUE TO WATER FROM COMBUSTION OF HYDROGEN = 11.942

HEAT LOSS DUE TO COMBUSTIBLE IN THE BOTTOM ASH = 0.000

HEAT LOSS DUE TO RADIATION AND UNACCOUNTED FOR LOSSES HAS BEEN ASSIGNED A VALUE OF 2%

CASE I EFFICIENCY IS 67.413

WHAT IS THE EXPECTED % OXYGEN AFTER INSTALLING COMBUSTION CONTROLS?
5.0

CASE II CARBON DIOXIDE = 9.000

CASE II NITROGEN = 85.990

PROGRAM ASSUMES CASE II OXYGEN CAN BE ACHIEVED WITHOUT SIGNIFICANT CARBON MONOXIDE FORMATION

CASE II % EXCESS AIR = 28.210

CASE II STACK TEMPERATURE DROP IS: 67.400 DEG. F

PROGRAM RECOGNIZES A REDUCTION IN STACK TEMPERATURE WITH A REDUCTION IN % O2

CASE II CALCULATED STACK TEMPERATURE IS 722.600 DEG. F

POUNDS OF DRY AIR PER POUND AS FIRED FUEL FOR CASE II = 19.973

POUNDS OF MOISTURE IN FLUE GAS PER POUND OF FIRED FUEL FOR CASE II = 2.279

POUNDS OF DRY GAS PER POUND OF FIRED FUEL FOR CASE II = 18.994

POUNDS OF WET GAS PER POUND AS FIRED FUEL FOR CASE II = 21.273

HEAT LOSS EFFICIENCY CALCULATIONS FOR CASE II ARE AS FOLLOWS:

HEAT LOSS DUE TO DRY GAS = 12.691

ENTHALPY OF VAPOR AT STACK TEMPERATURE = 1393.622

HEAT LOSS DUE TO MOISTURE IN THE FUEL = 0.000

HEAT LOSS DUE TO WATER FROM COMBUSTION OF HYDROGEN = 11.671

HEAT LOSS DUE TO COMBUSTIBLE IN THE BOTTOM ASH = 0.000

HEAT LOSS DUE TO RADIATION HAS BEEN ASSIGNED A VALUE OF 2%

CASE II EFFICIENCY IS 73.638

YEARLY COST FOR CASE I = $8.452.352.00

YEARLY COST FOR CASE II = $7.737.804.00

TOTAL DOLLAR SAVINGS PER YEAR BY IMPROVING COMBUSTION EFFICIENCY WITH COMBUSTION CONTROLS IS:

$714,548.00 PER YEAR

IF FORCED-DRAFT AND/OR INDUCED-DRAFT FANS ARE ASSOCIATED WITH THIS UNIT, ADDITIONAL SAVINGS ARE REALIZED BY THE REDUCTION IN HORSEPOWER REQUIREMENTS TO DELIVER THE LOWER PERCENTAGE OF EXCESS AIR. THESE SAVINGS CAN BE DOLLARIZED BY REVIEWING THE FAN CURVES AND ASSIGNING A DOLLAR FIGURE FOR EACH SAVED HORSEPOWER HOUR.

VARIOUS METHODS ARE AVAILABLE FOR CALCULATING THE WORK, AIR HORSEPOWER AND POWER OF FANS.

VARIABLE SPEED FANS ARE GOVERNED BY THE FOLLOWING STATEMENTS:

(1) CAPACITY IS DIRECTLY PROPORTIONAL TO SPEED.
(2) HEAD IS DIRECTLY PROPORTIONAL TO SPEED CUBED.
(3) POWER INPUT IS DIRECTLY PROPORTIONAL TO SPEED CUBED

THESE STATEMENTS RELATE A REDUCTION IN EXCESS AIR TO A DIRECT REDUCTION IN HORSEPOWER. FOR A CONSTANT SPEED MOTOR A RELATIONSHIP OF REDUCED HORSEPOWER TO REDUCED EXCESS AIR CAN BE DEVELOPED WHERE THE NO-LOAD POWER REQUIREMENTS ARE ASSUMED TO BE 0.25 OF FULL-LOAD POWER, THUS LEAVING 0.75 OF FULL-LOAD POWER TO ACTUALLY MOVE AIR AT FULL LOAD. FOR THIS CASE ASSUMING A CONSTANT SPEED MOTOR THE SAVINGS IN HORSEPOWER GOING FROM CASE I TO CASE II EXCESS AIR IS:

18.286%

ACTUAL HORSEPOWER SAVED DEPENDS ON THE SPECIFIC MEANS EMPLOYED AT THIS SITE TO VARY THE FAN OUTPUT AND THE NUMBER OF FANS INVOLVED. IN MOST CASES THE REDUCTION IN ENERGY CONSUMED BY BOILER OR HEATER FANS IS SIGNIFICANT AND SHOULD BE CONSIDERED IN OVERALL PROJECT SAVINGS.

* The author wishes to thank J. Shriver of The Foxboro Company for his work on Table 2.

Advanced control benefit analyses are most meaningful when calculations use recorded data. Test run data can be used to compare "base" operation with average daily operation. This establishes the minimum performance enhancement attainable. Computer programs for selected applications are available for operation performance and projected improved control benefit calculations.

Configurations of control systems are numerous. Upgrading a control system to perform advanced control can be as simple as panelboard controller replacement. Vendor selection qualifications should include reputation in the industry and ability to provide turnkey services.

References

1. Badavas, P. C., and Rys, R. A., "Advanced control techniques," *Chemical Engineering*, Nos. 8, 9, and 10 in series on process automation.
2. Stout, T. M., and Cline, R. P., "Control system justification," *Intech*, September 1976.
3. Hall, G. F., "Enhance refinery control: use microprocessor-based controllers," *Hydrocarbon Processing*, September 1983.
4. Ryskamp, C. J., "Seven ways to up crude unit productivity," *Oil & Gas Journal*, Feb. 21, 1977.
5. ASME Power Test Codes, "Steam generating units."

54

How Spreadsheets Aid Process and Control Engineering

T. S. Johnson

This chapter describes how a company that specializes in process control and instrumentation engineering has successfully incorporated the personal business computer into its organization. Emphasis will be placed on describing several useful engineering functions developed by using an electronic spreadsheet package.

The personal computer has proven to be an excellent tool in allowing companies to furnish clients with cost-effective engineering services. After a brief review of some engineering applications, details will be given on using electronic spreadsheets for process and control engineering applications.

A statistical analysis program has been used to generate curve-fits of engineering data, including analyzer calibration and certain simple process control models. A graphics package with limited CAD-type functions supplements a larger CAD system and is very good for generating simple loop diagrams, control room layouts and process flow diagrams. The large CAD system is also run on the P.C.

Applications specific to instrumentation engineering include programs that allow programming and configuration of distributed control and programmable logic controller equipment via the personal computer. These programs are generally faster and easier to use than previous configuration methods and allow offline generation and debugging of the software before hardware equipment is available.

SPECIFIC SPREADSHEET ENGINEERING APPLICATIONS

The electronic spreadsheet is the most frequently used program in engineering analyses. Although originally developed for business functions, the latest generation of electronic spreadsheets has enough calculation power to handle many sophisticated engineering problems.

The ease of programming and formating of data within the spreadsheet is another positive factor I believe contributes to its use as a engineering tool. For many engineering applications, programming expertise is not required. An engineer can program his requirements, enter data, and create documents formated for publication after only a few days of on-the-job training. This fact is important for the "working" manager who does not have the time nor desire to develop the conventional computer programs he may require.

At the same time, the spreadsheet has sophisticated programming capabilities approaching that of BASIC or FORTRAN, if the user desires to use them. Even when this sophistication is required, the spreadsheet has the advantages of "live data entry," ease of report formating, and very quick copying of code from one area of the spreadsheet to another.

I have found the "what-if" capability of the spreadsheet, that is, the ability to change certain variables to see their effect on the calculation results, to be another advantage in engineering applications.

Several examples of the electronic spreadsheet in engineering analysis are presented. I believe these examples demonstrate the flexibility and power of this approach.

Engineering Calculations

A spreadsheet can be programmed quickly to calculate many straightforward engineering calculations. The originally programmed formulas can be easily copied to adjacent columns of the spreadsheet with only a few keystrokes. The data then can be modified in each column to represent various conditions to be examined. Once entered, the calculated values are instantly available for comparison and evaluation.

For example, pressure drops of fluids through various pipe diameters can be presented on one spreadsheet for easy comparison and selection. Only the fluid physical properties are required as input data. Other applications such as orifice or control valve sizing, exchanger rating, etc., can be programmed in a similar nature.

Plant Data Analysis

When studying historical plant laboratory or operating data, I have found that the spreadsheet's built-in functions such as data averaging, determination of minimum, maximum, standard deviation, etc., and its ability to present data in tabular and graphic formats allow quick evaluation of trends or correlations. Also, these data can be directly transferred from the spreadsheet into finished reports (Fig. 1).

A useful feature of the spreadsheet is its ability to move data easily from one location to another within the same spreadsheet or to other files stored on disk. A process engineer responsible for tracking daily unit results could enter the data into a spreadsheet to perform calculations such as exchanger duties, furnace efficiencies, plant yields, etc., on a daily basis. He could then transfer the results for storage with previously entered results. From these data, graphs or tabulation of weekly, monthly, or even longer trends could be generated easily.

Process Sensitivity Study

A unique engineering application of the spreadsheet involves data analysis. In one recent study, a process model developed to provide an on-line feedforward correction to a unit's feed rate was used to analyze the current unit variability. This approach provided an insight into which variables were responsible for the variability and helped in efforts to improve unit control.

First, the model was created and a base set of conditions were entered into the spreadsheet. The model was copied to adjacent columns, and each variable was individually reduced to change the predicted productivity by a constant amount. The percent change in each variable was calculated and displayed for rapid comparison (Fig. 2). The sensitivity of production to each process variable was easily pointed out for discussion and further evaluation.

Plant operating data were entered into another spreadsheet, and simple statistics were calculated as described earlier. From these data, actual variability in each of the model parameters was determined. The spreadsheet model was then set up with each variable reduced by an amount shown to be typical (Fig. 3). In this way, the effect of each parameter's actual variability on production was documented. This information was used to prioritize control improvement recommendations.

Relative Gain Analysis

I have often used relative gain analysis to perform evaluations of distillation tower control stability. This technique has been developed by Bristol, Shinskey, and others and has been described in the literature.[1,2,3] Gordon's application of Shinskey's approach to tower composition control is of special interest.[3] An extension of the spreadsheet programming approach described previously has been implemented to provide fast, easy evaluation of the relative gain factors at various conditions that could affect the tower.

The relative gain technique is not discussed in detail. However, an overview is presented to allow the reader to understand its purpose and the advantages of placing the calculation on the electronic spreadsheet.

Relative gain analysis compares the ratio of a proposed control loop gain when other loops in the system are in manual (open-loop) to its gain when the others are in automatic (closed-loop). If there is no interaction from the other loops in the system, a relative gain value of 1.0 will result. As values digress from 1.0, increasing

GAS CONCENTRATION UNIT DEBUTANIZER

TRAY 4 TEMPERATURE VS OH COMPOSITION

TIME 20 MIN CYCLES	DEBUTZR FEED MBPD	DEBUTZR OH MBPD	FEED TEMP DEG F	BTM TEMP DEG F	OH TEMP DEG F	TRAY 4 TEMP DEG F	OH IC5 %	
1	29.15	16.80	230	342	142	172	0.60	MIN OVHD % IC5
2	25.35	14.66	231	348	143	171	0.65	TRAY 4 MINIMUM TEMP
3	28.73	14.88	230	349	143	173	0.68	
4	27.88	13.69	233	349	145	177	0.75	
5	27.50	14.21	233	351	147	178	0.72	
6	26.82	14.17	235	352	148	180	0.80	
7	26.65	14.25	236	355	149	182	1.01	TRAY 4 MAXIMUM TEMP
8	27.22	14.51	236	355	149	181	0.95	
9	28.01	14.31	235	354	148	180	1.10	MAX OVHD % IC5
10	28.42	14.65	232	352	147	178	0.90	
11	27.95	14.37	234	353	147	179	0.95	
12	28.86	14.44	233	352	147	179	0.88	
NO OF POINTS	12.00	12.00	12.00	12.00	12.00	12.00	6.00	
MINIMUM	25.35	13.69	230.00	342.00	142.00	170.00	0.16	
MAXIMUM	29.15	16.80	236.00	355.00	149.00	182.00	1.29	
RANGE	3.80	3.11	6.00	13.00	7.00	12.00	1.13	
AVERAGE	27.71	14.58	233.17	351.00	146.25	177.50	0.67	
STANDARD DEVIATION	1.04	0.73	2.03	3.49	2.31	3.50	0.41	

Figure 1. Plant data analysis.

SENSITIVITY STUDY
EFFECT ON PREDICTED PRODUCT DUE TO 10% REDUCTION IN EACH VARIABLE

	BASE	< SG A	< SP B	< A PUR	< TEMP	< GPM A	< GPM B	< B PUR
ONE TIME INPUTS								
SP GRAVITY COMPONENT A	1.4	1.26	1.4	1.4	1.4	1.4	1.4	1.4
SP GRAVITY COMPONENT B	1.17	1.17	1.053	1.17	1.17	1.17	1.17	1.17
COMPONENT A % PURITY	50	50	50	45	50	50	50	50
ONLINE INPUTS								
TEMPERATURE RISE	30	30	30	30	27	30	30	30
GPM COMPONENT A	25	25	25	25	25	22.5	25	25
GPM COMPONENT B	350	350	350	350	350	350	315	350
LABORATORY INPUTS								
COMPONENT B PURITY	190	190	190	190	190	190	190	171
PREDICTED PRODUCT	3315.3	3178.1	3155.9	2983.7	3277.0	3278.7	3348.9	3683.6
DELTA FROM BASE	0.0	137.2	159.4	331.5	38.2	36.6	-33.6	-368.4
% DEVIATION	0.0%	-4.1%	-4.8%	-10.0%	-1.2%	-1.1%	1.0%	11.1%

Figure 2. Production sensitivity to process variables.

SENSITIVITY STUDY
EFFECT ON PREDICTED PRODUCT DUE TO TYPICAL VARIABILITY

	BASE	< SG A	< SP B	< A PUR	< TEMP	< GPM A	< GPM B	< B PUR
ONE TIME INPUTS								
SP GRAVITY COMPONENT A	1.4	1.325	1.4	1.4	1.4	1.4	1.4	1.4
SP GRAVITY COMPONENT B	1.17	1.17	1.155	1.17	1.17	1.17	1.17	1.17
COMPONENT A % PURITY	50	50	50	47.5	50	50	50	50
ONLINE INPUTS								
TEMPERATURE RISE	30	30	30	30	24	30	30	30
GPM COMPONENT A	25	25	25	25	25	23	25	25
GPM COMPONENT B	350	350	350	350	350	350	328	350
LABORATORY INPUTS								
COMPONENT B PURITY	190	190	190	190	190	190	190	188
PREDICTED PRODUCT	3315.3	3243.9	3293.9	3149.5	3230.0	3286.5	3336.3	3350.5
DELTA FROM BASE	0.0	71.3	21.4	165.8	85.3	28.7	-21.1	-35.3
'% CHANGE IN VARIABLE	0.0%	5.4%	1.3%	5.0%	20.0%	8.0%	6.3%	1.1%

Figure 3. Study with each variable reduced by a typical amount.

loop interaction is indicated. Values near 0.5 or greater than 5.0 indicate very strong interaction and poor control stability.

The top and bottom composition control loops on a distillation column always have a certain amount of interaction. The purpose of the relative gain analysis is to find the best manipulative variables to control the compositions with the minimum interaction. Gordon[3] states that gains slightly greater than 1.0 are preferred over gains slightly less than 1.0 due to better stability of the bottom composition loop.

The technique uses only five column parameters to evaluate a number of possible control arrangements. The manipulative varibles considered include conventional control of reflux flow and reboiler heat input and certain ratios of column parameters. An example would be manipulating the column reflux ratio (L/D) or boilup to bottoms flow (V/B) to control composition. The input parameters are listed below:

- Feed composition
- Overhead purity specification
- Fraction light key loss in bottoms
- Number of theoretical trays
- Reflux ratio

From these data, correlations are made that indicate the degree of steady-state interaction between the overhead and bottom quality controllers for the many various combinations of manipulative variables. By choosing pairings that result in relative gain values near 1.0, a stable control system can be specificd.

In applying this concept to existing columns, I have found that variability in the operating data will occasionally indicate more than one control scheme is suggested. For example, a debutanizer receiving feed from several refinery sources may have feed composition changes great enough to totally change the relative gain results from one set of data to the next.

For this reason, the spreadsheet approach is very beneficial. As seen in Fig. 4, the various variables are entered as a base case in the first column of the spreadsheet. Each column thereafter changes one of the variables in a positive or negative direction. The results can be scanned quickly to determine if a conflict is noted.

Even more sophisticated calculations can be programmed effectively into the spreadsheet. For example, physical properties of petroleum fractions (specific gravity, density, enthalpy) have been calculated via the spreadsheet, allowing quick access to these data when required. I find it faster to call up the program and enter the few pieces of data required than to thumb through a data book to find a table or graph containing the information. Extensions of the physical property program include a multi-draw tower heat and material balance, heat exchanger duty calculations, and integration of the

07-Jan-85 RELATIVE GAIN MATRIX SENSITIVITY ANALYSIS

RELATIVE GAIN ANALYSIS AND SENSITIVITY TO VARYING TOWER CONDITIONS

	BASE CASE	<D >B	>D <B	<Y	>Y	<X	>X	<X <Y	>X >Y	<X >Y	>X <Y
DISTILLATE FRACTION(D)	0.6	0.480	0.720	0.600	0.600	0.600	0.600	0.600	0.600	0.600	0.600
BOTTOMS FRACTION(B)	0.400	0.520	0.280	0.400	0.400	0.400	0.400	0.400	0.400	0.400	0.400
OVHD PURITY FRACTION(Y)	0.985	0.9850	0.9850	0.9775	0.9925	0.9850	0.9850	0.9775	0.9925	0.9925	0.9775
BTM PURITY FRACTION(X)	0.035	0.0350	0.0350	0.0350	0.0350	0.0175	0.0525	0.0175	0.0525	0.0175	0.0525
THEORETICAL TRAYS(nE)	25	25.0	25.0	25.0	25.0	25.0	25.0	25.0	25.0	25.0	25.0
REFLUX RATIO(L/D)	2.5	2.50	2.50	2.50	2.50	2.50	2.50	2.50	2.50	2.50	2.50
FRACTION LT IN FEED(Z)	0.605	0.491	0.719	0.601	0.610	0.598	0.612	0.594	0.617	0.603	0.608
MATERIAL BALANCE FACTORS											
YDS(YBS)	0.396	0.288	0.529	0.494	0.248	0.563	0.308	0.657	0.183	0.394	0.399
YSB(YSD)	0.604	0.712	0.471	0.506	0.752	0.437	0.692	0.343	0.817	0.606	0.601
OTHER VARIATIONS											
YDV	0.462	0.375	0.575	0.577	0.289	0.610	0.384	0.713	0.228	0.426	0.498
YDV/B	0.560	0.455	0.693	0.701	0.350	0.680	0.498	0.796	0.295	0.475	0.647
YSV	4.261	3.067	5.912	3.532	5.360	5.694	3.504	4.415	4.178	7.992	3.015
YLV	13.760	9.796	19.254	10.921	18.039	18.405	11.304	13.671	14.054	26.906	9.307
YLL/B	4.078	4.613	3.365	3.240	5.340	5.204	3.482	3.868	4.323	7.603	2.871
YLV/B	2.670	2.812	2.319	2.123	3.493	3.284	2.344	2.442	2.908	4.795	1.935
YSL/B	2.652	2.426	2.604	2.201	3.333	3.381	2.267	2.623	2.699	4.742	1.953
YSV/B	2.067	1.937	1.994	1.716	2.595	2.540	1.817	1.972	2.161	3.561	1.567
YDL/B	0.513	0.407	0.646	0.642	0.321	0.647	0.444	0.756	0.263	0.451	0.576
YLB	0.557	0.650	0.438	0.447	0.723	0.403	0.638	0.303	0.785	0.583	0.530
YDL	0.443	0.350	0.562	0.553	0.277	0.597	0.362	0.697	0.215	0.417	0.470
YSL	5.724	4.008	8.088	4.743	7.204	7.797	4.629	6.044	5.523	10.946	3.981
SEPR	1810.5	1810.5	1810.5	1197.8	3648.6	3686.7	1135.1	2439.1	2388.3	7429.6	784.1
SYMM	2.33	2.33	2.33	1.56	4.67	1.17	3.50	0.78	7.00	2.33	2.33

NOTE: S REPRESENTS CONTROL OF SEPARATION VIA D/V, L/D, OR L/V

GAINS BETWEEN 0.9 AND 5.0 ACCEPTABLE SLIGHTLY ABOVE 1.0 PREFERRED OVER THOSES SLIGHTLY LESS THAN 1.0

Figure 4. Relative gain matrix sensitivity analysis.

previously mentioned pressure drop correlation, all of which use calculated physical property data.

CONCLUSION

I believe the "user-friendliness" and flexibility of the personal computer and electronic spreadsheet are a powerful combination that can be used effectively in providing engineers with better tools with which to work. Hopefully, the reader will extrapolate from these examples and find many useful applications for the electronic spreadsheet in carrying out his job functions.

References

1. Bristol, E. H., "On a New Measure of Interaction for Multivariable Process Control," *IEEE Transactions on Automatic Control,* 1966, pp. 133–134.
2. Shinskey, F. G., "From Process Knowledge to Computer Control: A Methodology," The Foxboro Company, Foxboro, Mass.
3. Gordon, L. M., "Practical evaluation of relative gains: The key to designing dual composition controls," *Hydrocarbon Processing,* December 1982, p. 87. The Foxboro Company, Foxboro, Mass.

55
Process Control Models

T. L. Blevins and K. Langley

Development of operating policies and design of appropriate control systems for process plants should be easy. On paper, there are several approaches that can be taken, any of which should lead to success. In reality, there seem to be roadblocks at almost every turn. This chapter presents a procedure successfully used to develop a technology (design or model) for ammonia plant control.

ANALYTICAL PROCEDURES

Some control specialists favor analytical techniques. These involve writing the equations that describe a process and deciding on the form of a well-behaved solution, then inserting terms or factors into the expressions to yield the desired results. The elements added to the equations are the functions contributed by the controllers—and implementation involves finding the corresponding hardware and software.

Almost any competent process engineer, given enough time, can write the process equations. Likewise, almost any control systems engineer can select a combination of analog, discrete, and digital controllers to implement particular transfer functions. The problem is that except in trivial cases, there are too many interacting equations to yield meaningful solutions.

EMPIRICAL METHODS

To avoid mathematical complexity, control specialists may turn to an empirical or plant study approach. The general procedure is to tweak and probe the process from all sides, under a variety of operating conditions, and to log the responses. The data then indicate what inputs affect the important process parameters, and in what manner. With this casual information, it is possible to devise compensations for disturbances to hold inputs at desired levels, and to establish the manipulative actions necessary to change process outputs.

One problem with this approach is that processes are rarely available for the types of tests it would be most desirable to perform. Even in circumstances where natural phenomena, such as feedstream fluctuations can be monitored, or where it is permitted to introduce low-intensity disturbances, the level of process noise can be greater than the expected responses, so the results are difficult to interpret. Plant testing has several other shortcomings. Because variables are coupled, effects of particular inputs are difficult to isolate. The number of experimental conditions of interest may be large, so comprehensive tests may involve more effort and the instrumenting of more points than could be justified. And responses may be so slow, or subject to such long delays, that they cannot be properly correlated with stimuli.

MODELING

Any description of a process—for example a diagram or a set of equations—is a model, and could conceivably be used as a tool for control system design. Modeling, however, has come to imply the use of computers to simulate process operation. The technique represents a compromise between analytical and empirical approaches to control system design. It first became popular in the 1960s, when analog computers were found to be effective modeling tools. Digital computation has added impetus to the concept, although it has also led to abuses by users confusing the precision of the machine with the accuracy of the model.

Modeling borrows from the analytical and empirical domains. Variables are identified and equations written to formulate the model, while results are obtained by testing.

The advantage over purely analytical methods is in avoiding the need to solve differential equations in the traditional ways. It is, therefore, possible not only to obtain solutions to complex problems, but to experiment with many solutions and to do so under the control of a program that systematically seeks some form of optimum behavior.

The benefit of simulation relative to plant testing is that the experimental conditions can be easily and precisely manipulated. As a result, stimuli can be applied individually to achieve decoupling, conditions can be varied under program control, and time-scaling can be used to obtain results rapidly. Moreover, because such systems can be made self-documenting, no measurements are needed for data acquisition, and records can be captured in a form suitable for sophisticated analysis.

In specifying simulation conditions, it should be remembered that a model is only as valid as the assumptions used in its formulation, so there is a limit to the amount and reliability of information to be gleaned. In particular, constraints that might govern a real process are often absent. Moreover, simulated conditions may only approximate those encountered in practice, or gross phenomena might conform with real behavior while details differ significantly.

SIMULATION GOALS

The goals of simulation should be to obtain trends and qualitative rather than quantitative information. It is particularly beneficial to use models for determining whether there exist circumstances under which process performance exhibits some maxima, globally, or at plant constraints, and if so, how sensitive the operation is to small changes in the various conditions. The parameters used in establishing criteria need not be restricted to measurable quantities, since it is relatively simple to calculate indices such as production cost or efficiency for use in the evaluations. Likewise it is practical to investigate variables in combination as well as singly, to thoroughly study process interactions.

A complete model-oriented study therefore, begins by investigating process sensitivities to variables, singly and in combinations, to determine which quantities seem to affect performance enough to warrant consideration in control system design. Particular emphasis should be given to operation near plant constraints, because thruput, efficiency and similar performance maxima often occur when the plant is limited in some way.

APPLICATION

A detailed simulation was used in developing a technology for ammonia plant control. The model was formulated especially for this application, on the basis of the various unit operations in the Kellogg process (Fig. 1), although the general structure was consistent with published work in the area.[1,2] Conclusions drawn from the model have been employed in production plants, and have yielded improvements in profitability consistent with those predicted.

Makeup feed, the gas flowing to the synthesis compressor from the suction drum, has a significant effect on production. Simulation showed that, with all other loop parameters constant, ammonia output in mols/hr would increase monotonically with this variable (Fig. 2).

In many ammonia plants, makeup feed is limited by front end restrictions such as insufficient synthesis compressor power or absorber capacity. The simulation showed the importance of a control strategy that could monitor front end restrictions, and regulate makeup feed to maintain the front end at its operating constraints. This ensures maximum makeup under given operating conditions.

The *hydrogen/nitrogen ratio* (loop H/N) at the synthesis converter inlet was also found to be an important variable. For constant makeup rates, the simulation showed that production increased as H/N decreased (Fig. 3), with the effect becoming more pronounced at high flows. The data suggested that close control over H/N ratio would be advantageous, particularly for high-rate feed-limited conditions. In addition, as H/N is reduced, synthesis compressor loading increases. The lower operating limit on the ratio is therefore often established by compressor capacity.

Synthesis loop H/N ratio was also studied with constant H_2 process gas in the makeup feed (Fig. 4). This condition occurs when production is limited by natural gas feedstock availability or primary reformer capacity. Over the simulation range, production for constant process gas rate was found to be greatest at low H/N ratios. As before, the effect was most significant at high makeup feed rates, with the synthesis converter near capacity.

Synthesis loop H/N ratio was investigated with constant N_2 in the makeup flow (Fig. 5). This condition occurs when the plant is air-limited. Over the range of simulation, maximum production occurred at an H/N ratio of approximately 2.9 for constant air flow, but sensitivity was low and minor variations about this value had only a small effect. As a result, production increase obtainable with closer H/N control appeared to be negligible for an air-limited plant. Makeup flow control for the air-limited case, however, can provide significant production increases. It is, therefore, advantageous to provide automatic adjustment of makeup flow to the highest value which still allows all available air to be used. This degree of automation is possible only if H/N is under control.

The *effect of synthesis loop pressure* variations on production, with other process parameters constant, is shown in Fig. 6. Over the simulation range, production was found to increase with loop pressure. This suggested controlling the purge system to maintain the loop just below the design pressure limit, when compressor ca-

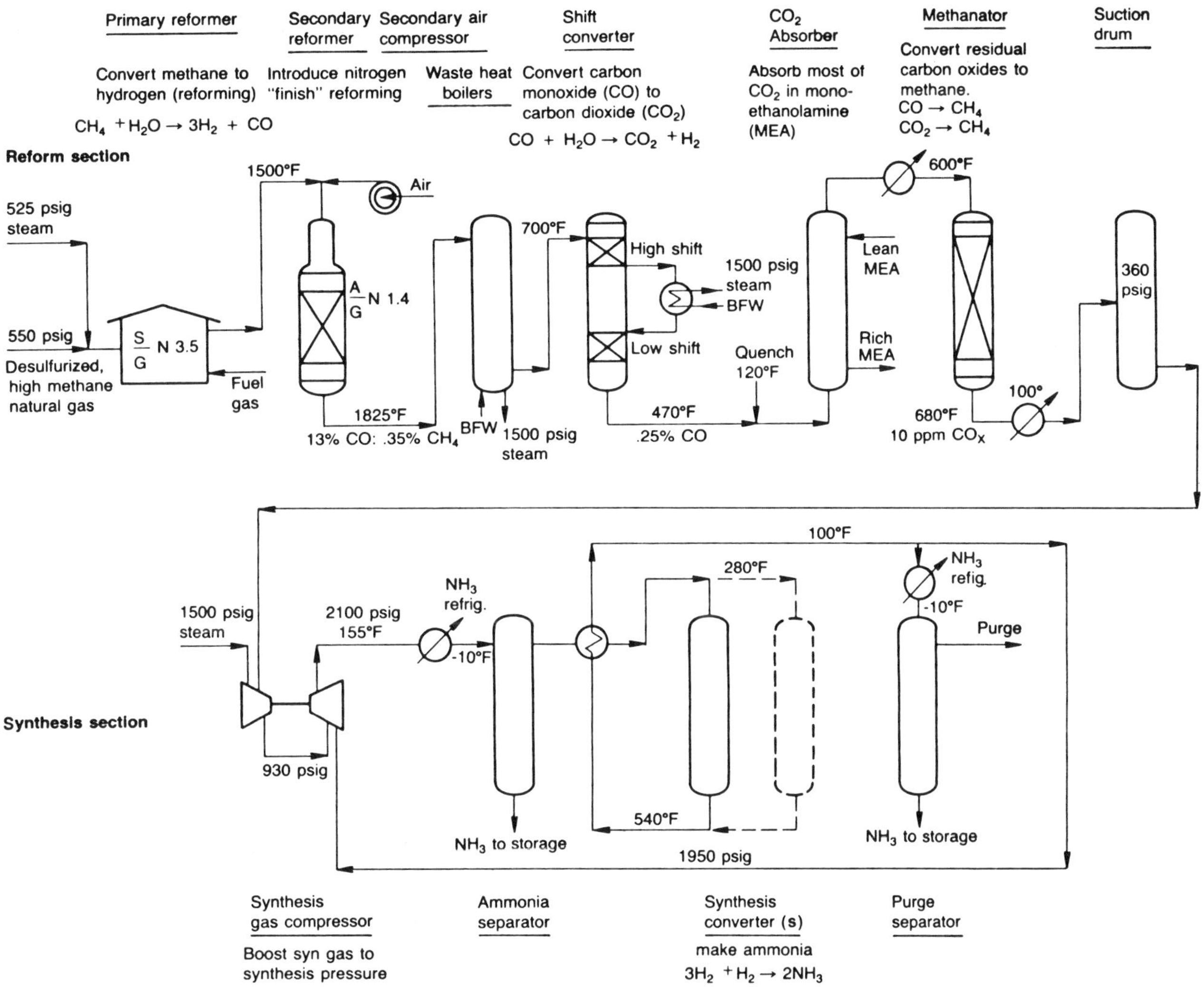

Figure 1. Kellogg ammonia process.

pacity is sufficient to maintain this pressure at maximum makeup feed rate. Under such circumstances, close purge control could be highly beneficial if variations in loop pressure would otherwise occur due to normal disturbances.

Loop pressure was also studied with constant synthesis compressor loading. This condition occurs in plants lacking sufficient compressor power to simultaneously utilize front end capacity and maintain maximum loop pressure.

Consequences of pressure variations depended on feed rate (Fig. 7). In the simulation, output increased at low makeup feed rates when loop pressure dropped; at makeup feed rates approaching synthesis converter limits, however, lowering loop pressure also decreased production.

The simulation revealed that synthesis compressor suction could be monitored to indicate load limits. The purge controller could be designed to either raise or lower loop pressure depending on operating conditions.

Simulated production appeared to be exceptionally sensitive to *temperature at the inlet of the first synthesis converter bed* (Fig. 8). Over the simulation range, for a given set of loop conditions, there is an optimum temperature and a sharp production peak. The temperature of the first synthesis converter bed inlet should, therefore, be closely controlled.

Methane leakage, the CH_4 in the makeup feed gas, can occur because of factors such as improper primary reformer firing, poor absorber efficiency and inactive or insufficient reformer catalyst. Varying amounts of methane leakage for constant process gas flow are simulated. Figure 9 predicts that production will increase if makeup gas methane content is decreased. The control strategy suggested was to improve primary reformer temperature control.

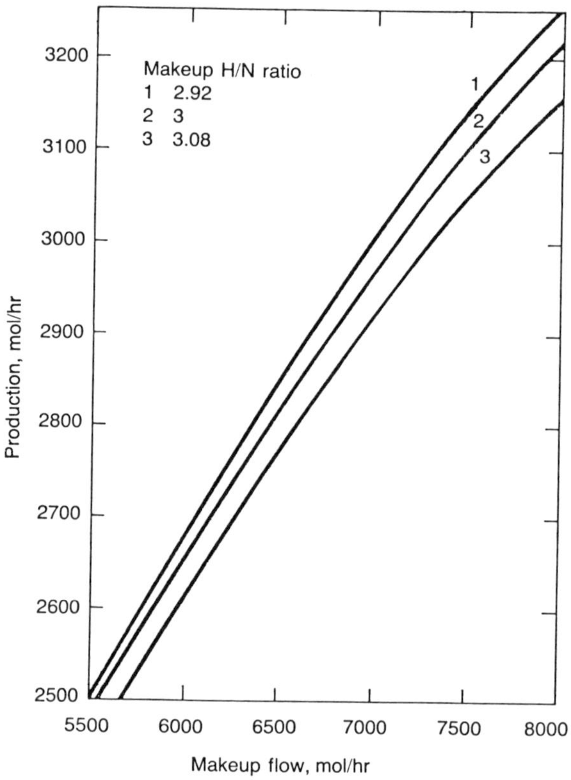

Figure 2. Production vs. makeup flow with constant H/N ratio.

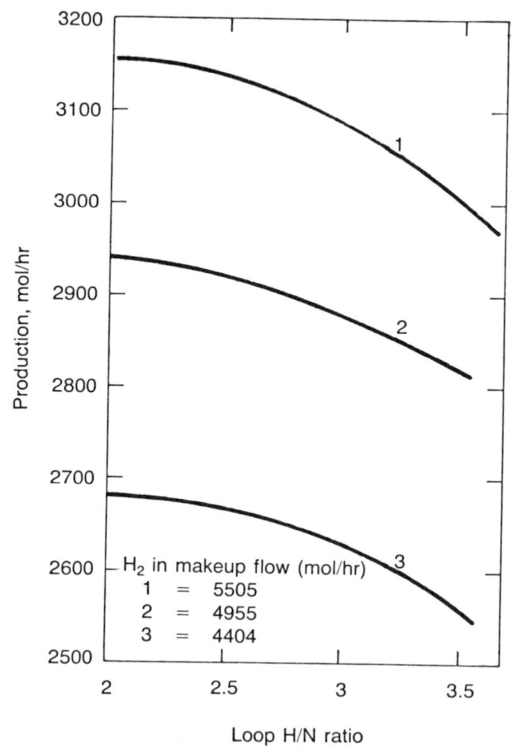

Figure 4. Production vs. H/N ratio with constant hydrogen.

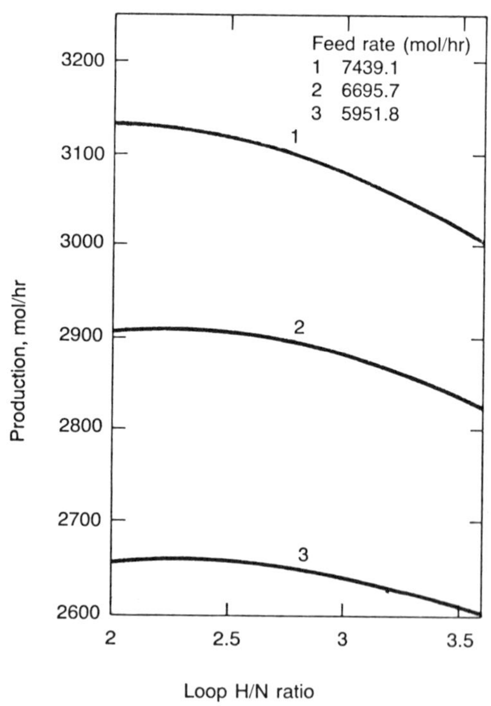

Figure 3. Production vs. H/N ratio with constant makeup flow.

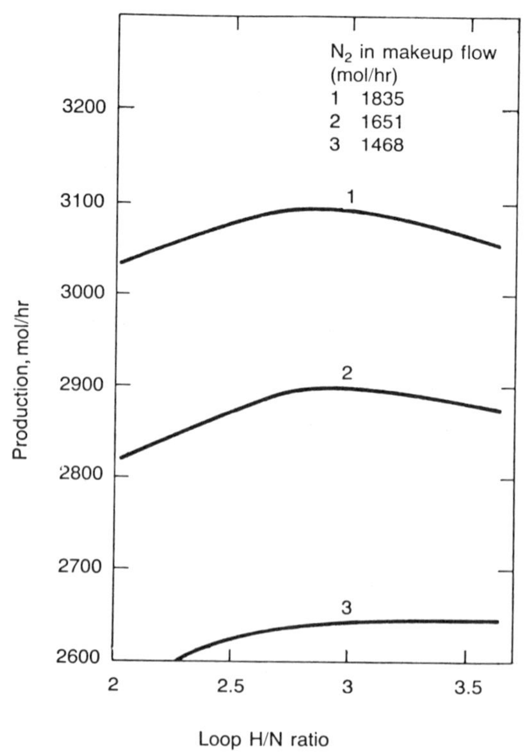

Figure 5. Production vs. H/N ratio with constant nitrogen.

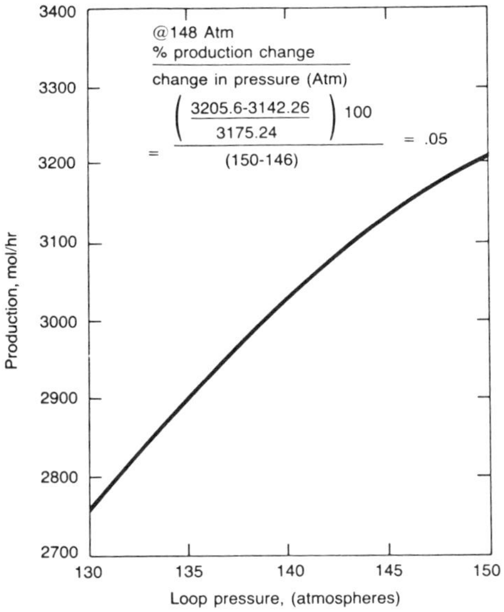

Figure 6. Production vs. loop pressure.

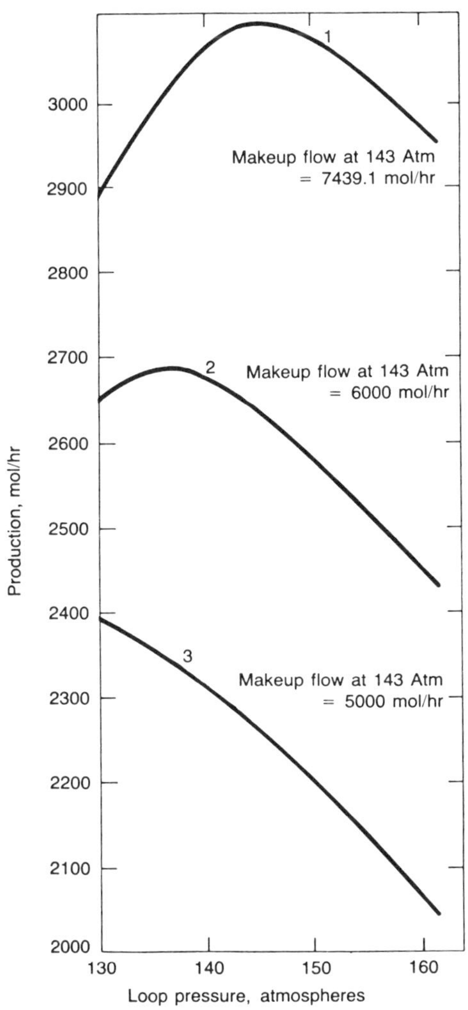

Figure 7. Production vs. loop pressure for various makeup flow rates.

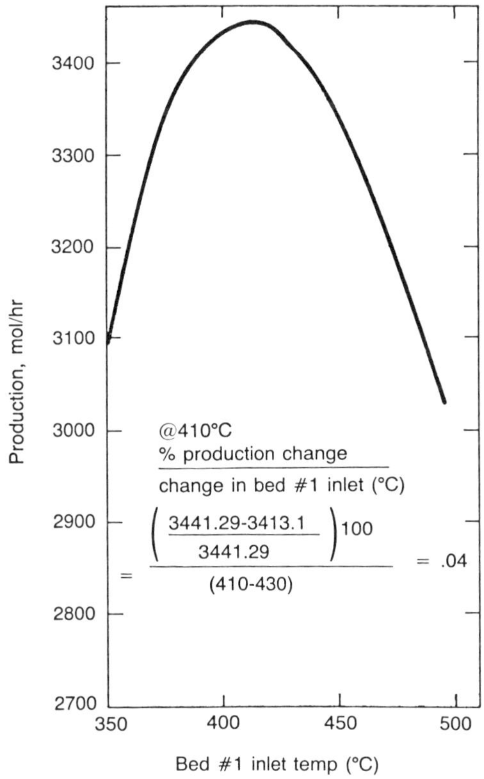

Figure 8. Production vs. first converter bed inlet temperature.

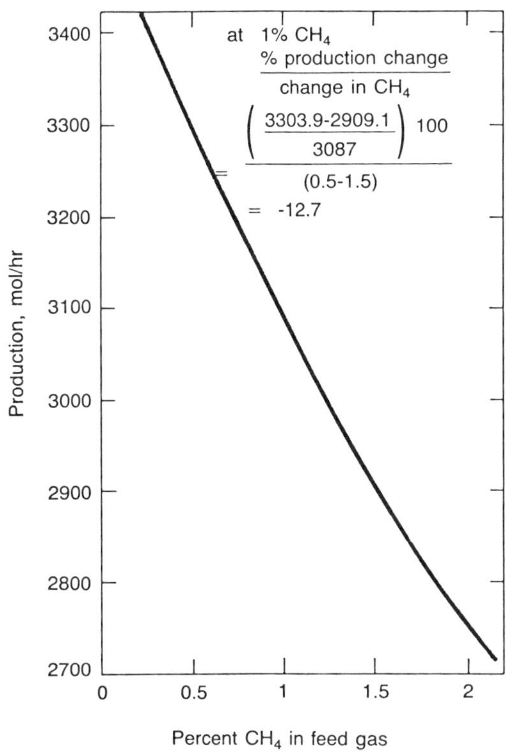

Figure 9. Production vs. methane leakage.

MODEL JUSTIFICATION

Conclusions drawn from the ammonia plant model have been put into practice at a number of installations. These have consistently led to greatly improved productivity and conversion efficiency.

Some controls resulting from the simulation were developed for individual units, when the need to minimize variations was indicated. The details of the loops in such cases were studied using simulations of the structures of the units, when it was not immediately clear how the systems should be implemented. The strategy for adjusting the reformer temperature was an example of a technique developed in this manner.

Thruput has also been maximized with production limited by operating constraints, based on conclusions drawn from the model. This involved calculating setpoints for a number of loops using measurements of operating conditions, then applying the setpoint values as commands to the controllers.

In general, modeling has proved advantageous for exploring process interactions, and investigating whether improvements in the control of certain variables can have a significant effect on performance. The technique is particularly advantageous in complex processes, because it shows behavior in a unified graphic manner.

References

1. Ferraris, G. Buzzi "An Investigation of Kinetic Models for Ammonia Synthesis," *Chemical Engineering Science,* Vol. 29, 1974, pp. 1621–1627.
2. Hays, G. E. "Ammonia Synthesis Plant Kinetics," *Chemical Engineering Progress,* Vol. 60, No. 1, January 1964, pp. 61–65.
3. Shah, Manesh J. "Control Simulation in Ammonia Production," *Industrial and Engineering Chemistry,* Vol. 59, No. 1, January 1967, pp. 72–83.

56

Algorithm for Multivariable Control

G. R. Arulalan and P. B. Deshpande

This chapter describes an algorithm for multivariable control and how it is applied to a simulated side-draw distillation column. The algorithm uses readily available step responses for implementation. It inherently compensates for deadtime and process interactions, and inverse response behavior poses no special problems. The order of the process is not at all important and constraints on inputs/outputs can be accommodated. The robustness issues can be addressed in the design phase. The algorithm is simple to understand and easy to implement, and the memory requirements of the digital control algorithm are modest. The only requirement for application is that the process be open-loop stable. The results of the simulation study show the excellent capability of the algorithm for multivariable control.

PRESENT MULTIVARIABLE CONTROL TECHNIQUES

Many control loops in process plants are single-input single-output systems satisfactorily controlled by standard PID-type controllers. A small number of commercially important loops, however, are multivariable. Some examples include distillation columns, fluid catalytic crackers, exothermic polymerization reactors, etc. Automatic control of multivariable processes with PID-type controllers poses difficulties due to the presence of deadtime, process interactions, and constraints.

Several multivariable control techniques are available. They include:

- Relative gain arrays/decoupling
- Nyquist arrays
- Optimal control with state-space methods
- Singular value decomposition
- Model predictive control

Edgar[1] surveyed the multivariable control methods available up to 1976. A monograph describing the more recent methods is under preparation.[2]

The focus in this chapter is on model predictive control. In the late seventies two multivariable control techniques, dynamic matrix control (DMC) (Cutler and Ramaker[3]) and model algorithmic control (MAC) (Richalet, et al.[4]) appeared in the control literature. They were developed independently in industry and had already been applied successfully by the time they appeared in the literature. These algorithms did not result from a new theory but rather had a heuristic basis. Each requires process impulse responses derived from open-loop step responses for implementation. Interaction and deadtime compensation is inherent in these algorithms. Other important features include the facility to handle inverse response behavior with equal ease, specification of robustness as a design criterion, and the ability to handle constraints. The principal difference between MAC and DMC is that they use different optimization procedures.

In a series of recent papers, Morari and Garcia[5,6,7] showed that DMC and MAC indeed had a theoretical basis and were a subset of a new methodology which they referred to as Internal Model Control (IMC).

In the following section a new methodology for multivariable control is described and then applied to a simulated 3 × 3 distillation column. Although it can be examined in the context of IMC, it is much easier to design and implement than other model predictive control techniques. It also retains the essential features of model predictive control algorithms mentioned earlier.

THE NEW METHOD

We refer to this algorithm as SMPC, for simplified model predictive control. The mathematical details of algorithm development are given in Arulalan and Deshpande[9] and will not be repeated here. Instead a philosophical basis will be presented.

The algorithm is based on the premise that it should always be possible to design a control algorithm that results in closed-loop setpoint responses that are at least as good as the normalized open-loop responses. This implies that the open-loop transfer function matrix and the closed-loop transfer function matrix may be equated giving for a 3 × 3 system (chosen to illustrate the method):

$$M_1^n = \sum_{i=1}^{3} k_{1i}E_i^n + \sum_{i=1}^{3} k_{1i} \sum_{j=1}^{N}(h_{11}^j M_1^{n-j} + h_{12}^j M_2^{n-j} + h_{13}^j M_3^{n-j}) \quad (1a)$$

$$M_2^n = \sum_{i=1}^{3} k_{2i}E_i^n + \sum_{i=1}^{3} k_{2i} \sum_{j=1}^{N}(h_{21}^j M_1^{n-j} + h_{22}^j M_2^{n-j} + h_{23}^j M_3^{n-j}) \quad (1b)$$

$$M_3^n = \sum_{i=1}^{3} k_{3i}E_i^n + \sum_{i=1}^{3} k_{3i} \sum_{j=1}^{N}(h_{31}^j M_1^{n-j} + h_{32}^j M_2^{n-j} + h_{33}^j M_3^{n-j}) \quad (1c)$$

where M_1, M_2, M_3 = manipulated variables
E_1, E_2, E_3 = errors, (setpoint measurement)
h_{ij} = impulse response coefficients
k_{ij} = elements of process gain inverse matrix
n = sampling instants

Equations (1a), (1b), and (1c) will yield closed-loop responses that are identical to the normalized open-loop responses. The algorithm can be speeded up by replacing k_{ij} with a set of α_{ij} which are the turning constants of the algorithm giving:

$$M_1^n = \sum_{i=1}^{3} \alpha_{1i}E_i^n + \sum_{i=1}^{3} k_{1i} \sum_{j=1}^{N}(h_{11}^j M_1^{n-j} + h_{12}^j M_2^{n-j} + h_{13}^j M_3^{n-j}) \quad (2a)$$

$$M_2^n = \sum_{i=1}^{3} \alpha_{2i}E_i^n + \sum_{i=1}^{3} k_{2i} \sum_{j=1}^{N}(h_{21}^j M_1^{n-j} + h_{22}^j M_2^{n-j} + h_{23}^j M_3^{n-j}) \quad (2b)$$

$$M_3^n = \sum_{i=1}^{3} \alpha_{3i}E_i^n + \sum_{i=1}^{3} k_{3i} \sum_{j=1}^{N}(h_{31}^j M_1^{n-j} + h_{32}^j M_2^{n-j} + h_{33}^j M_3^{n-j}) \quad (2c)$$

The α_{ij} are determined through offline simulation by an optimization program such that a suitable performance index, say of the type,

$$J = \int_0^t |E_1|\,dt + \int_0^t |E_2|\,dt \quad (3)$$

is minimized.

Robustness of the system can be ensured by inserting a filter matrix in the feedback path and constraints can be accommodated during the optimization step. The filter constraints can serve as adjustable parameters that address the degree of plant-model mismatch.

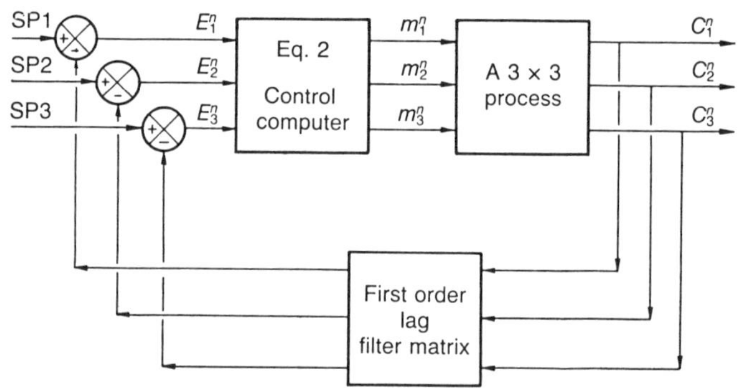

Figure 1. SMPC system for 3 × 3 process.

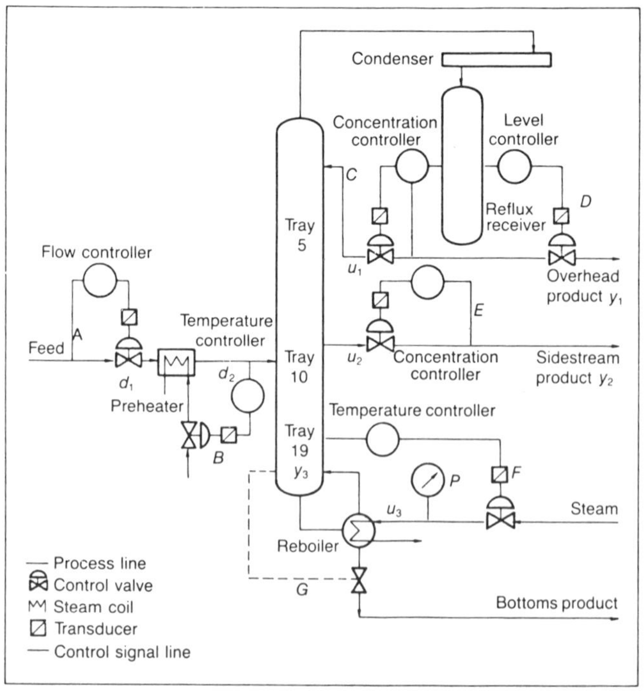

Figure 2. Conventional feedback control scheme.

IMPLEMENTATION AND TESTING

A block diagram of the SMPC control system for a 3×3 process is shown in Fig. 1. The control algorithm has been tested on a simulated 3×3 distillation system reported in the literature (Ogunnaike et al.[8]). A schematic of the distillation column and the associated control system is shown in Fig. 2. A flowchart of the computer program used to implement SMPC is shown in Fig. 3. The steady state operating data for the column and the associated transfer functions reported in the literature are shown in Table 1. The step responses required for implementation of SMPC are obtained from these transfer functions. Table 1 also shows the tuning constants, α_{ij}, for various tests.

The results of the simulation are shown in Figs. 4 through 16. Figures 4 through 10 show setpoint responses and the associated manipulated moves. Figures 11 through 16 show the responses to load upsets in feed flow and feed temperature and the associated manipulated variables moves. Throughout this study, perfect modeling is assumed and therefore the filter matrix is

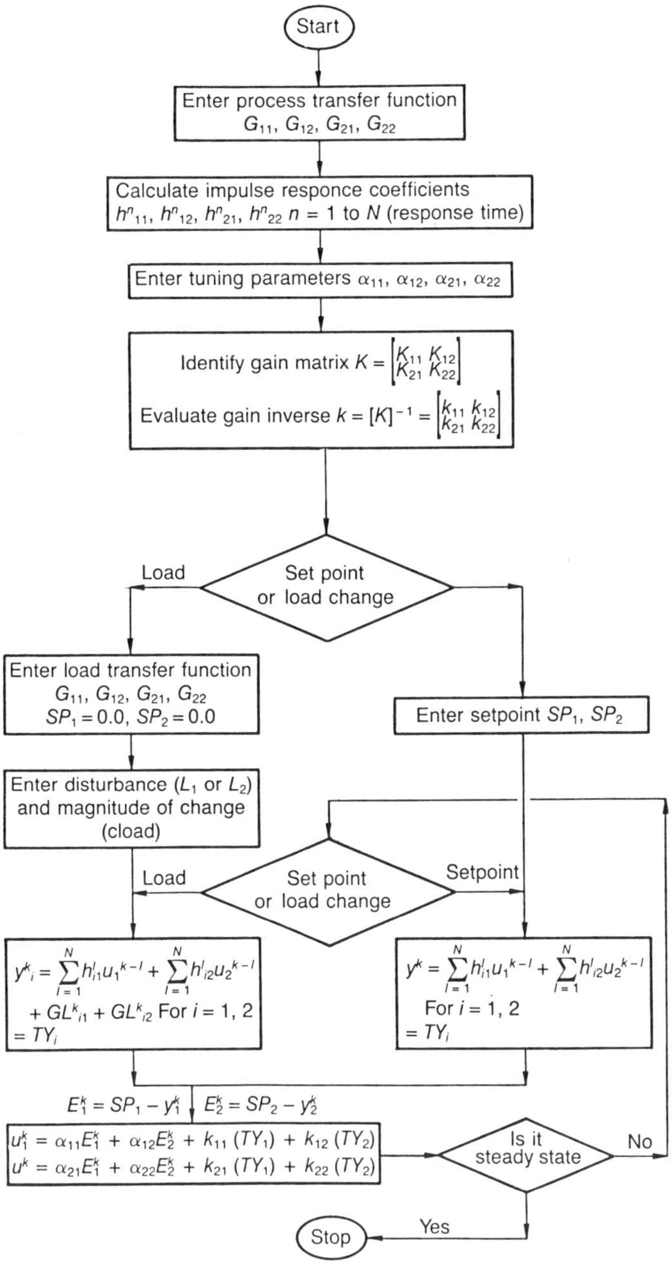

Figure 3. Flowchart for SMPC simulation (3×3 process).

Table 1
Column Data and Tuning Constants

(a) Steady state operating conditions

Control variables		Controlled variables	
Reflux flowrate	0.18 gpm (0.01 l/s)	Overhead mole fraction ethanol	0.7
Side stream flowrate	0.046 gpm (0.003 l/s)	Side stream mole fraction ethanol	0.52
Reboiler steam pressure	20 psig (137.89 kPa)	Temperature on tray 19	92°C

(b) Transfer functions*

$$\begin{bmatrix} Y_1(s) \\ Y_2(s) \\ Y_3(s) \end{bmatrix} = \begin{bmatrix} \dfrac{0.66e^{-2.6s}}{(6.7s+1)} & \dfrac{-0.61e^{-3.5s}}{(8.64s+1)} & \dfrac{-0.0049e^{-s}}{(9.06s+1)} \\ \dfrac{1.11e^{-6.5s}}{(3.25s+1)} & \dfrac{-2.36e^{-3s}}{(5.0s+1)} & \dfrac{-0.012e^{-1.2s}}{(7.09s+1)} \\ \dfrac{-34.68e^{-9.2s}}{(8.15s+1)} & \dfrac{46.2e^{-9.4s}}{(10.9s+1)} & \dfrac{0.87(11.61s+1)e^{-s}}{(3.89s+1)(18.8s+1)} \end{bmatrix} \begin{bmatrix} U_1(s) \\ U_2(s) \\ U_3(s) \end{bmatrix}$$

$$+ \begin{bmatrix} \dfrac{0.14e^{-12s}}{(6.2s+1)} & \dfrac{-0.0011(26.32s+1)e^{-2.66s}}{(7.85s+1)(14.63s+1)} \\ \dfrac{0.53e^{-10.5s}}{(6.9s+1)} & \dfrac{-0.0032(19.62s+1)e^{-3.44s}}{(7.29s+1)(8.94s+1)} \\ \dfrac{-11.54e^{-0.6s}}{(7.01s+1)} & \dfrac{0.32e^{-2.6s}}{(7.76s+1)} \end{bmatrix} \begin{bmatrix} d_1(s) \\ d_2(s) \end{bmatrix}$$

* See Fig. 2 for description of variables.

(c) Process gain matrix and inverse

$$K = \begin{bmatrix} 0.66 & -0.61 & -0.0049 \\ 1.11 & -2.36 & -0.012 \\ -34.68 & 46.2 & 0.87052 \end{bmatrix}$$

$$k = K^{-1} = \begin{bmatrix} 2.97218 & -0.60361 & 0.00841 \\ 1.09001 & -0.80170 & -0.00492 \\ 60.55808 & 18.50102 & 1.74465 \end{bmatrix}$$

(d) Tuning parameters, α

Type of change	Magnitude change		α		Minimum IAE
setpoint	$y_1 = -0.05$	2.9496	−0.4388	−0.0056	
	$y_2 = -0.08$	1.5093	−0.4794	−0.0042	8.0477
	$y_3 = 0.0$	0.3718	−0.9184	1.9722	
setpoint	$y_1 = 0.0$	0.8484	−1.1243	0.0192	
	$y_2 = 0.0$	0.0760	−0.9945	−0.0016	28.78
	$y_3 =$ 5°C	−0.2240	−0.5029	2.9045	
load	$d_1 = 0.2$	2.6677	−0.7609	−0.1108	
	$d_2 = 0.0$	0.8745	−1.2673	−0.0593	13.985
		−0.1811	−1.1977	3.1230	
load	$d_1 = 0.0$	0.6144	−1.0873	−0.0279	
	$d_2 =$ −20°C	−0.5831	−1.0810	−0.0035	37.08
		0.5797	−1.0904	2.9359	

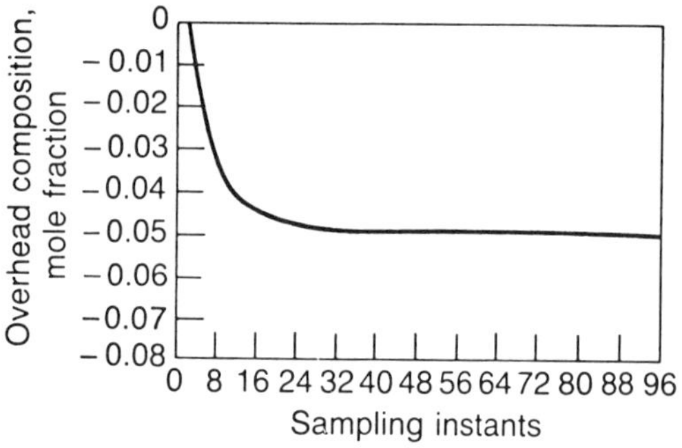

Figure 4. Response of overhead composition, y_1, to simultaneous setpoint change in y_1 and y_2.

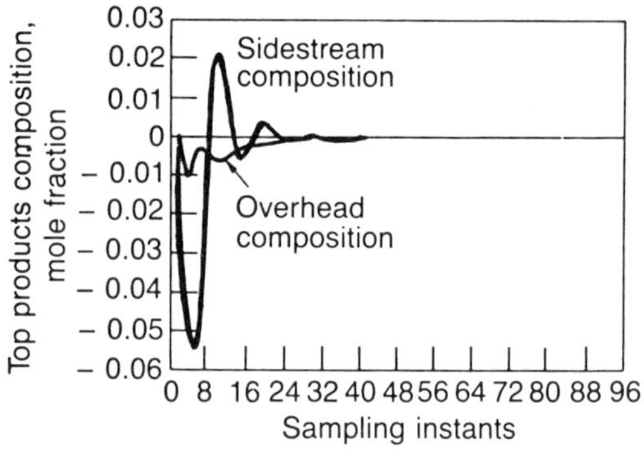

Figure 8. Response of overhead and sidestream product compositions to a setpoint change in Tray 19 temperature.

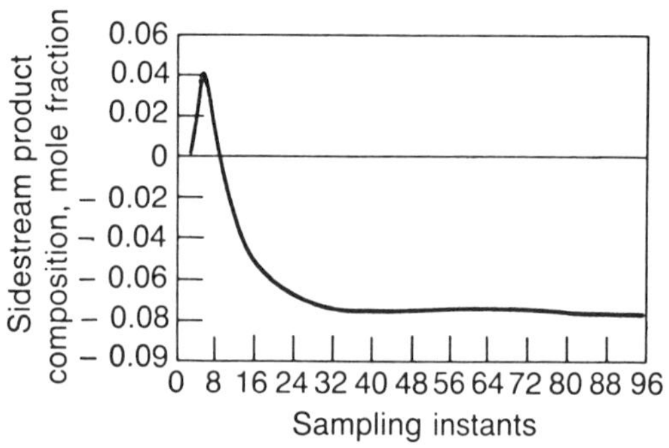

Figure 5. Response of sidestream product composition, y_2, to simultaneous setpoint change in y_1 and y_2.

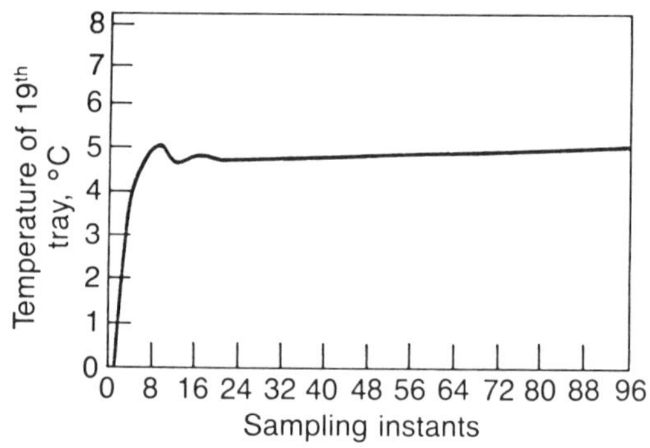

Figure 9. Response of Tray 19 temperature to a setpoint change.

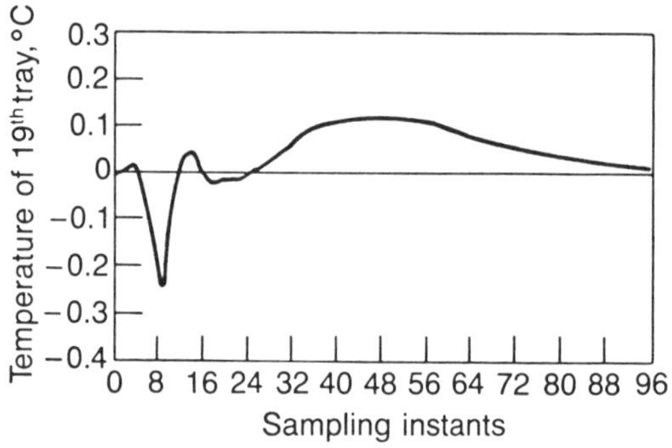

Figure 6. Response of temperature on Tray 19, y_3, to simultaneous setpoint change in y_1 and y_2.

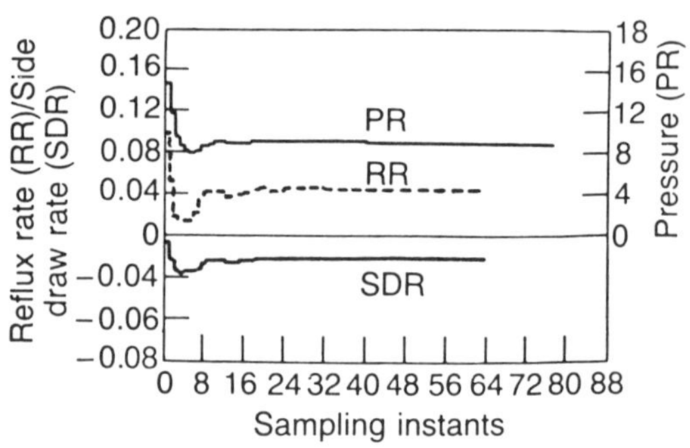

Figure 10. Manipulated variable moves for temperature setpoint test.

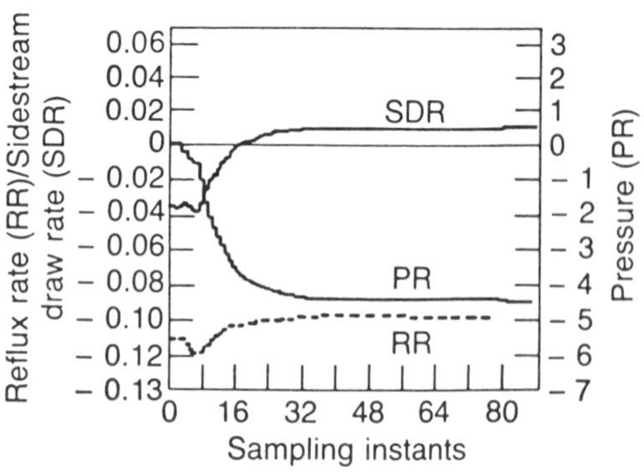

Figure 7. Manipulated variable moves for setpoint changes in y_1 and y_2.

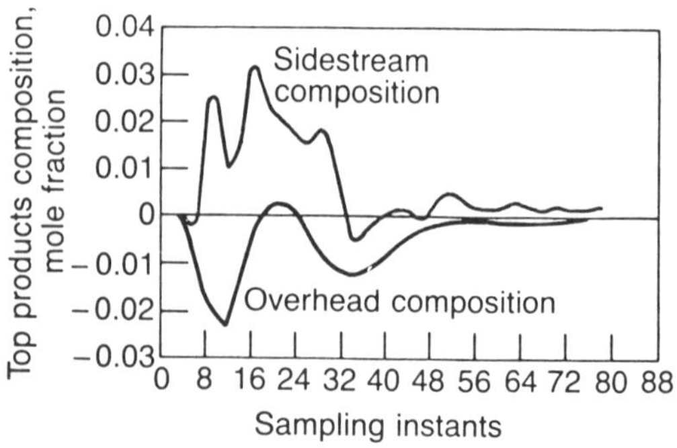

Figure 11. Response of top product compositions to feed flow disturbance of 0.2 gpm.

not used. The results clearly show the excellent capability of SMPC for multivariable control.

In conclusion, the algorithm is simple and easy to implement. It retains the features of the more complex model predictive control algorithms, and should be of interest to control engineers who are concerned with designing and implementing multivariable control systems.

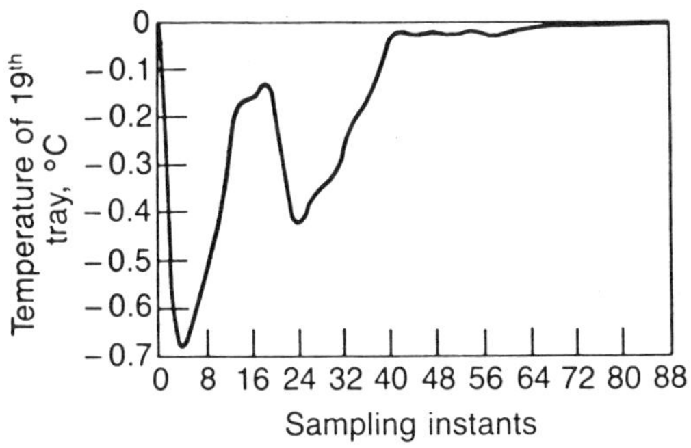

Figure 12. Response of Tray 19 temperature to feed flow disturbance of 0.2 gpm.

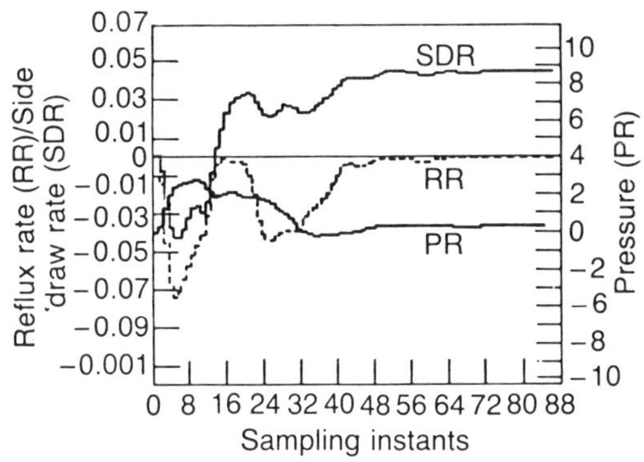

Figure 13. Manipulated variable moves for feed flow disturbance test.

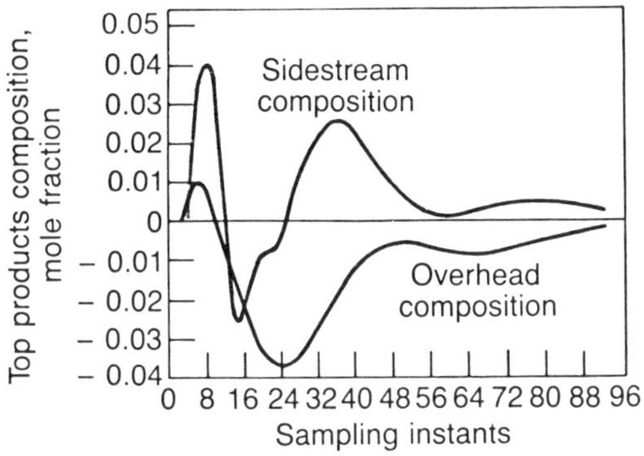

Figure 14. Response of top product compositions to feed temperature disturbance of − 20°C.

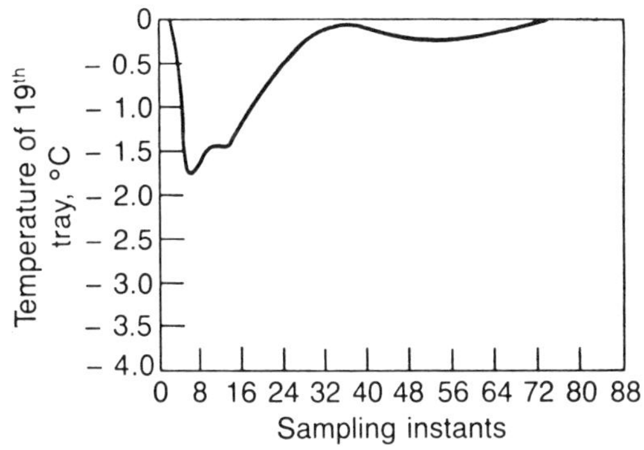

Figure 15. Response of Tray 19 temperature to feed temperature disturbance of − 20°C.

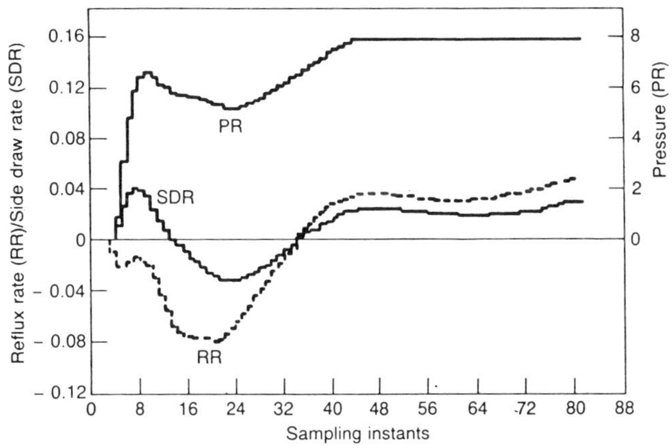

Figure 16. Manipulated variable moves for feed temperature disturbance test.

References

1. Edgar, T. J., *Proceedings,* Chemical Process Control I, Asilomar, Calif., January 1976.
2. Deshpande, P. B., "Multivariable Control Methods," Monograph accepted for publication by Instrument Society of America, 1987.
3. Cutler, C. R., Ramaker, B. I., Paper No. WP-8, JACC, 1980.
4. Richalet, J. A., Rault, A., Testud, J. D., Papon, J., *Automatica,* 14, p. 413, 1978.
5. Garcia, C. E., Morari, M., *Ind. Eng. Chem. Proc. Des. Dev.,* 21, 2, p. 308, 1982.
6. Garcia, C. E., Morari, M., *Ind. Eng. Chem. Proc. Des. Dev.,* 24, 2, p. 472.
7. Garcia, C. E., Morari, M., *Ind. Eng. Chem. Proc. Des. Dev.,* 24, 2, p. 484, 1985.
8. Ogunnaike, B. A., et al., "Advanced Multivariable Control of a Pilot Plant Distillation," *AIChE J.,* 29, 4, pp. 632–640, July 1983.
9. Arulalan, G. R., Deshpande, P. B., "Simplified Model Predictive Control," Paper submitted to *I & E C Proc. Des. Dev.*

57
Optimizing Software Systems

B. D. Campbell

By approaching supervisory control projects in a disciplined manner, it is possible to implement optimizing programs effectively. Agreement on required behavior of a system is necessary because of the potential impact on plant operations. In fact, with the ability to describe optimizing programs, the initial risk of these kinds of projects can be lowered significantly.

At the present time, these programs are viewed as a specialist activity, divorced from the main lines of software development. This need not be the case. It is crucial to the development of these programs that their operation fit into the established pattern of production control. In fact, should it be found that a model cannot be agreed upon for a particular plant situation, the proposed software system ought not to be built.

If the system is built anyway, experience has often shown that it later falls into disuse or interferes with ongoing production more than it helps increase plant productivity. With the appropriate definition of software requirements, a project to create optimizing software systems will have a high probability of success. It will also deliver the ability to handle cost volatilities, the reason that the project was initiated in the first place.

The history of applying optimization to real time chemical process control problems has been spotty. The cost of maintaining and installing these specialized programs is often thought to outweigh any possible benefit.

In recent years, however, an approach to an operationally useful optimizing program has been developed. It uses "structured analysis" to put together the kinds of expertise necessary for the success of software projects of this type.[1-4,9]

Optimizing programs use mathematical, linear, and nonlinear programming techniques to determine optimal solutions to supervisory control problems. Supervisory control problems involve choosing one set from among a number of physically possible set points consisting of those that do not violate mass and energy balance. Thus, if the process physics can be described by a series of equations representing significant mass and energy balance, then chances are that a useful optimization can be performed.

Most of the discussion in the open literature has been concerned with the algorithmic issues inherent in the computation of a correct answer. In practice, software design issues[5,6] that deal with the creation of modules in a specified computing environment are also important.

In an optimizing program, such a problem would be the design of the software module that represents the process mass and energy balances. The module should be designed so that as it is used by the optimizing program in an iterative manner, the selection of the "best" answer occurs rapidly.

Usually, the optimizing program is a part of the production planning process. The program itself figures out the best way to obtain the necessary levels of production. The actual levels are determined by production planning systems. Thus, the interface between the supervisory control system and the production planning system must be clearly understood.

PRODUCTION CONTROL

Production control is the process of determining how much of a given product is to be produced and which machinery will be used. Optimizing programs in a chemical plant environment ought to work well with existing or planned production control software and also with existing regulatory controls. Many existing regulatory and production control packages have complex interfaces, which means that the required behavior of the software programs must be defined before establishing the technology to implement the required behavior.

Keeping track of the scope and projected costs of a software project is difficult, partly a result of not understanding software systems. Since the personnel involved in a software project usually are from varied backgrounds, a common understanding of the required behavior of the hardware and software systems is necessary. This is another way of saying that the analysis phase of a project must be done diligently so that a given system will function practically in a working environment.

SOFTWARE ENGINEERING

Software engineering is the art and practice of applying software technologies to meet human needs. The disciplines of structured design, structured analysis, simulation and numerical analysis are all branches.

Five different phases of a software project have been identified: analysis, design, implementation, testing and maintenance. This chapter will focus on the analysis document which provides the means to getting a project under way. Further, the bridge to the design stage will be described.

"Analysis" is the description of a particular application need without reference to the technology being used. It is of increasing importance in areas where the kinds and varieties of devices being employed change rapidly. *"Design"* is the process of taking these technology-free specifications and applying a particular kind of software and hardware to the problem being solved.

In building control systems, it was found that the process of understanding an application was so helpful that standard methods were developed for describing requirements. In like manner, the process of developing a computational system can make use of symbology for defining the goal of the computation. If the decision about the technology to be used can be decided apart from the problem to be solved, the development of supervisory control systems will occur at a predictable and reasonable pace.

OPTIMIZING PROGRAM EXAMPLE

Suppose that the ethylene glycol process shown in Fig. 1 were to be instrumented. In addition to the regulatory controls, it is proposed that an optimization routine be created to calculate the most profitable recycle and feed rates.[7,8]

First, the problem should be described as is shown in Fig. 2. This will allow the users, designers and specialists involved in the software project to understand the scope of the effort being undertaken. Four entities external to the program are identified: a production control software system, a supervisory terminal, an operator terminal and the process itself.

The diagram also scopes the size of the problem to be solved. If, for instance, the whole problem includes a larger portion than that shown in Fig. 1, the size of the

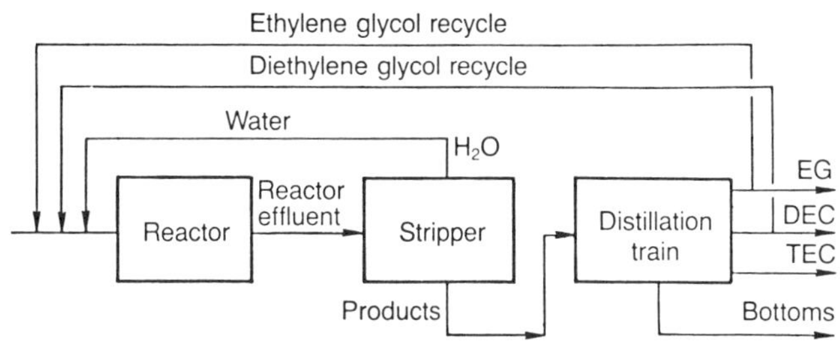

Figure 1. Ethylene glycol process.

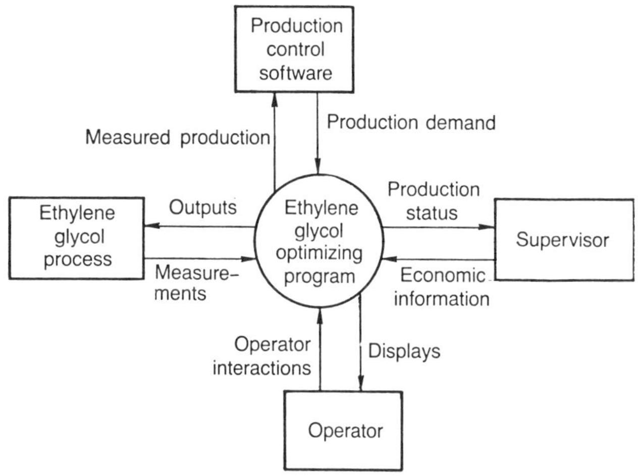

Figure 2. Context of ethylene glycol optimizing program.

optimizer will increase, as will the risk that an acceptable optimizing program cannot be finished. Industry figures state that 15% of all software projects begun in earnest never deliver anything at all.[9]

Figure 3 describes the two portions of the optimizing program operation—the data flow and state descriptions. The data flow description tells what will be done, as is illustrated by the processes numbered 1 through 5. Process 1 reformats the output from a production planning system to the required production for process 3, the computation of optimal set points. Process 2 uploads actual production figures from measurements and optimal answers computed by process 3, which is the computation of optimal set points. The interface to the regulatory controls is shown as a ramp set point function, and the regulatory controls are represented by process 5.

The state description tells when these functions are performed, when they are active, that there is a run state for the ethylene glycol process, and that there is a system upload and a system download.

The state and data flow diagrams are connected through the activate and deactivate descriptions on the transfer from one state to the other. The condition that causes the transition is shown above the line on the tran-

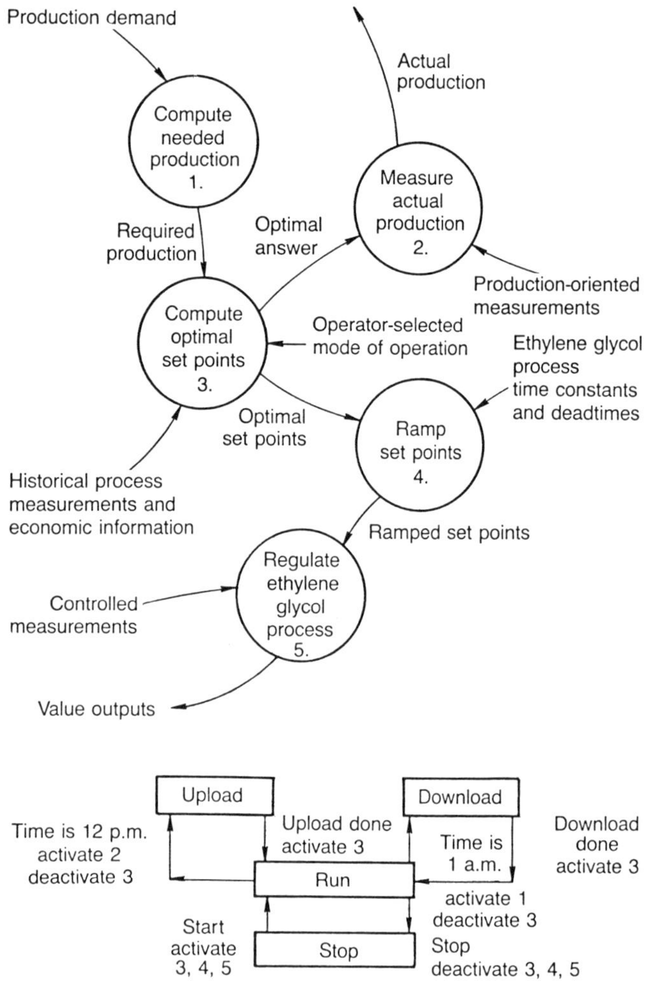

Figure 3. Operation of optimizing program.

sition. For instance, at 12 p.m. a transition to the upload state occurs activating the process that measures actual production. The computation of optimal set points is deactivated at that time as well.

Studying process 3 further, the operator selects the mode of operation in which the optimal set points are computed. This computation requires historical process measurements, economic information and the required production. At the same time, the process provides a summary of its operation for system upload. The optimal set points are approached through a ramping function. Each ramp rate is selected so that it is larger than the fundamental period of the process. If the ramp rate is too quick, the process is likely to be disturbed, negating potential optimization benefits.

These ramped set points are used by the regulatory control system to compute valve positions. The focus of this chapter is on the computation of optimal set points. That calculation is detailed in Fig. 4. Note that beginning with a measured starting point and running to meet a defined production demand, a nonlinear programming package is utilized with an ethylene glycol process model.

The computer must recalibrate the model for changes from its steady-state operation, which can be done with the data utilized from the historical process measurements. This is the first use of the storage symbol. The two lines surrounding the name of the information to be stored indicate a time delay between when the information is entered and when it is recalled.

If a numerical criterion is not available to determine whether a given fitting is "good enough," then perhaps model calibration should be done manually. This could be diagrammed by taking the calibrated model out of the system to be automated and discussing how it would work with a model calibration system external to the optimizing program.

The optimization has several modes of operation, described in Fig. 4. In each, one or several of the data flows is activated. Further, the operator will have control of these modes of operation. This describes the operational aspects of the supervisory control scheme.

These user-recognizable "modes of operation" describe how the optimizing program would work for the actual day-to-day users of the system. It is represented

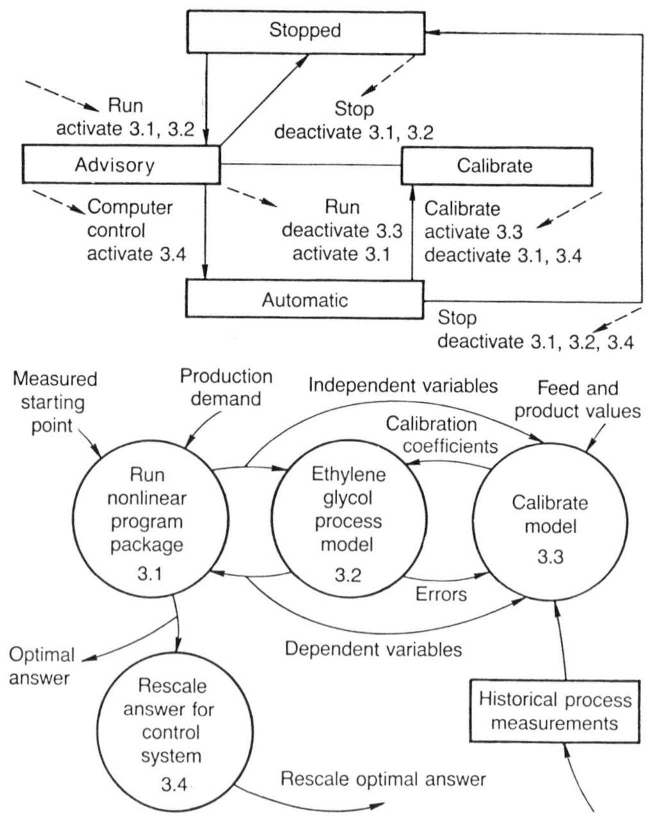

Figure 4. Computing optimal set points (3, Fig. 3).

as a state machine. Flaws in the scheme can be worked out with people who have experience in operating similar systems; a knowledge of the algorithms being used to compute the optimal answers is not required. This assures that the designed system will be under the control of operations personnel.

The ethylene glycol example contains four modes of operation: stopped, advisory, calibrate and automatic. As the algorithm moves from the stopped to the advisory mode, the package is run and answers are computed. However, the answers are not passed on to the control system until the automatic mode is invoked. This is shown by the activation of 3.4, rescale answer for control system, when the operator places the computer in control on the transfer from advisory to automatic.

In the calibrate mode, the calibrate model function is activated along with the ethylene glycol process model. The nonlinear programming package cannot run in this mode. This is shown by the activate and deactivate statements on the state transfer. The calibration function adjusts the calibration coefficients so that model errors are minimized.

The major difficulty in performing any optimizing project is determining what model to use. Figure 5 and Table 2 show how a given static model can be described using data flow diagrams.

Figure 5 breaks the problem down into the reactor, splitter and distillation train. It also shows the calculation of the optimization criterion, process profit, which is the basis upon which the nonlinear programming package chooses between the many possible selections of independent variables.

Table 1 shows that the reactor effluent has four components. The amount of each is determined by the ratio of water to ethylene glycol in the mixture, the amount of recycle, and the average residence time in the reactor. Table 1 also shows that the functions will be represented by polynomials and that the average residence time is proportional to the constant coefficient of the polynomial. Physically, this means that the curve is biased up and down by the residence time in the reactor.

With Figs. 2 through 5 all discussing the same system in increasing detail, the possibility of confusion exists in relating one level to the next. The data structure dictionary (Table 2) indicates how one level of diagram relates to the next. The = sign is read as "is"; the + sign is read as "and"; and the statements begun with asterisks are read as comments.

Following this discipline provides a way of precisely defining what a program should do. It can also be viewed as a language for describing problems precisely. For example, one could discuss what would be the correct model for the proposed ethylene glycol process with experts in reaction kinetics.

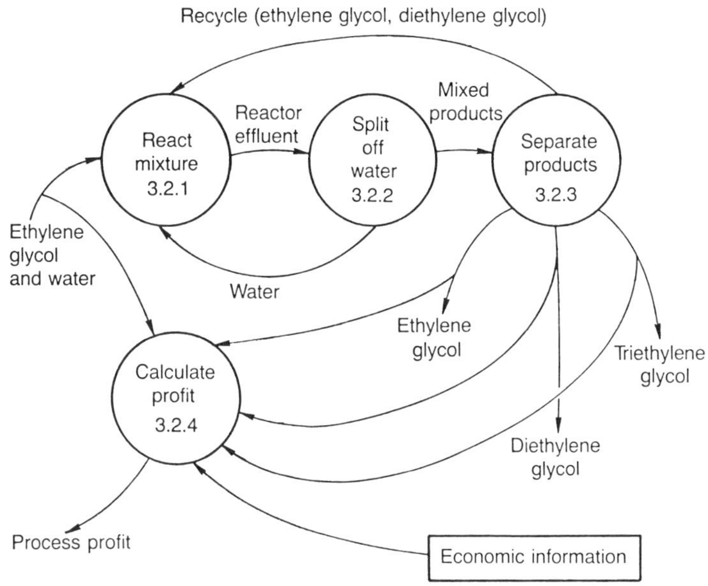

Figure 5. Static model of ethylene glycol process (3.2, Fig. 4).

Table 1
Specification of a Reactor Model which Determines Composition of Effluent (3.2.1, Fig. 5)

Reactor effluent	=	EG + DEG + TEG + Bottoms
EG	=	f_1 (EG/H_2O, recycle, average residence time)
DEG	=	f_2 (EG/H_2O, recycle, average residence time)
TEG	=	f_3 (EG/H_2O, recycle, average residence time)
Bottoms	=	f_4 (EG/H_2O, recycle, average residence time)
where:		
$f_j (X_1, X_2, X_3)$	=	$\sum_{k=1}^{2} \sum_{l=0}^{3} \text{coefficient}_{jlk} X_k^l$
and coefficient_{jok}	=	*Average resident time* constant
and where:		
EG	=	the amount (Vol 1 unit time) of ethylene glycol
DEG	=	the amount (Vol 1 unit time) of diethylene glycol
TEG	=	the amount (Vol 1 unit time) of triethylene glycol
Bottoms	=	the amount (Vol 1 unit time) of bottoms
Constant	=	an adjustable constant for model tuning

Other languages for describing requirements exist, but this language is useful because it is simple and easy to learn. The use of only five symbols (terminator, process, data flow, storage of information and state symbol) makes the diagrams easy to read.

The terminator is the square symbol used in Fig. 2, indicating the entities that are to provide and receive information. The storage of information is shown by two

Table 2
A Partial Data Structure Dictionary

Independent - variables	=	Ethylene glycol and water + water + recycle
	=	** Connection between Figs. 4 and 5
Operator-selected mode of operation	=	stop | run | computer-control | calibrate
	=	** Connection between Figs. 3 and 4
Recycle	=	Diethylene-glycol-recycle + ethylene-glycol-recycle
	=	** Further defines Fig. 5

lines, as in Fig. 4 (historical process measurements). The state symbols used in Figs. 3 and 4 indicate when processes are to be activated. Most other requirement languages are much more ornate.

ANALYSIS VS. DESIGN ISSUES

In the early phases of a project where functional diagramming occurs, several issues should be addressed. First and foremost is whether an optimization ought to be performed on this particular process. If, for example, a high volatility in the cost per unit feed or value per unit product exists, then an optimizing program may be of some assistance in determining the set points of a regulatory control scheme.

It also ought to be determined that the control scheme, as configured, can respond to changes in the supervisory control system. That is, that it can respond to the set point changes inherent in computing the optimal set points to the controllers.

Another level of concern is in the area of determining how "real time" the optimizer should be. It is possible that the optimizer can make use of real time or manually entered data. It is also possible for it to be plugged directly into the regulatory control system or to provide advisory information to operators.

Operational detail is what can make or break a given optimizer installation. This means the way that operations people will manipulate the supervisory calculation. In the ethylene glycol program, several modes of operation were defined. There is no standard solution to the problem; operations people can be consulted to determine what's best for them. The correct answer partially depends on how well the given model represents the dominant physics of the process being studied.

The major issues in analyzing the optimizing program will be the approach to calibrating the model, the nature of the static process model, and the operational character of the optimizing program. Most practical optimizing programs use simple models and make use of process measurements for calibration.

Models based on obscure physical phenomena are rarely useful; simple energy and material balances are all that is usually required. These balances can then be calibrated by actual measurements to show their validity. One must, however, contend with averaging out process dynamics before computing the mass and energy balances. Also, a physical properties package must be used to convert volumetric flow to mass flow. It is also needed for the computation of energy balances.

Design issues, by contrast, are concerned with putting a particular optimizer together in a particular computer. For example, it has been found that certain formulation strategies for nonlinear programming models help speed computation.[5,6] The particular strategy depends on the algorithms employed.

Other design issues include the selection of a commercially available code to perform the optimization and whether to use a linear programming package or a nonlinear programming code for the problem at hand. Linear codes can be used with nonlinear models through the process of model linearization. If a model were linearized, another issue is whether the errors from steady-state operation are acceptable. Having defined the problem using a structured analysis, it then becomes much easier to address these design concerns. Since using a structured requirements definition takes some of the magic out of supervisory control schemes, it is hoped that they can become a part of the maintenance of control systems design.

References

1. DeMarco, T., *Structured Analysis and System Specification,* New York, Yourdon Press, 1982.
2. Morrow, I. & Robinson, W. R., "Software Engineering for Real-Time Systems," *Proceedings* of ISA/83, Houston, Oct. 1983.
3. Morrow, I. & Robinson, W. R., "An Analysis Language for Real-Time Control Systems," *Proceedings* of 1983 American Automatic Control Conference, San Francisco, June 1983.
4. Ward, P. & Campbell, B. D., "Requirements Definition for Process Control Systems," *Proceedings* of ISA/83, Houston, Oct. 1983.
5. Campbell, B. D., "Grasping the Concepts Behind Optimization," *Control Engineering,* November 1979.
6. Gill, T. E., Murray, W., Saunders, M. A., Wright, M. H., "Aspects of mathematical modeling related to optimization," *Applied Mathematical Modeling,* April 1981, pp. 71–83.
7. Epperly, A., Personal communication, Sept. 1983.
8. Gunkler, A., Personal notes, Jan. 1984.
9. DeMarco, T., Controlling Software Projects, New York, Yourdon Press, 1984.

58

Uncontrollable Processes and What To Do About Them

F. G. Shinskey

I have had the responsibility for commissioning and evaluating control systems on many processes that proved to be uncontrollable. Typically, operators and process engineers place the blame for control failures on the control system rather than on the process, which is probably correct more often than not. However, in situations where the process simply cannot be controlled without changing its design, the burden of proof falls upon the control engineer.

Solutions to these problems range from additional controls to relatively minor changes in process piping. However, in certain cases, the only remedy may be a decrease in production rate or other major departure in operating conditions.

Several factors may contribute to uncontrollability.

NEGATIVE RESISTANCE

If a process has no steady state stability, it cannot be controlled. Unless equilibrium exists at the operating point, control is impossible regardless of the dynamic characteristics of the process. Steady-state instability is identified by negative resistance. In an electrical component, this property would cause current to fall as voltage is raised; negative resistance is often used to make an oscillator.

Negative resistance is a transitional regime between two stable operating modes. For example, a pool-boiling curve (Fig. 1) is characterized by two stable modes separated by a transitional mode of negative resistance.

At low heat transfer rates, nuclear boiling exists, where vapor is generated at many individual sites on the heat transfer surface. As the driving temperature difference is raised, more sites develop and bubbles expand. Eventually, vapor generation can increase enough to blanket the heat transfer surface with a vapor film having a low coefficient. This is the film-boiling regime, identified by the lower positive slope. Between these two stable modes, heat flux falls as temperature rises: negative resistance. Heat flux cannot remain in this transitional regime but will drop from the peak reached in the nucleate regime to the valley in the film-boiling regime.

If an excessively hot heat transfer medium is provided for a reboiler, the transitional regime could be entered during an upset condition. Then once film-boiling has become established, the system will be locked in that mode until the controls are transferred to manual and the temperature difference is reduced enough to restore nucleate boiling.

Negative resistance can also be found in steam-jet ejectors (Fig. 2). Normally, pumping rate will increase with suction pressure. When the negative resistance region is entered, the slope reverses sign and control becomes impossible.

The most common means of pressure control is the admission of air or gas into the ejector suction. Where resistance is positive, opening the control valve increases the load on the ejector and therefore raises pressure proportionately. However, when the slope becomes negative, pumping rate will fall so that pressure rises spontaneously. Since there is no way to position the control valve to counteract this tendency, pressure will continue

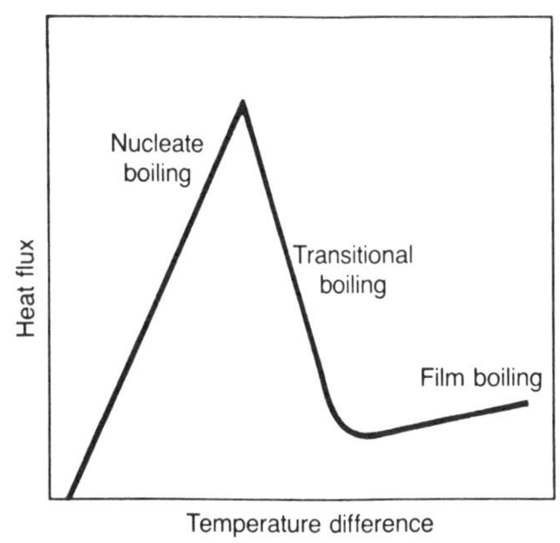

Figure 1. Pool-boiling curve.

to rise to the point where pumping rate again matches the flow admitted by the valve. Any attempt to control pressure within the negative resistance region will result in limit cycling between the two adjacent stable regimes, as traced by the dashed lines in Fig. 2.

The only way to solve this problem is to alter the characteristics of the jet either by changing steam pressure or nozzle size, or by rearranging a configuration of multiple jets. It is likely to be encountered when suction pressure must be controlled over a very wide range.

EXOTHERMIC REACTIONS

An exothermic reaction has the potential for instability in that its rate of heat evolution increases with temperature. For a continuous reactor, this plots as the S-shaped curve in Fig. 3. Heat transfer, however, is a self-regulating process and is capable of stabilizing a reactor. The same increase in reaction temperature that raises heat evolution will also raise heat transfer to the cooling system.

If an increase in reaction temperature raises heat transfer rate more than heat evolution, temperature will fall and a stable equilibrium will be established. The straight line in Fig. 3, however, describes a cooling system with insufficient heat transfer surface to bring about stability. Here, an increase in temperature above the control point will result in a greater rise in heat release than in heat transfer, causing temperature to increase further. The control point then represents an *unstable* equilibrium. Reactor temperature will tend to rise to the stable intersection near 100% conversion or fall to the other stable intersection near zero conversion.

It is possible to control a reactor having this steady-state instability if its dynamic properties are favorable. In essence, the reactor can be stabilized if enough negative feedback can be applied by a temperature controller to overcome the positive feedback of the exotherm.[1] But the gain of the temperature controller is limited by the ratio of the capacitive time-constant of the reactor to its dead time. If the reactor is dominated by dead time (as is common with longitudinal, plug-flow processes), it may be uncontrollable. The temperature will cycle in a saw-tooth pattern between the lower and upper stable intersections of Fig. 3, regardless of the settings of the temperature controller.

This problem is not easy to correct. To increase the slope of the heat transfer line requires a larger surface area or coefficient; boiling water systems are recommended for reactor cooling, since they are more isothermal than sensible-heat cooling. If such alterations are not possible, production rate must be reduced to the point where the slope of the heat evolution curve matches that of the heat transfer line.

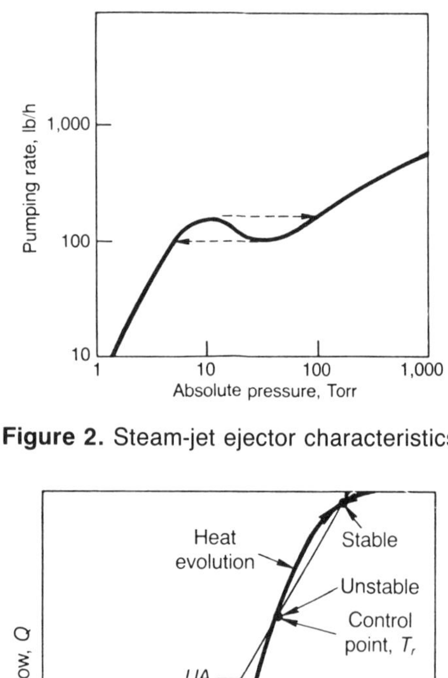

Figure 2. Steam-jet ejector characteristics.

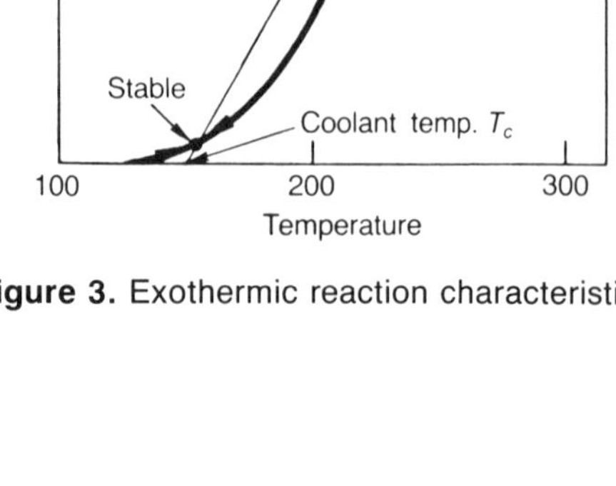

Figure 3. Exothermic reaction characteristics.

STIRRED TANK REACTORS

A stirred tank reactor such as that shown in Fig. 4 typically has sufficient heat capacity to permit a high gain in its temperature controller. If it should be unstable without control, it nonetheless may still be controllable.

However, the performance of the controls depends on uniform velocities on both sides of the heat transfer surface. Inside the vessel, this is achieved by agitation. But uniform velocity is also necessary through the jacket, and this can only be provided by a circulating pump. Without circulation, coolant velocity will be low during times of low heat transfer and high at high rates, causing the dead time in the control loop to vary.

When dead time varies with the output of the controller (coolant flow in this case), a nonlinear cycle is developed. A rising temperature will cause the controller to increase flow, making the temperature drop sharply. However, a falling temperature will cause flow to be reduced and response to be retarded. A cycle of sharp peaks and broad valleys is developed that cannot be

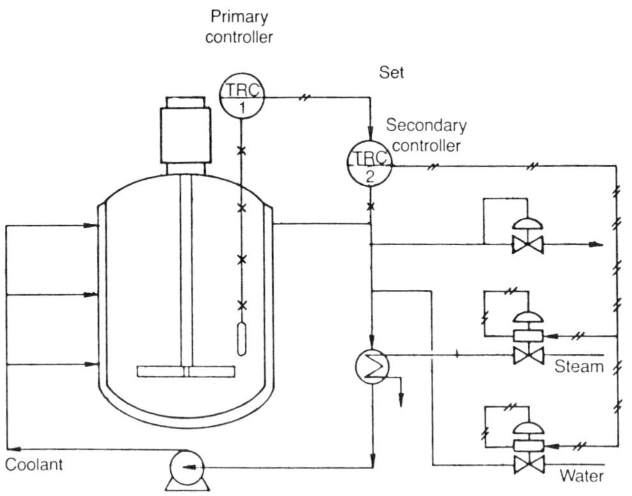

Figure 4. Stirred tank reactor.

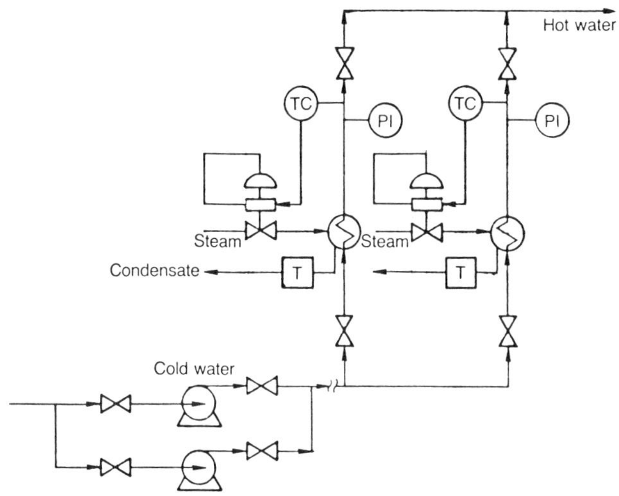

Figure 5. Parallel heat transfer system.

dampened and whose period varies inversely with heat load. A circulating pump solves the problem. In addition, cascading reactor temperature to coolant *outlet* temperature (as shown) greatly improves dynamic response.

PARALLEL UNITS

Instability is often caused by interaction between parallel unit operations. In the heat transfer system of Fig. 5, water is raised to 99°C to be used for sterilizing purposes. In the event of an upset, however, the discharge temperatures would begin a cycle that could not be stopped without placing *both* controllers in manual and closing both steam valves.

Again, controllers were blamed for causing the cycles, but the culprit was the process itself. Whenever a momentary reduction in water flow caused one temperature to rise above 100°C, a vapor bubble would form. Its expansion would develop a back pressure, reducing water flow and forcing the balance into the other exchanger. This decreased the load on the first exchanger even more, further promoting boiling. Eventually, when all the liquid was expelled, pressure drop across the first exchanger would fall. This would draw liquid from the second exchanger, cooling the first and overheating the second. Then a vapor bubble would develop in the second exchanger, initiating the second half of the cycle.

The two exchangers would continue to cycle out of phase until an operator intervened. The period of the cycle was about 20 seconds, the same as the period for each individual temperature loop. This interaction can be overcome by forcing the flows to stay equal. A single flow controller with one water flow signal as measurement and the other as set point could do this by manipulating a single three-way valve combining the streams that leave the two exchangers. Alternately, each exchanger could have its own feed pump.

A similar cycle was observed in parallel sulfur dioxide scrubbers where pH controllers were drawing lime slurry from a single, unregulated source. An increase in lime demand in one scrubber would rob reagent from the other, developing a sympathetic cycle. It could be eliminated by the use of separate lime pumps or by installing a cascade flow controller on each supply from the single pump.

INVERSE RESPONSE

Inverse response is a dynamic phenomenon caused by concurrent reactions having opposite signs. One reaction to an input will have a steady-state component and a delay or time lag, while the other will have a time lead of the opposite sign. The combination gives a response that starts in one direction but later reverses direction (Fig. 6).

Inverse response is common to processes having two-phase flow with variable vapor content. Consider a distillation column with a forced circulation reboiler in which partial vaporization occurs. An increase in heat transfer will generate more vapor, which displaces liquid from the vapor return line into the base of the column. The displaced liquid raises the level in the base of the column, but the increased vaporization ultimately draws the liquid inventory downward. The time elapsed between the change in the heat input and the crossing of the original position of the base level is the inversion time, roughly equivalent to dead time in its effect on a control loop.[2]

Inverse response can also occur within the column itself. Each tray contains a mixture of liquid and vapor.

An increase in vapor flow can displace some liquid from each tray, causing it to overflow to the tray below. Thus, each tray will begin to spill some of its liquid inventory on an increase in boilup, although the effect is distributed across all the trays. The excess from the first tray is immediately felt at the base. However, that of the others is successively delayed by the transport hydraulics. Consequently, the overall response may not invert completely but simply appear as a dead time equal to the transport time of liquid from the top of the column to the base.

Inverse response is not observed in columns with sieve trays operating fully loaded. Because their pressure drop tends to rise steeply with vapor flow in this regime, a change in tray inventory is accommodated by an opposite change in downcomer inventory. This leaves base inventory constant. At lower vapor rates, there is less change in pressure drop with boilup, and inverse response may be observed. Because valve trays regulate their pressure drop, inverse response is common to them throughout their throttling range. It is enough to prevent base level from being controlled by heat input.

Inverse response is also found in the drum-level loop of a steam boiler, again because of partial vaporization. Adding cold feedwater to the drum can collapse steam bubbles, causing level in the drum to fall at the same time that liquid inventory is rising. This makes liquid-level control especially difficult and has brought on the need for feedforward (three-element) controls.

SUPPLY SIDE CAPACITY

The worst place for a capacitive time lag in a control loop is at the manipulated variable. Yet one is often found in pH control systems, which already have enough problems. Consider the lime-feeding system shown in Fig. 7. The flow of dry lime is manipulated into a slurry tank, which overflows into the neutralization vessel. As built, the system is typically supplied with a constant flow of water so that slurry concentration must be raised or lowered to change lime flow into the neutralization vessel.

If the pH is too low, the controller will increase feeder speed, but response is delayed while the lime concentration rises. Similarly, when the pH is too high, the controller will stop the feeder, but concentrated slurry will continue to overflow until the lime is depleted. This causes very slow control that cannot keep up with even moderate load changes.

This situation may be corrected by installing a control valve on the water line to the slurry tank. Then the pH controller can change the overflow rate directly, so that the concentration of the slurry remains virtually constant. The performance of the system with and without the water valve is shown in Fig. 8.

The waste being treated came from a metal-plating shop, which began discharge at about 6 a.m. At that time, the pH in the neutralization vessel fell sharply, and the controller began feeding lime. The upper record shows the controlled pH cycling between about 3 and 11, with a period of about an hour. Since the effluent discharged from the downstream settling tank violated the limits of pH 6 to 9, performance was unacceptable.

In the lower chart, water was fed in proportion to the lime. The pH controller had been left in manual until 6:30 a.m., but once recovered from the initial upset, it controlled with a cycle of about 7 minutes. Excursions at 9:30 and 11:30 were caused by reduced waste production during coffee and lunch breaks. Treatment plant effluent remained in the pH 6 to 9 range throughout.

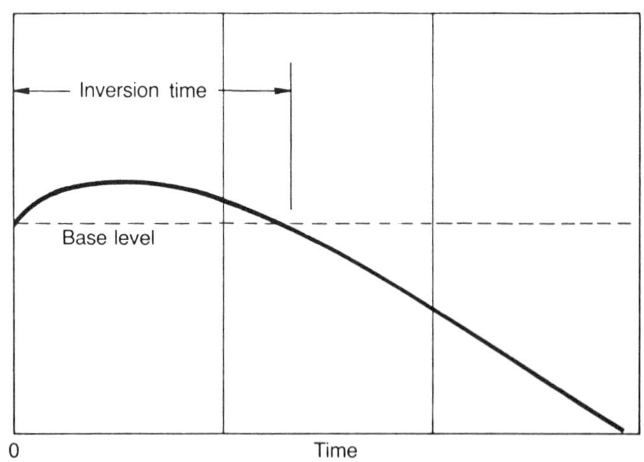

Figure 6. Inverse response in a distillation column.

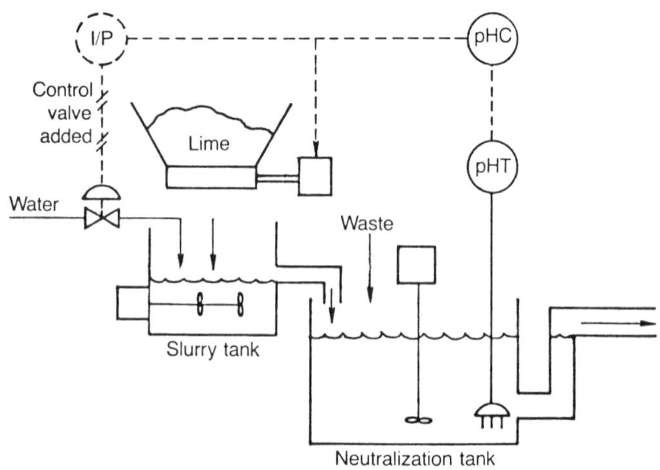

Figure 7. Lime feeding system.

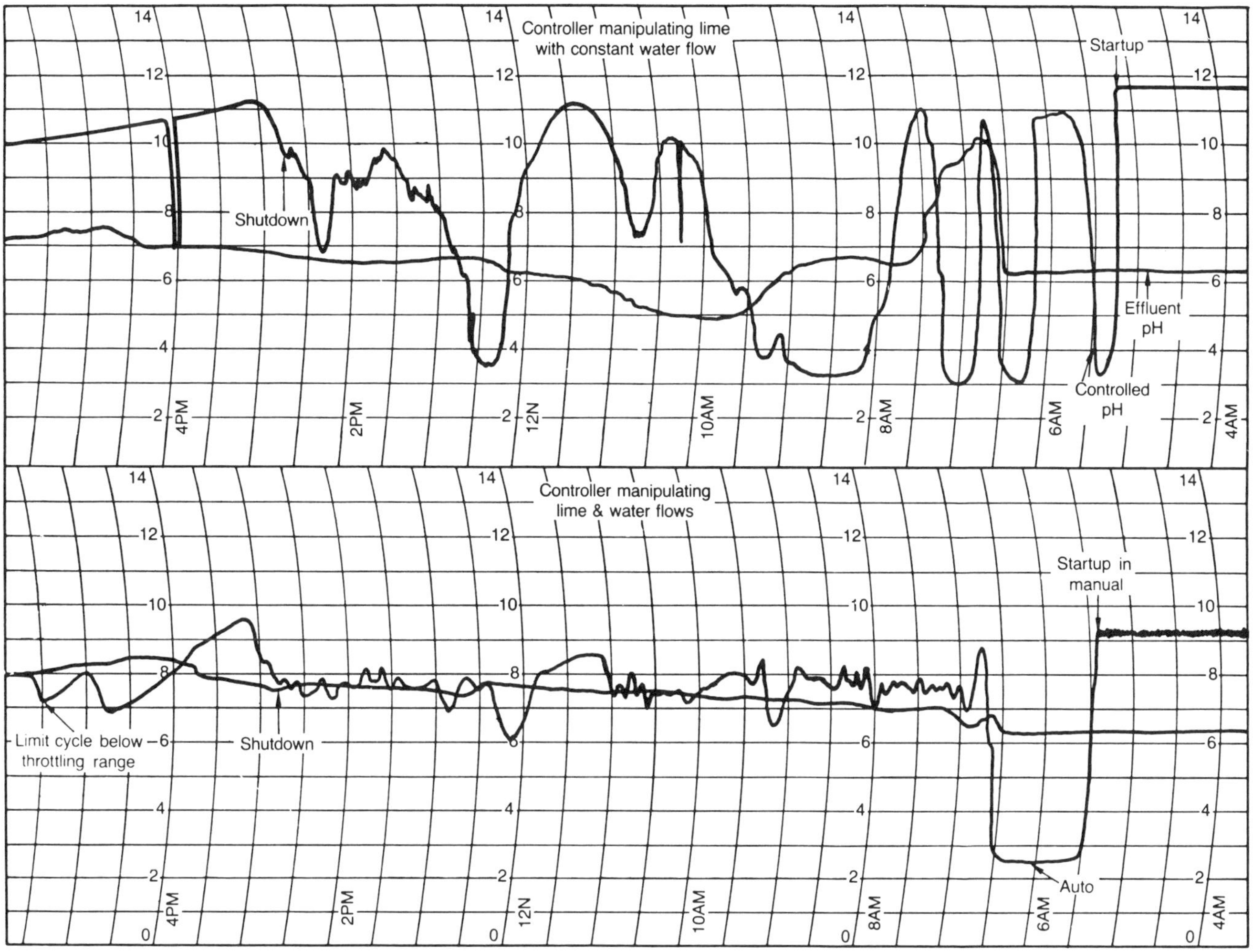

Figure 8. Performance of pH system with and without water valve.

Supply-side lags can also appear when liquid reagents are used if the delivery line downstream of the reagent control valve is allowed to fill and drain. This is especially common when delivering sulfuric acid, which because of its high concentration, tends to have low flow rates and high time constants. The delivery line should be kept full at all times by installing a loop seal at the discharge end.

CONCLUSION

While there are countless examples of faulty process designs causing control failures, those cited here seem to be the most common. In every case, it became the responsibility of the control engineer to correct or to work around the problem. A better approach would be to avoid the problem in the first place, by having the process designer either work with a control engineer or learn some process control himself.

References

1. Shimskey, F. G., *Process Control Systems,* 2nd Ed., pp. 253–257, McGraw-Hill Book Company, New York, N.Y., 1979.
2. Ibid, pp. 236, 237.

59

Avoiding Partial Optimization

G. L. Funk, C. C. Kania, and R. D. Terhune

For all situations an ethylene plant might encounter, there is no single "best" way to operate. Each new situation has a different "best" way to operate. In earlier times it was acceptable for plant operations and management to just keep on producing olefins. If you kept the plant running and periodically set new production records, management was called successful. This is not the case now. Production is not the supreme arbiter of success, efficiency now is taking its place along-side production. The "best" way to run an ethylene plant is now influenced by factors such as relative costs and prices, political policies, and feedstock availability. With these new factors comes the problem to plant operations and management of where and how to run the plant equipment so as to take full advantage of the changing marketplace. Even captive suppliers are feeling the need to respond to "efficiency and external factors" in plant operations. Being able to respond to a changing situation is valuable to management. Our experience indicates that optimal operation of a one billion pound per year ethylene plant is easily worth an additional $8 to $20 million (U.S.) per year (based on U.S. Gulf Coast values.) This benefit is derived from knowing where to operate and then shifting the plant operation in response to the factors mentioned earlier. The interactive nature of the major process units makes it necessary to examine the "TOTAL" plant to determine the best conditions.

The economic benefits due to optimizing (the operation and control) an entire ethylene plant are tremendous. This is true of all large industrial plants. "Optimization" and "Control" are two distinct features in the operation of a plant. To obtain maximum benefits both must be used and there definitely is a synergism between the two features. Each feature can be developed independently and the two individual benefits are less than the sum of the combined feature "Operation and Control."

Our development of a kinetic based furnace yield model has allowed us to simulate and study the interactions of entire ethylene plants for various clients. As a result of this extensive experience, we have discovered several facts that question the truth of several popular beliefs and common practices in the optimization of ethylene plants. We will demonstrate some of these "controversies" by using a single gas feedstock ethylene plant. However, similar problems with the popular beliefs exist for liquid feedstocks ethylene plants. We have found in practice and will demonstrate that:

1. The value for optimal operating conditions depends upon how much area of the plant is being included in the system, economic values, physical contraints and the specific optimization criteria. The benefits increase as more of the plant is included in the optimization.
2. Minimum energy does not ensure maximum profit.
3. Optimizing distinct areas of the plant and then combining the results is *not* equivalent to optimizing the "total plant."
4. The best feed composition for maximum profit is not the best ethylene producing feedstock.
5. False economic values lead to incorrect optimum conditions.
6. Minimum impurity in the recycle streams is not always good.

MODEL DESCRIPTION

Figure 1 is a simplified schematic of a typical ethylene plant. For the discussion we created a detailed model of a typical ethylene plant containing heat exchangers, flashes, stream mixers, stream dividers, fractionators, compressors, special equipment and cracking furnaces. The model contained over 66 process modules and 88 streams. The simulation uses Applied Automation's Proprietary Process Modeling System (APMS).

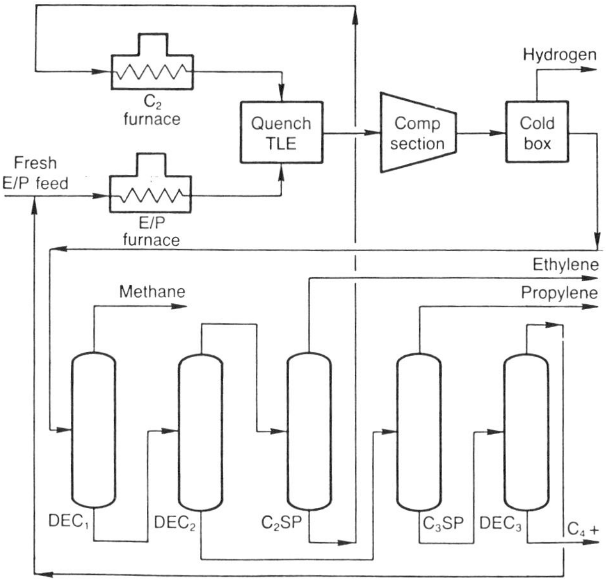

Figure 1. Typical ethylene plant (light feedstock).

Table 2 contains the typical constraints on the plant which were used in this model. Most are maximum constraint limits having to do with typical equipment limitations and product specifications. Table 1 contains the independent variables in the model. These were optimized in various combinations:

- Furnaces only
- Fractionators only
- Furnaces plus fractionators
- Furnaces plus fractionators plus feed

The four different combinations are also run for three different objective functions:

- Maximize profit
- Minimize production cost
- Minimize production cost per pound of olefin production

OPTIMIZATION RESULTS

The optimal operating conditions can be quite different from what is classically expected and the optimal operating conditions are continually changing. Tables 4 through 7 summarize the optimization results for the ethane-propane feedstock based ethylene plant shown in Fig. 1. The simulation model used 13 distinct hydrocarbon components to characterize each process stream. Table 8 summarizes the typical 1983 U.S. Gulf Coast prices used in generating economics for these examples.

Table 1
Major Independent Variables for Each Ethylene Plant

Conversion (severity) at each furnace
Steam-to-hydrocarbon in each furnace
Fractionator overhead separation (each fractionator)
Fractionator bottom separation (each fractionator)
Ethane-propane feed rate (or liquid rate to each plant)
Feed composition (or characterization)
Feed allocation to each furnace
Cracked gas compressor suction pressure (discharge pressure)
Hydrogenation temperature

Table 2
Major Typical Plant Constraints

Constraint	Type
Tube skin temperature	Max
Coil outlet temperature	Max
Furnace heat input limit	Max
Total cracked gas compressor horsepower	Max
Ethylene refrigeration limit	Max
Propylene refrigeration limit	Max
Fractionator condenser duty (for each fractionator)	Max
Fractionator reboiler duty (for each fractionator)	Max
Ethylene purity	Min
Propylene purity	Min
Total olefin production	Min
Ethylene production	Min, Max
Propylene production	Min, Max
Feed rate	Max

Three common different performance optimization functions have been used:

1. Maximum Profit = All Product Streams Values − all raw material costs − all utility costs + utility credits + credits for byproducts
2. Minimum "Cost of Production" = All raw material costs + utility costs − utility credits − credits for byproducts
3. Minimum "Cost of Production" per pound of olefins produced = (cost of production)/(Pounds of olefin product stream).

While these three criteria are similar, optimizing these three performance measures does not result in operating the plant at the same conditions. Many other good criteria for operating the plant exist. Tables 4 through 7 show that for the "total" plant, the feed rates, furnace operation, and fractionation train are all operated at different values for each of the three criteria.

Four different optimization cases were executed to show the affects of individual or partial equipment optimization compared to a total plant optimization. The four cases are as follows:

1. The fractionators' heavy key and light key were free to move over specified ranges subject to plant constraints, but the furnace conversions, steam to hydrocarbon ratio, EP feed rate and EP composition were fixed at the base case values.
2. The furnace conversions and steam to hydrocarbon ratio were free to move over specified ranges subject to plant constraint but the fractionators' heavy key and light key and the EP feed rate and composition were fixed at the base case values.
3. The furnace conversions, steam to hydrocarbon ratio, and the fractionators' heavy key and light key are free to move over specified ranges subject to plant constraints. The EP feed rate and composition were fixed at base case values.
4. The furnace conversions, steam to hydrocarbon ratios, fractionators' heavy key and light key, EP feed rate and EP composition were free to move over specified ranges subject to plant constraints.

Each of the four cases was run with the three performance criteria. The discussion of the performance criteria for each case are in the following sections.

MAXIMUM PROFIT OBJECTIVE FUNCTION

Tables 3–7 summarize the results obtained for running the four cases with the maximum profit objective function. The following observation can be made for those tables along with Fig. 2.

1. As more sections of the plant are optimized, the olefins profit increased from .0285 to .0402 $/lb.
2. Just optimizing the fractionator area by itself generated much less profit than other areas or combination of areas. However the fractionator optimization was still worth an additional $739,000 per year for a billion pound per year plant.
3. As more sections are optimized, the ratio of ethylene to propylene production increased. This is interesting when the fixed rate cases are compared since the total olefins rate remain nearly the same. The optimization shifted propylene production to ethylene production thus increasing this ratio.
4. When the fractionation area was added to the furnace area, it allowed the furnaces to generate more savings by moving the heavy key and light key to keep purities while the furnaces moved conversions. The combination meant more profits than just adding the two individual furnace and fractionation case profits. This synergist effect results in an additional 13.5% or $1.2 million per year more than the individual cases.
5. When compared to the results from the other two objective functions, the solution from optimization of the fractionator area does not give a minimum

Table 3
Profit Summary (Percent Change from Base Case in M $/yr) for the Objective Function

Objective function	Case A Fractionators	Case B Furnaces	Case C Fractionators & furnaces	Case D Fractionators furnaces & feed
Max profit, %	1.88	20.58	25.51	59.90
Min cost prod, %	1.32	2.90	4.16	14.73
Min cost prod/lb, %	0.42	3.42	4.19	7.28

Table 4
Furnace Summary

Objective function	Case B Furnaces only	Case C Furnaces & fractionators	Case D Fractionators & furnaces & feed
Max profit	C_2 conv 0.8000 EP conv 0.7859	C_2 conv 0.8000 EP conv 0.8104	C_2 conv 0.8000 EP conv 0.8430
Min cost prod	C_2 conv 0.8000 EP conv 0.7859	C_2 conv 0.8000 EP conv 0.8023	C_2 conv 0.8000 EP conv 0.8340
Min cost prod/lb	C_2 conv 0.8000 EP conv 0.7859	C_2 conv 0.8000 EP conv 0.8080	C_2 conv 0.8000 EP conv 0.8500

Table 5
Fractionator Summary

Tower	Objective function		Fractionator only	Furnaces & fractionators	Fractionators & furnaces & feed
DeC_1	$/lb	HK*	0.00720	0.00470	0.00590
		LK*	0.00022	0.00024	0.00021
	Min	HK	0.01000	0.01000	0.01000
	Cost	LK	0.00020	0.00021	0.00023
	Max	HK	0.00640	0.00540	0.00750
	Profit	LK	0.00020	0.00021	0.00025
DeC_2	$/lb	HK	0.00420	0.01100	0.00100
		LK	0.00056	0.00035	0.00014
	Min	HK	0.01000	0.00510	0.00830
	Cost	LK	0.00045	0.00064	0.00059
	Max	HK	0.00200	0.00100	0.00250
	Profit	LK	0.00058	0.00069	0.00010
C_2Sp	$/lb	HK	0.00010	0.00010	0.00010
		LK	0.00190	0.01200	0.01200
	Min	HK	0.00010	0.00010	0.00010
	Cost	LK	0.01500	0.01500	0.01500
	Max	HK	0.00010	0.00010	0.00010
	Profit	LK	0.00400	0.00140	0.00150

* HK = Heavy Key separation; LK = Light Key separation

Table 6
Feed Summary

Objective function	Feed Rate (% change)	C_3 Comp
Max profit	+ 19.49%	0.95
Min cost	− 4.39%	0.95
Min cost/lb ole	+ 17.90%	0.95

Table 7
Ethylene to Propylene Ratio

Objective Func.	Fractionators	Furnaces	Furnaces & fractionators	Fractionators & furnaces & feed
Max profit	1.876	1.999	2.108	2.139
Min prod cost	1.927	1.999	2.094	1.880
Min cost/lb	1.892	1.999	2.108	2.160

Table 8
Ethylene Unit Computer Control Economic Evaluation

Ethylene H.P.	19.3¢/lb
Propylene H.P.	18.3¢/lb
Propylene L.P.	17.3¢/lb
Butadiene mix	19.0¢/lb
Aromatic distillate	12.0¢/lb
Fuel	$3.9/10^6 Btu
Hydrogen	$3.9/10^6 Btu
FCC gas	$3.9/10^6 Btu
Ethane	9.3¢/lb
Propane	10.5¢/lb
Butane	10.8¢/lb
Electric power	$.09/kWh
Steam 800 #	$11.00/M #
600 #	$10.50/M #
150 #	$ 7.50/M #
50 #	$ 6.00/M #
Condensate	$ 1.00/M Gal
Cooling water	$ .10/M Gal

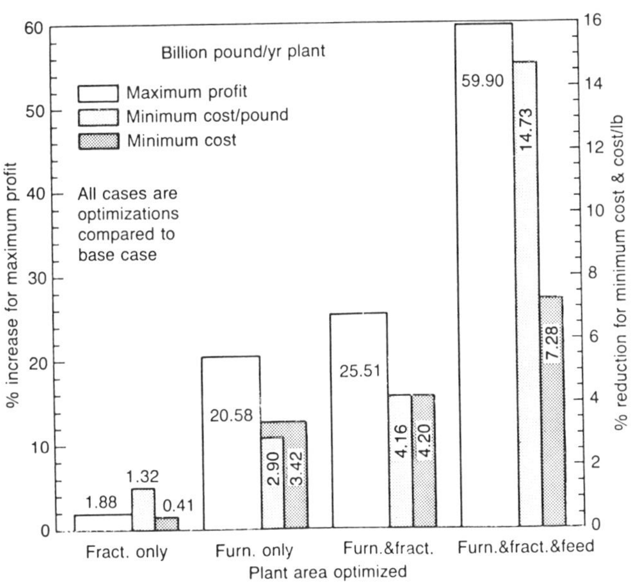

Figure 2. Comparison of benefits from partial and total optimization.

energy answer. This is due to the pricing structure of the ethylene and propylene products. Since the optimizer could make more profits by recycling ethane and propane to produce ethylene and propylene, the fractionators did not go to minimum energy. They found a point that balanced increased production with the increased energy consumption.

Figure 2 (and Table 3) shows the basic affect that adding more areas to the optimization would have. For a billion pounds per year plant, case A made 1.88% or $739,000 per year. Case B made 20.58% or $8.10 million per year above the base case. Case C made 25.51% more profit or $10.04 million per year. Case D made 59.90% additional profit or $23.61 million per year over the base case.

MINIMUM COST OBJECTIVE FUNCTION

The ethylene plant modeled was optimized using an objective function of minimum cost. This was achieved by giving ethylene and propylene products a *zero value.* Credit was given for byproducts such as butanes, aromatics and steam.

From Tables 3 through 7, the following observations can be made:

1. As more sections of the plant were included in the optimization, the greater the reduction in cost per pound of olefin. This would be expected since the more degrees of freedom the optimizer has normally the better the answer.
2. The ratio of ethylene to propylene increased for each case with the exception of Case D. With a fixed feed (Case A, B and C) the optimizer favored production of ethylene over propylene.
3. When feed rate and composition were allowed to move in Case D, the composition shifted to 95% propane with heavier cracking in the furnaces. This combination took advantage of the by-product pricing given to the C_4^+ compounds.
4. The greatest variation in the fractionator optimal separation values was in the deethanizer. This was a key column since it made the split between ethane/ethylene overhead and the propane/propylene bottoms.

Each selected a slightly different optimum. Case A operated the fractionators in a classical minimum energy solution and shifted splits on the columns to produce more specification ethylene. Case B operated the ethane furnace at the maximum allowed conversion. The mixed-feed furnace was operated to mid-range conversion. Case C was able to increase the ethylene production due to the dual freedom of increasing the mixed-feed furnace conversion and shifting the column specifications. Case D operated in a different manner.

The ethane-propane feed rate was decreased to the minimum allowed value. The feed composition was also increased to maximum allowed propane. The mixed-feed conversion increased until the minimum propylene product limit stopped the increase. The columns were operated close to the minimum energy solution used in Case A. Everything Case D did was designed to decrease operating costs and increase profit by taking advantage of byproduct values.

As shown in Fig. 2 based on a billion pound per year ethylene plant, Case A saved 1.32% or $798,000 per year, Case B saved 2.90% or $1.747 million per year, Case C saved 4.161% or $2.507 million per year, and Case D saved 14.731% or $8.877 million per year. Table 3 contains the economic benefit summary.

MINIMUM COST OF PRODUCTION PER POUND OF OLEFIN

Examining Tables 3 through 7 and Fig. 2 shows that for the same optimizing objective:

1. The "optimal" operating conditions change as the scope of the optimization increases from one area to the entire plant. Optimal conversion on the mixed feed furnace varies from 78.6% to 85% depending upon how much of the plant is optimized (Case A–D). Similarly the separation on the demethanizer, deethanizer, and C_2 splitter are sensitive to what part of the plant is optimized.
2. When just the fractionators are optimized (Case A), the ethylene recycled in the ethane recycle stream is decreased from 1.14% to 0.19%, but if the whole plant is considered the optimal amount of ethylene in the recycle is raised to 1.23%.
3. The benefit received increases as more and more of the plant is included in the optimization. Optimizing just the fractionators, Case A, results in reducing the cost by 0.41%, while considering the whole plant results in reducing the cost by 4.2% (a fixed feed rate and composition). This variation represents a cost savings of between $910,000/year and $9,400,000/year.
4. If there is a market for the additional olefins, then Case a D shows the cost per pound of olefins can be reduced by 7.3%. This represents a savings of $11.7 million on the billion pound per year plant.
5. The benefits from optimizing the furnace and the fractionators simultaneously exceed the benefit from optimizing them separately. The additional savings due the furnaces and fractionators amounts to about 0.37% of the base case cost per pound. For a billion pound per year ethylene plant this is an added savings of $880,000 per year.
6. Maximum conversion on the furnaces is not always an optimal policy.
7. The pricing shows that the value of the heavy byproducts is greater than the cost of feedstock and cost of utilities. Thus the cost per pound of olefins can be reduced by increasing feed rate up to a constraint (ethylene refrigeration).
8. As the price of the byproducts is reduced the optimal operating conditions (furnaces, fractionation, feed) all change. Therefore optimal plant operation depends upon relative prices. For example, with no credit for by-product butane and heavier compounds, the optimal conditions go to the minimum feed flow to meet production, only 72% conversion in the mixed feed furnace and 60% conversion in the ethane furnace.

As a general rule, the furnace conversions were increased as more of the plant was included in the optimization. Typically better operation resulted in higher ethylene to propylene ratio even though zero credit was given for olefin production. This is due to higher conversions yielding both more olefins and more by-product heavy hydrocarbons.

It is easy to understand why the optimum shifts for some situations while for the other cases it is not obvious. We have seen that a minimum cost criterion seeks a minimum feed rate, and a maximum profit criterion seeks a maximum feed rate, and a minimum cost per pound of olefin seeks a high feed rate dependent upon the value of by-products. Similarly the best operating conditions vary considerably. For a single plant and a single optimizing criterion, optimal operation of a fractionator can call for shifting of the separation from 0.002 to 0.012 (a change of 600%) and moving a furnace conversion level from 78.6% to 81% (a change of over 3%) when there is a change in the plant units involved. These are significant changes.

Generalizations are dangerous. For this plant and economics, optimization favors using propane as a feedstock and not a mixture of propane and ethane. This is not always true. At a fixed feed rate and composition, using optimal minimum fractionation energy set points when trying to produce at a minimum cost per pound of olefin results in a benefit reduced by $0.0003 per pound. This benefit is $417,000 per year less than what it could have been had the fractionator been operated for optimal minimum cost per pound. Similarly, using minimum cost set point for the fractionator results in $1.9 millions per year less profits when it is desired to actually maximize profits.

It is commonly believed that optimizing different parts of a plant and then combining the results "optimal" set points will result in a truly optimal total plant operation. There is a fallacy in this kind of thinking. There is always a positive synergistic effect as more independent variables are introduced to the optimization. As more and more of the plant is included in the scope of the optimizer, the final performance index can only be equal to or greater than the sum of the performance indices for the individual plant areas. The spread between the total performance index and the sum of the individual performance indices is a measure of the synergism (interaction) between the plant area with respect to the performance index. The value varies. For example, comparing maximum profit at a fixed feed rate for Case C versus the sum of cases A and B, we see Case C results in a gain of three percent in additional profit. Comparing the minimum cost of production cases shows the sums are practically equal (sum of cases A and B versus Case C).

In this chapter we have demonstrated that there are serious pitfalls to considering an ethylene plant "optimized" when only parts of the plant have been optimized and the results combined. Practice confirms these pitfalls. A true optimal set of operating conditions is dependent upon (1) the total equipment of the plant, (2) what each plant area is doing, (3) the performance, economic, and market conditions surrounding the plant, and (4) what is the criterion being used. Thus, even for the exact same situation the optimal operating conditions changes as (1) the performance criterion shifts (Case D, all three criteria) and as (2) the amounts of the plant being considered (Cases A, B, C and D) at any given criterion.

60

Dynamic Simulation of Cascade Control*

R. O. Brantley, K. W. Leffew, and P. B. Deshpande

Dynamic simulation can often serve as a very useful tool in assessing potential improvements with cascade control. While there are some guidelines available for assessing potential benefits of cascade control in specific applications, they are not complete. This chapter presents an application of dynamic simulation to a cascade control system. Based on results of dynamic simulations, cascade control is implemented on an experimental system. Closed-loop response of the experimental system with disturbances in the manipulated variable shows substantial improvement and thus, justify dynamic simulation in assessment of potential benefits of cascade control. Also presented is a commercial application of the cascade control technique.

A schematic of the process unit for this study is shown in Fig. 1. The control loops shown are typical of those found in the process industry. The objective of the feedback control system is to maintain the liquid temperature at set point, but because of frequent disturbances in the manipulated variable (steam supply-pressure upsets in this example), this is not possible. Such disturbances frequently occur in industrial plants having many process units each requiring varying amounts of heating or cooling. A sudden change in demand from one or more of the units can cause disturbances in many controlled variables. Of course, the feedback control system eventually brings the controlled variables back to their set points, but in the meantime the process is perturbed. If the disturbances are frequent, the controlled variables do not remain at set point very long.

Cascade control can frequently improve the response of a feedback control system which is subject to disturbances in the manipulated variable. The schematic of the process unit with the cascade control and arrangement is shown in Fig. 2. With cascade control, distur-

* The authors thank E.I. duPont de Nemours and company for their permission to publish this work.

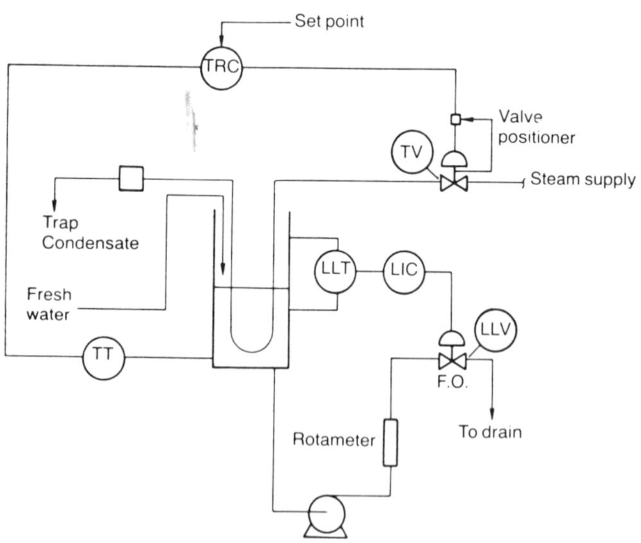

Figure 1. Typical feedback control system.

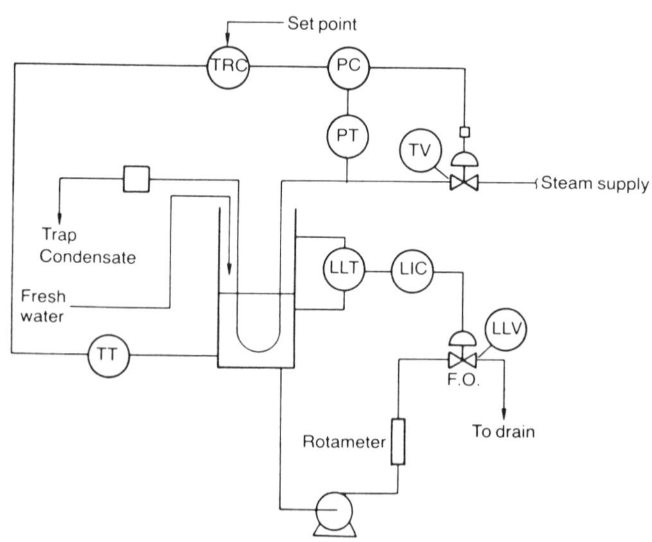

Figure 2. Cascade control of temperature.

bances in the manipulated variable are corrected by the action of the pressure controller before they upset the controlled variable; temperature.

Intuitively, it should be clear that the inner loop must be at least as fast as the outer loop if the cascade control system is to be effective. However, quantitative information is needed to properly assess potential benefits of cascade control. Franks and Worley[1] presented a design graph (Fig. 3) which shows improvement possible with cascade control for certain types of processes. Fig. 3 does not apply to processes that exhibit apparent deadtime characteristics.

Dynamic simulation has proven useful in design and analysis of process control systems. The simulation technique can be applied to a control system to find out if cascade control is likely to improve response of the feedback system. This article presents the results of a simulation study applied to the thermal process in Fig. 1. Results of the simulation are compared with the experimental response of the same system. Also presented

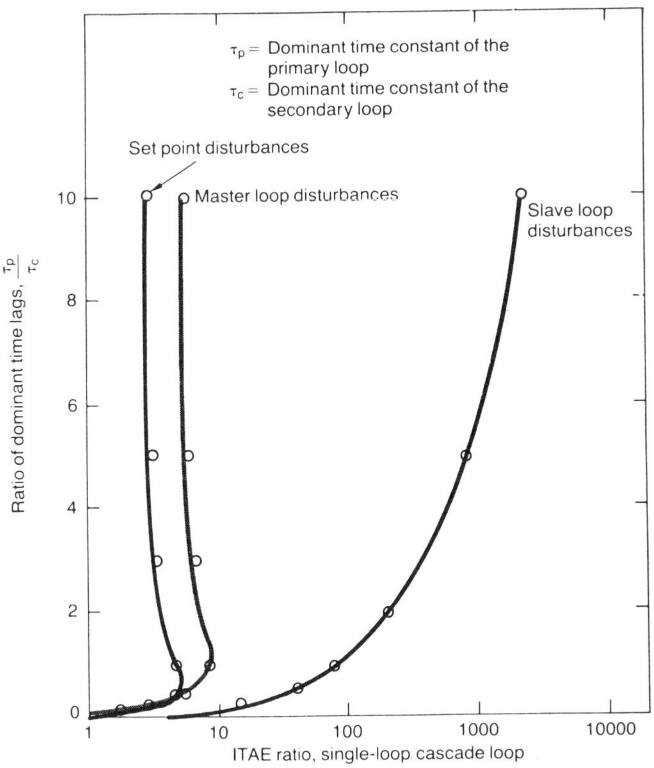

Notes: 1. ITAE $= \int_0^t t\,|e|\,dt$
where: E = error = set point-measurement
2. Primary process, $G_{p_2}(s) = \dfrac{1}{(\tau_a s + 1)(0.1\,\tau_a s + 1)}$
where: τ_a = 0, 2, 5, 10, 30, 50 and 100
3. Secondary process, $G_{p_1}(s) = \dfrac{1}{(10s + 1)(s + 1)^2}$

Figure 3. Relative performance of cascade and single loop systems.

is a commercial application of cascade control to an exothermic polymerization reactor.

DYNAMIC SIMULATION STUDIES

The dynamic mathematical model of the thermal process is presented in the Appendix. Development of the model is based on the following assumptions.

- The process is described by a set of linearized equations. Since in control work we are usually interested in the behavior of the system in the vicinity of some steady-state, this approximation is usually valid. Use of linearized equations facilitates their solution by Laplace transforms.
- Dynamics of the level loop are much faster than those of the temperature loop. Thus, the liquid level remains constant at all times.
- Temperature of the liquid in the stirred tank is uniform.

The system equations can be combined to obtain the block diagrams of the feedback and cascade control systems as shown in Figs. 4a and b. The equations are solved on a digital computer using the CSMP language[2]. Figure 5 shows response of the simulated feedback and cascade systems to a disturbance in the manipulated variable. Figure 6 shows response of the two systems to changes in set point.

Figure 5 shows that cascade control has resulted in substantial improvement in dynamic response with disturbances in the manipulated variable. Figure 6 shows some improvement in set-point response with cascade control.

The simulation work clearly favors installation of a cascade control system for temperature control. Closed-loop response of the experimental system shown in Fig. 1 with and without cascade control is shown in Fig. 7. This figure shows that with disturbances in the manipulated variable, the cascade system fares better than the feedback system.

Comparison of the simulation results (Fig. 5) and the experimental response (Fig. 7) shows that the response curves are not identical. Differences arise due to the nonlinearities of the system. However, the useful role of dynamic simulation in assessing potential application for cascade control is clearly demonstrated.

COMMERCIAL APPLICATION

To further demonstrate the benefits of cascade control, we present a commercial application of a cascade system to temperature control of an exothermic polymerization reactor[3]. The block diagram of the reactor

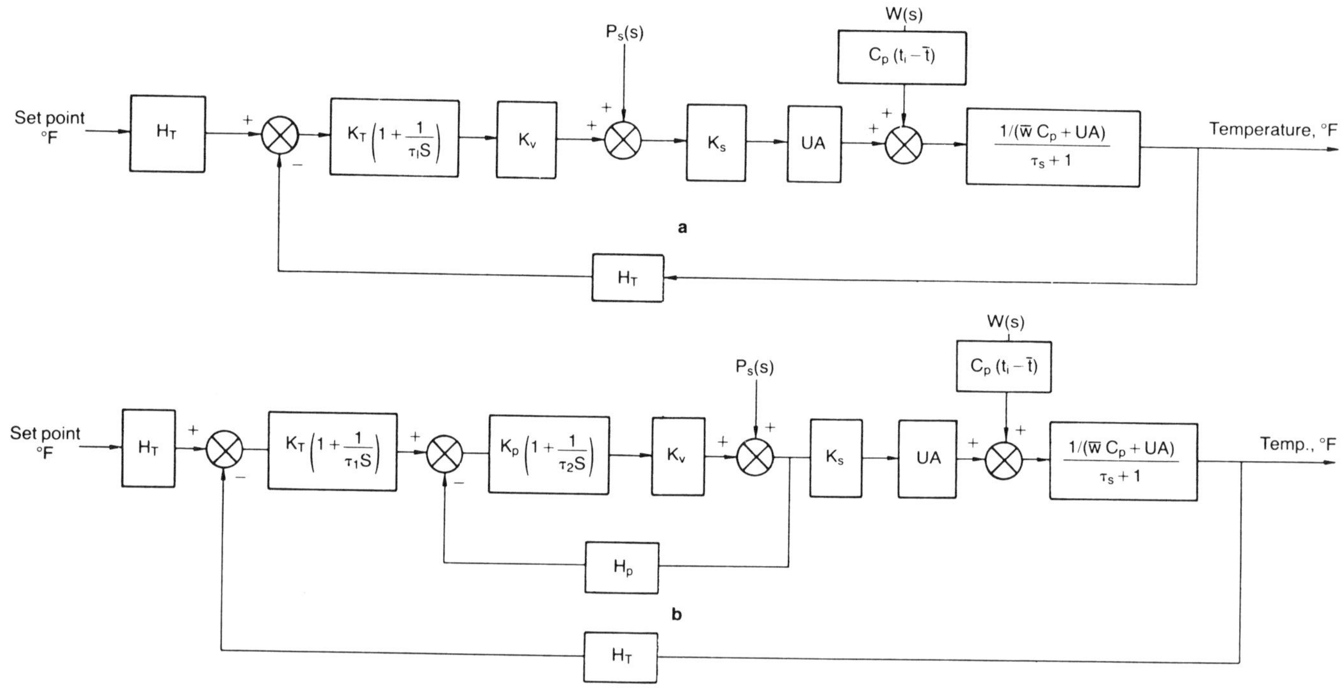

Figure 4. Block diagrams of feedback and cascade control loops.

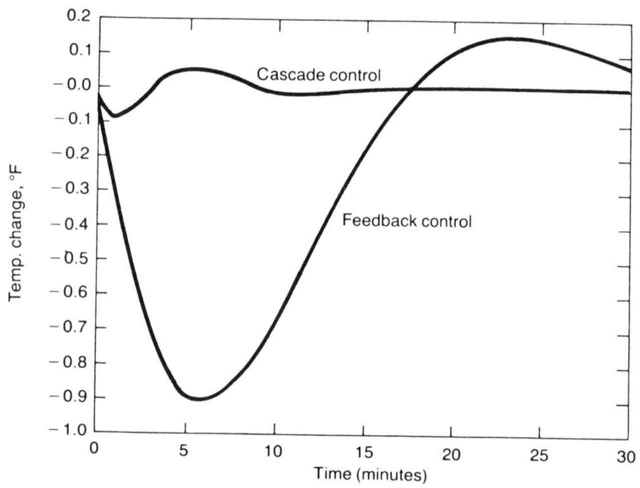

Figure 5. Response of simulated systems to 10 psi drop in steam supply pressure.

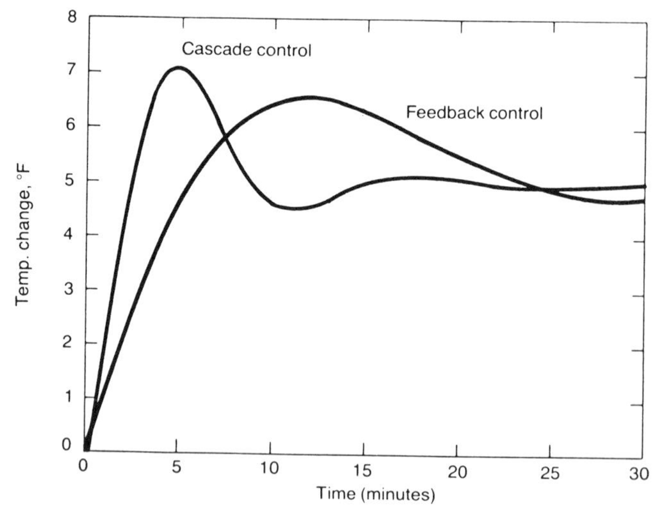

Figure 6. Response of simulated systems to 5 psi increase in set point.

feedback control system is shown in Fig. 8. Closed-loop response of the feedback control system (Fig. 9) reveals that variations in the manipulated variable may be causing deterioration in response of the reactor temperature. Thus, the system is ideal for exploring cascade control for improving response.

The block diagram of the cascade control system for temperature control is shown in Fig. 10. Closed-loop response of the cascade control system is shown in Fig. 11. This figure shows a substantial improvement in response with cascade control.

APPENDIX—DEVELOPMENT OF THE DYNAMIC MODEL

The dynamic mathematical model of the thermal process shown in Figs. 1 and 2 is based on the following unsteady-state energy balance:

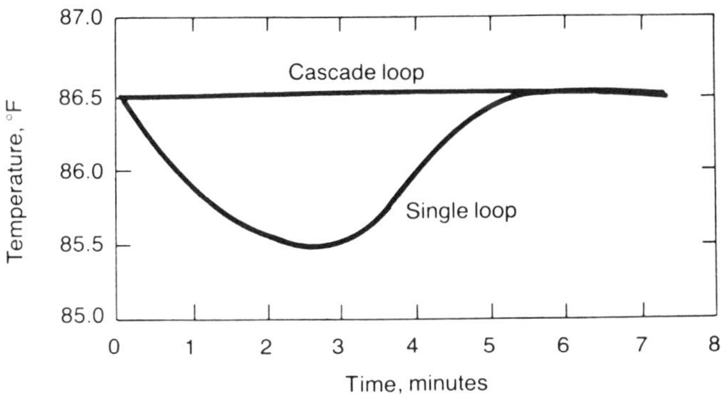

Figure 7. Response of experimental system to 10 psi drop in steam supply pressure.

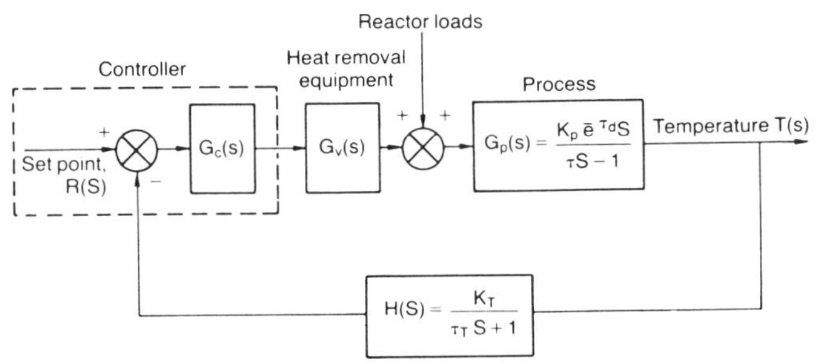

Figure 8. Block diagram of reactor control system.

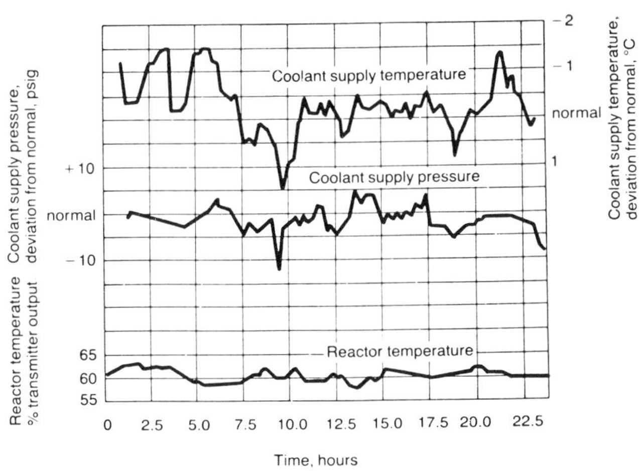

Figure 9. Response of reactor feedback control system.

Energy added due to flow and heat transfer	−	Energy removed due to flow	=	Accumulation of energy in the tank

$$wc_p(t_i - t_r) + UA(t_s - t) - wc_p(t - t_r) = dMc_pt/d\theta \quad (1)$$

As usual, this equation is linearized then expressed in terms of deviation variables and finally Laplace transformed to give:

$$T(s) = \frac{c_p(t_i - \bar{t})}{UA + \overline{w}c_p}\,\frac{W(s)}{\tau s + 1} + \frac{UA}{UA + \overline{w}c_p} + \frac{T_s(s)}{\tau s + 1} \quad (2)$$

where τ = time constant, $Mc_p/(UA + c_p\overline{w})$
$= 154.5/(4.92 + 51.5) = 2.7$ min.

Equation 2 relates the liquid temperature to steam temperature and water flow.

The temperature transmitter has an input range of 0 to 200°F and an output range of 3 to 15 psig. Thus, gain of the transmitter is

$$H_T = (15 - 3)/200 = 0.06 \text{ psi/°F} \quad (3)$$

The temperature controller is the proportional plus integral type, and may be described by the equation:

$$\text{c.o.} = K_TE(s) + (K_TE(s)/\tau_Is) \quad (4)$$

The controller tuning constants used in this study are:

$$K_T = 37.5 \text{ psi/psi}$$
$$\tau_I = 4.76 \text{ minutes} \quad (5)$$

In the feedback control system the temperature controller output is the air-top pressure to the steam control valve. Experimental data on this process suggests the following relationship between the air-top pressure and coil-side steam pressure (at a supply pressure of 50 psig):

Coil-side pressure = 2.65 times air-top pressure in psig − 10.99

Therefore, the valve gain is:

$$K_V = 2.65 \text{ psi/psi} \quad (6)$$

Coil-side pressure is related to steam temperature through the following experimental relationship:

Steam temperature, °F + 1.024 times coil-side pressure in psia + 245.4

Therefore, gain relating coil-side pressure to steam temperature is:

$$K_s = 1.024\text{°F/psi} \quad (7)$$

Steady-state operating data with the experimental unit suggests that the overall heat transfer coefficient is:

$$UA = 4.92 \text{ Btu/hr} - \text{°F} \quad (8)$$

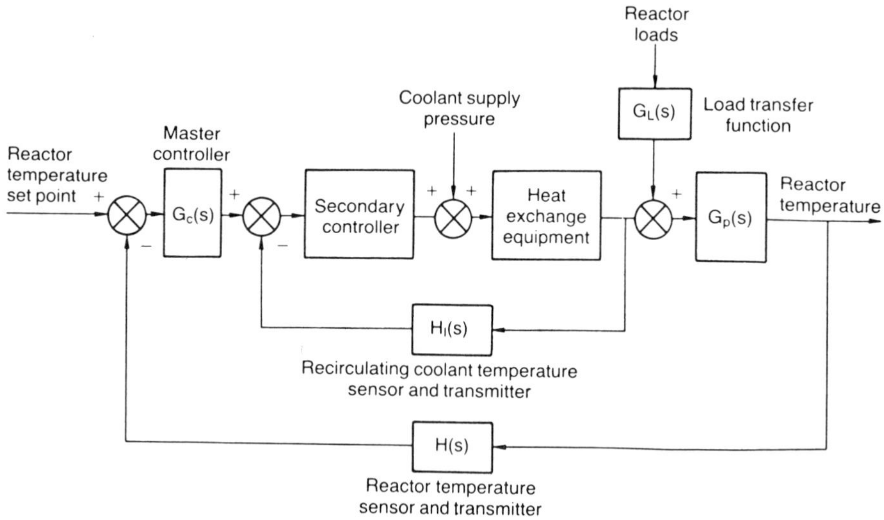

Figure 10. Block diagram of cascade control system for reactor temperature control.

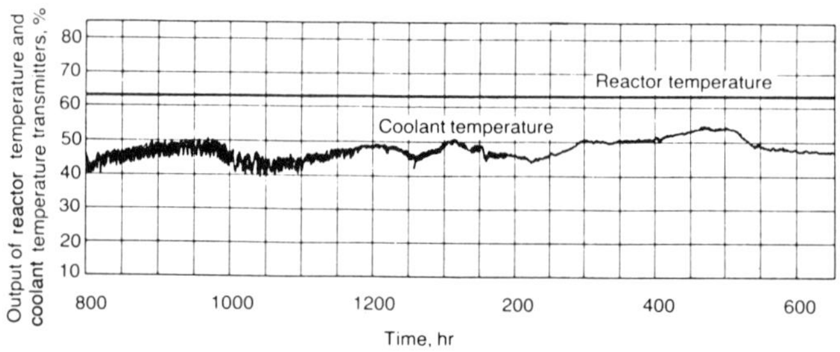

Figure 11. Response of cascade control system for reactor temperature control.

Equations 1 through 8 represent the dynamic mathematical model of the feedback control system. These equations may be combined to obtain a block diagram of the control system as shown in Fig. 4(a).

The dynamic model of the cascade control system requires an additional controller and a pressure transmitter. The pressure transmitter gain is known to be

$$H_p = 0.2545 \text{ psi/psi} \quad (9)$$

Tuning constants of the primary and the secondary controllers of the cascade system are:

Primary (temperature) $K_T = 64.83$ psi/psi $\tau_I = 2.08$ min. controller

Secondary (pressure) $K_p = 0.667$ psi/psi $\tau_I = 0.2$ min. controller (10)

These elements and Eqs. 9 and 10, are incorporated to obtain the block diagram of the cascade control system as shown in Fig. 4(b).

Nomenclature

c_p	specific heat of water, Btu/lb – °F
E	error, set point - measurement
H_p	pressure transmitter gain, psi/psi
H_T	temperature transmitter gain, psi/°F
K_p	proportional gain for pressure controller, psi/psi
K_s	steam temperature to coil-side pressure gain, °F/psi
K_T	proportional gain for temperature controller, psi/psi
K_v	valve gain, psi/psi
M	mass of water in tank, lb
p_s	steady-state steam supply pressure, 20 psig
T	temperature deviation from set point, °F
t	water temperature in tank, °F
t_i	temperature of inlet water, 70°F
t_r	reference temperature, °F
$\bar{t}$	steady-state water temperature, 86.5°F
$\bar{w}$	steady-state flowrate of water, lb/hr
θ	time
τ	time constant min
τ_I	integral time, min
τ_1, τ_2	integral time of cascade system controllers, min

References

1. Franks, R. G., Worley, C. W., Quantitative Analysis of Cascade Control, *Ind. Eng. Chem.*, 48, 1956, p. 1074.
2. Continuous System Modeling Program, Tymshare Tymcom - IX Manual Reference Series, Palo Alto, California, May 1972.
3. Deshpande, P. B. and Ash, R. H., "Elements of Computer Process Control with Advanced Control Applications," Instrument Society of America, 1980.

Index

B–C

D

H

I-K

L-N

O

P-Q

R

S

T-Z